MATLAB®
Programming
for Engineers

MATLAB®
Programming
for Engineers

Fifth Edition

Stephen J. Chapman
BAE Systems Australia

CENGAGE
Learning·

Australia • Brazil • Mexico • Singapore • United Kingdom • United States

CENGAGE
Learning®

**MATLAB Programming for Engineers,
Fifth Edition**
Stephen J. Chapman

Product Director, Global Engineering:
Timothy L. Anderson

Senior Content Developer: Mona Zeftel

Media Assistant: Ashley Kaupert

Product Assistant: Alexander Sham

Product Assistant: Teresa Versaggi

Marketing Manager: Kristin Stine

Director, Content and Media Production:
Sharon L. Smith

Senior Content Project Manager:
Colleen A. Farmer

Production Service: RPK Editorial Services, Inc.

Copyeditor: Harlan James

Proofreader: Martha McMaster

Indexer: Shelly Gerger-Knechtl

Compositor: MPS Limited

Senior Art Director: Michelle Kunkler

Internal Designer: Carmela Pereira

Cover Designer: Andrew Adams

Cover Image: © Polushkina Svetlana/
Dreamstime.com

Intellectual Property

   Analyst: Christine Myaskovsky

   Project Manager: Sarah Shainwald

Text and Image Permissions Researcher:
Kristiina Paul

Senior Manufacturing Planner: Doug Wilke

For product information and technology assistance, contact us at
Cengage Learning Customer & Sales Support, 1-800-354-9706.

For permission to use material from this text or product,
submit all requests online at **www.cengage.com/permissions**.
Further permissions questions can be emailed to
permissionrequest@cengage.com.

Library of Congress Control Number: 2015937331

ISBN: 978-1-111-57671-4

Cengage Learning
20 Channel Center Street
Boston, MA 02210
USA

Cengage Learning is a leading provider of customized learning solutions with employees residing in nearly 40 different countries and sales in more than 125 countries around the world. Find your local representative at **www.cengage.com**.

Cengage Learning products are represented in Canada by Nelson Education Ltd.

To learn more about Cengage Learning Solutions, visit **www.cengage.com/engineering**.

Purchase any of our products at your local college store or at our preferred online store **www.cengagebrain.com**.

MATLAB is a registered trademark of The MathWorks, Inc.,
3 Apple Hill Road, Natick, MA.

Printed in the United States of America
Print Number: 01   Print Year: 2015

This book is dedicated with love to my youngest daughter Devorah.

Preface

MATLAB (short for MATrix LABoratory) is a special-purpose computer program optimized to perform engineering and scientific calculations. It started life as a program designed to perform matrix mathematics, but over the years it has grown into a flexible computing system capable of solving essentially any technical problem.

The MATLAB program implements the MATLAB language and provides a very extensive library of pre-defined functions to make technical programming tasks easier and more efficient. This extremely wide variety of functions makes it much easier to solve technical problems in MATLAB than in other languages such as Fortran or C. This book introduces the MATLAB language as it is implemented in version R2014b, and shows how to use it to solve typical technical problems.

This book teaches MATLAB as a technical programming language showing students how to write clean, efficient, and documented programs. It makes no pretense at being a complete description of all of MATLAB's hundreds of functions. Instead, it teaches the student how to use MATLAB as a computer language and how to locate any desired function with MATLAB's extensive on-line help facilities.

The first eight chapters of the text are designed to serve as the text for an "Introduction to Programming/Problem Solving" course for freshman engineering students. This material should fit comfortably into a nine-week, three-hour course. The remaining chapters cover advanced topics such as I/O, Object-Oriented Programming, and Graphical User Interfaces. These chapters may be covered in a longer course or used as a reference by engineering students or practicing engineers who use MATLAB as a part of their coursework or employment.

Changes in the Fifth Edition

The fifth edition of this book is specifically devoted to MATLAB R2014b. Release 2014b is the first edition of MATLAB to enable the new H2 Graphics System, which produces higher-quality outputs. Graphic components are now MATLAB objects with handles returning properties. In addition, the MATLAB implementation of objects and object-oriented programming has matured since the last edition of this book and deserves to be covered in detail. This book has been expanded to cover MATLAB classes and objects, which work very closely with the new handle graphics system.

The major changes in this edition of the book include:

- Reduced chapter size in earlier chapters. Branches and loops now each have a chapter to themselves and the discussion of functions is split over two chapters. This change helps students to absorb the material in more bite-sized chunks.
- A new Chapter 3 is totally dedicated to 2D plots, collecting all of the plotting information in a single place.
- Chapter 8 has more extensive coverage of 3D plots and Chapter 13 now has a section devoted to animations.
- Chapter 12 is a totally new discussion of MATLAB classes and object-oriented programming.
- Chapter 13 has been rewritten to cover the new H2 handle graphics, where handles are now MATLAB objects instead of numbers.

The Advantages of MATLAB for Technical Programming

MATLAB has many advantages compared to conventional computer languages for technical problem solving. Among them are:

1. **Ease of Use**

   MATLAB is an interpreted language, like many versions of Basic. Like Basic, it is very easy to use. The program can be used as a scratch pad to evaluate expressions typed at the command line or it can be used to execute large pre-written programs. Programs may be easily written and modified with the built-in integrated development environment and debugged with the MATLAB debugger. Because the language is so easy to use, it is ideal for educational use and for the rapid prototyping of new programs.

   Many program development tools are provided to make the program easy to use. They include an integrated editor/debugger, online documentation and manuals, a workspace browser, and extensive demos.

2. **Platform Independence**

   MATLAB is supported on many different computer systems, providing a large measure of platform independence. At the time of this writing, the language is supported on Windows 7/8, Linux, and Macintosh. Programs written on any platform will run on all of the other platforms, and data files written on any platform may be read transparently on any other platform. As a result, programs written in MATLAB can migrate to new platforms when the needs of the user change and can easily be shared.

3. **Pre-defined Functions**

   MATLAB comes complete with an extensive library of pre-defined functions that provide tested and pre-packaged solutions to many basic technical tasks.

For example, suppose that you are writing a program that must calculate the statistics associated with an input data set. In most languages, you would need to write your own subroutines or functions to implement calculations such as the arithmetic mean, standard deviation, median, etc. These and hundreds of other functions are built right into the MATLAB language, making your job much easier.

In addition to the large library of functions built into the basic MATLAB language, there are many special-purpose toolboxes available to help solve complex problems in specific areas. For example, you can buy standard toolboxes to solve problems in Signal Processing, Control Systems, Communications, Image Processing, and Neural Networks, among many others.

4. **Device-Independent Plotting**

Unlike other computer languages, MATLAB has many integral plotting and imaging commands. The plots and images can be displayed on any graphical output device supported by the computer on which MATLAB is running. This capability makes MATLAB an outstanding tool for visualizing technical data.

5. **Graphical User Interface**

MATLAB includes tools that allow a programmers to interactively construct a Graphical User Interface (GUI) for their programs. With this capability, programmers can design sophisticated data analysis programs that can be operated by relatively inexperienced users.

Features of this Book

Many features of this book are designed to emphasize the proper way to write reliable MATLAB programs. These features should serve a student well as he or she is first learning MATLAB, but should also be useful to the practitioner on the job. They include:

1. **Emphasis on Top-Down Design Methodology**

The book introduces a top-down design methodology in Chapter 4 and then uses it consistently throughout the rest of the book. This methodology encourages a student to think about the proper design of a program *before* beginning to code. The book emphasizes the importance of clearly defining the problem to be solved and the required inputs and outputs before any other work is begun. Once the problem is properly defined, the book teaches students to employ stepwise refinement to break tasks down into successively smaller sub-tasks, and to implement the subtasks as separate subroutines or functions. Finally, students are taught the importance of testing at all stages of the process, both unit testing of the component routines and exhaustive testing of the final product.

The formal design process taught by the book may be summarized as follows:

1. *Clearly state the problem that you are trying to solve.*
2. *Define the inputs required by the program and the outputs to be produced by the program.*

3. *Describe the algorithm that you intend to implement in the program.* This step involves top-down design and stepwise decomposition using pseudocode or flow charts.
4. *Turn the algorithm into MATLAB statements.*
5. *Test the MATLAB program.* This step includes unit testing of specific functions as well as exhaustive testing of the final program with many different data sets.

2. Emphasis on Functions

The book emphasizes the use of functions to logically decompose tasks into smaller subtasks. It teaches the advantages of functions for data hiding. It also emphasizes the importance of unit testing functions before they are combined into the final program. In addition, students learn about the common mistakes made with functions and how to avoid them.

3. Emphasis on MATLAB Tools

The book teaches the proper use of MATLAB's built-in tools to make programming and debugging easier. The tools covered include the Editor/ Debugger, Workspace Browser, Help Browser, and GUI design tools.

4. Good Programming Practice Boxes

These boxes highlight good programming practices as they are introduced for the student's convenience. In addition, the good programming practices introduced in a chapter are summarized at the end of each chapter. An example of a Good Programming Practice Box is shown below.

 Good Programming Practice

Always indent the body of an `if` construct by 2 or more spaces to improve the readability of the code.

5. Programming Pitfalls Boxes

These boxes highlight common errors so that they can be avoided. An example Programming Pitfalls Box is shown below.

 Programming Pitfalls

Make sure that your variable names are unique in the first 31 characters. Otherwise, MATLAB will not be able to tell the difference between them.

6. **Emphasis on Data Structures**

   Chapter 10 contains a detailed discussion of MATLAB data structures, including sparse arrays, cell arrays, and structure arrays. The proper use of these data structures is illustrated in the chapters on Handle Graphics and Graphical User Interfaces.

7. **Emphasis on Object-Oriented MATLAB**

   Chapter 12 includes an introduction to object-oriented programming (OOP) and describes the MATLAB implementation of OOP in detail.

Pedagogical Features

The first eight chapters of this book are specifically designed to be used in a freshman "Introduction to Program/Problem Solving" course. It is possible to cover this material comfortably in a nine-week, three-hour per week course. If there is insufficient time to cover all of the material in a particular engineering program, Chapter 8 may be deleted as the remaining material still teaches the fundamentals of programming and using MATLAB to solve problems. This feature should appeal to harassed engineering educators trying to cram ever more material into a finite curriculum.

The remaining chapters cover advanced material that is useful to the engineer and engineering students as they progress in their careers. This material includes advanced I/O, object-oriented programming, and the design of Graphical User Interfaces for programs.

The book includes several features designed to aid student comprehension. A total of 17 quizzes appear scattered throughout the chapters, with answers to all questions included in Appendix B. These quizzes can serve as a useful self-test of comprehension. In addition, there are approximately 180 end-of-chapter exercises. Answers to all exercises are included in the Instructor's Manual. Good programming practices are highlighted in all chapters with special Good Programming Practice boxes and common errors are highlighted in Programming Pitfalls boxes. End-of-chapter materials include Summaries of Good Programming Practice and Summaries of MATLAB Commands and Functions.

The book is accompanied by an Instructor's Manual, containing the solutions to all end-of-chapter exercises. The source code for all examples in the book is available from the book's web site, and the source code for all solutions in the Instructor's Manual is available separately to instructors.

MindTap Online Course

This textbook is also available online through Cengage Learning's MindTap, a personalized learning program that can be purchased as an addition to the book. Students who purchase the MindTap have access to the book's electronic Reader and are able to complete homework and assessment material online, on their desktops, laptops, or iPads. Instructors who use a Learning Management System (such as Blackboard or Moodle) for tracking course content, assignments, and grading, can seamlessly access the MindTap suite of content and assessments for this course.

With MindTap, instructors can:

- Personalize the Learning Path to match the course syllabus by rearranging content or appending original material to the online content
- Connect a Learning Management System portal to the online course and Reader
- Customize online assessments and assignments
- Track student progress and comprehension
- Promote student engagement through interactivity and exercises

Additionally, students can listen to the text through ReadSpeaker, take notes, study from or create their own Flashcards, highlight content for easy reference, and check their understanding of the material through practice quizzes and gradable homework.

A Final Note to the User

No matter how hard I try to proofread a document like this book, it is inevitable that some typographical errors slip through and appear in print. If you should spot any such errors, please drop me a note via the publisher, and I will do my best to get them eliminated from subsequent printings and editions. Thank you very much for your help in this matter.

I will maintain a complete list of errata and corrections at the book's web site, which is available through www.cengage.com. Please check that site for any updates and/or corrections.

Acknowledgments

I would like to thank all my friends at Cengage Learning for the support they have given me in getting this book to market.

I would like to thank these reviewers who offered their helpful suggestions for this edition:

David Eromon	Georgia Southern University
Arlene Guest	Naval Postgraduate School
Mary M. Hofle	Idaho State University
Mark Hutchenreuther	California Polytechnic State University
Mani Mina	Iowa State Univesity

In addition, I would like to thank my wife, Rosa, and our children, Avi, David, Rachel, Aaron, Sarah, Naomi, Shira, and Devorah for their help and encouragement.

Stephen J. Chapman
Melbourne, Australia

Contents

Chapter 4   Branching Statements and Program Design   127

Chapter 5   Loops and Vectorization   175

Chapter 12 User-Defined Classes and Object-Oriented Programming   461

Chapter 13 Handle Graphics and Animation 531

Chapter 14 Graphical User Interfaces 571

Chapter 1

Introduction to MATLAB

MATLAB (short for MATrix LABoratory) is a special-purpose computer program optimized to perform engineering and scientific calculations. It started life as a program designed to perform matrix mathematics, but over the years it has grown into a flexible computing system capable of solving essentially any technical problem.

The MATLAB program implements the MATLAB programming language and provides a very extensive library of predefined functions to make technical programming tasks easier and more efficient. This book introduces the MATLAB language as it is implemented in MATLAB Version 2014B and shows how to use it to solve typical technical problems.

MATLAB is a huge program, with an incredibly rich variety of functions. Even the basic version of MATLAB without any toolkits is much richer than other technical programming languages. There are more than 1000 functions in the basic MATLAB product alone, and the toolkits extend this capability with many more functions in various specialties. Furthermore, these functions often solve very complex problems (solving differential equations, inverting matrices, and so forth) in a *single step*, saving large amounts of time. Doing the same thing in another computer language usually involves writing complex programs yourself or buying a third-party software package (such as IMSL or the NAG software libraries) that contains the functions.

The built-in MATLAB functions are almost always better than anything that an individual engineer could write on his or her own because many people have worked on them, and they have been tested against many different data sets. These functions are also robust, producing sensible results for wide ranges of input data and gracefully handling error conditions.

This book makes no attempt to introduce the user to all of MATLAB's functions. Instead, it teaches a user the basics of how to write, debug, and optimize good MATLAB programs, plus a subset of the most important functions used to solve

1

common scientific and engineering problems. Just as importantly, it teaches the scientist or engineer how to use MATLAB's own tools to locate the right function for a specific purpose from the enormous list of choices available. In addition, it teaches how to use MATLAB to solve many practical engineering problems, such as vector and matrix algebra, curve fitting, differential equations, and data plotting.

The MATLAB program is a combination of a procedural programming language, an integrated development environment (IDE) including an editor and debugger, and an extremely rich set of functions to perform many types of technical calculations.

The MATLAB language is a procedural programming language, meaning that the engineer writes *procedures*, which are effectively mathematical recipes for solving a problem. This makes MATLAB very similar to other procedural languages such as C, Basic, Fortran, and Pascal. However, the extremely rich list of predefined functions and plotting tools makes it superior to these other languages for many engineering analysis applications.

1.1 The Advantages of MATLAB

MATLAB has many advantages compared to conventional computer languages for technical problem solving. Among them are:

1. **Ease of Use**

   MATLAB is an interpreted language, like many versions of Basic. Like Basic, it is very easy to use. The program can be used as a scratch pad to evaluate expressions typed at the command line, or it can be used to execute large prewritten programs. Programs may be easily written and modified with the built-in integrated development environment and debugged with the MATLAB debugger. Because the language is so easy to use, it is ideal for the rapid prototyping of new programs.

   Many program development tools are provided to make the program easy to use. They include an integrated editor/debugger, online documentation and manuals, a workspace browser, and extensive demos.

2. **Platform Independence**

   MATLAB is supported on many different computer systems, providing a large measure of platform independence. At the time of this writing, the language is supported on Windows XP/Vista/7, Linux, Unix, and the Macintosh. Programs written on any platform will run on all of the other platforms, and data files written on any platform may be read transparently on any other platform. As a result, programs written in MATLAB can migrate to new platforms when the needs of the user change.

3. **Predefined Functions**

   MATLAB comes complete with an extensive library of predefined functions that provide tested and prepackaged solutions to many basic technical tasks. For example, suppose that you are writing a program that must calculate the statistics associated with an input data set. In most languages, you would need to write your own subroutines or functions to implement calculations such as the arithmetic mean, standard deviation, median, and so forth. These and hundreds of other functions are built right into the MATLAB language, making your job much easier.

In addition to the large library of functions built into the basic MATLAB language, there are many special-purpose toolboxes available to help solve complex problems in specific areas. For example, a user can buy standard toolboxes to solve problems in signal processing, control systems, communications, image processing, and neural networks, among many others. There is also an extensive collection of free user-contributed MATLAB programs that are shared through the MATLAB website.

4. **Device-Independent Plotting**
Unlike most other computer languages, MATLAB has many integral plotting and imaging commands. The plots and images can be displayed on any graphical output device supported by the computer on which MATLAB is running. This capability makes MATLAB an outstanding tool for visualizing technical data.

5. **Graphical User Interface**
MATLAB includes tools that allow an engineer to interactively construct a Graphical User Interface (GUI) for his or her program. With this capability, the engineer can design sophisticated data analysis programs that can be operated by relatively inexperienced users.

6. **MATLAB Compiler**
MATLAB's flexibility and platform independence is achieved by compiling MATLAB programs into a device-independent p-code, and then interpreting the p-code instructions at run-time. This approach is similar to that used by Microsoft's Visual Basic or by Java. Unfortunately, the resulting programs can sometimes execute slowly because the MATLAB code is interpreted rather than compiled. Recent versions of MATLAB have partially overcome this problem by introducing just-in-time (JIT) compiler technology. The JIT compiler compiles portions of the MATLAB code as it is executed to increase overall speed.

A separate MATLAB compiler is also available. This compiler can compile a MATLAB program into a standalone executable that can run on a computer without a MATLAB license. It is a great way to convert a prototype MATLAB program into an executable suitable for sale and distribution to users.

1.2   Disadvantages of MATLAB

MATLAB has two principal disadvantages. The first is that it is an interpreted language and therefore can execute more slowly than compiled languages. This problem can be mitigated by properly structuring the MATLAB program to maximize the performance of vectorized code and by the use of the JIT compiler.

The second disadvantage is cost: a full copy of MATLAB is five to ten times more expensive than a conventional C or Fortran compiler. This relatively high cost is more than offset by the reduced time required for an engineer or scientist to create a working program, so MATLAB is cost-effective for businesses. However, it is too expensive for most individuals to consider purchasing. Fortunately, there is also an inexpensive student edition of MATLAB, which is a great tool for students wishing to learn the language. The student edition of MATLAB is essentially identical to the full edition.[1]

1.3 The MATLAB Environment

The fundamental unit of data in any MATLAB program is the **array**. An array is a collection of data values organized into rows and columns and known by a single name. Individual data values within an array can be accessed by including the name of the array followed by subscripts in parentheses, which identify the row and column of the particular value. Even scalars are treated as arrays by MATLAB—they are simply arrays with only one row and one column. We will learn how to create and manipulate MATLAB arrays in Section 1.4.

When MATLAB executes, it can display several types of windows that accept commands or display information. The three most important types of windows are the Command Window, where commands may be entered; figure windows, which display plots and graphs; and edit windows, which permit a user to create and modify MATLAB programs. We will see examples of all three types of windows in this section.

In addition, MATLAB can display other windows that provide help and that allow the user to examine the values of variables defined in memory. We will examine some of these additional windows here and examine the others when we discuss how to debug MATLAB programs.

1.3.1 The MATLAB Desktop

When you start MATLAB Version 2014B, a special window called the MATLAB desktop appears. The desktop is a window that contains other windows showing MATLAB data, plus toolbars and a "Toolstrip" or "Ribbon Bar" similar to that used by Microsoft Office. By default, most MATLAB tools are docked to the desktop, so that they appear inside the desktop window. However, the user can choose to undock any or all tools, making them appear in windows separate from the desktop.

The default configuration of the MATLAB desktop is shown in Figure 1.1. It integrates many tools for managing files, variables, and applications within the MATLAB environment.

The major tools within or accessible from the MATLAB desktop are:

- Command Window
- Toolstrip

---

[1]There are also some free software programs that are largely compatible with MATLAB, such as GNU Octave and FreeMat.

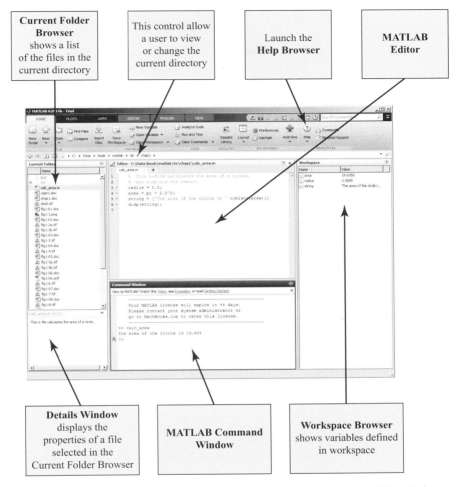

Figure 1.1 The default MATLAB desktop. The exact appearance of the desktop may differ slightly on different types of computers.

- Documents Window, including the Editor/Debugger and Array Editor
- Figure Windows
- Workspace Browser
- Current Folder Browser, with the Details Window
- Help Browser
- Path Browser
- Popup Command History Window

The functions of these tools are summarized in Table 1.1. We will discuss them in later sections of this chapter.

Table 1.1: Tools and Windows Included in the MATLAB Desktop

Tool	Description
Command Window	A window where the user can type commands and see immediate results
Toolstrip	A strip across the top of the desktop containing icons to select functions and tools, arranged in tabs and sections of related functions
Command History Window	A window that displays recently used commands, accessed by clicking the up arrow when typing in the Command Window
Document Window	A window that displays MATLAB files and allows the user to edit or debug them
Figure Window	A window that displays a MATLAB plot
Workspace Browser	A window that displays the names and values of variables stored in the MATLAB Workspace
Current Folder Browser	A window that displays the names of files in the current directory. If a file is selected in the Current Folder Browser, details about the file will appear in the Details Window
Help Browser	A tool to get help for MATLAB functions, accessed by clicking the Help button
Path Browser	A tool to display the MATLAB search path, accessed by clicking the Set Path button

1.3.2 The Command Window

The bottom center of the default MATLAB desktop contains the **Command Window**. A user can enter interactive commands at the command prompt (») in the Command Window, and they will be executed on the spot.

As an example of a simple interactive calculation, suppose that you want to calculate the area of a circle with a radius of 2.5 m. This can be done in the MATLAB Command Window by typing:

```
» area = pi * 2.5^2
area =
   19.6350
```

MATLAB calculates the answer as soon as the Enter key is pressed, and stores the answer in a variable (really a 1×1 array) called `area`. The contents of the variable are displayed in the Command Window as shown in Figure 1.2, and the variable can be used in further calculations. (Note that π is predefined in MATLAB, so we can just use `pi` without first declaring it to be 3.141592....)

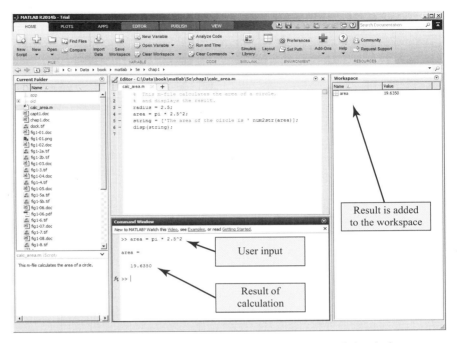

Figure 1.2 The Command Window appears in the center of the desktop. Users enter commands and see responses here.

If a statement is too long to type on a single line, it may be continued on successive lines by typing an **ellipsis** (. . .) at the end of the first line, and then continuing on the next line. For example, the following two statements are identical.

```
x1 = 1 + 1/2 + 1/3 + 1/4 + 1/5 + 1/6
```

and

```
x1 = 1 + 1/2 + 1/3 + 1/4 ...
    + 1/5 + 1/6
```

Instead of typing commands directly in the Command Window, a series of commands can be placed into a file, and the entire file can be executed by typing its name in the Command Window. Such files are called **script files**. Script files (and functions, which we will see later) are also known as **M-files**, because they have a file extension of ".m".

1.3.3 The Toolstrip

The Toolstrip (see Figure 1.3) is a bar of tools that appears across the top of the desktop. The controls on the Toolstrip are organized into related categories of functions, first by tabs and then by groups. For example, the tabs visible in Figure 1.3 are Home, Plots, Apps, Editor, and so forth. When one of the tabs is selected, a series of controls

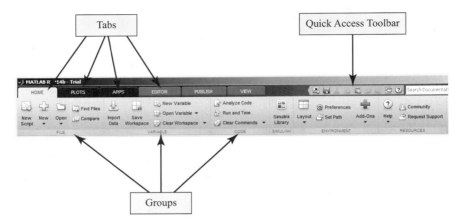

Figure 1.3 The Toolstrip, which allows a user to select from a wide variety of MATLAB tools and commands.

grouped into sections is displayed. In the Home tab, the sections are File, Variable, Code, and so forth. With practice, the logical grouping of commands helps the user to quickly locate any desired function.

In addition, the upper right-hand corner of the Toolstrip contains the Quick Access Toolbar, which is a place where the user can customize the interface and display the most commonly used commands and functions at all times. To customize the functions displayed there, right-click on the toolbar and select the Customize option from the popup menu.

1.3.4 The Command History Window

The Command History window displays a list of the commands that a user has previously entered in the Command Window. The list of commands can extend back to previous executions of the program. Commands remain in the list until they are deleted. To display the Command History window, press the up arrow key while typing in the Command Window. To re-execute any command, simply double-click it with the left mouse button. To delete one or more commands from the Command History window, select the commands and right-click them with the mouse. A popup menu will be displayed that allows the user to delete the items (see Figure 1.4).

1.3.5 The Document Window

A **Document Window** (also called an **Edit/Debug Window**) is used to create new M-files or modify existing ones. An Edit Window is created automatically when you create a new M-file or open an existing one. You can create a new M-file with the New Script command from the File group on the Toolstrip (Figure 1.5*a*), or by clicking the New icon and selecting Script from the popup menu (Figure 1.5*b*). You

Figure 1.4 The Command History Window, showing two commands being deleted.

can open an existing M-file file with the Open command from the File section on the Toolstrip.

An Edit Window displaying a simple M-file called `calc_area.m` is shown in Figure 1.5. This file calculates the area of a circle given its radius and displays the result. By default, the Edit Window is docked to the desktop, as shown in Figure 1.5c. The Edit Window can also be undocked from the MATLAB desktop. In that case, it appears within a container called the Documents Window, as shown in Figure 1.5d. We will learn how to dock and undock a window later in this chapter.

The Edit Window is essentially a programming text editor, with the MATLAB language's features highlighted in different colors. On screen, comments in an M-file file appear in green, variables and numbers appear in black, complete character strings appear in magenta, incomplete character strings appear in red, and language keywords appear in blue. [See color insert.]

After an M-file is saved, it may be executed by typing its name in the Command Window. For the M-file in Figure 1.5, the results are:

```
» calc_area
The area of the circle is 19.635
```

The Edit Window also doubles as a debugger, as we shall see in Chapter 2.

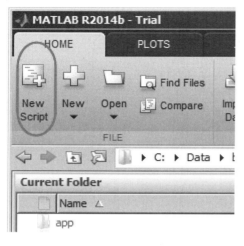

(a)

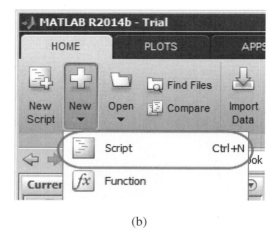

(b)

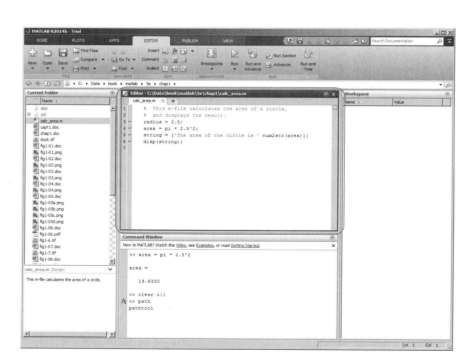

(c)

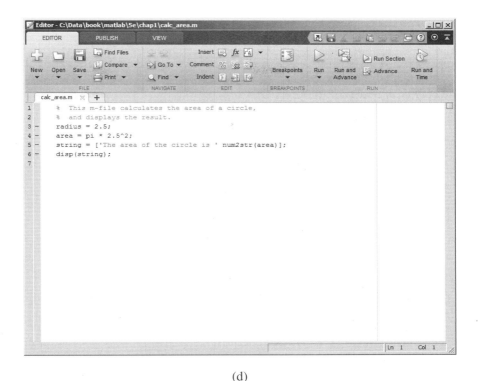

(d)

Figure 1.5 (a) Creating a new M-file with the New Script command.
(b) Creating a new M-file with the New >> Script popup menu. (c) The MATLAB Editor, docked to the MATLAB desktop. (d) The MATLAB Editor, displayed as an independent window. [See color insert.]

1.3.6 Figure Windows

A **figure window** is used to display MATLAB graphics. A figure can be a two- or three-dimensional plot of data, an image, or a graphical user interface (GUI). A simple script file that calculates and plots the function sin x is shown below:

```
% sin_x.m: This M-file calculates and plots the
% function sin(x) for 0 <= x <= 6.
x = 0:0.1:6
y = sin(x)
plot(x,y)
```

If this file is saved under the name `sin_x.m`, then a user can execute the file by typing "`sin_x`" in the Command Window. When this script file is executed, MATLAB opens a figure window and plots the function sin x in it. The resulting plot is shown in Figure 1.6.

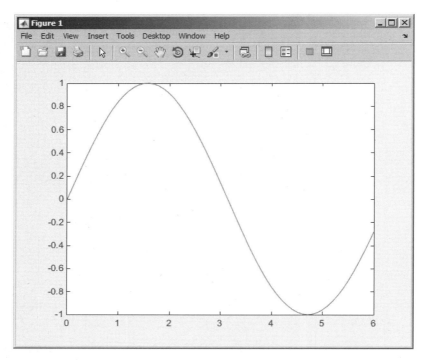

Figure 1.6 MATLAB plot of sin *x* versus *x*.

1.3.7 Docking and Undocking Windows

MATLAB windows such as the Command Window, the Edit Window, and Figure Windows can either be *docked* to the desktop, or they can be *undocked*. When a window is docked, it appears as a pane within the MATLAB desktop. When it is undocked, it appears as an independent window on the computer screen separate from the desktop. When a window is docked to the desktop, it can be undocked by selecting the small down arrow in the upper right-hand corner and selecting the Undock option from the popup menu (see Figure 1.7). When the window is an independent window, the upper right-hand corner contains a small button with an arrow pointing down and to the right (⬒). If this button is clicked, then the window will be re-docked with the desktop. The Dock button is visible in the upper right hand corner of Figure 1.6.

1.3.8 The MATLAB Workspace

A statement like

```
z = 10
```

creates a variable named z, stores the value 10 in it, and saves it in a part of computer memory known as the **workspace**. A workspace is the collection of all the variables and arrays that can be used by MATLAB when a particular command, M-file, or function is executing. All commands executed in the Command Window (and all

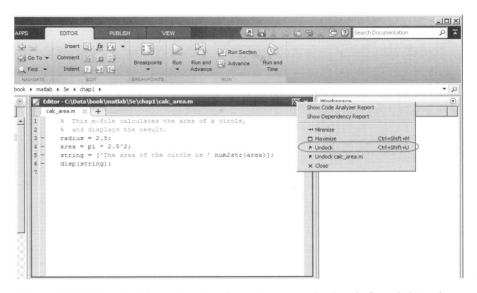

Figure 1.7 Select the Undock option from the menu displayed after clicking the small down arrow in the upper right-hand corner of a pane.

script files executed from the Command Window) share a common workspace, so they can all share variables. As we will see later, MATLAB functions differ from script files in that each function has its own separate workspace.

A list of the variables and arrays in the current workspace can be generated with the `whos` command. For example, after M-files `calc_area` and `sin_x` are executed, the workspace contains the following variables.

```
» whos

Name        Size       Bytes     Class       Attributes

area        1x1            8     double
radius      1x1            8     double
string      1x32          64     char
x           1x61         488     double
y           1x61         488     double
```

Script file `calc_area` created variables `area`, `radius`, and `string`, while script file `sin_x` created variables `x` and `y`. Note that all of the variables are in the same workspace, so if two script files are executed in succession, the second script file can use variables created by the first script file.

The contents of any variable or array may be determined by typing the appropriate name in the Command Window. For example, the contents of `string` can be found as follows:

```
» string
string =
The area of the circle is 19.635
```

A variable can be deleted from the workspace with the `clear` command. The `clear` command takes the form

```
clear var1 var2 ...
```

where `var1` and `var2` are the names of the variables to be deleted. The command `clear variables`, or simply `clear`, deletes all variables from the current workspace.

1.3.9 The Workspace Browser

The contents of the current workspace can also be examined with a GUI-based Workspace Browser. The Workspace Browser appears by default in the right-hand side of the desktop. It provides a graphic display of the same information as the `whos` command, and it also shows the actual contents of each array if the information is short enough to fit within the display area. The Workspace Browser is dynamically updated whenever the contents of the workspace change.

A typical Workspace Browser window is shown in Figure 1.8. As you can see, it displays the same information as the `whos` command. Double-clicking on any variable in the window will bring up the Array Editor, which allows the user to modify the information stored in the variable.

One or more variables may be deleted from the workspace by selecting them in the Workspace Browser with the mouse and pressing the delete key or by right-clicking with the mouse and selecting the delete option.

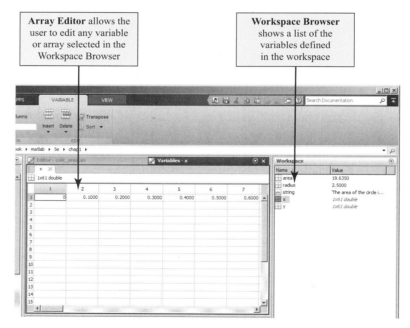

Figure 1.8 The Workspace Browser and Array Editor. The Array Editor is invoked by double-clicking a variable in the Workspace Browser. It allows a user to change the values contained in a variable or array.

1.3.10 The Current Folder Browser

The Current Folder Browser is displayed on the upper left-hand side of the desktop. It shows all the files in the currently selected folder, and allows the user to edit or execute any desired file. You can double-click on any M-file to open it in the MATLAB editor, or you can right-click it and select Run to execute it. The Current Folder Browser is shown in Figure 1.9. A toolbar above the browser is used to select the current folder to display.

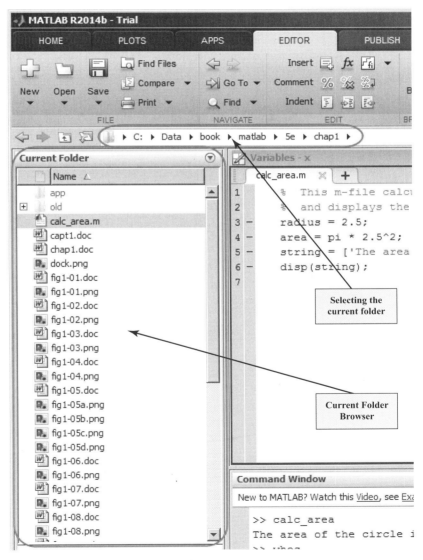

Figure 1.9 The Current Folder Browser.

1.3.11 Getting Help

There are three ways to get help in MATLAB. The preferred method is to use the Help Browser. The Help Browser can be started by selecting the ? icon from the Toolstrip or by typing `helpdesk` or `helpwin` in the Command Window. A user can get help by browsing the MATLAB documentation, or he or she can search for the details of a particular command. The Help Browser is shown in Figure 1.10.

There are also two command-line oriented ways to get help. The first way is to type `help` or `help` followed by a function name in the Command Window. If you just type `help`, MATLAB will display a list of possible help topics in the Command Window. If a specific function or a toolbox name is included, help will be provided for that particular function or toolbox.

The second way to get help is the `lookfor` command. The `lookfor` command differs from the `help` command in that the `help` command searches for an exact function name match, while the `lookfor` command searches the quick summary information in each function for a match. This makes `lookfor` slower than `help`,

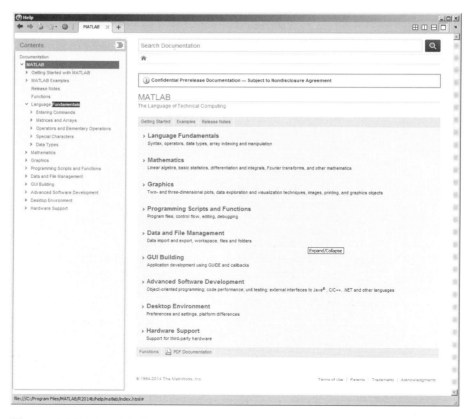

Figure 1.10 The Help Browser.

but it improves the chances of getting back useful information. For example, suppose that you were looking for a function to take the inverse of a matrix. Since MATLAB does not have a function named `inverse`, the command `help inverse` will produce nothing. On the other hand, the command `lookfor inverse` will produce the following results:

```
» lookfor inverse
ifft          - Inverse discrete Fourier transform.
ifft2         - Two-dimensional inverse discrete Fourier transform.
ifftn         - N-dimensional inverse discrete Fourier transform.
ifftshift     - Inverse FFT shift.
acos          - Inverse cosine, result in radians.
acosd         - Inverse cosine, result in degrees.
acosh         - Inverse hyperbolic cosine.
acot          - Inverse cotangent, result in radian.
acotd         - Inverse cotangent, result in degrees.
acoth         - Inverse hyperbolic cotangent.
acsc          - Inverse cosecant, result in radian.
acscd         - Inverse cosecant, result in degrees.
acsch         - Inverse hyperbolic cosecant.
asec          - Inverse secant, result in radians.
asecd         - Inverse secant, result in degrees.
asech         - Inverse hyperbolic secant.
asin          - Inverse sine, result in radians.
asind         - Inverse sine, result in degrees.
asinh         - Inverse hyperbolic sine.
atan          - Inverse tangent, result in radians.
atan2         - Four quadrant inverse tangent.
atan2d        - Four quadrant inverse tangent, result in degrees.
atand         - Inverse tangent, result in degrees.
atanh         - Inverse hyperbolic tangent.
invhilb       - Inverse Hilbert matrix.
ipermute      - Inverse permute array dimensions.
inv           - Matrix inverse.
pinv          - Pseudoinverse.
betaincinv    - Inverse incomplete beta function.
erfcinv       - Inverse complementary error function.
erfinv        - Inverse error function.
gammaincinv   - Inverse incomplete gamma function.
acde          - Inverse of cd elliptic function.
asne          - Inverse of sn elliptic function.
icceps        - Inverse complex cepstrum.
idct          - Inverse discrete cosine transform.
ifwht         - Fast Inverse Discrete Walsh-Hadamard Transform.
unshiftdata   - The inverse of SHIFTDATA.
```

From this list, we can see that the function of interest is named `inv`.

1.3.12 A Few Important Commands

If you are new to MATLAB, a few demonstrations may help to give you a feel for its capabilities. To run MATLAB's built-in demonstrations, type `demo` in the Command Window, or select demos from the Start button.

The contents of the Command Window can be cleared at any time using the `clc` command, and the contents of the current figure window can be cleared at any time using the `clf` command. The variables in the workspace can be cleared with the `clear` command. As we have seen, the contents of the workspace persist between the executions of separate commands and M-files, so it is possible for the results of one problem to have an effect on the next one that you may attempt to solve. To avoid this possibility, it is a good idea to issue the `clear` command at the start of each new independent calculation.

Another important command is the **abort** command. If an M-file appears to be running for too long, it may contain an infinite loop, and it will never terminate. In this case, the user can regain control by typing control-c (abbreviated ^c) in the Command Window. This command is entered by holding down the control key while typing a "c". When MATLAB detects a ^c, it interrupts the running program and returns a command prompt.

There is also an auto-complete feature in MATLAB. If a user starts to type a command and then presses the Tab key, a popup list of recently typed commands and MATLAB functions that match the string will be displayed (see Figure 1.11). The user can complete the command by selecting one of the items from the list.

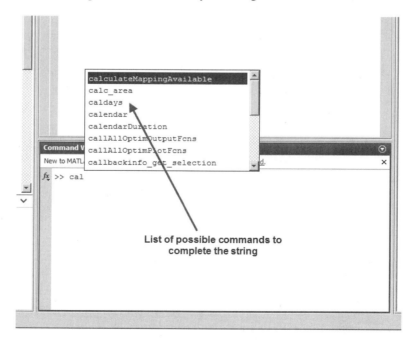

Figure 1.11 If a user types a partial command and then hits the Tab key, MATLAB will pop up a window of suggested commands or functions that match the string.

The exclamation point (!) is another important special character. Its special purpose is to send a command to the computer's operating system. Any characters after the exclamation point will be sent to the operating system and executed as though they had been typed at the operating system's command prompt. This feature lets you embed operating system commands directly into MATLAB programs.

Finally, it is possible to keep track of everything done during a MATLAB session with the **diary** command. The form of this command is

```
diary filename
```

After this command is typed, a copy of all input and most output typed in the Command Window is echoed in the diary file. This is a great tool for recreating events when something goes wrong during a MATLAB session. The command "diary off" suspends input into the diary file, and the command "diary on" resumes input again.

1.3.13 The MATLAB Search Path

MATLAB has a search path, which it uses to find M-files. MATLAB's M-files are organized in directories on your file system. Many of these directories of M-files are provided along with MATLAB, and users may add others. If a user enters a name at the MATLAB prompt, the MATLAB interpreter attempts to find the name as follows:

1. It looks for the name as a variable. If it is a variable, MATLAB displays the current contents of the variable.
2. It checks to see if the name is an M-file in the current directory. If it is, MATLAB executes that function or command.
3. It checks to see if the name is an M-file in any directory in the search path. If it is, MATLAB executes that function or command.

Note that MATLAB checks for variable names first, so *if you define a variable with the same name as a MATLAB function or command, that function or command becomes inaccessible*. This is a common mistake made by novice users.

///

☒ Programming Pitfalls

Never use a variable with the same name as a MATLAB function or command. If you do so, that function of command will become inaccessible.

///

Also, if there is more than one function or command with the same name, the *first* one found on the search path will be executed, and all of the others will be inaccessible. This is a common problem for novice users, since they sometimes create M-files files with the same names of standard MATLAB functions, making them inaccessible.

///

[x] ## Programming Pitfalls

Never create an M-file with the same name as a MATLAB function or command.

///

MATLAB includes a special command (`which`) to help you find out just which version of a file is being executed and where it is located. This can be useful in finding filename conflicts. The format of this command is `which` *functionname*, where *functionname* is the name of the function that you are trying to locate. For example, the cross-product function `cross.m` can be located as follows:

```
» which cross
C:\Program
Files\MATLAB\R2014b\toolbox\matlab\specfun\cross.m
```

The MATLAB search path can be examined and modified at any time by selecting the Set Path tool from the Environment section of the Home tab on the Toolstrip, or by typing `editpath` in the Command Window. The Path Tool is shown in Figure 1.12. It allows a user to add, delete, or change the order of directories in the path.

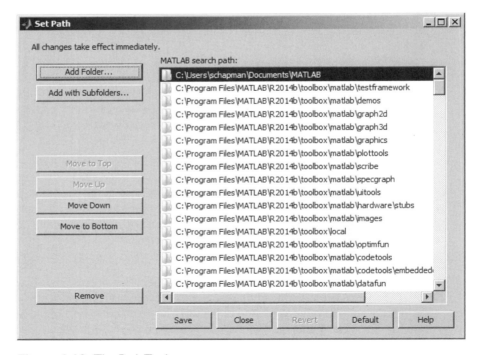

Figure 1.12 The Path Tool.

Other path-related functions include:

- `addpath`   Add directory to MATLAB search path.
- `path`   Display MATLAB search path.
- `path2rc`   Adds current directory to MATLAB search path.
- `rmpath`   Remove directory from MATLAB search path.

1.4 Using MATLAB as a Calculator

In its simplest form, MATLAB can be used as a calculator to perform mathematical calculations. The calculations to be performed are typed directly into the Command Window, using the symbols +, -, *, /, and ^ for addition, subtraction, multiplication, division, and exponentiation respectively. After an expression is typed, the results of the expression will be automatically calculated and displayed. If an equal sign is used in the expression, then the result of the calculation is saved in the variable name to the left of the equal sign.

For example, suppose that we would like to calculate the volume of a cylinder of radius r and length l. The area of the circle at the base of the cylinder is given by the equation

$$A = \pi r^2 \tag{1.1}$$

and the total volume of the cylinder will be

$$V = Al \tag{1.2}$$

If the radius of the cylinder is 0.1 m and the length is 0.5 m, then the volume of the cylinder can be found using the MATLAB statements (user inputs are shown in bold face):

```
» A = pi * 0.1^2
A =
    0.0314
» V = A * 0.5
V =
    0.0157
```

Note that `pi` is predefined to be the value 3.141592....

When the first expression is typed, the area at the base of the cylinder is calculated, stored in variable `A`, and displayed to the user. When the second expression is typed, the volume of the cylinder is calculated, stored in variable `V`, and displayed to the user. Note that the value stored in `A` was saved by MATLAB and re-used when we calculated `V`.

If an expression *without an equal sign* is typed into the Command Window, MATLAB will evaluate it, store the result in a special variable called `ans`, and display the result.

```
» 200 / 7
ans =
    28.5714
```

The value in ans can be used in later calculations, but be careful! Every time a new expression without an equal sign is evaluated, the value saved in ans will be overwritten.

```
» ans * 6
ans =
   171.4286
```

The value stored in ans is now 171.4286, not 28.5714.

If you want to save a calculated value and reuse it later, be sure to assign it to a specific name instead of using the default name ans.

//

Programming Pitfalls

If you want to reuse the result of a calculation in MATLAB, be sure to include a variable name to store the result. Otherwise, the result will be overwritten the next time that you perform a calculation.

//

Quiz 1.1

This quiz provides a quick check to see if you have understood the concepts introduced in Chapter 1. If you have trouble with the quiz, reread the sections, ask your instructor, or discuss the material with a fellow student. The answers to this quiz are found in the back of the book.

1. What is the purpose of the MATLAB Command Window? The Edit Window? The Figure Window?
2. List the different ways that you get help in MATLAB.
3. What is a workspace? How can you determine what is stored in a MATLAB workspace?
4. How can you clear the contents of a workspace?
5. The distance traveled by a ball falling in the air is given by the equation

$$x = x_0 + v_0 t + \frac{1}{2}at^2$$

Use MATLAB to calculate the position of the ball at time $t = 5$ s if $x_0 = 10$ m, $v_0 = 15$ m/s, and $a = -9.81$ m/sec^2.
6. Suppose that $x = 3$ and $y = 4$. Use MATLAB to evaluate the following expression:

$$\frac{x^2 y^3}{(x - y)^2}$$

The following questions are intended to help you become familiar with MATLAB tools.

7. Execute the M-files `calc_area.m` and `sin_x.m` in the Command Window (these M-files are available from the book's website). Then use the Workspace Browser to determine what variables are defined in the current workspace.

8. Use the Array Editor to examine and modify the contents of variable `x` in the workspace. Then type the command `plot(x,y)` in the Command Window. What happens to the data displayed in the Figure Window?

1.5 Summary

In this chapter, we learned about the MATLAB integrated development environment (IDE). We learned about basic types of MATLAB windows, the workspace, and how to get online help. The MATLAB desktop appears when the program is started. It integrates many of the MATLAB tools in a single location. These tools include the Command Window, the Command History Window, the Toolstrip, the Document Window, the Workspace Browser and Array Editor, and the Current Folder viewer. The Command Window is the most important of the windows. It is the one in which all commands are typed and results are displayed.

The Document Window (or Edit/Debug Window) is used to create or modify M-files. It displays the contents of the M-file with the contents of the file color-coded according to function: comments, keywords, strings, and so forth.

The Figure Window is used to display graphics.

A MATLAB user can get help by either using the Help Browser or the command-line help functions `help` and `lookfor`. The Help Browser allows full access to the entire MATLAB documentation set. The command-line function `help` displays help about a specific function in the Command Window. Unfortunately, you must know the name of the function in order to get help about it. The function `lookfor` searches for a given string in the first comment line of every MATLAB function and displays any matches.

When a user types a command in the Command Window, MATLAB searches for that command in the directories specified in the MATLAB path. It will execute the *first* M-file in the path that matches the command, and any further M-files with the same name will never be found. The Path Tool can be used to add, delete, or modify directories in the MATLAB path.

1.5.1 MATLAB Summary

The following summary lists all of the MATLAB special symbols described in this chapter, along with a brief description of each one.

Special Symbols

+	Addition
-	Subtraction
*	Multiplication
/	Division
^	Exponentiation

1.6 Exercises

1.1 The following MATLAB statements plot the function $y(x) = 2e^{-0.2x}$ for the range $0 \leq x \leq 10$.

```
x = 0:0.1:10;
y = 2 * exp( -0.2 * x);
plot(x,y);
```

Use the MATLAB Edit Window to create a new empty M-file, type these statements into the file, and save the file with the name `test1.m`. Then, execute the program by typing the name `test1` in the Command Window. What result do you get?

1.2 Get help on the MATLAB function `exp` using: (a) The `help exp` command typed in the Command Window, and (b) the Help Browser.

1.3 Use the `lookfor` command to determine how to take the base-10 logarithm of a number in MATLAB.

1.4 Suppose that $u = 1$ and $v = 3$. Evaluate the following expressions using MATLAB.

(a) $\dfrac{4u}{3v}$

(b) $\dfrac{2v^{-2}}{(u + v)^2}$

(c) $\dfrac{v^3}{v^3 - u^3}$

(d) $\dfrac{4}{3}\pi v^2$

1.5 Suppose that $x = 2$ and $y = -1$. Evaluate the following expressions using MATLAB.

(a) $\sqrt[4]{2x^3}$

(b) $\sqrt[4]{2y^3}$

Note that MATLAB evaluates expressions with complex or imaginary answers transparently.

1.6 Type the following MATLAB statements into the Command Window:

```
4 * 5
a = ans * pi
b = ans / pi
ans
```

What are the results in a, b, and ans? What is the final value saved in ans? Why was that value retained during the subsequent calculations?

1.7 Use the MATLAB Help Browser to find the command required to show MATLAB's current directory. What is the current directory when MATLAB starts up?

1.8 Use the MATLAB Help Browser to find out how to create a new directory from within MATLAB. Then, create a new directory called mynewdir under the current directory. Add the new directory to the top of MATLAB's path.

1.9 Change the current directory to mynewdir. Then open an Edit Window and add the following lines:

```
% Create an input array from -2*pi to 2*pi
t = -2*pi:pi/10:2*pi;

% Calculate |sin(t)|
x = abs(sin(t));

% Plot result
plot(t,x);
```

Save the file with the name test2.m, and execute it by typing test2 in the Command Window. What happens?

1.10 Close the Figure Window, and change back to the original directory that MATLAB started up in. Next, type "test2" in the Command Window. What happens, and why?

Chapter 2

MATLAB Basics

In this chapter, we will introduce some basic elements of the MATLAB language. By the end of the chapter, you will be able to write simple but functional MATLAB programs.

2.1 Variables and Arrays

The fundamental unit of data in any MATLAB program is the **array**. An array is a collection of data values organized into rows and columns and is known by a single name (See Figure 2.1). Individual data values within an array are accessed by including the name of the array followed by subscripts in parentheses, which identify the row and column of the particular value. Even scalars are treated as arrays by MATLAB—they are simply arrays with only one row and one column.

Arrays can be classified as either **vectors** or **matrices**. The term "vector" is usually used to describe an array with only one dimension, while the term "matrix" is usually used to describe an array with two or more dimensions. In this text, we will use the term "vector" when discussing one-dimensional arrays and the term "matrix" when discussing arrays with two or more dimensions. If a particular discussion applies to both types of arrays, we will use the generic term "array."

The **size** of an array is specified by the number of rows and the number of columns in the array, with the number of rows mentioned first. The total number of elements in the array will be the product of the number of rows and the number of columns. For example, the sizes of the following arrays are

Array	Size
$a = \begin{bmatrix} 1 & 2 \\ 3 & 4 \\ 5 & 6 \end{bmatrix}$	This is a 3 × 2 matrix, containing 6 elements.
$b = \begin{bmatrix} 1 & 2 & 3 & 4 \end{bmatrix}$	This is a 1 × 4 array containing 4 elements, known as a **row vector**.
$c = \begin{bmatrix} 1 \\ 2 \\ 3 \end{bmatrix}$	This is a 3 × 1 array containing 3 elements, known as a **column vector**.

Individual elements in an array are addressed by the array name followed by the row and column of the particular element. If the array is a row or column vector, then only one subscript is required. For example, in the above arrays a(2,1) is 3 and c(2) = 2.

A MATLAB **variable** is a region of memory containing an array and is known by a user-specified name. The contents of the array may be used or modified at any time by including its name in an appropriate MATLAB command.

MATLAB variable names must begin with a letter, followed by any combination of letters, numbers, and the underscore (_) character. Only the first 63 characters are significant; if more than 63 are used, the remaining characters will be ignored. If two variables are declared with names that only differ in the 64th character, MATLAB will treat them as the same variable. MATLAB will issue a warning if it has to truncate a long variable name to 63 characters.

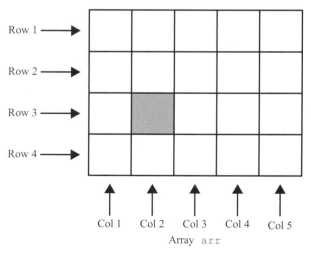

Figure 2.1 An array is a collection of data values organized into rows and columns.

▣ Programming Pitfalls

Make sure that your variable names are unique in the first 63 characters. Otherwise, MATLAB will not be able to tell the difference between them.

When writing a program, it is important to pick meaningful names for the variables. Meaningful names make a program *much* easier to read and to maintain. Names such as `day`, `month`, and `year` are quite clear even to a person seeing a program for the first time. Since spaces cannot be used in MATLAB variable names, underscore characters can be substituted to create meaningful names. For example, *exchange rate* might become `exchange_rate`.

⟨/⟩ Good Programming Practice

Always give your variables descriptive and easy-to-remember names. For example, a currency exchange rate could be given the name `exchange_rate`. This practice will make your programs clearer and easier to understand.

It is also important to include a **data dictionary** in the header of any program that you write. A data dictionary lists the definition of each variable used in a program. The definition should include both a description of the contents of the item and the units in which it is measured. A data dictionary may seem unnecessary while the program is being written, but it is invaluable when you or another person has to go back and modify the program at a later time.

⟨/⟩ Good Programming Practice

Create a data dictionary for each program to make program maintenance easier.

The MATLAB language is case-sensitive, which means that uppercase and lowercase letters are not the same. Thus the variables `name`, `NAME`, and `Name` are all different in MATLAB. You must be careful to use the same capitalization every time that variable name is used.

⟨/⟩ Good Programming Practice

Be sure to capitalize a variable exactly the same way each time that it is used. It is good practice to use only lowercase letters in variable names.

Many MATLAB programmers use the convention that variable names use all lowercase letters, with underscores between words. The variable `exchange_rate` mentioned previously is an example of this convention. It is used in this book.

Other MATLAB programmers use the convention common in Java and C++, where underscores are not used, the first word is all lowercase, and all subsequent words are capitalized. The same variable written in this convention would be `exchangeRate`. Either convention is fine, but be consistent throughout your programs.

Good Programming Practice

Adopt a standard naming and capitalization convention, and use it consistently throughout your programs.

The most common types of MATLAB variables are `double` and `char`. Variables of type `double` consist of scalars or arrays of 64-bit double-precision floating-point numbers. They can hold real, imaginary, or complex values. The real and imaginary components of each variable can be positive or negative numbers in the range 10^{-308} to 10^{308}, with 15 to 16 significant decimal digits of accuracy, plus the number zero. They are the principal numerical data type in MATLAB.

A variable of type `double` is automatically created whenever a numerical value is assigned to a variable name. The numerical values assigned to `double` variables can be real, imaginary, or complex. A real value is just a number. For example, the following statement assigns the real value 10.5 to the `double` variable `var`:

```
var = 10.5
```

An imaginary number is defined by appending the letter `i` or `j` to a number[1]. For example, `10i` and `-4j` are both imaginary values. The following statement assigns the imaginary value $4i$ to the `double` variable `var`:

```
var = 4i
```

A complex value has both a real and an imaginary component. It is created by adding a real and an imaginary number together. For example, the following statement assigns the complex value $10 + 10i$ to variable `var`:

```
var = 10 + 10i
```

Variables of type `char` consist of scalars or arrays of 16-bit values, each representing a single character. Arrays of this type are used to hold character strings. They are automatically created whenever a single character or a character

---

[1]An imaginary number is a number multiplied by $\sqrt{-1}$. The letter i is the symbol for $\sqrt{-1}$ used by most mathematicians and scientists. The letter j is the symbol for $\sqrt{-1}$ used by electrical engineers, because the letter i is usually reserved for currents in that discipline.

string is assigned to a variable name. For example, the following statement creates a variable of type `char` whose name is `comment` and stores the specified string in it. After the statement is executed, `comment` will be a 1×26 character array.

```
comment = 'This is a character string'
```

In a language such as C, the type of every variable must be explicitly declared in a program before it is used. These languages are said to be **strongly typed**. In contrast, MATLAB is a **weakly typed** language. Variables may be created at any time by simply assigning values to them, and the type of data assigned to the variable determines the type of variable that is created.

2.2 Creating and Initializing Variables in MATLAB

MATLAB variables are automatically created when they are initialized. There are three common ways to initialize a variable in MATLAB:

1. Assign data to the variable in an assignment statement.
2. Input data into the variable from the keyboard.
3. Read data from a file.

The first two ways will be discussed here, and the third approach will be discussed in Section 2.6.

2.2.1 Initializing Variables in Assignment Statements

The simplest way to initialize a variable is to assign it one or more values in an **assignment statement**. An assignment statement has the general form

```
var = expression;
```

where `var` is the name of a variable, and `expression` is a scalar constant, an array, or combination of constants, other variables, and mathematical operations ($+$, $-$, and so forth). The value of the expression is calculated using the normal rules of mathematics, and the resulting values are stored in the named variable. The semicolon at the end of the statement is optional. If the semicolon is absent, the value assigned to `var` will be echoed in the Command Window. If it is present, nothing will be displayed in the Command Window, even though the assignment has occurred.

Simple examples of initializing variables with assignment statements include

```
var = 40i;
var2 = var / 5;
x = 1; y = 2;
array = [1 2 3 4];
```

The first example creates a scalar variable of type `double`, and stores the imaginary number $40i$ in it. The second example creates a scalar variable and stores the result of the expression `var/5` in it. The third example shows that multiple assignment statements can be placed on a single line, provided that they are separated by semicolons or commas. The fourth example creates a variable and stores a 4-element row vector in it. Note that if any of the variables had already existed when the statements were executed, then their old contents would have been lost.

The last example shows that variables can also be initialized with arrays of data. Such arrays are constructed using brackets ([ ]) and semicolons. All of the elements of an array are listed in **row order**. In other words, the values in each row are listed from left to right, with the topmost row first and the bottommost row last. Individual values within a row are separated by blank spaces or commas, and the rows themselves are separated by semicolons or new lines. The following expressions are all legal arrays that can be used to initialize a variable:

`[3.4]`	This expression creates a 1×1 array (a scalar) containing the value 3.4. The brackets are not required in this case.
`[1.0  2.0  3.0]`	This expression creates a 1×3 array containing the row vector $[1\ 2\ 3]$.
`[1.0; 2.0; 3.0]`	This expression creates a 3×1 array containing the column vector $\begin{bmatrix} 1 \\ 2 \\ 3 \end{bmatrix}$.
`[1, 2, 3; 4, 5, 6]`	This expression creates a 2×3 array containing the matrix $\begin{bmatrix} 1 & 2 & 3 \\ 4 & 5 & 6 \end{bmatrix}$.
`[1, 2, 3` ` 4, 5, 6]`	This expression creates a 2×3 array containing the matrix $\begin{bmatrix} 1 & 2 & 3 \\ 4 & 5 & 6 \end{bmatrix}$. The end of the first line terminates the first row.
`[]`	This expression creates an **empty array**, which contains no rows and no columns. (Note that this is not the same as an array containing zeros.)

The number of elements in every row of an array must be the same, and the number of elements in every column must be the same. An expression such as

```
[1 2 3; 4 5];
```

is illegal because row 1 has three elements while row 2 has only two elements.

Programming Pitfalls

The number of elements in every row of an array must be the same, and the number of elements in every column must be the same. Attempts to define an array with different numbers of elements in its rows or different numbers of elements in its columns will produce an error when the statement is executed.

The expressions used to initialize arrays can include algebraic operations and all or portions of previously defined arrays. For example, the assignment statements

```
a = [0 1+7];
b = [a(2) 7 a];
```

will define an array a = [0   8] and an array b = [8   7   0   8].

Also, not all of the elements in an array must be defined when it is created. If a specific array element is defined and one or more of the elements before it are not, then the earlier elements will automatically be created and initialized to zero. For example, if c is not previously defined, the statement

```
c(2,3) = 5;
```

will produce the matrix $c = \begin{bmatrix} 0 & 0 & 0 \\ 0 & 0 & 5 \end{bmatrix}$. Similarly, an array can be extended by specifying a value for an element beyond the currently defined size. For example, suppose that array d = [1   2]. Then the statement

```
d(4) = 4;
```

will produce the array d = [1   2   0   4], as previously explained.

The semicolon at the end of each assignment statement shown above has a special purpose: it *suppresses the automatic echoing of values* that normally occurs whenever an expression is evaluated in an assignment statement. If an assignment statement is typed without the semicolon, the result of the statement is automatically displayed in the Command Window:

```
» e = [1, 2, 3; 4, 5, 6]
e =
   1  2  3
   4  5  6
```

If a semicolon is added at the end of the statement, the echoing disappears. Echoing is an excellent way to quickly check your work, but it seriously slows down the execution of MATLAB programs. For that reason, we normally suppress echoing at all times by ending each line with a semicolon.

However, echoing the results of calculations makes a great quick-and-dirty debugging tool. If you are not certain what the results of a specific assignment statement are, just leave off the semicolon from that statement, and the results will be displayed in the Command Window as the statement is executed.

Good Programming Practice

Use a semicolon at the end of all MATLAB assignment statements to suppress echoing of assigned values in the Command Window. This greatly speeds program execution.

Good Programming Practice

If you need to examine the results of a statement during program debugging, you may remove the semicolon from that statement only so that its results are echoed in the Command Window.

2.2.2 Initializing with Shortcut Expressions

It is easy to create small arrays by explicitly listing each term in the array, but what happens when the array contains hundreds or even thousands of elements? It is just not practical to write out each element in the array separately!

MATLAB provides a special shortcut notation for these circumstances using the **colon operator**. The colon operator specifies a whole series of values by specifying the first value in the series, the stepping increment, and the last value in the series. The general form of a colon operator is

```
first:incr:last
```

where `first` is the first value in the series, `incr` is the stepping increment, and `last` is the last value in the series. If the increment is one, it may be omitted. This expression will generate an array containing the values `first`, `first+incr`, `first+2*incr`, `first+3*incr`, and so forth as long as the values are less than or equal to `last`. The list stops when the next value in the series is greater than the value of `last`.

For example, the expression 1:2:10 is a shortcut for a 1×5 row vector containing the values 1, 3, 5, 7, and 9. The next value in the series would be 11, which is greater than 10, so the series terminates at 9.

```
» x = 1:2:10
x =
    1   3   5   7   9
```

With colon notation, an array can be initialized to have the hundred values $\frac{\pi}{100}, \frac{2\pi}{100}, \frac{3\pi}{100}, ..., \pi$ as follows:

```
angles = (0.01:0.01:1.00) * pi;
```

Shortcut expressions can be combined with the **transpose operator** (') to initialize column vectors and more complex matrices. The transpose operator swaps the row and columns of any array that it is applied to. Thus the expression

```
f = [1:4]' ;
```

generates a 4-element row vector [1   2   3   4], and then transposes it into the

4-element column vector $f = \begin{bmatrix} 1 \\ 2 \\ 3 \\ 4 \end{bmatrix}$. Similarly, the expressions

```
g = 1:4;
h = [g' g'];
```

will produce the matrix $h = \begin{bmatrix} 1 & 1 \\ 2 & 2 \\ 3 & 3 \\ 4 & 4 \end{bmatrix}$.

2.2.3 Initializing with Built-in Functions

Arrays can also be initialized using built-in MATLAB functions. For example, the function `zeros` can be used to create an all-zero array of any desired size. There are several forms of the `zeros` function. If the function has a single scalar argument, it will produce a square array using the single argument as both the number of rows and the number of columns. If the function has two scalar arguments, the first argument will be the number of rows, and the second argument will be the number of columns. Since the `size` function returns two values containing the number of rows and columns in an array, it can be combined with the `zeros` function to generate an array of zeros that is the same size as another array. Some examples using the `zeros` function follow:

```
a = zeros(2);
b = zeros(2,3);
c = [1 2; 3 4];
d = zeros(size(c));
```

These statements generate the following arrays:

$$a = \begin{bmatrix} 0 & 0 \\ 0 & 0 \end{bmatrix} \qquad b = \begin{bmatrix} 0 & 0 & 0 \\ 0 & 0 & 0 \end{bmatrix}$$

$$c = \begin{bmatrix} 1 & 2 \\ 3 & 4 \end{bmatrix} \qquad d = \begin{bmatrix} 0 & 0 \\ 0 & 0 \end{bmatrix}$$

Similarly, the `ones` function can be used to generate arrays containing all ones, and the `eye` function can be used to generate arrays containing **identity matrices**, in which all on-diagonal elements are one, while all off-diagonal elements are zero. Table 2.1 contains list of common MATLAB functions useful for initializing variables.

2.2.4 Initializing Variables with Keyboard Input

It is also possible to prompt a user and initialize a variable with data that he or she types directly at the keyboard. This option allows a script file to prompt a user for

Table 2.1: MATLAB Functions Useful for Initializing Variables

Function	Purpose
`zeros(n)`	Generates an n × n matrix of zeros.
`zeros(m,n)`	Generates an m × n matrix of zeros.
`zeros(size(arr))`	Generates a matrix of zeros of the same size as `arr`.
`ones(n)`	Generates an n × n matrix of ones.
`ones(m,n)`	Generates an m × n matrix of ones.
`ones(size(arr))`	Generates a matrix of ones of the same size as `arr`.
`eye(n)`	Generates an n × n identity matrix.
`eye(m,n)`	Generates an m × n identity matrix.
`length(arr)`	Returns the length of a vector, or the longest dimension of a 2-D array.
`size(arr)`	Returns two values specifying the number of rows and columns in `arr`.

input data values while it is executing. The `input` function displays a prompt string in the Command Window and then waits for the user to type in a response. For example, consider the following statement:

```
my_val = input('Enter an input value:');
```

When this statement is executed, MATLAB prints out the string `'Enter an input value:'` and then waits for the user to respond. If the user enters a single number, it may just be typed in. If the user enters an array, it must be enclosed in brackets. In either case, whatever is typed will be stored in variable `my_val` when the return key is entered. If only the return key is entered, then an empty matrix will be created and stored in the variable.

If the `input` function includes the character `'s'` as a second argument, then the input data is stored in the returned variable as a character string. Thus, the statement

```
» in1 = input('Enter data: ');
Enter data: 1.23
```

stores the value 1.23 into `in1`, while the statement

```
» in2 = input('Enter data: ','s');
Enter data: 1.23
```

stores the character string `'1.23'` into `in2`.

Quiz 2.1

This quiz provides a quick check to see if you have understood the concepts introduced in Sections 2.1 and 2.2. If you have trouble with the quiz, reread the sections, ask your instructor, or discuss the material with a fellow student. The answers to this quiz are found in the back of the book.

1. What is the difference between an array, a matrix, and a vector?
2. Answer the following questions for the array shown below.

$$c = \begin{bmatrix} 1.1 & -3.2 & 3.4 & 0.6 \\ 0.6 & 1.1 & -0.6 & 3.1 \\ 1.3 & 0.6 & 5.5 & 0.0 \end{bmatrix}$$

   (a) What is the size of `c`?
   (b) What is the value of `c(2,3)`?
   (c) List the subscripts of all elements containing the value 0.6.

3. Determine the size of the following arrays. Check you answers by entering the arrays into MATLAB and using the `whos` command or the Workspace Browser. Note that the later arrays may depend on the definitions of arrays defined earlier in this exercise.

   (a) `u = [10 20*i 10+20];`
   (b) `v = [-1; 20; 3];`
   (c) `w = [1 0 -9; 2 -2 0; 1 2 3];`
   (d) `x = [u' v];`
   (e) `y(3,3) = -7;`
   (f) `z = [zeros(4,1) ones(4,1) zeros(1,4)'];`
   (g) `v(4) = x(2,1);`

4. What is the value of `w(2,1)` above?
5. What is the value of `x(2,1)` above?
6. What is the value of `y(2,1)` above?
7. What is the value of `v(3)` after statement (g) is executed?

2.3 Multidimensional Arrays

As we have seen, MATLAB arrays can have one or more dimensions. One-dimensional arrays can be visualized as a series of values laid out in a row or column, with a single subscript used to select the individual array elements (Figure 2.2a). Such arrays are useful to describe data that is a function of one independent variable, such as a series of temperature measurements made at fixed intervals of time.

Some types of data are functions of more than one independent variable. For example, we might wish to measure the temperature at five different locations at four different times. In this case, our 20 measurements could logically be grouped into five different columns of four measurements each, with a separate column for each location (Figure 2.2b). In this case, we will use two subscripts to access a given element in the array: the first one to select the row and the second one to select the column. Such arrays are called **two-dimensional arrays**. The number of elements in a two-dimensional array will be the product of the number of rows and the number of columns in the array.

MATLAB allows us to create arrays with as many dimensions as necessary for any given problem. These arrays have one subscript for each dimension, and an individual element is selected by specifying a value for each subscript. The total

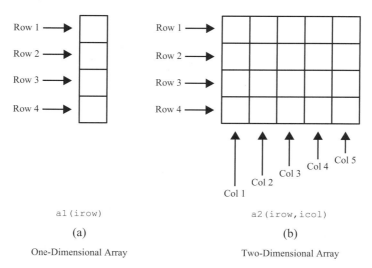

al(irow)

(a)

One-Dimensional Array

a2(irow,icol)

(b)

Two-Dimensional Array

Figure 2.2 Representations of one- and two-dimensional arrays.

number of elements in the array will be the product of the maximum value of each subscript. For example, the following two statements create a $2 \times 3 \times 2$ array c:

```
» c(:,:,1)=[1 2 3; 4 5 6];
» c(:,:,2)=[7 8 9; 10 11 12];
» whos c
```

```
Name     Size      Bytes     Class       Attributes
c        2x3x2        96      double
```

This array contains 12 elements ($2 \times 3 \times 2$). Its contents can be displayed just like any other array.

```
» c
c(:,:,1)  =
            1        2        3
            4        5        6
c(:,:,2)  =
            7        8        9
           10       11       12
```

2.3.1 Storing Multidimensional Arrays in Memory

A two-dimensional array with m rows and n columns will contain m × n elements, and these elements will occupy m × n successive locations in the computer's memory. How are the elements of the array arranged in the computer's memory? MATLAB always allocates array elements in **column major order**. That is, MATLAB allocates the first column in memory, then the second, then the third, etc., until all of the columns have been allocated. Figure 2.3 illustrates this memory allocation scheme

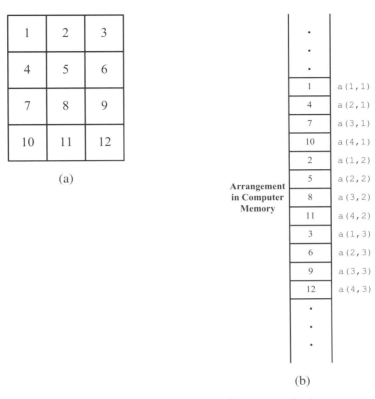

Figure 2.3 (a) Data values for array a. (b) Layout of values in memory for array a.

for a 4 × 3 array a. As we can see, element a(1,2) is really the fifth element allocated in memory. The order that elements are allocated in memory will become important when we discuss single-subscript addressing in the next section, and low-level I/O functions in Chapter 8.

This same allocation scheme applies to arrays with more than two dimensions. The first array subscript is incremented most rapidly, the second subscript is incremented less rapidly, and so forth, and the last subscript in incremented most slowly. For example, in a 2 × 2 × 2 array, the elements would be allocated in the following order: (1,1,1), (2,1,1), (1,2,1), (2,2,1), (1,1,2), (2,1,2), (1,2,2), (2,2,2).

2.3.2 Accessing Multidimensional Arrays with One Dimension

One of MATLAB's peculiarities is that it will permit a user or programmer to treat a multidimensional array as though it were a one-dimensional array whose length is equal to the number of elements in the multidimensional array. If a multidimensional array is addressed with a single dimension, then the elements will be accessed in the order in which they were allocated in memory.

For example, suppose that we declare the 4×3 element array `a` as follows:

```
» a = [1 2 3; 4 5 6; 7 8 9; 10 11 12]
a =
         1         2         3
         4         5         6
         7         8         9
        10        11        12
```

Then the value of `a(5)` will be 2, which is the value of element `a(1,2)`, because `a(1,2)` was allocated fifth in memory.

Under normal circumstances, you should never use this feature of MATLAB. Addressing multidimensional arrays with a single subscript is a recipe for confusion.

⟨/⟩ Good Programming Practice

Always use the proper number of dimensions when addressing a multidimensional array.

//

2.4 Subarrays

It is possible to select and use subsets of MATLAB arrays as though they were separate arrays. To select a portion of an array, just include a list of all of the elements to be selected in the parentheses after the array name. For example, suppose array `arr1` is defined as follows:

```
    arr1 = [1.1 -2.2 3.3 -4.4 5.5];
```

Then `arr1(3)` is just 3.3, `arr1([1  4])` is the array `[1.1  -4.4]`, and `arr1(1:2:5)` is the array `[1.1 3.3 5.5]`.

For a two-dimensional array, a colon can be used in a subscript to select all of the values of that subscript. For example, suppose

```
    arr2 = [1 2 3; -2 -3 -4; 3 4 5];
```

This statement creates an array `arr2` containing the values $\begin{bmatrix} 1 & 2 & 3 \\ -2 & -3 & -4 \\ 3 & 4 & 5 \end{bmatrix}$.

With this definition, the subarray `arr2(1,:)` is `[1  2  3]`, and the subarray `arr2(:,1:2:3)` is $\begin{bmatrix} 1 & 3 \\ -2 & -4 \\ 3 & 5 \end{bmatrix}$.

2.4.1   The end Function

MATLAB includes a special function named end, which is very useful for creating array subscripts. When used in an array subscript, end *returns the highest value taken on by that subscript*. For example, suppose that array arr3 is defined as follows:

```
arr3 = [1 2 3 4 5 6 7 8];
```

Then arr3(5:end) would be the array [5 6 7 8], and arr3(end) would be the value 8.

The value returned by end is always the *highest value* of a given subscript. If end appears in different subscripts, it can return *different* values within the same expression. For example, suppose that the 3×4 array arr4 is defined as follows:

```
arr4 = [1 2 3 4; 5 6 7 8; 9 10 11 12];
```

Then the expression arr4(2:end,2:end) would return the array $\begin{bmatrix} 6 & 7 & 8 \\ 10 & 11 & 12 \end{bmatrix}$.

Note that the first end returned the value 3, while the second end returned the value 4!

2.4.2   Using Subarrays on the Left-hand Side of an Assignment Statement

It is also possible to use subarrays on the left-hand side of an assignment statement to update only some of the values in an array, as long as the **shape** (the number of rows and columns) of the values being assigned matches the shape of the subarray. If the shapes do not match, then an error will occur. For example, suppose that the 3×4 array arr4 is defined as follows:

```
» arr4 = [1 2 3 4; 5 6 7 8; 9 10 11 12]
arr4 =
     1     2     3     4
     5     6     7     8
     9    10    11    12
```

Then the following assignment statement is legal, since the expressions on both sides of the equal sign have the same shape (2×2):

```
» arr4(1:2,[1 4]) = [20 21; 22 23]
arr4 =
    20     2     3    21
    22     6     7    23
     9    10    11    12
```

Note that the array elements (1,1), (1,4), (2,1), and (2,4) were updated. In contrast, the following expression is illegal because the two sides do not have the same shape.

```
» arr5(1:2,1:2) = [3 4]
??? In an assignment A(matrix,matrix) = B, the number
of rows in B and the number of elements in the A row
index matrix must be the same.
```

⊠ Programming Pitfalls

For assignment statements involving subarrays, the *shapes of the subarrays on either side of the equal sign must match.* MATLAB will produce an error if they do not match.

There is a major difference in MATLAB between assigning values to a subarray and assigning values to an array. If values are assigned to a subarray, *only those values are updated, while all other values in the array remain unchanged.* On the other hand, if values are assigned to an array, *the entire contents of the array are deleted and replaced by the new values.* For example, suppose that the 3×4 array arr4 is defined as follows:

```
» arr4 = [1 2 3 4; 5 6 7 8; 9 10 11 12]
arr4 =
     1     2     3     4
     5     6     7     8
     9    10    11    12
```

Then the following assignment statement replaces the *specified elements* of arr4:

```
» arr4(1:2,[1 4]) = [20 21; 22 23]
arr4 =
    20     2     3    21
    22     6     7    23
     9    10    11    12
```

In contrast, the following assignment statement replaces the *entire contents* of arr4 with a 2×2 array:

```
» arr4 = [20 21; 22 23]
arr4 =
    20    21
    22    23
```

</> Good Programming Practice

Be sure to distinguish between assigning values to a subarray and assigning values to an array. MATLAB behaves differently in these two cases.

2.4.3 Assigning a Scalar to a Subarray

A scalar value on the right-hand side of an assignment statement always matches the shape specified on the left-hand side. The scalar value is copied into every element specified on the left-hand side of the statement. For example, assume that the 3×4 array `arr4` is defined as follows:

```
arr4 = [1 2 3 4; 5 6 7 8; 9 10 11 12];
```

Then the expression shown below assigns the value one to four elements of the array.

```
» arr4(1:2,1:2) = 1
arr4 =
     1     1     3     4
     1     1     7     8
     9    10    11    12
```

2.5 Special Values

MATLAB includes a number of predefined special values. These predefined values may be used at any time in MATLAB without initializing them first. A list of the most common predefined values is given in Table 2.2.

Table 2.2: Predefined Special Values

Function	Purpose
pi	Contains π to 15 significant digits.
i, j	Contain the value i ($\sqrt{-1}$).
Inf	This symbol represents machine infinity. It is usually generated as a result of a division by 0.
NaN	This symbol stands for Not-a-Number. It is the result of an undefined mathematical operation, such as the division of zero by zero.
clock	This special variable contains the current date and time in the form of a 6-element row vector containing the year, month, day, hour, minute, and second.
date	Contains the current data in a character string format, such as 24-Nov-1998.
eps	This variable name is short for "epsilon". It is the smallest difference between two numbers that can be represented on the computer.
ans	A special variable used to store the result of an expression if that result is not explicitly assigned to some other variable.

These predefined values are stored in ordinary variables, so they can be overwritten or modified by a user. If a new value is assigned to one of the predefined variables, then that new value will replace the default one in all later calculations. For example, consider the following statements that calculate the circumference of a circle with a radius of 10 cm:

```
circ1 = 2 * pi * 10
pi = 3;
circ2 = 2 * pi * 10
```

In the first statement, `pi` has its default value of 3.14159..., so `circ1` is 62.8319, which is the correct circumference. The second statement redefines `pi` to be 3, so in the third statement `circ2` is 60. Changing a predefined value in the program has created an incorrect answer and also introduced a subtle and hard-to-find bug. Imagine trying to locate the source of such a hidden error in a 10,000-line program!

///

▣ Programming Pitfalls

Never redefine the meaning of a predefined variable in MATLAB. It is a recipe for disaster, producing subtle and hard-to-find bugs.

///

Quiz 2.2

This quiz provides a quick check to see if you have understood the concepts introduced in Sections 2.3 through 2.5. If you have trouble with the quiz, reread the sections, ask your instructor, or discuss the material with a fellow student. The answers to this quiz are found in the back of the book.

1. Assume that array c is defined as shown, and determine the contents of the following sub-arrays:

$$
c = \begin{bmatrix} 1.1 & -3.2 & 3.4 & 0.6 \\ 0.6 & 1.1 & -0.6 & 3.1 \\ 1.3 & 0.6 & 5.5 & 0.0 \end{bmatrix}
$$

    (a) `c(2,:)`
    (b) `c(:,end)`
    (c) `c(1:2,2:end)`
    (d) `c(6)`
    (e) `c(4:end)`
    (f) `c(1:2,2:4)`
    (g) `c([1 3],2)`
    (h) `c([2 2],[3 3])`

2. Determine the contents of array a after the following statements are executed.

   (a) ```
   a = [1 2 3; 4 5 6; 7 8 9];
   a([3 1],:) = a([1 3],:);
   ```

   (b) ```
   a = [1 2 3; 4 5 6; 7 8 9];
   a([1 3],:) = a([2 2],:);
   ```

   (c) ```
   a = [1 2 3; 4 5 6; 7 8 9];
   a = a([2 2],:);
   ```

3. Determine the contents of array a after the following statements are executed.

   (a) ```
   a = eye(3,3);
   b = [1 2 3];
   a(2,:) = b;
   ```

   (b) ```
   a = eye(3,3);
   b = [4 5 6];
   a(:,3) = b';
   ```

   (c) ```
   a = eye(3,3);
   b = [7 8 9];
   a(3,:) = b([3 1 2]);
   ```

2.6 Displaying Output Data

There are several ways to display output data in MATLAB. The simplest way is one we have already seen—just leave the semicolon off of the end of a statement and it will be echoed to the Command Window. We will now explore a few other ways to display data.

2.6.1 Changing the Default Format

When data is echoed in the Command Window, integer values are always displayed as integers, character values are displayed as strings, and other values are printed using a **default format**. The default format for MATLAB shows four digits after the decimal point, and it may be displayed in scientific notation with an exponent if the number is too large or too small. For example, the statements

```
x = 100.11
y = 1001.1
z = 0.00010011
```

produce the following output

```
x =
   100.1100

y =
   1.0011e+003

z =
   1.0011e-004
```

This default format can be changed in one of two ways: from the main MATLAB Window menu or using the **format** command. You can change the format by selecting the Preferences icon on the Toolstrip. This option will pop up the preferences window (see Figure 2.4), and the format can be selected from the Command Window item in the preferences list.

Alternately, a user can use the `format` command to change the preferences. The format command changes the default format according to the values given in Table 2.3. The default format can be modified to display more significant digits of data, force the display to be in scientific notation, display data to two decimal digits, or eliminate extra line feeds to get more data visible in the Command Window at a single time. Experiment with the commands in Table 2.3 for yourself.

Which of these ways to change the data format is better? If you are working directly at the computer, it is probably easier to use the Toolbar. On the other hand, if you are writing programs, it is probably better to use the `format` command, because it can be embedded directly into a program.

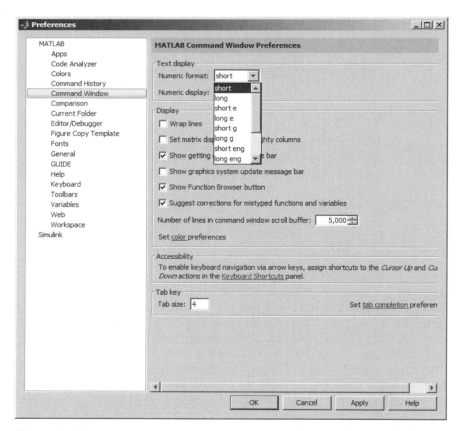

Figure 2.4 Selecting the desired numeric format within the Command Window preferences.

Table 2.3: Output Display Formats

Format Command	Results	Example[1]
format short	4 digits after decimal (default format)	12.3457
format long	14 digits after decimal	12.34567890123457
format short e	5 digits plus exponent	1.2346e+001
format short g	5 total digits with or without exponent	12.346
format long e	15 digits plus exponent	1.234567890123457e+001
format long g	15 total digits with or without exponent	12.3456789012346
format bank	"dollars and cents" format	12.35
format hex	hexadecimal display of bits	4028b0fcd32f707a
format rat	approximate ratio of small integers	1000/81
format compact	suppress extra line feeds	
format loose	restore extra line feeds	
format +	Only signs are printed	+

[1]The data value used for the example is 12.345678901234567 in all cases.

2.6.2 The disp Function

Another way to display data is with the disp function. The disp function accepts an array argument and displays the value of the array in the Command Window. If the array is of type char, then the character string contained in the array is printed out.

This function is often combined with the functions num2str (convert a number to a string) and int2str (convert an integer to a string) to create messages to be displayed in the Command Window. For example, the following MATLAB statements will display "The value of pi = 3.1416" in the Command Window. The first statement creates a string array containing the message, and the second statement displays the message.

```
str = ['The value of pi = ' num2str(pi)];
disp (str);
```

2.6.3 Formatted Output with the fprintf Function

An even more flexible way to display data is with the fprintf function. The fprintf function displays one or more values together with related text and lets the programmer control the way that the displayed value appears. The general form of this function when it is used to print to the Command Window is:

```
fprintf(format,data)
```

where format is a string describing the way the data is to be printed, and data is one or more scalars or arrays to be printed. The format is a character string containing text to be printed plus special characters describing the format of the data. For example, the function

```
fprintf('The value of pi is %f \n',pi)
```

will print out `'The value of pi is 3.141593'` followed by a line feed. The characters `%f` are called **conversion characters**; they indicate that the value in the data list should be printed out in floating point format at that location in the format string. The characters `\n` are **escape characters**; they indicate that a line feed should be issued so that the following text starts on a new line. There are many types of conversion characters and escape characters that may be used in an `fprintf` function. A few of them are listed in Table 2.4, and a complete list can be found in Chapter 11.

It is also possible to specify the width of the field in which a number will be displayed and the number of decimal places to display. This is done by specifying the the width and precision after the `%` sign and before the `f`. For example, the function

```
fprintf('The value of pi is %6.2f \n',pi)
```

will print out `'The value of pi is 3.14'` followed by a line feed. The conversion characters `%6.2f` indicate that the first data item in the function should be printed out in floating point format in a field six characters wide, including two digits after the decimal point.

The `fprintf` function has one very significant limitation: *it only displays the real portion of a complex value*. This limitation can lead to misleading results when calculations produce complex answers. In those cases, it is better to use the `disp` function to display answers.

For example, the following statements calculate a complex value x and display it using both `fprintf` and `disp`.

```
x = 2 * ( 1 - 2*i )^3;
str = ['disp:    x = ' num2str(x)];
disp(str);
fprintf('fprintf: x = %8.4f\n',x);
```

Table 2.4: Common Special Characters in `fprintf` Format Strings

Format String	Results
%d	Display value as an integer.
%e	Display value in exponential format.
%f	Display value in floating point format.
%g	Display value in either floating point or exponential format, whichever is shorter.
\n	Skip to a new line.

The results printed out by these statements are

```
disp:    x = -22+4i
fprintf: x = -22.0000
```

Note that the `fprintf` function ignored the imaginary part of the answer.

 ## Programming Pitfalls

The `fprintf` function only displays the *real* part of a complex number, which can produce misleading answers when working with complex values.

2.7 Data Files

There are many ways to load and save data files in MATLAB, most of which will be addressed in Chapter 11. For the moment, we will consider only the **load** and **save** commands, which are the simplest ones to use.

The **save** command saves data from the current MATLAB workspace into a disk file. The most common form of this command is

```
save filename var1 var2 var3
```

where `filename` is the name of the file where the variables are saved, and `var1`, `var2`, and so forth are the variables to be saved in the file. By default, the file name will be given the extension "`mat`", and such data files are called MAT-files. If no variables are specified, then the entire contents of the workspace are saved.

MATLAB saves MAT-files in a special compact format which preserves many details, including the name and type of each variable, the size of each array, and all data values. A MAT-file created on any platform (PC, Mac, Unix, or Linux) can be read on any other platform, so MAT-files are a good way to exchange data between computers if both computers run MATLAB. Unfortunately, the MAT-file is in a format that cannot be read by other programs. If data must be shared with other programs, then the `-ascii` option should be specified, and the data values will be written to the file as ASCII character strings separated by spaces. However, the special information such as variable names and types are lost when the data is saved in ASCII format, and the resulting data file will be much larger.

For example, suppose the array x is defined as

```
x =[1.23 3.14 6.28; -5.1 7.00 0];
```

Then the command "`save x.dat x -ascii`" will produce a file named `x.dat` containing the following data:

```
  1.2300000e+000   3.1400000e+000   6.2800000e+000
 -5.1000000e+000   7.0000000e+000   0.0000000e+000
```

This data is in a format that can be read by spreadsheets or by programs written in other computer languages, so it makes it easy to share data between MATLAB programs and other applications.

Good Programming Practice

If data must be exchanged between MATLAB and other programs, save the MATLAB data in ASCII format. If the data will only be used in MATLAB, save the data in MAT-file format.

MATLAB doesn't care what file extension is used for ASCII files. However, it is better for the user if a consistent naming convention is used, and an extension of "dat" is a common choice for ASCII files.

Good Programming Practice

Save ASCII data files with a "dat" file extension to distinguish them from MAT-files, which have a "mat" file extension.

The **load** command is the opposite of the save command. It loads data from a disk file into the current MATLAB workspace. The most common form of this command is

```
load filename
```

where filename is the name of the file to be loaded. If the file is a MAT-file, then all of the variables in the file will be restored, with the names and types the same as before. If a list of variables is included in the command, then only those variables will be restored. If the given filename has no extension, or if the file extension is .mat, then the load command will treat the file as a MAT-file.

MATLAB can load data created by other programs in comma- or space-separated ASCII format. If the given filename has any file extension other than .mat, then the load command will treat the file as an ASCII file. The contents of an ASCII file will be converted into a MATLAB array having the same name as the file (without the file extension) that the data was loaded from. For example, suppose that an ASCII data file named x.dat contains the following data:

```
    1.23    3.14    6.28
    -5.1    7.00    0
```

Then the command "load x.dat" will create a 2 × 3 array named x in the current workspace, containing these data values.

The load statement can be forced to treat a file as a MAT-file by specifying the −mat option. For example, the statement

```
load −mat x.dat
```

would treat file x.dat as a MAT-file even though its file extension is not .mat. Similarly, the load statement can be forced to treat a file as an ASCII file by specifying the −ascii option. These options allow the user to load a file properly even if its file extension doesn't match the MATLAB conventions.

Quiz 2.3

This quiz provides a quick check to see if you have understood the concepts introduced in Sections 2.6 and 2.7. If you have trouble with the quiz, reread the sections, ask your instructor, or discuss the material with a fellow student. The answers to this quiz are found in the back of the book.

1. How would you tell MATLAB to display all real values in exponential format with 15 significant digits?
2. What do the following sets of statements do? What is the output from them?

    (a)
    ```
    radius = input('Enter circle radius:\n');
    area = pi * radius^2;
    str = ['The area is ' num2str(area)];
    disp(str);
    ```

    (b)
    ```
    value = int2str(pi);
    disp(['The value is ' value '!']);
    ```

3. What do the following sets of statements do? What is the output from them?

    ```
    value = 123.4567e2;
    fprintf('value = %e\n',value);
    fprintf('value = %f\n',value);
    fprintf('value = %g\n',value);
    fprintf('value = %12.4f\n',value);
    ```

2.8 Scalar and Array Operations

Calculations are specified in MATLAB with an assignment statement, whose general form is

```
variable_name = expression;
```

The assignment statement calculates the value of the expression to the right of the equal sign and *assigns* that value to the variable named on the left of the equal sign. Note that the equal sign does not mean equality in the usual sense of the word. Instead, it means: *store the value of* `expression` *into location* `variable_name`. For this reason, the equal sign is called the **assignment operator**. A statement like

```
ii = ii + 1;
```

is complete nonsense in ordinary algebra, but makes perfect sense in MATLAB. It means: take the current value stored in variable `ii`, add one to it, and store the result back into variable `ii`.

2.8.1 Scalar Operations

The expression to the right of the assignment operator can be any valid combination of scalars, arrays, parentheses, and arithmetic operators. The standard arithmetic operations between two scalars are given in Table 2.5.

Parentheses may be used to group terms whenever desired. When parentheses are used, the expressions inside the parentheses are evaluated before the expressions outside the parentheses. For example, the expression `2 ^ ((8+2)/5)` is evaluated as shown below

```
2 ^ ((8 + 2)/5) = 2 ^ (10/5)
                = 2 ^ 2
                = 4
```

2.8.2 Array and Matrix Operations

MATLAB supports two types of operations between arrays, known as *array operations* and *matrix operations*. **Array operations** are operations performed between arrays on an **element-by-element basis**. That is, the operation is performed on corresponding elements in the two arrays. For example, if $a = \begin{bmatrix} 1 & 2 \\ 3 & 4 \end{bmatrix}$ and $b = \begin{bmatrix} -1 & 3 \\ -2 & 1 \end{bmatrix}$, then $a + b = \begin{bmatrix} 0 & 5 \\ 1 & 5 \end{bmatrix}$. Note that for these operations to work, *the*

Table 2.5: Arithmetic Operations between Two Scalars

Operation	Algebraic Form	MATLAB Form
Addition	$a + b$	`a + b`
Subtraction	$a - b$	`a - b`
Multiplication	$a \times b$	`a * b`
Division	$\dfrac{a}{b}$	`a / b`
Exponentiation	a^b	`a ^ b`

number of rows and columns in both arrays must be the same. If not, MATLAB will generate an error message.

Array operations may also occur between an array and a scalar. If the operation is performed between an array and a scalar, the value of the scalar is applied to every element of the array. For example, if $a = \begin{bmatrix} 1 & 2 \\ 3 & 4 \end{bmatrix}$, then $a + 4 = \begin{bmatrix} 5 & 6 \\ 7 & 8 \end{bmatrix}$.

In contrast, **matrix operations** follow the normal rules of linear algebra, such as matrix multiplication. In linear algebra, the product $c = a \times b$ is defined by the equation

$$c(i,j) = \sum_{k=1}^{n} a(i,k)\, b(k,j)$$

where n is the number of columns on matrix a and the number of rows in matrix b.

For example, if $a = \begin{bmatrix} 1 & 2 \\ 3 & 4 \end{bmatrix}$ and $b = \begin{bmatrix} -1 & 3 \\ -2 & 1 \end{bmatrix}$, then $a \times b = \begin{bmatrix} -5 & 5 \\ -11 & 13 \end{bmatrix}$.

Note that for matrix multiplication to work, *the number of columns in matrix* a *must be equal to the number of rows in matrix* b.

MATLAB uses a special symbol to distinguish array operations from matrix operations. In the cases where array operations and matrix operations have a different definition, MATLAB uses a period before the symbol to indicate an array operation (for example, .*). A list of common array and matrix operations is given in Table 2.6.

Table 2.6: Common Array and Matrix Operations

Operation	MATLAB Form	Comments
Array Addition	a + b	Array addition and matrix addition are identical.
Array Subtraction	a - b	Array subtraction and matrix subtraction are identical.
Array Multiplication	a .* b	Element-by-element multiplication of a and b. Both arrays must be the same shape, or one of them must be a scalar.
Matrix Multiplication	a * b	Matrix multiplication of a and b. The number of columns in a must equal the number of rows in b.
Array Right Division	a ./ b	Element-by-element division of a and b: a(i,j) / b(i,j). Both arrays must be the same shape, or one of them must be a scalar.
Array Left Division	a .\ b	Element-by-element division of a and b, but with b in the numerator: b(i,j) / a(i,j). Both arrays must be the same shape, or one of them must be a scalar.
Matrix Right Division	a / b	Matrix division defined by a * inv(b), where inv(b) is the inverse of matrix b.

(continued)

Table 2.6: Common Array and Matrix Operations (*Continued*)

Operation	MATLAB Form	Comments
Matrix Left Division	a \ b	Matrix division defined by inv(a) * b, where inv(a) is the inverse of matrix a.
Array Exponentiation	a .^ b	Element-by-element exponentiation of a and b: a(i,j) ^ b(i,j). Both arrays must be the same shape, or one of them must be a scalar.

New users often confuse array operations and matrix operations. In some cases, substituting one for the other will produce an illegal operation, and MATLAB will report an error. In other cases, both operations are legal, and MATLAB will perform the wrong operation and come up with a wrong answer. The most common problem happens when working with square matrices. Both array multiplication and matrix multiplication are legal for two square matrices of the same size, but the resulting answers are totally different. Be careful to specify exactly what you want!

☒ Programming Pitfalls

Be careful to distinguish between array operations and matrix operations in your MATLAB code. It is especially common to confuse array multiplication with matrix multiplication.

►Example 2.1—Assume that a, b, c, and d are defined as follows

$$a = \begin{bmatrix} 1 & 0 \\ 2 & 1 \end{bmatrix} \qquad b = \begin{bmatrix} -1 & 2 \\ 0 & 1 \end{bmatrix}$$

$$c = \begin{bmatrix} 3 \\ 2 \end{bmatrix} \qquad d = 5$$

What is the result of each of the following expressions?

(a) a + b

(b) a .* b

(c) a * b

(d) a * c

(e) a + c

(f) a + d

(g) a .* d

(h) a * d

Solution

(a) This is array or matrix addition: $a + b = \begin{bmatrix} 0 & 2 \\ 2 & 2 \end{bmatrix}$

(b) This is element-by-element array multiplication: $a \cdot * \ b = \begin{bmatrix} -1 & 0 \\ 0 & 1 \end{bmatrix}$

(c) This is matrix multiplication: $a \ * \ b = \begin{bmatrix} -1 & 2 \\ -2 & 5 \end{bmatrix}$

(d) This is matrix multiplication: $a \ * \ c = \begin{bmatrix} 3 \\ 8 \end{bmatrix}$

(e) This operation is illegal, since a and c have different numbers of columns.

(f) This is addition of an array to a scalar: $a + d = \begin{bmatrix} 6 & 5 \\ 7 & 6 \end{bmatrix}$

(g) This is array multiplication: $a \cdot * \ d = \begin{bmatrix} 5 & 0 \\ 10 & 5 \end{bmatrix}$

(h) This is matrix multiplication: $a \ * \ d = \begin{bmatrix} 5 & 0 \\ 10 & 5 \end{bmatrix}$

The matrix left division operation has a special significance that we must understand. A 3×3 set of simultaneous linear equations takes the form

$$a_{11}x_1 + a_{12}x_2 + a_{13}x_3 = b_1$$
$$a_{21}x_1 + a_{22}x_2 + a_{23}x_3 = b_2 \qquad (2.1)$$
$$a_{31}x_1 + a_{32}x_2 + a_{33}x_3 = b_3$$

which can be expressed as

$$Ax = B \qquad (2.2)$$

where $A = \begin{bmatrix} a_{11} & a_{12} & a_{13} \\ a_{21} & a_{22} & a_{23} \\ a_{31} & a_{32} & a_{33} \end{bmatrix}$, $B = \begin{bmatrix} b_1 \\ b_2 \\ b_3 \end{bmatrix}$, and $x = \begin{bmatrix} x_1 \\ x_2 \\ x_3 \end{bmatrix}$.

Equation (2.2) can be solved for x using linear algebra. If A is a non-singular (*i.e.*, invertible) matrix, the result is

$$x = A^{-1}B \qquad (2.3)$$

Since the left division operator $A \setminus B$ is defined to be `inv(A) * B`, the left division operator solves a system of simultaneous equations in a single statement!

Good Programming Practice:

Use the left division operator to solve systems of simultaneous equations.

2.9 Hierarchy of Operations

Often, many arithmetic operations are combined into a single expression. For example, consider the equation for the distance traveled by an object starting from rest and subjected to a constant acceleration:

```
distance = 0.5 * accel * time ^ 2
```

There are two multiplications and an exponentiation in this expression. In such an expression, it is important to know the order in which the operations are evaluated. If exponentiation is evaluated before multiplication, this expression is equivalent to

```
distance = 0.5 * accel * (time ^ 2)
```

But if multiplication is evaluated before exponentiation, this expression is equivalent to

```
distance = (0.5 * accel * time) ^ 2
```

These two equations have different results, and we must be able to unambiguously distinguish between them.

To make the evaluation of expressions unambiguous, MATLAB has established a series of rules governing the hierarchy or order in which operations are evaluated within an expression. The rules generally follow the normal rules of algebra. The order in which the arithmetic operations are evaluated is given in Table 2.7.

Table 2.7: Hierarchy of Arithmetic Operations

Precedence	Operation
1	The contents of all parentheses are evaluated, starting from the innermost parentheses and working outward.
2	All exponentials are evaluated, working from left to right.
3	All multiplications and divisions are evaluated, working from left to right.
4	All additions and subtractions are evaluated, working from left to right.

►**Example 2.2—Variables a, b, c, and d have been initialized to the following values**

```
      a = 3;         b = 2;         c = 5;         d = 3;
```

Evaluate the following MATLAB assignment statements:

(a) `output = a*b+c*d;`
(b) `output = a*(b+c)*d;`
(c) `output = (a*b)+(c*d);`
(d) `output = a^b^d;`
(e) `output = a^(b^d);`

Solution

(a) Expression to evaluate:

output = a*b+c*d;

Fill in numbers:

output = 3*2+5*3;

First, evaluate multiplications
and divisions from left to right:

output = 6 +5*3;

output = 6 + 15;

Now evaluate additions:

output = 21

(b) Expression to evaluate:

output = a*(b+c)*d;

Fill in numbers:

output = 3*(2+5)*3;

First, evaluate parentheses:

output = 3*7*3;

Now, evaluate multiplications
and divisions from left to right:

output = 21*3;

output = 63;

(c) Expression to evaluate:

output = (a*b)+(c*d);

Fill in numbers:

output = (3*2)+(5*3);

First, evaluate parentheses:

output = 6 + 15;

Now evaluate additions:

output = 21

(d) Expression to evaluate:

output = a^b^d;

Fill in numbers:

output = 3^2^3;

Evaluate exponentials
from left to right:

output = 9^3;

output = 729;

(e) Expression to evaluate:

output = a^(b^d);

Fill in numbers:

output = 3^(2^3);

First, evaluate parentheses:

output = 3^8;

Now, evaluate exponential:

output = 6561;

◀

As we see above, the order in which operations are performed has a major effect on the final result of an algebraic expression.

It is important that every expression in a program be made as clear as possible. Any program of value must not only be written but also be maintained and modified when necessary. You should always ask yourself: "Will I easily understand this expression if I come back to it in six months? Can another programmer look at my code and easily understand what I am doing?" If there is any doubt in your mind, use extra parentheses in the expression to make it as clear as possible.

Good Programming Practice

Use parentheses as necessary to make your equations clear and easy to understand.

///

If parentheses are used within an expression, then the parentheses must be balanced. That is, there must be an equal number of open parentheses and close parentheses within the expression. It is an error to have more of one type than the other.

Errors of this sort are usually typographical, and they are caught by the MATLAB interpreter when the command is executed. For example, the expression

$$(2 + 4) / 2)$$

produces an error when the expression is executed.

Quiz 2.4

This quiz provides a quick check to see if you have understood the concepts introduced in Sections 2.8 and 2.9. If you have trouble with the quiz, reread the sections, ask your instructor, or discuss the material with a fellow student. The answers to this quiz are found in the back of the book.

1. Assume that a, b, c, and d are defined as follows, and calculate the results of the following operations if they are legal. If an operation is, explain why it is illegal.

$$a = \begin{bmatrix} 2 & 1 \\ -1 & 2 \end{bmatrix} \qquad b = \begin{bmatrix} 0 & -1 \\ 3 & 1 \end{bmatrix}$$

$$c = \begin{bmatrix} 1 \\ 2 \end{bmatrix} \qquad d = -3$$

   (a) result = a .* c;
   (b) result = a * [c c];
   (c) result = a .* [c c];
   (d) result = a + b * c;
   (e) result = a + b .* c;

2. Solve for x in the equation $Ax = B$, where $A = \begin{bmatrix} 1 & 2 & 1 \\ 2 & 3 & 2 \\ -1 & 0 & 1 \end{bmatrix}$

   and $B = \begin{bmatrix} 1 \\ 1 \\ 0 \end{bmatrix}$.

2.10 Built-in MATLAB Functions

In mathematics, a **function** is an expression that accepts one or more input values and calculates a single result from them. Scientific and technical calculations usually require functions that are more complex than the simple addition, subtraction, multiplication, division, and exponentiation operations that we have discussed so far. Some of these functions are very common and are used in many different technical disciplines. Others are rarer and specific to a single problem or a small number of problems. Examples of very common functions are the trigonometric functions, logarithms, and square roots. Examples of rarer functions include the hyperbolic functions, Bessel functions, and so forth. One of MATLAB's greatest strengths is that it comes with an incredible variety of built-in functions ready for use.

2.10.1 Optional Results

Unlike mathematical functions, MATLAB functions can return *more than one result* to the calling program. The function max is an example of such a function. This function normally returns the maximum value of an input vector, but it can also return a second argument containing the location in the input vector where the maximum value was found. For example, the statement

```
maxval = max ([1 -5 6 -3])
```

returns the result maxval = 6. However, if two variables are provided to store results in, the function returns *both* the maximum value *and* the location of the maximum value.

```
[maxval, index] = max ([1 -5 6 -3])
```

produces the results maxval = 6 and index = 3.

2.10.2 Using MATLAB Functions with Array Inputs

Many MATLAB functions are defined for one or more scalar inputs and produce a scalar output. For example, the statement y = sin(x) calculates the sine of x and stores the result in y. If these functions receive an array of input values, then they will calculate an array of output values on an element-by-element basis. For example, if x = [0 pi/2 pi 3*pi/2 2*pi], then the statement

```
y = sin(x)
```

will produce the result y = [0 1 0 -1 0] .

2.10.3 Common MATLAB Functions

A few of the most common and useful MATLAB functions are shown in Table 2.8. These functions will be used in many examples and homework problems. If you need to locate a specific function not on this list, you can search for the function alphabetically or by subject using the MATLAB Help Browser.

Note that, unlike most computer languages, many MATLAB functions work correctly for both real and complex inputs. MATLAB functions automatically calculate the correct answer, even if the result is imaginary or complex. For example, the function sqrt(-2) will produce a runtime error in languages such as C++, Java, or Fortran. In contrast, MATLAB correctly calculates the imaginary answer:

```
» sqrt(-2)
ans =
    0 + 1.4142i
```

Table 2.8: Common MATLAB Functions

Function	Description
Mathematical functions	
vabs(x)	Calculates $|x|$.
acos(x)	Calculates $\cos^{-1}x$ (results in radians).
acosd(x)	Calculates $\cos^{-1}x$ (results in degrees).
angle(x)	Returns the phase angle of the complex value x, in radians.
asin(x)	Calculates $\sin^{-1}x$ (results in radians).
asind(x)	Calculates $\sin^{-1}x$ (results in degrees).
atan(x)	Calculates $\tan^{-1}x$ (results in radians).
atand(x)	Calculates $\tan^{-1}x$ (results in degrees).
atan2(y,x)	Calculates $\theta = \tan^{-1}\dfrac{y}{x}$ over all four quadrants of the circle, taking into account the boundaries between the quadrants (results in radians in the range $-\pi \le \theta \le \pi$).
atan2d(y,x)	Calculates $\theta = \tan^{-1}\dfrac{y}{x}$ over all four quadrants of the circle, taking into account the boundaries between the quadrants (results in degrees in the range $-180° \le \theta \le 180°$).
cos(x)	Calculates $\cos x$, with x in radians.
cosd(x)	Calculates $\cos x$, with x in degrees.
exp(x)	Calculates e^x.
log(x)	Calculates the natural logarithm $\log_e x$
[value,index] = max(x)	Returns the maximum value in vector x, and optionally the location of that value.
[value,index] = min(x)	Returns the minimum value in vector x, and optionally the location of that value.
mod(x,y)	Remainder or modulo function.
sin(x)	Calculates $\sin x$, with x in radians.
sind(x)	Calculates $\sin x$, with x in degrees.
sqrt(x)	Calculates the square root of x.
tan(x)	Calculates $\tan x$, with x in radians.
tand(x)	Calculates $\tan x$, with x in degrees.
Rounding functions	
ceil(x)	Rounds x to the nearest integer towards positive infinity: ceil(3.1) = 4 and ceil(-3.1) = -3.
fix(x)	Rounds x to the nearest integer towards zero: fix(3.1) = 3 and fix(-3.1) = -3.
floor(x)	Rounds x to the nearest integer towards minus infinity: floor(3.1) = 3 and floor(-3.1) = -4.
round(x)	Rounds x to the nearest integer.

(continued)

Table 2.8: Common **MATLAB** Functions (*Continued*)

String conversion functions	
char(x)	Converts a matrix of numbers into a character string. For ASCII characters the matrix should contain numbers ≤ 127.
double(x)	Converts a character string into a matrix of numbers.
int2str(x)	Converts x into an integer character string.
num2str(x)	Converts x into a character string.
str2num(s)	Converts character string s into a numeric array.

2.11   Introduction to Plotting

MATLAB's extensive, device-independent plotting capabilities are one of its most powerful features. They make it very easy to plot any data at any time. To plot a data set, just create two vectors containing the x and y values to be plotted, and use the plot function.

For example, suppose that we wish to plot the function $y = x^2 - 10x + 15$ for values of x between 0 and 10. It takes only three statements to create this plot. The first statement creates a vector of x values between 0 and 10 using the colon operator. The second statement calculates the y values from the equation (note that we are using array operators here so that this equation is applied to each x value on an element-by-element basis). Finally, the third statement creates the plot.

```
x = 0:1:10;
y = x.^2 - 10.*x + 15;
plot(x,y);
```

When the plot function is executed, MATLAB opens a Figure Window and displays the plot in that window. The plot produced by these statements is shown in Figure 2.5.

2.11.1   Using Simple *xy* Plots

As we saw above, plotting is *very* easy in MATLAB. Any pair of vectors can be plotted versus each other as long as both vectors have the same length. However, the result is not a finished product, since there are no titles, axis labels, or grid lines on the plot.

Titles and axis labels can be added to a plot with the title, xlabel, and ylabel functions. Each function is called with a string containing the title or label to be applied to the plot. Grid lines can be added or removed from the plot with the grid command: grid on turns on grid lines, and grid off turns off grid lines. For example, the statements that follow generate a plot of the function $y = x^2 - 10x + 15$ with titles, labels, and gridlines. The resulting plot is shown in Figure 2.6.

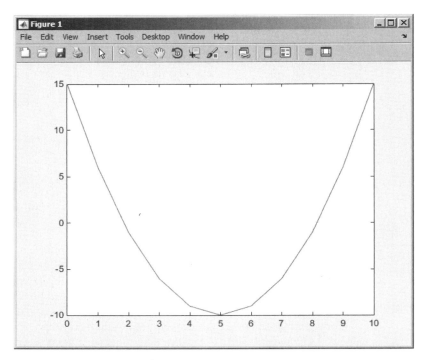

Figure 2.5 Plot of $y = x^2 - 10x + 15$ from 0 to 10.

```
x = 0:1:10;
y = x.^2 - 10.*x + 15;
plot(x,y);
title ('Plot of y = x.^2 - 10.*x + 15');
xlabel ('x');
ylabel ('y');
grid on;
```

2.11.2 Printing a Plot

Once created, a plot may be printed on a printer with the `print` command, by clicking on the "print" icon in the Figure Window, or by selecting the File/Print menu option in the Figure Window.

The `print` command is especially useful because it can be included in a MATLAB program, allowing the program to automatically print graphical images. The form of the `print` command is:

```
print <options> <filename>
```

If no filename is included, this command prints a copy of the current figure on the system printer. If a filename is specified, the command prints a copy of the current figure to the specified file.

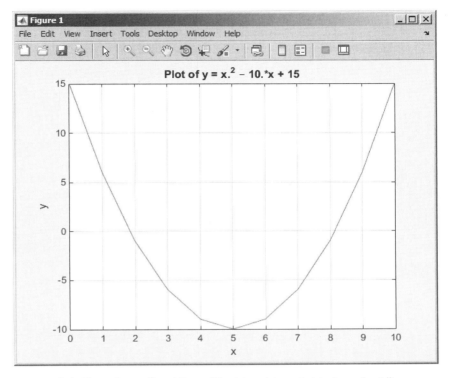

Figure 2.6 Plot of $y = x^2 - 10x + 15$ with a title, axis labels, and gridlines.

2.11.3 Exporting a Plot as a Graphical Image

The `print` command can be used to save a plot as a graphical image by specifying appropriate options and a file name.

```
print <options> <filename>
```

There are many different options that specify the format of the output sent to a file. One very important option is `-dpng`. This option specifies that the output will be to a file in Portable Network Graphics Format (PNG). Since this format can be imported into all of the important word processors on PC, Mac, Unix, and Linux platforms, it is a great way to include MATLAB plots in a document. The following command will create a 300 dot-per-inch PNG image of the current figure and store it in a file called `my_image.png`:

```
print -dpng -r300 my_image.png
```

Note that the `-png` specifies that the image should be in PNG format, and the `-r300` specifies that the resolution should be 300 dots per inch.

Other options allow image files to be created in other formats. Some of the most important image file formats are given in Table 2.9.

Table 2.9: `print` Options to Create Graphics Files

Option	Description
`-deps`	Creates a monochrome encapsulated postscript image.
`-depsc`	Creates a color encapsulated postscript image.
`-djpeg`	Creates a JPEG image.
`-dpng`	Creates a Portable Network Graphic color image.
`-dtiff`	Creates a compressed TIFF image.

In addition, the File/Save As menu option on the Figure Window can be used to save a plot as a graphical image. In this case, the user selects the file name and the type of image from a standard dialog box (see Figure 2.7).

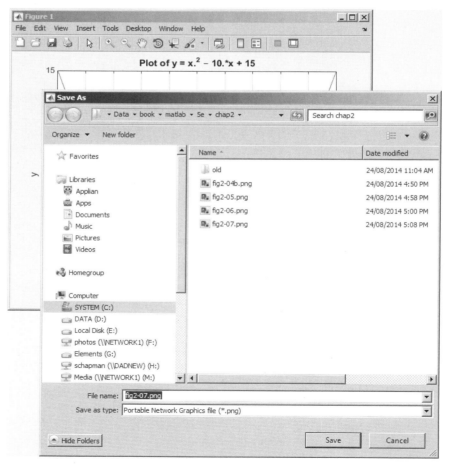

Figure 2.7 Exporting a plot as an image file using the File/Save As menu item.

2.11.4 Multiple Plots

It is possible to plot multiple functions on the same graph by simply including more than one set of (x, y) values in the plot function. For example, suppose that we wanted to plot the function $f(x) = \sin 2x$ and its derivative on the same plot. The derivative of $f(x) = \sin 2x$ is:

$$\frac{d}{dt} \sin 2x = 2 \cos 2x \qquad (2.4)$$

To plot both functions on the same axes, we must generate a set of x values and the corresponding y values for each function. Then to plot the functions, we would simply list both sets of (x, y) values in the plot function as shown below.

```
x = 0:pi/100:2*pi;
y1 = sin(2*x);
y2 = 2*cos(2*x);
plot(x,y1,x,y2);
```

The resulting plot is shown in Figure 2.8.

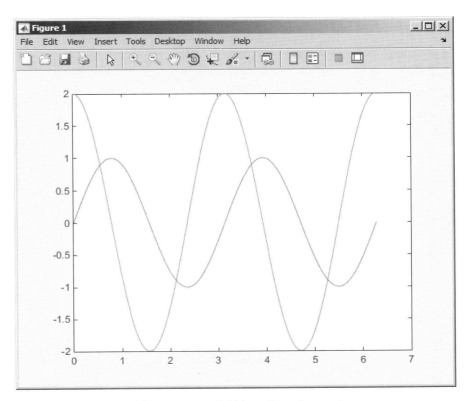

Figure 2.8 Plot of $f(x) = \sin 2x$ and $f(x) = 2 \cos 2x$ on the same axes.

2.11.5 Line Color, Line Style, Marker Style, and Legends

MATLAB allows a programmer to select the color of a line to be plotted, the style of the line to be plotted, and the type of marker to be used for data points on the line. These traits may be selected using an attribute character string after the x and y vectors in the plot function.

The attribute character string can have up to three characters, with the first character specifying the color of the line, the second character specifying the style of the marker, and the last character specifying the style of the line. The characters for various colors, markers, and line styles are shown in Table 2.10.

The attribute characters may be mixed in any combination, and more than one attribute string may be specified if more than one pair of (x, y) vectors are included in a single `plot` function call. For example, the following statements will plot the function $y = x^2 - 10x + 15$ with a dashed red line and include the actual data points as blue circles (see Figure 2.9).

```
x = 0:1:10;
y = x.^2 - 10.*x + 15;
plot(x,y,'r--',x,y,'bo');
```

Legends may be created with the `legend` function. The basic form of this function is

```
legend('string1','string2',..., pos)
```

Table 2.10: Table of Plot Colors, Marker Styles, and Line Styles

Color		Marker Style		Line Style	
y	yellow	.	point	–	solid
m	magenta	o	circle	:	dotted
c	cyan	x	x-mark	-.	dash-dot
r	red	+	plus	--	dashed
g	green	*	star	<none>	no line
b	blue	s	square		
w	white	d	diamond		
k	black	v	triangle (down)		
		^	triangle (up)		
		<	triangle (left)		
		>	triangle (right)		
		p	pentagram		
		h	hexagram		
		<none>	no marker		

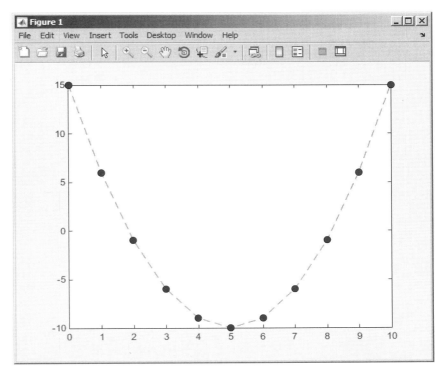

Figure 2.9 Plot of the function $y = x^2 - 10x + 15$ with a dashed line, showing the actual data points as circles.

where string1, string2, and so forth are the labels associated with the lines plotted, and pos is a string specifying where to place the legend. The possible values for pos are given in Table 2.11, and are shown graphically in Figure 2.10.

The command legend off will remove an existing legend.

An example of a complete plot is shown in Figure 2.11, and the statements to produce that plot are shown below. They plot the function $f(x) = \sin 2x$ and its derivative $f'(x) = 2\cos 2x$ on the same axes, with a solid black line for $f(x)$ and a dashed line for its derivative. The plot includes a title, axis labels, a legend in the top left corner of the plot, and grid lines.

```
x = 0:pi/100:2*pi;
y1 = sin(2*x);
y2 = 2*cos(2*x);
plot(x,y1,'k-',x,y2,'b--');
title ('Plot of f(x) = sin(2x) and its derivative');
xlabel ('x');
ylabel ('y');
legend ('f(x)','d/dx f(x)','Location','NW')
grid on;
```

Table 2.11: Values of `pos` in the `legend` Command

Value	Legend Location
'NW'	Above and to the left
'NL'	Above top left corner
'NC'	Above center of top edge
'NR'	Above top right corner
'NE'	Above and to right
'TW'	At top and to left
'TL'	Top left corner
'TC'	At top center
'TR'	Top right corner
'TE'	At top and to right
'MW'	At middle and to left
'ML'	Middle left edge
'MC'	Middle and center
'MR'	Middle right edge
'ME'	At middle and to right
'BW'	At bottom and to left
'BL'	Bottom left corner
'BC'	At bottom center
'BR'	Bottom right corner
'BE'	At bottom and to right
'SW'	Below and to left
'SL'	Below bottom left corner
'SC'	Below center of bottom edge
'SR'	Below bottom right corner
'SE'	Below and to right

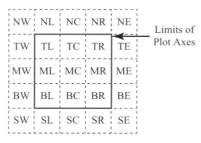

Figure 2.10 Possible locations for a plot legend.

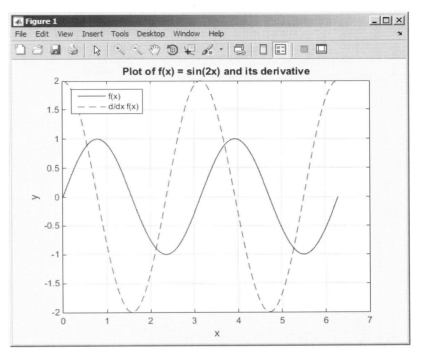

Figure 2.11 A complete plot with title, axis labels, legend, grid, and multiple line styles.

2.11.6 Logarithmic Scales

It is possible to plot data on logarithmic scales as well as linear scales. There are four possible combinations of linear and logarithmic scales on the *x*- and *y*- axes, and each combination is produced by a separate function.

1. The `plot` function plots both *x* and *y* data on linear axes.
2. The `semilogx` function plots *x* data on logarithmic axes and *y* data on linear axes.
3. The `semilogy` function plots *x* data on linear axes and *y* data on logarithmic axes.
4. The `loglog` function plots both *x* and *y* data on logarithmic axes.

All of these functions have identical calling sequences—the only difference is the type of axis used to plot the data. Examples of each plot are shown in Figure 2.12.

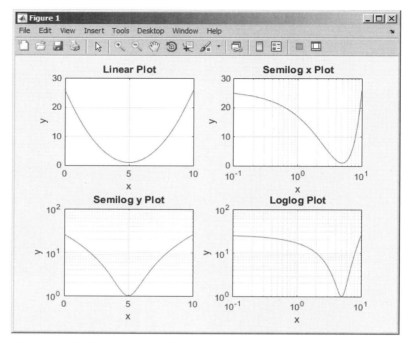

Figure 2.12 Comparison of linear, semilog *x*, semilog *y*, and log-log plots.

2.12 Examples

The following examples illustrate problem solving with MATLAB.

▶Example 2.3—Temperature Conversion

Design a MATLAB program that reads an input temperature in degrees Fahrenheit, converts it to an absolute temperature in kelvin, and writes out the result.

Solution The relationship between temperature in degrees Fahrenheit (°F) and temperature in kelvins (K) can be found in any physics textbook. It is

$$T \text{ (in kelvin)} = \left[\frac{5}{9} T \text{ (in °F)} - 32.0 \right] + 273.15 \qquad (2.5)$$

The physics books also give us sample values on both temperature scales, which we can use to check the operation of our program. Two such values are:

The boiling point of water	212° F	373.15 K
The sublimation point of dry ice	-110° F	194.26 K

Our program must perform the following steps:

1. Prompt the user to enter an input temperature in °F.
2. Read the input temperature.

3. Calculate the temperature in kelvin from Equation (2.5).
4. Write out the result and stop.

We will use function `input` to get the temperature in degrees Fahrenheit and function `fprintf` to print the answer. The resulting program is shown below.

```
%   Script file: temp_conversion
%
%   Purpose:
%     To convert an input temperature from degrees
%     Fahrenheit to an output temperature in kelvins.
%
%   Record of revisions:
%       Date        Programmer         Description of change
%       ====        ==========         =====================
%     01/03/14     S. J. Chapman      Original code
%
% Define variables:
%    temp_f     -- Temperature in degrees Fahrenheit
%    temp_k     -- Temperature in kelvins

% Prompt the user for the input temperature.
temp_f = input('Enter the temperature in degrees Fahrenheit:');

% Convert to kelvin.
temp_k = (5/9) * (temp_f - 32) + 273.15;

% Write out the result.
fprintf('%6.2f degrees Fahrenheit = %6.2f kelvins.\n', ...
        temp_f,temp_k);
```

To test the completed program, we will run it with the known input values given above. Note that user inputs appear in bold face below.

```
» temp_conversion
Enter the temperature in degrees Fahrenheit: 212
212.00 degrees Fahrenheit = 373.15 kelvins.
» temp_conversion
Enter the temperature in degrees Fahrenheit: -110
-110.00 degrees Fahrenheit = 194.26 kelvins.
```

The results of the program match the values from the physics book.

◄

In the above program, we echoed the input values and printed the output values together with their units. The results of this program only make sense if the units (degrees Fahrenheit and kelvins) are included together with their values. As a general rule, the units associated with any input value should always be printed along with the prompt that requests the value, and the units associated with any output value should always be printed along with that value.

The above program exhibits many of the good programming practices that we have described in this chapter. It includes a data dictionary defining the meanings of all of the variables in the program. It also uses descriptive variable names, and appropriate units are attached to all printed values.

►Example 2.4—Electrical Engineering: Maximum Power Transfer to a Load

Figure 2.13 shows a voltage source $V = 120$ V with an internal resistance R_S of 50 Ω supplying a load of resistance R_L. Find the value of load resistance R_L that will result in the maximum possible power being supplied by the source to the load. How much power will be supplied in this case? Also, plot the power supplied to the load as a function of the load resistance R_L.

Solution In this program, we need to vary the load resistance R_L and compute the power supplied to the load at each value of R_L. The power supplied to the load resistance is given by the equation

$$P_L = I^2 R_L \tag{2.6}$$

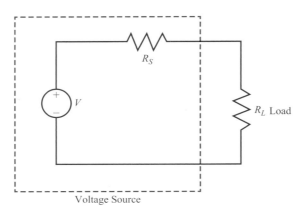

Figure 2.13 A voltage source with a voltage V and an internal resistance R_S supplying a load of resistance R_L.

where I is the current supplied to the load. The current supplied to the load can be calculated by Ohm's Law:

$$I = \frac{V}{R_{\text{TOT}}} = \frac{V}{R_S + R_L} \tag{2.7}$$

The program must perform the following steps:

1. Create an array of possible values for the load resistance R_L. The array will vary R_L from 1 Ω to 100 Ω in 1 Ω steps.
2. Calculate the current for each value of R_L.
3. Calculate the power supplied to the load for each value of R_L.
4. Plot the power supplied to the load for each value of R_L and determine the value of load resistance resulting in the maximum power.

The final MATLAB program is shown below.

```
%  Script file: calc_power.m
%
%  Purpose:
%     To calculate and plot the power supplied to a load as
%     a function of the load resistance.
%
%  Record of revisions:
%       Date       Programmer        Description of change
%       ====       ==========        =====================
%     01/03/14   S. J. Chapman   Original code
%
% Define variables:
%     amps         -- Current flow to load (amps)
%     pl           -- Power supplied to load (watts)
%     rl           -- Resistance of the load (ohms)
%     rs           -- Internal resistance of the power source
%                     (ohms)
%     volts        -- Voltage of the power source (volts)

% Set the values of source voltage and internal resistance
volts = 120;
rs = 50;

% Create an array of load resistances
rl = 1:1:100;

% Calculate the current flow for each resistance
amps = volts ./ ( rs + rl );

% Calculate the power supplied to the load
pl = (amps .^ 2) .* rl;
```

```
% Plot the power versus load resistance
plot(rl,pl);
title('Plot of power versus load resistance');
xlabel('Load resistance (ohms)');
ylabel('Power (watts)');
grid on;
```

When this program is executed, the resulting plot is shown in Figure 2.14. From this plot, we can see that the maximum power is supplied to the load when the load's resistance is 50 Ω. The power supplied to the load at this resistance is 72 watts.

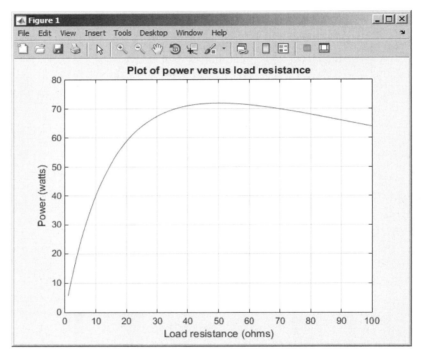

Figure 2.14 Plot of power supplied to load versus load resistance.

Note the use of the array operators `.*`, `.^`, and `./` in the above program. These operators cause the arrays `amps` and `pl` to be calculated on an element-by-element basis.

►Example 2.5—Carbon-14 Dating

A radioactive isotope of an element is a form of the element that is not stable. Instead, it spontaneously decays into another element over a period of time. Radioactive decay is an exponential process. If Q_0 is the initial quantity of a radioactive substance at time $t = 0$, then the amount of that substance which will be present at any time t in the future is given by

$$Q(t) = Q_0 e^{-\lambda t} \tag{2.8}$$

where λ is the radioactive decay constant.

Because radioactive decay occurs at a known rate, it can be used as a clock to measure the time since the decay started. If we know the initial amount of the radioactive material Q_0 present in a sample and the amount of the material Q left at the current time, we can solve for t in Equation (2.8) to determine how long the decay has been going on. The resulting equation is

$$t_{\text{decay}} = -\frac{1}{\lambda} \log_e \frac{Q}{Q_0} \tag{2.9}$$

Equation (2.9) has practical applications in many areas of science. For example, archaeologists use a radioactive clock based on carbon-14 to determine the time that has passed since a once-living thing died. Carbon-14 is continually taken into the body while a plant or animal is living, so the amount of it present in the body at the time of death is assumed to be known. The decay constant λ of carbon-14 is well known to be 0.00012097/year, so if the amount of carbon-14 remaining now can be accurately measured, then Equation (2.9) can be used to determine how long ago the living thing died. The amount of carbon-14 remaining as a function of time is shown in Figure 2.15.

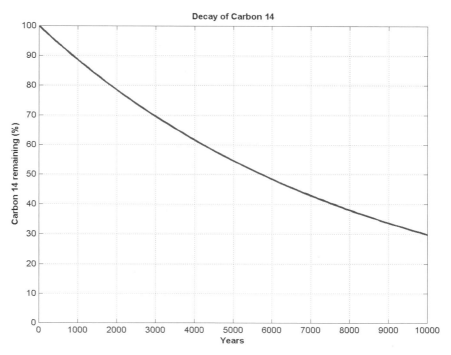

Figure 2.15 The radioactive decay of carbon-14 as a function of time. Notice that 50 percent of the original carbon-14 is left after about 5730 years have elapsed.

Write a program that reads the percentage of carbon-14 remaining in a sample, calculates the age of the sample from it, and prints out the result with proper units.

Solution Our program must perform the following steps:

1. Prompt the user to enter the percentage of carbon-14 remaining in the sample.
2. Read in the percentage.
3. Convert the percentage into the fraction $\dfrac{Q}{Q_0}$.
4. Calculate the age of the sample in years using Equation (2.9).
5. Write out the result and stop.

The resulting code is shown below.

```
% Script file: c14_date.m
%
% Purpose:
%   To calculate the age of an organic sample from the
%   percentage of the original carbon-14 remaining in
%   the sample.
%
% Record of revisions:
%     Date       Programmer        Description of change
%     ====       ==========        =====================
%   01/05/14    S. J. Chapman     Original code
%
% Define variables:
%   age        -- The age of the sample in years
%   lambda     -- The radioactive decay constant for
%                 carbon-14, in units of 1/years.
%   percent    -- The percentage of carbon-14 remaining
%                 at the time of the measurement
%   ratio      -- The ratio of the carbon-14 remaining at
%                 the time of the measurement to the
%                 original amount of carbon-14.

% Set decay constant for carbon-14
lambda = 0.00012097;

% Prompt the user for the percentage of C-14 remaining.
percent = input('Enter the percentage of carbon-14 remaining:\n');

% Perform calculations
ratio = percent / 100;        % Convert to fractional ratio
age = (-1.0 / lambda) * log(ratio); % Get age in years
```

```
% Tell the user about the age of the sample.
string = ['The age of the sample is' num2str(age) 'years.'];
disp(string);
```

To test the completed program, we will calculate the time it takes for half of the carbon-14 to disappear. This time is known as the *half-life* of carbon-14.

```
» c14_date
Enter the percentage of carbon-14 remaining:
50
The age of the sample is 5729.9097 years.
```

The *CRC Handbook of Chemistry and Physics* states that the half-life of carbon-14 is 5730 years, so the output of the program agrees with the reference book.

2.13 Debugging MATLAB Programs

There is an old saying that the only sure things in life are death and taxes. We can add one more certainty to that list: If you write a program of any significant size, it won't work the first time you try it! Errors in programs are known as **bugs**, and the process of locating and eliminating them is known as **debugging**. Given that we have written a program and it is not working, how do we debug it?

Three types of errors are found in MATLAB programs. The first type of error is a **syntax error**. Syntax errors are errors in the MATLAB statement itself, such as spelling errors or punctuation errors. These errors are detected by the MATLAB compiler the first time that an M-file is executed. For example, the statement

```
x = (y + 3) / 2);
```

contains a syntax error because it has unbalanced parentheses. If this statement appears in an M-file named test.m, the following message appears when test is executed.

```
» test
??? x = (y + 3) / 2)
                    |
Missing operator, comma, or semi-colon.

Error in ==> d:\book\matlab\chap1\test.m
On line 2  ==>
```

The second type of error is the **run-time error**. A run-time error occurs when an illegal mathematical operation is attempted during program execution (for example, attempting to divide by 0). These errors cause the program to return `Inf` or `NaN`, which is then used in further calculations. The results of a program that contains calculations using `Inf` or `NaN` are usually invalid.

The third type of error is a **logical error**. Logical errors occur when the program compiles and runs successfully but produces the wrong answer.

The most common mistakes made during programming are *typographical errors*. Some typographical errors create invalid MATLAB statements. These errors produce syntax errors that are caught by the compiler. Other typographical errors occur in variable names. For example, the letters in some variable names might have been transposed, or an incorrect letter might be typed. The result will be a new variable, and MATLAB simply creates the new variable the first time that it is referenced. MATLAB cannot detect this type of error. Typographical errors can also produce logical errors. For example, if variables `vel1` and `vel2` are both used for velocities in the program, then one of them might be inadvertently used instead of the other one at some point. You must check for that sort of error by manually inspecting the code.

Sometimes, a program will start to execute, but run-time errors or logical errors occur during execution. In this case, there is either something wrong with the input data or something wrong with the logical structure of the program. The first step in locating this sort of bug should be to *check the input data to the program*. Either remove semicolons from input statements or add extra output statements to verify that the input values are what you expect them to be.

If the variable names seem to be correct and the input data is correct, then you are probably dealing with a logical error. You should check each of your assignment statements.

1. If an assignment statement is very long, break it into several smaller assignment statements. Smaller statements are easier to verify.
2. Check the placement of parentheses in your assignment statements. It is a very common error to have the operations in an assignment statement evaluated in the wrong order. If you have any doubts as to the order in which the variables are being evaluated, add extra sets of parentheses to make your intentions clear.
3. Make sure that you have initialized all of your variables properly.
4. Be sure that any functions you use are in the correct units. For example, the input to trigonometric functions must be in units of radians, not degrees.

If you are still getting the wrong answer, add output statements at various points in your program to see the results of intermediate calculations. If you can locate the point where the calculations go bad, then you know just where to look for the problem, which is 95% of the battle.

If you still cannot find the problem after all of the above steps, explain what you are doing to another student or to your instructor, and let them look at the code. It is very common for a person to see just what he or she expects to see when they look at their own code. Another person can often quickly spot an error that you have overlooked time after time.

Good Programming Practice

To reduce your debugging effort, make sure that during your program design you:
1. Initialize all variables.
2. Use parentheses to make the functions of assignment statements clear.

MATLAB includes a special debugging tool called a *symbolic debugger*, which is embedded into the Edit/Debug Window. A symbolic debugger is a tool that allows you to walk through the execution of your program one statement at a time and to examine the values of any variables at each step along the way. Symbolic debuggers allow you to see all of the intermediate results without having to insert a lot of output statements into your code. We will learn how to use MATLAB's symbolic debugger in Chapter 3.

2.14 Summary

In this chapter, we have presented many of the fundamental concepts required to write functional MATLAB programs. We learned about the basic types of MATLAB windows, the workspace, and how to get online help.

We introduced two data types: `double` and `char`. We also introduced assignment statements, arithmetic calculations, intrinsic functions, input/output statements, and data files.

The order in which MATLAB expressions are evaluated follows a fixed hierarchy, with operations at a higher level evaluated before operations at lower levels. The hierarchy of operations is summarized in Table 2.12.

The MATLAB language includes an extremely large number of built-in functions to help us solve problems. This list of functions is *much* richer than the list of functions found in other languages like Fortran or C, and it includes device-independent plotting capabilities. A few of the common intrinsic functions are summarized in Table 2.8, and many others will be introduced throughout the remainder of the book. A complete list of all MATLAB functions is available through the online Help Desk.

Table 2.12: Hierarchy of Operations

Precedence	Operation
1	The contents of all parentheses are evaluated, starting from the innermost parentheses and working outward.
2	All exponentials are evaluated, working from left to right.
3	All multiplications and divisions are evaluated, working from left to right.
4	All additions and subtractions are evaluated, working from left to right.

2.14.1 Summary of Good Programming Practice

Every MATLAB program should be designed so that another person who is familiar with MATLAB can easily understand it. This is very important, since a good program may be used for a long period of time. Over that time, conditions will change, and the program will need to be modified to reflect the changes. The program modifications may be done by someone other than the original programmer. The programmer making the modifications must understand the original program well before attempting to change it.

It is much harder to design clear, understandable, and maintainable programs than it is to simply write programs. To do so, a programmer must develop the discipline to properly document his or her work. In addition, the programmer must be careful to avoid known pitfalls along the path to good programs. The following guidelines will help you to develop good programs:

1. Use meaningful variable names whenever possible. Use names that can be understood at a glance, like day, month, and year.
2. Create a data dictionary for each program to make program maintenance easier.
3. Use only lowercase letters in variable names, so that there won't be errors due to capitalization differences in different occurrences of a variable name.
4. Use a semicolon at the end of all MATLAB assignment statements to suppress echoing of assigned values in the Command Window. If you need to examine the results of a statement during program debugging, you may remove the semicolon from that statement only.
5. If data must be exchanged between MATLAB and other programs, save the MATLAB data in ASCII format. If the data will only be used in MATLAB, save the data in MAT-file format.
6. Save ASCII data files with a "dat" file extension to distinguish them from MAT-files, which have a "mat" file extension.
7. Use parentheses as necessary to make your equations clear and easy to understand.
8. Always include the appropriate units with any values that you read or write in a program.

2.14.2 MATLAB Summary

The following summary lists all of the MATLAB special symbols, commands, and functions described in this chapter, along with a brief description of each one.

Special Symbols

[ ]	Array constructor
( )	Forms subscripts
' '	Marks the limits of a character string
'	1. Separates subscripts or matrix elements
	2. Separates assignment statements on a line

(continued)

,	Separates subscripts or matrix elements
;	1. Suppresses echoing in Command Window
	2. Separates matrix rows
	3. Separates assignment statements on a line
%	Marks the beginning of a comment
:	Colon operator, used to create shorthand lists
+	Array and matrix addition
−	Array and matrix subtraction
.*	Array multiplication
*	Matrix multiplication
./	Array right division
.\	Array left division
/	Matrix right division
\	Matrix left division
.^	Array exponentiation
'	Transpose operator

Commands and Functions

...	Continues a MATLAB statement on the following line.
abs(x)	Calculates the absolute value of *x*.
ans	Default variable used to store the result of expressions not assigned to another variable.
acos(x)	Calculates the inverse cosine of *x*. The resulting angle is in radians between 0 and π.
acosd(x)	Calculates the inverse cosine of *x*. The resulting angle is in degrees between 0° and 180°.
asin(x)	Calculates the inverse sine of *x*. The resulting angle is in radians between $-\pi/2$ and $\pi/2$.
asind(x)	Calculates the inverse sine of *x*. The resulting angle is in degrees between −90° and 90°.
atan(x)	Calculates the inverse tangent of *x*. The resulting angle is in radians between $-\pi/2$ and $\pi/2$.
atand(x)	Calculates the inverse tangent of *x*. The resulting angle is in radians between −90° and 90°.
atan2(y,x)	Calculates the inverse tangent of *y/x*, taking into account the boundaries between the quadrants. The resulting angle is in radians between $-\pi$ and π.
atan2d(y,x)	Calculates the inverse tangent of *y/x*, taking into account the boundaries between the quadrants. The resulting angle is in degrees between −180° and 180°.
ceil(x)	Rounds *x* to the nearest integer towards positive infinity: floor(3.1) = 4 and floor(-3.1) = -3.

(continued)

Commands and Functions (*Continued*)

char	Converts a matrix of numbers into a character string. For ASCII characters the matrix should contain numbers ≤ 127.
clock	Current time.
cos(x)	Calculates cosine of x, where x is in radians.
cosd(x)	Calculates cosine of x, where x is in degrees.
date	Current date.
disp	Displays data in Command Window.
doc	Open HTML Help Desk directly at a particular function description.
double	Converts a character string into a matrix of numbers.
eps	Represents machine precision.
exp(x)	Calculates e^x.
eye(m,n)	Generates an identity matrix.
fix(x)	Rounds x to the nearest integer towards zero: fix (3.1) = 3 and fix(-3.1) = -3.
floor(x)	Rounds x to the nearest integer towards minus infinity: floor(3.1) = 3 and floor(-3.1) = -4.
format +	Prints + and − signs only.
format bank	Prints in "dollars and cents" format.
format compact	Suppresses extra linefeeds in output.
format hex	Prints hexadecimal display of bits.
format long	Prints with 14 digits after the decimal.
format long e	Prints with 15 digits plus exponent.
format long g	Prints with 15 digits with or without exponent.
format loose	Prints with extra linefeeds in output.
format rat	Prints as an approximate ratio of small integers.
format short	Prints with 4 digits after the decimal.
format short e	Prints with 5 digits plus exponent.
format short g	Prints with 5 digits with or without exponent.
fprintf	Prints formatted information.
grid	Adds/removes a grid from a plot.
i	$\sqrt{-1}$.
Inf	Represents machine infinity (∞).
input	Writes a prompt and reads a value from the keyboard.
int2str	Converts x into an integer character string.
j	$\sqrt{-1}$.
legend	Adds a legend to a plot.
length(arr)	Returns the length of a vector, or the longest dimension of a 2-D array.
load	Load data from a file.

(*continued*)

`log(x)`	Calculates the natural logarithm of x.
`loglog`	Generates a log-log plot.
`lookfor`	Look for a matching term in the one-line MATLAB function descriptions.
`max(x)`	Returns the maximum value in vector x, and optionally the location of that value.
`min(x)`	Returns the minimum value in vector x, and optionally the location of that value.
`mod(m,n)`	Remainder or modulo function.
`NaN`	Represents not-a-number.
`num2str(x)`	Converts x into a character string.
`ones(m,n)`	Generates an array of ones.
`pi`	Represents the number π.
`plot`	Generates a linear xy plot.
`print`	Prints a Figure Window.
`round(x)`	Rounds x to the nearest integer.
`save`	Saves data from workspace into a file.
`semilogx`	Generates a log-linear plot.
`semilogy`	Generates a linear-log plot.
`sin(x)`	Calculates sine of x, where x is in radians.
`sind(x)`	Calculates sine of x, where x is in degrees.
`size`	Get number of rows and columns in an array.
`sqrt`	Calculates the square root of a number.
`str2num`	Converts a character string into a number.
`tan(x)`	Calculates tangent of x, where x is in radians.
`tand(x)`	Calculates tangent of x, where x is in degrees.
`title`	Adds a title to a plot.
`zeros(m,n)`	Generate an array of zeros.

2.15 Exercises

2.1 Answer the following questions for the array shown below.

$$\text{array1} = \begin{bmatrix} 0.0 & 0.5 & 2.1 & -3.5 & 6.0 \\ 0.0 & -1.1 & -6.6 & 2.8 & 3.4 \\ 2.1 & 0.1 & 0.3 & -0.4 & 1.3 \\ 1.1 & 5.1 & 0.0 & 1.1 & -2.0 \end{bmatrix}$$

(a) What is the size of `array1`?
(b) What is the value of `array1(1,4)`?
(c) What is the size and value of `array1(:,1:2:5)`?
(d) What is the size and value of `array1([1 3],end)`?

2.2 Are the following MATLAB variable names legal or illegal? Why?

(a) `dog1`
(b) `1dog`
(c) `Do_you_know_the_way_to_san_jose`
(d) `_help`
(e) `What 's_up?`

2.3 Determine the size and contents of the following arrays. Note that the later arrays may depend on the definitions of arrays defined earlier in this exercise.

(a) `a = 2:3:8;`
(b) `b = [a' a' a'];`
(c) `c = b(1:2:3,1:2:3);`
(d) `d = a + b(2,:);`
(e) `w = [zeros(1,3) ones(3,1)' 3:5'];`
(f) `b([1 3],2) = b([3 1],2);`
(g) `e = 1:-1:5;`

2.4 Assume that array `array1` is defined as shown, and determine the contents of the following sub-arrays:

$$
\text{array1} = \begin{bmatrix}
1.1 & 0.0 & -2.1 & -3.5 & 6.0 \\
0.0 & -3.0 & -5.6 & 2.8 & 4.3 \\
2.1 & 0.3 & 0.1 & -0.4 & 1.3 \\
-1.4 & 5.1 & 0.0 & 1.1 & -3.0
\end{bmatrix}
$$

(a) `array1(3,:)`
(b) `array1(:,3)`
(c) `array1(1:2:3,[3 3 4])`
(d) `array1([1 1],:)`

2.5 Assume that `value` has been initialized to 10π, and determine what is printed out by each of the following statements.

```
disp (['value = ' num2str(value)]);
disp (['value = ' int2str(value)]);
fprintf('value = %e\n',value);
fprintf('value = %f\n',value);
fprintf('value = %g\n',value);
fprintf('value = %12.4f\n',value);
```

2.6 Assume that a, b, c, and d are defined as follows, and calculate the results of the following operations if they are legal. If an operation is illegal, explain why.

$$
a = \begin{bmatrix} 2 & 1 \\ -1 & 4 \end{bmatrix} \qquad b = \begin{bmatrix} -1 & 3 \\ 0 & 2 \end{bmatrix}
$$

$$
c = \begin{bmatrix} 2 \\ 1 \end{bmatrix} \qquad d = \text{eye}(2)
$$

```
(a) result = a + b;
(b) result = a * d;
(c) result = a .* d;
(d) result = a * c;
(e) result = a .* c;
(f) result = a \ b;
(g) result = a .\ b;
(h) result = a .^ b;
```

2.7 Evaluate each of the following expressions.

```
(a) 11 / 5 + 6
(b) (11 / 5) + 6
(c) 11 / (5 + 6)
(d) 3 ^ 2 ^ 3
(e) v3 ^ (2 ^ 3)
(f) (3 ^ 2) ^ 3
(g) round(-11/5) + 6
(h) ceil(-11/5) + 6
(i) floor(-11/5) + 6
```

2.8 Use MATLAB to evaluate each of the following expressions.

(a) $(3 - 4i)(-4 + 3i)$

(b) $\cos^{-1}(1.2)$

2.9 Solve the following system of simultaneous equations for x:

```
-2.0 x₁ + 5.0 x₂ + 1.0 x₃ + 3.0 x₄ + 4.0 x₅ - 1.0 x₆ =  0.0
 2.0 x₁ - 1.0 x₂ - 5.0 x₃ - 2.0 x₄ + 6.0 x₅ + 4.0 x₆ =  1.0
-1.0 x₁ + 6.0 x₂ - 4.0 x₃ - 5.0 x₄ + 3.0 x₅ - 1.0 x₆ = -6.0
 4.0 x₁ + 3.0 x₂ - 6.0 x₃ - 5.0 x₄ - 2.0 x₅ - 2.0 x₆ = 10.0
-3.0 x₁ + 6.0 x₂ + 4.0 x₃ + 2.0 x₄ - 6.0 x₅ + 4.0 x₆ = -6.0
 2.0 x₁ + 4.0 x₂ + 4.0 x₃ + 4.0 x₄ + 5.0 x₅ - 4.0 x₆ = -2.0
```

2.10 **Position and Velocity of a Ball** If a stationary ball is released at a height h_0 above the surface of the Earth with a vertical velocity v_0, the position and velocity of the ball as a function of time will be given by the equations

$$h(t) = \frac{1}{2}gt^2 + v_0 t + h_0 \tag{2.10}$$

$$v(t) = gt + v_0 \tag{2.11}$$

where g is the acceleration due to gravity (-9.81 m/s^2), h is the height above the surface of the Earth (assuming no air friction), and v is the vertical component of velocity. Write a MATLAB program that prompts a user for the initial height of the ball in meters and the vertical velocity of the ball in meters per second, and plots the height and vertical velocity as a function of time. Be sure to include proper labels in your plots.

2.11 The distance between two points (x_1, y_1) and (x_2, y_2) on a Cartesian coordinate plane is given by the equation

$$d = \sqrt{(x_1 - x_2)^2 + (y_1 - y_2)^2} \qquad (2.12)$$

(See Figure 2-16). Write a program to calculate the distance between any two points (x_1, y_1) and (x_2, y_2) specified by the user. Use good programming practices in your program. Use the program to calculate the distance between the points $(-3, 2)$ and $(3, -6)$.

2.12 The distance between two points (x_1, y_1, z_1) and (x_2, y_2, z_2) in a three-dimensional Cartesian coordinate system is given by the equation

$$d = \sqrt{(x_1 - x_2)^2 + (y_1 - y_2)^2 + (z_1 - z_2)^2} \qquad (2.13)$$

Write a program to calculate the distance between any two points (x_1, y_1, z_1) and (x_2, y_2, z_2) specified by the user. Use good programming practices in your program. Use the program to calculate the distance between the points $(-3, 2, 5)$ and $(3, -6, -5)$.

2.13 **Decibels** Engineers often measure the ratio of two power measurements in *decibels*, or dB. The equation for the ratio of two power measurements in decibels is

$$dB = 10 \log_{10} \frac{P_2}{P_1} \qquad (2.14)$$

where P_2 is the power level being measured, and P_1 is some reference power level.

(a) Assume that the reference power level P_1 is 1 milliwatt, and write a program that accepts an input power P_2 and converts it into dB with respect to the 1 mW reference level. (Engineers have a special unit for dB power levels with respect to a 1 mW reference: dBm.) Use good programming practices in your program.

(b) Write a program that creates a plot of power in watts versus power in dBm with respect to a 1 mW reference level. Create both a linear *xy* plot and a log-linear *xy* plot.

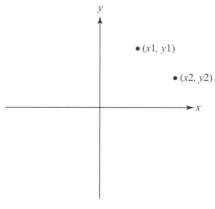

Figure 2.16 Distance between two points on a Cartesian plane.

2.14 **Power in a Resistor** The voltage across a resistor is related to the current flowing through it by Ohm's law

$$V = IR \qquad (2.15)$$

and the power consumed in the resistor is given by the equation

$$P = IV \qquad (2.16)$$

Write a program that creates a plot of the power consumed by a 1000 Ω resistor as the voltage across it is varied from 1 V to 200 V. Create two plots, one showing power in watts, and one showing power in dBW (dB power levels with respect to a 1 W reference).

2.15 A three dimensional vector can be represented in either rectangular coordinates (x, y, z) or the spherical coordinates (r, θ, ϕ), as shown in Figure 2.18[2]. The relationships among these two sets of coordinates are given by the following equations:

$$x = r \cos \phi \cos \theta \qquad (2.17)$$

$$y = r \cos \phi \sin \theta \qquad (2.18)$$

$$z = r \sin \phi \qquad (2.19)$$

$$r = \sqrt{x^2 + y^2 + z^2} \qquad (2.20)$$

$$\theta = \tan^{-1} \frac{y}{x} \qquad (2.21)$$

$$\phi = \tan^{-1} \frac{z}{\sqrt{x^2 + y^2}} \qquad (2.22)$$

Use the MATLAB help system to look up function `atan2`, and use that function in answering the questions below.

(a) Write a program that accepts a 3D vector in rectangular coordinates and calculates the vector in spherical coordinates, with the angles θ and ϕ expressed in degrees.

(b) Write a program that accepts a 3D vector in spherical coordinates (with the angles θ and ϕ in degrees) and calculates the vector in rectangular coordinates.

2.16 MATLAB includes two functions `cart2sph` and `sph2cart` to convert back and forth between Cartesian and spherical coordinates. Look these functions up

Figure 2.17 Voltage and current in a resistor.

---

[2]These definitions of the angles in spherical coordinates are non-standard according to international usage, but match the definitions employed by the MATLAB program.

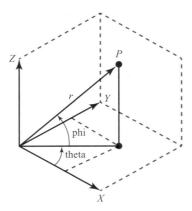

Figure 2.18 A three-dimensional vector v can be represented in either rectangular coordinates (x, y, z) or spherical coordinates (r, θ, ϕ).

in the MATLAB help system and re-write the programs in Exercise 2.15 using these functions. How do the answers compare between the programs written using Equations (2.17) through (2.22) and the programs written using the built-in MATLAB functions?

2.17 Hyperbolic cosine The hyperbolic cosine function is defined by the equation

$$\cosh x = \frac{e^x + e^{-x}}{2} \tag{2.23}$$

Write a program to calculate the hyperbolic cosine of a user-supplied value x. Use the program to calculate the hyperbolic cosine of 3.0. Compare the answer that your program produces to the answer produced by the MATLAB intrinsic function `cosh(x)`. Also, use MATLAB to plot the function `cosh(x)`. What is the smallest value that this function can have? At what value of x does it occur?

2.18 Energy Stored in a Spring The force required to compress a linear spring is given by the equation

$$F = kx \tag{2.24}$$

where F is the force in newtons and k is the spring constant in newtons per meter. The potential energy stored in the compressed spring is given by the equation

$$E = \frac{1}{2} kx^2 \tag{2.25}$$

where E is the energy in joules. The following information is available for four springs:

	Spring 1	Spring 2	Spring 3	Spring 4
Force (N)	20	30	25	20
Spring constant k (N/m)	200	250	300	400

Determine the compression of each spring and the potential energy stored in each spring. Which spring has the most energy stored in it?

2.19 Radio Receiver A simplified version of the front end of an AM radio receiver is shown in Figure 2.19. This receiver consists of an *RLC* tuned circuit containing a resistor, capacitor, and an inductor connected in series. The *RLC* circuit is connected to an external antenna and ground, as shown in the picture.

The tuned circuit allows the radio to select a specific station out of all the stations transmitting on the AM band. At the resonant frequency of the circuit, essentially all of the signal V_0 appearing at the antenna appears across the resistor, which represents the rest of the radio. In other words, the radio receives its strongest signal at the resonant frequency. The resonant frequency of the LC circuit is given by the equation

$$f_0 = \frac{1}{2\pi\sqrt{LC}} \tag{2.26}$$

where L is inductance in henrys (H) and C is capacitance in farads (F). Write a program that calculates the resonant frequency of this radio set given specific values of L and C. Test your program by calculating the frequency of the radio when $L = 0.25$ mH and $C = 0.10$ nF.

2.20 Radio Receiver The average (rms) voltage across the resistive load in Figure 2.18 varies as a function of frequency according to Equation (2.27).

$$V_R = \frac{R}{\sqrt{R^2 + \left(\omega L - \dfrac{1}{\omega C}\right)^2}} V_0 \tag{2.27}$$

where $\omega = 2\pi f$ and f is the frequency in hertz. Assume that $L = 0.25$ mH, $C = 0.10$ nF, $R = 50\ \Omega$, and $V_0 = 10$ mV.

(a) Plot the rms voltage on the resistive load as a function of frequency. At what frequency does the voltage on the resitive load peak? What is the voltage on the load at this frequency? This frequency is called the resonant frequency f_0 of the circuit.

(b) If the frequency is changed to 10% greater than the resonant frequency, what is the voltage on the load? How selective is this radio receiver?

(c) At what frequencies will the voltage on the load drop to half of the voltage at the resonant frequency?

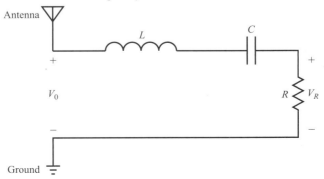

Figure 2.19 A simplified version of the front end of an AM radio receiver.

2.21 Suppose two signals were received at the antenna of the radio receiver described in the previous problem. One signal has a strength of 1 V at a frequency of 1000 kHz, and the other signal has a strength of 1 V at 950 kHz. Calculate the voltage V_R that will be received for each of these signals. How much power will the first signal supply to the resistive load R? How much power will the second signal supply to the resistive load R? Express the ratio of the power supplied by signal 1 to the power supplied by signal 2 in decibels (see Problem 2.13 above for the definition of a decibel). How much is the second signal enhanced or suppressed compared to the first signal? *(Note:* The power supplied to the resistive load can be calculated from the equation $P = V_R^2/R$.)

2.22 Aircraft Turning Radius An object moving in a circular path at a constant tangential velocity v is shown in Figure 2.20. The radial acceleration required for the object to move in the circular path is given by the Equation (2.28)

$$a = \frac{v^2}{r} \tag{2.28}$$

where a is the centripetal acceleration of the object in m/s², v is the tangential velocity of the object in m/s, and r is the turning radius in meters. Suppose that the object is an aircraft, and answer the following questions about it:

(a) Suppose that the aircraft is moving at Mach 0.85, or 85% of the speed of sound. If the centripetal acceleration is 2 g, what is the turning radius of the aircraft? (Note: For this problem, you may assume that Mach 1 is equal to 340 m/s, and that 1 g = 9.81 m/s²).

(b) Suppose that the speed of the aircraft increases to Mach 1.5. What is the turning radius of the aircraft now?

(c) Plot the turning radius as a function of aircraft speed for speeds between Mach 0.5 and Mach 2.0, assuming that the acceleration remains 2 g.

(d) Suppose that the maximum acceleration that the pilot can stand is 7 g. What is the minimum possible turning radius of the aircraft at Mach 1.5?

(e) Plot the turning radius as a function of centripetal acceleration for accelerations between 2 g and 8 g, assuming a constant speed of Mach 0.85.

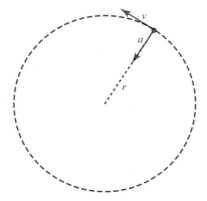

Figure 2.20 An object moving in uniform circular motion due to the centripetal acceleration a.

Chapter | 3

Two-Dimensional Plots

One of the most powerful features of MATLAB is the ability to easily create plots that visualize the information that an engineer is working with. In other programming languages used by engineers (such as C++, Java, Fortran, and so forth), plotting is a major task involving either a lot of effort or additional software packages that are not a part of the basic language. In contrast, MATLAB is ready to create high-quality plots with minimal effort right out of the box.

We introduced a few simple plotting commands in Chapter 2 and used them to display a variety of data on linear and logarithmic scales in various examples and exercises.

Because the ability to create plots is so important, we will devote this entire chapter to learning how to make good two-dimensional plots of engineering data. Three-dimensional plots will be addressed later in Chapter 8.

3.1 Additional Plotting Features for Two-Dimensional Plots

This section describes additional features that improve the simple two-dimensional plots introduced in Chapter 2. These features permit us to control the range of x and y values displayed on a plot, lay multiple plots on top of each other, create multiple figures, create multiple subplots within a figure, and provide greater control of the plotted lines and text strings. In addition, we will learn how to create polar plots.

3.1.1 Logarithmic Scales

It is possible to plot data on logarithmic scales as well as linear scales. There are four possible combinations of linear and logarithmic scales on the x- and y-axes, and each combination is produced by a separate function.

1. The `plot` function plots both *x* and *y* data on linear axes.
2. The `semilogx` function plots *x* data on a logarithmic axis and *y* data on a linear axis.
3. The `semilogy` function plots *x* data on a linear axis and *y* data on a logarithmic axis.
4. The `loglog` function plots both *x* and *y* data on logarithmic axes.

All of these functions have identical calling sequences—the only difference is the type of axis used to plot the data.

To compare these four types of plots, we will plot the function $y(x) = 2x^2$ over the range 0 to 100 with each type of plot. The MATLAB code to do this is:

```
x = 0:0.2:100;
y = 2 * x.^2;

% For the linear / linear case
plot(x,y);
title('Linear / linear Plot');
xlabel('x');
ylabel('y');
grid on;

% For the log / linear case
semilogx(x,y);
title('Log / linear Plot');
xlabel('x');
ylabel('y');
grid on;

% For the linear / log case
semilogy(x,y);
title('Linear / log Plot');
xlabel('x');
ylabel('y');
grid on;

% For the log / log case
loglog(x,y);
title('Log / log Plot');
xlabel('x');
ylabel('y');
grid on;
```

Examples of each plot are shown in Figure 3.1.

It is important to consider the type of data being plotted when selecting linear or logarithmic scales. In general, if the range of the data being plotted covers many orders of magnitude, a logarithmic scale will be more appropriate, because on a linear scale the very small part of the data set will be invisible. If the data being plotted covers a relatively small dynamic range, then linear scales work very well.

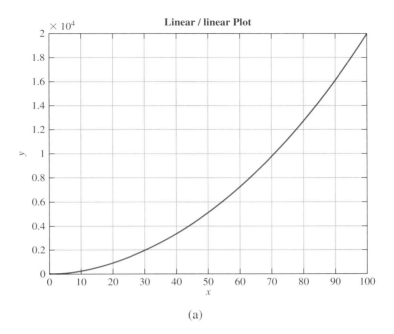

(a)

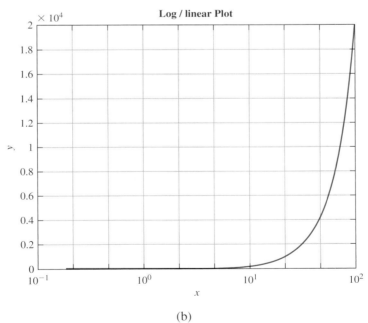

(b)

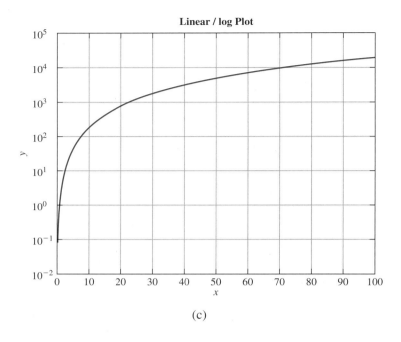

(c)

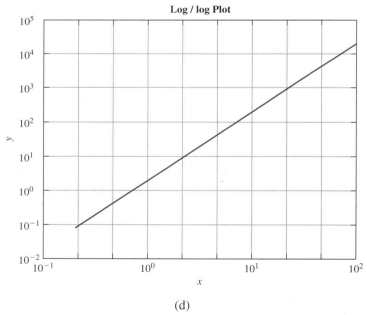

(d)

Figure 3.1 Comparison of linear, semilog *x*, semilog *y*, and log-log plots.

Good Programming Practice

If the range of the data to plot covers many orders of magnitude, use a logarithmic scale to represent the data properly. If the range of the data to plot is an order of magnitude or less, then use a linear scale.

Also, be careful of trying to plot data with zeros or negative values on a logarithmic scale. The logarithm of zero or a negative number is undefined for real numbers, so those negative points will never be plotted. MATLAB issues a warning and ignores those negative values.

Programming Pitfalls

Do not attempt to plot negative data on a logarithmic scale. The data will be ignored.

3.1.2 Controlling *x*- and *y*-axis Plotting Limits

By default, a plot is displayed with *x*- and *y*-axis ranges wide enough to show every point in an input data set. However, it is sometimes useful to display only the subset of the data that is of particular interest. This can be done using the **axis** command/function (see the Sidebar about the relationship between MATLAB commands and functions).

Command/Function Duality

Some items in MATLAB seem to be unable to make up their minds whether they are commands (words typed out on the command line) or functions (with arguments in parentheses). For example, sometimes axis seems to behave like a command and sometimes it seems to behave like a function. Sometimes we treat it as a command: axis on, and other times we might treat it as a function: axis ([0 20 0 35]). How is this possible?

The short answer is that MATLAB commands are really implemented by functions, and the MATLAB interpreter is smart enough to substitute the function call whenever it encounters the command. It is always possible to call the command directly as a function instead of using the command syntax. Thus the following two statements are identical:

```
axis on;
axis ('on');
```

Whenever MATLAB encounters a command, it forms a function from the command by treating each command argument as a character string and calling the equivalent function with those character strings as arguments. Thus MATLAB interprets the command

(continued)

```
garbage 1 2 3
```

as the following function call:

```
garbage('1','2','3')
```

Note that *only functions with character arguments can be treated as commands*. Functions with numerical arguments must be used in function form only. This fact explains why `axis` is sometimes treated as a command and sometimes treated as a function.

Some of the forms of the `axis` command/function are shown in Table 3.1 below. The two most important forms are shown in bold type—they let an engineer get the current limits of a plot and modify them. A complete list of all options can be found in the MATLAB online documentation.

To illustrate the use of `axis`, we will plot the function $f(x) = \sin x$ from -2π to 2π, and then restrict the axes to the region defined by $0 \leq x \leq \pi$ and $0 \leq y \leq 1$. The statements to create this plot are shown below, and the resulting plot is shown in Figure 3.2*a*.

```
x = -2*pi:pi/20:2*pi;
y = sin(x);
plot(x,y);
title ('Plot of sin(x) vs x');
grid on;
```

The current limits of this plot can be determined from the basic `axis` function.

```
» limits = axis
limits =
     -8      8     -1      1
```

These limits can be modified with the function call `axis([0 pi 0 1])`. After that function is executed, the resulting plot is shown in Figure 3.2*b*.

Table 3.1: Forms of the `axis` Function/Command

Command	Description
v = axis;	This function returns a 4-element row vector containing `[xmin xmax ymin ymax]`, where `xmin`, `xmax`, `ymin`, and `ymax` are the current limits of the plot.
axis ([xmin xmax ymin ymax]);	This function sets the *x* and *y* limits of the plot to the specified values.
axis equal	This command sets the axis increments to be equal on both axes.
axis square	This command makes the current axis box square.
axis normal	This command cancels the effect of axis equal and axis square.
axis off	This command turns off all axis labeling, tick marks, and background.
axis on	This command turns on all axis labeling, tick marks, and background (default case).

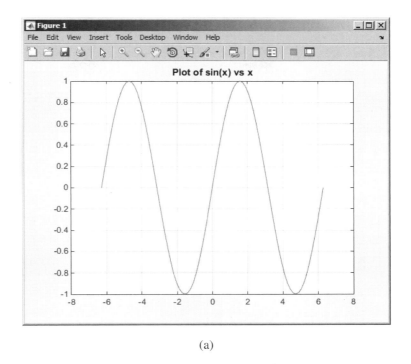

(a)

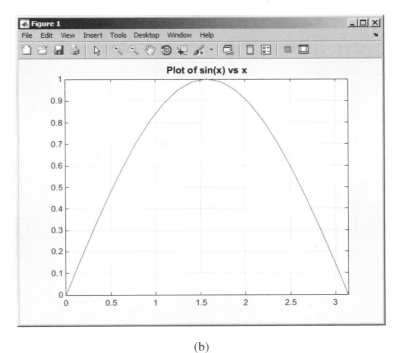

(b)

Figure 3.2 (a) Plot of sin x versus x. (b) Closeup of the region [0 π 0 1].

3.1.3 Plotting Multiple Plots on the Same Axes

Normally, a new plot is created each time that a `plot` command is issued, and the previous data displayed on the figure are lost. This behavior can be modified with the **hold** command. After a `hold on` command is issued, all additional plots will be laid on top of the previously existing plots. A `hold off` command switches plotting behavior back to the default situation, in which a new plot replaces the previous one.

For example, the following commands plot sin *x* and cos *x* on the same axes. The resulting plot is shown in Figure 3.3.

```
x = -pi:pi/20:pi;
y1 = sin(x);
y2 = cos(x);
plot(x,y1,'b-');
hold on;
plot(x,y2,'k--');
hold off;
legend ('sin x','cos x');
```

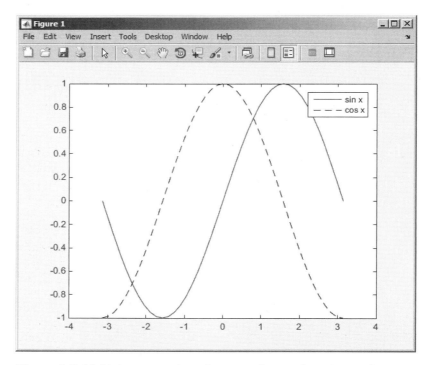

Figure 3.3 Multiple curves plotted on a single set of axes using the `hold` command.

3.1.4 Creating Multiple Figures

MATLAB can create multiple Figure Windows, with different data displayed in each window. Each Figure Window is identified by a *figure number*, which is a small positive integer. The first Figure Window is Figure 1, the second is Figure 2, and so forth. One of the Figure Windows will be the **current figure**, and all new plotting commands will be displayed in that window.

The current figure is selected with the **figure function**. This function takes the form "figure(n)", where n is a figure number[1]. When this command is executed, Figure n becomes the current figure and is used for all plotting commands. The figure is automatically created if it does not already exist. The current figure may also be selected by clicking on it with the mouse.

The function gcf returns a *handle* (a reference) to the current figure, so this function can be used by an M-file if it needs to know the current figure.

The following commands illustrate the use of the figure function. They create two figures, displaying e^x in the first figure and e^{-x} in the second one (see Figure 3.4).

```
figure(1)
x = 0:0.05:2;
y1 = exp(x);
plot(x,y1);
title(' exp(x)');
grid on;

figure(2)
y2 = exp(-x);
plot (x,y2);
title(' exp(-x)');
grid on;
```

3.1.5 Subplots

It is possible to place more than one set of axes on a single figure, creating multiple **subplots**. Subplots are created with a subplot command of the form

```
subplot(m,n,p)
```

This command divides the current figure into m × n equal-sized regions, arranged in m rows and n columns, and creates a set of axes at position p to receive all current plotting commands. The subplots are numbered from left to right and from top to bottom. For example, the command subplot(2,3,4) would divide the current figure into six regions arranged in two rows and three columns and create an axis in position 4 (the lower left one) to accept new plot data (see Figure 3.5).

If a subplot command creates a new set of axes that conflict with a previously existing set, then the older axes are automatically deleted.

---

[1]The figure function can also accept a figure handle, as will be explained further in Chapter 13.

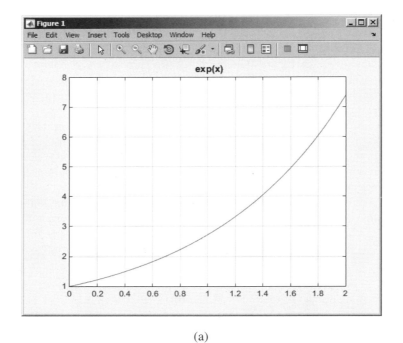

(a)

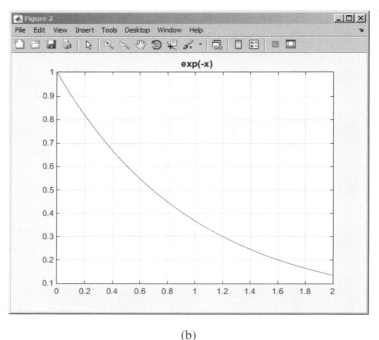

(b)

Figure 3.4 Creating multiple plots on separate figures using the `figure` function. (a) Figure 1; (b) Figure 2.

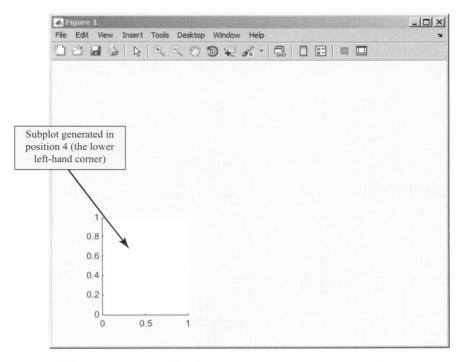

Figure 3.5 The axis created by the `subplot(2,3,4)` command.

The commands below create two subplots within a single window and display the separate graphs in each subplot. The resulting figure is shown in Figure 3.6.

```
figure(1)
subplot(2,1,1)
x = -pi:pi/20:pi;
y = sin(x);
plot(x,y);
title('Subplot 1 title');
subplot(2,1,2)
x = -pi:pi/20:pi;
y = cos(x);
plot(x,y);
title('Subplot 2 title');
```

3.1.6 Controlling the Spacing Between Points on a Plot

In Chapter 2, we learned how to create an array of values using the colon operator. The colon operator

```
start:incr:end
```

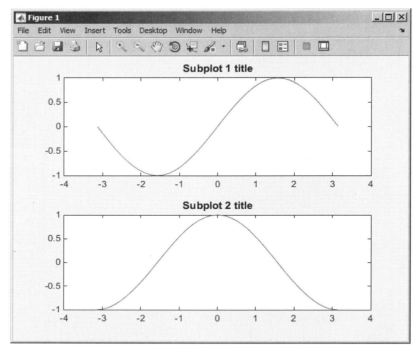

Figure 3.6 A figure with two subplots showing sin *x* and cos *x* respectively.

produces an array that starts at `start`, advances in increments of `incr`, and ends when the last point equals the value `end` or when the last point plus the increment would exceed the value `end`. The colon operator can be used to create an array, but it has two disadvantages in regular use:

1. It is not always easy to know how many points will be in the array. For example, can you tell how many points would be in the array defined by `0:pi:20`?
2. There is no guarantee that the last specified point will be in the array, since the increment could overshoot that point.

To avoid these problems, MATLAB includes two functions to generate an array of points where the user has full control of both the exact limits of the array and the number of points in the array. These functions are `linspace`, which produces a linear spacing between samples, and `logspace`, which produces a logarithmic spacing between samples.

The forms of the `linspace` function are:

```
y = linspace(start,end);
y = linspace(start,end,n);
```

where start is the starting value, end is the ending value, and n is the number of points to produce in the array. If only the start and end values are specified, linspace produces 100 equally spaced points starting at start and ending at end. For example, we can create an array of 10 evenly spaced points on a linear scale with the command

```
» linspace(1,10,10)
ans =
     1    2    3    4    5    6    7    8    9    10
```

The forms of the logspace function are:

```
y = logspace(start,end);
y = logspace(start,end,n);
```

where start is the *exponent* of the starting power of 10, end is the *exponent* of the ending power of 10, and n is the number of points to produce in the array. If only the start and end values are specified, logspace produces 50 points equally spaced on a logarithmic scale, starting at start and ending at end. For example, we can create an array of logarithmically spaced points starting at 1 ($= 10^0$) and ending at 10 ($= 10^1$) on a logarithmic scale with the command

```
» logspace(0,1,10)
ans =
     1.0000      1.2915      1.6681      2.1544      2.7826
     3.5938      4.6416      5.9948      7.7426     10.0000
```

The logspace function is especially useful for generating data to be plotted on a logarithmic scale, since the points on the plot will be evenly spaced.

▶ Example 3.1—Creating Linear and Logarithmic Plots

Plot the function

$$y(x) = x^2 - 10x + 25 \qquad (3.1)$$

over the range 0 to 10 on a linear plot using 21 evenly spaced points in one subplot and over the range 10^{-1} to 10^1 on a semi logarithmic plot using 21 evenly spaced points on a logarithmic x-axis in a second subplot. Put markers on each point used in the calculation so that they will be visible, and be sure to include a title and axis labels on each plot.

Solution To create these plots, we will use function linspace to calculate an evenly spaced set of 21 points on a linear scale and function logspace to calculate an evenly spaced set of 21 points on a logarithmic scale. Next, we will evaluate Equation (3.1) at those points and plot the resulting curves. The MATLAB code to do this is shown on the following page.

```
%   Script file: linear_and_log_plots.m
%
%   Purpose:
%     This program plots y(x) = x^2 - 10*x + 25
%     on linear and semilogx axes.
%
%   Record of revisions:
%     Date         Programmer          Description of change
%     ====         ==========          =====================
%     11/15/14     S. J. Chapman       Original code
%

% Create a figure with two subplots
subplot(2,1,1);

% Now create the linear plot
x = linspace(0, 10, 21);
y =  x.^2 - 10*x + 25;
plot(x,y,'b-');
hold on;
plot(x,y,'ro');
title('Linear Plot');
xlabel('x');
ylabel('y');
hold off;

% Select the other subplot
subplot(2,1,2);

% Now create the logarithmic plot
x = logspace(-1, 1, 21);
y =  x.^2 - 10*x + 25;
semilogx(x,y,'b-');
hold on;
semilogx(x,y,'ro');
title('Semilog x Plot');
xlabel('x');
ylabel('y');
hold off;
```

The resulting plot is shown in Figure 3.7. Note that the plot scales are different, but each plot includes 21 evenly spaced samples.

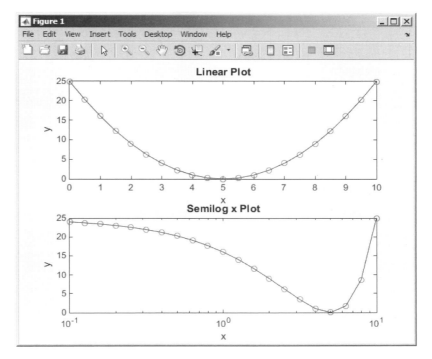

Figure 3.7 Plots of the function $y(x) = x^2 - 10x + 25$ on linear and semi-logarithmic axes.

3.1.7 Enhanced Control of Plotted Lines

In Chapter 2, we learned how to set the color, style, and marker type for a line. It is also possible to set four additional properties associated with each line:

- `LineWidth` – specifies the width of each line in points
- `MarkerEdgeColor` – specifies the color of the marker or the edge color for filled markers
- `MarkerFaceColor` – specifies the color of the face of filled markers.
- `MarkerSize` – specifies the size of the marker in points.

These properties are specified in the `plot` command after the data to be plotted in the following fashion:

```
plot(x,y,'PropertyName',value,...)
```

For example, the following command plots a 3-point-wide solid black line with 6-point-wide circular markers at the data points. Each marker has a red edge and a green center, as shown in Figure 3.8.

```
x = 0:pi/15:4*pi;
y = exp(2*sin(x));
```

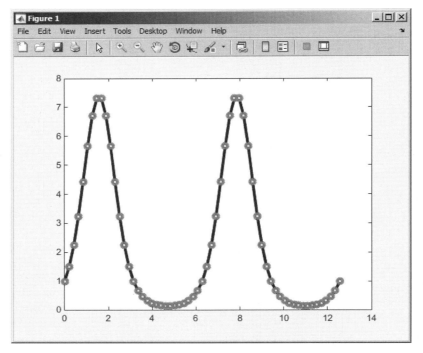

Figure 3.8 A plot illustrating the use of the `LineWidth` and `Marker` properties. [See color insert.]

```
plot(x,y,'-ko','LineWidth',3.0,'MarkerSize',6,...
    'MarkerEdgeColor','r','MarkerFaceColor','g')
```

3.1.8 Enhanced Control of Text Strings

It is possible to enhance plotted text strings (titles, axis labels, and so forth) with formatting such as bold face, italics, and so forth, and with special characters such as Greek and mathematical symbols.

The font used to display the text can be modified by **stream modifiers**. A stream modifier is a special sequence of characters that tells the MATLAB interpreter to change its behavior. The most common stream modifiers are:

- `\bf`—Bold face
- `\it`—Italics
- `\rm`—Removes stream modifiers, restoring normal font
- `\fontname{`*fontname*`}`—Specify the font name to use
- `\fontsize{`*fontsize*`}`—Specify font size
- `_{`xxx`}`—The characters inside the braces are subscripts
- `^{`xxx`}`—The characters inside the braces are superscripts

Once a stream modifier has been inserted into a text string, it will remain in effect until the end of the string or until cancelled. Any stream modifier can be followed by braces `{}`. If a modifier is followed by braces, only the text within the braces is affected.

Special Greek and mathematical symbols may also be used in text strings. They are created by embedding *escape sequences* into the text string. These escape sequences are the same as those defined in the TeX language. A sample of the possible escape sequences is shown in Table 3.2; the full set of possibilities is included in the MATLAB online documentation.

If one of the special escape characters \, {, }, _, or ^ must be printed, precede it by a backslash character.

The following examples illustrate the use of stream modifiers and special characters.

String	Result
`\tau_{ind} versus \omega_{\itm}`	τ_{ind} versus ω_m
`\theta varies from 0\circ to 90\circ`	θ varies from 0° to 90°
`\bf{B}_{\itS}`	$\mathbf{B}_S$

Good Programming Practice

Use stream modifiers to create effects such as bold, italics, superscripts, subscripts, and special characters in your plot titles and labels.

Table 3.2: Selected Greek and Mathematical Symbols

Character Sequence	Symbol	Character Sequence	Symbol	Character Sequence	Symbol
\alpha	α			\int	∫
\beta	β			\cong	≅
\gamma	γ	\Gamma	Γ	\sim	~
\delta	δ	\Delta	Δ	\infty	∞
\epsilon	ε			\pm	±
\eta	η			\leq	≤
\theta	θ			\geq	≥
\lambda	λ	\Lambda	Λ	\neq	≠
\mu	μ			\propto	∝
\nu	ν			\div	÷
\pi	π	\Pi	Π	\circ	°
\phi	φ			\leftrightarrow	↔
\rho	ρ			\leftarrow	←
\sigma	σ	\Sigma	Σ	\rightarrow	→
\tau	τ			\uparrow	↑
\omega	ω	\Omega	Ω	\downarrow	↓

►Example 3.2—Labeling Plots with Special Symbols

Plot the decaying exponential function

$$y(t) = 10e^{-t/\tau} \sin \omega t \tag{3.2}$$

where the time constant $\tau = 3$ s and the radial velocity $\omega = \pi$ rad/s over the range $0 \le t \le 10$ s. Include the plotted equation in the title of the plot, and label the x- and y-axes properly.

Solution To create this plot, we will use function `linspace` to calculate an evenly spaced set of 100 points between 0 and 10. Next, we will evaluate Equation (3.2) at those points and plot the resulting curve. Finally, we will use the special symbols in this chapter to create the title of the plot.

The title of the plot must include italic letters for $y(t)$, t/τ, and ωt, and it must set the $-t/\tau$ as a superscript. The string of symbols that will do this is

```
\it{y(t)} = \it{e}^{-\it{t / \tau}} sin \it{\omegat}
```

The MATLAB code that plots this function is shown below.

```
%  Script file: decaying_exponential.m
%
%  Purpose:
%    This program plots the function
%    y(t) = 10*EXP(-t/tau)*SIN(omega*t)
%    on linear and semilogx axes.
%
%  Record of revisions:
%    Date        Programmer        Description of change
%    ====        ==========        =====================
%    11/15/14   S. J. Chapman     Original code
%
% Define variables:
%    tau          -- Time constant, s
%    omega        -- Radial velocity, rad/s
%    t            -- Time (s)
%    y            -- Output of function

% Declare time constant and radial velocity
tau = 3;
omega = pi;

% Now create the plot
t = linspace(0, 10, 100);
y =  10 * exp(-t./tau) .* sin(omega .* t);
plot(t,y,'b-');
title('Plot of \it{y(t)} = \it{e}^{-\it{t / \tau}} sin \it{\omegat}');
```

```
xlabel('\it{t}');
ylabel('\it{y(t)}');
grid on;
```

The resulting plot is shown in Figure 3.9.

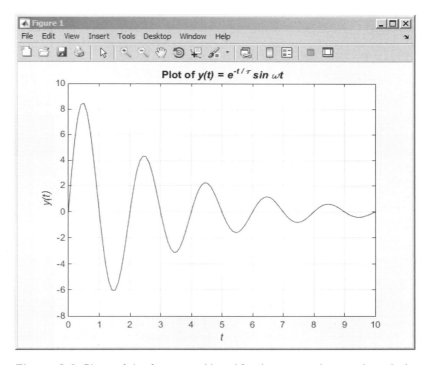

Figure 3.9 Plots of the function $y(t) = 10e^{-t/\tau} \sin \omega t$ with special symbols used to reproduce the equation in the title.

3.2 Polar Plots

MATLAB includes a special function called `polar`, which plots two-dimensional data in polar coordinates instead of rectangular coordinates. The basic form of this function is

```
polar(theta,r)
```

where `theta` is an array of angles in radians, and `r` is an array of distances from the center of the plot. The angle `theta` is the angle (in radians) of a point counterclockwise from the right-hand horizontal axis, and r is distance from the center of the plot to the point.

This function is useful for plotting data that is intrinsically a function of angle, as we will see in the next example.

►Example 3.3—Cardioid Microphone

Most microphones designed for use on a stage are directional microphones, which are specifically built to enhance the signals received from the singer in the front of the microphone while suppressing the audience noise from behind the microphone. The gain of such a microphone varies as a function of angle according to the equation

$$Gain = 2g(1 + \cos\theta) \tag{3.3}$$

where g is a constant associated with a particular microphone, and θ is the angle from the axis of the microphone to the sound source. Assume that g is 0.5 for a particular microphone, and make a polar plot the gain of the microphone as a function of the direction of the sound source.

Solution We must calculate the gain of the microphone versus angle and then plot it with a polar plot. The MATLAB code to do this is shown below.

```
%   Script file: microphone.m
%
%   Purpose:
%     This program plots the gain pattern of a cardioid
%     microphone.
%
%   Record of revisions:
%     Date          Engineer          Description of change
%     ====          ==========        =====================
%     01/05/14      S. J. Chapman     Original code
%
% Define variables:
%   g           -- Microphone gain constant
%   gain        -- Gain as a function of angle
%   theta       -- Angle from microphone axis (radians)

% Calculate gain versus angle
g = 0.5;
theta = linspace(0,2*pi,41);
gain = 2*g*(1+cos(theta));

% Plot gain
polar (theta,gain,'r-');
title ('\bfGain versus angle \it{\theta}');
```

The resulting plot is shown in Figure 3.10. Note that this type of microphone is called a "cardioid microphone" because its gain pattern is heart-shaped.

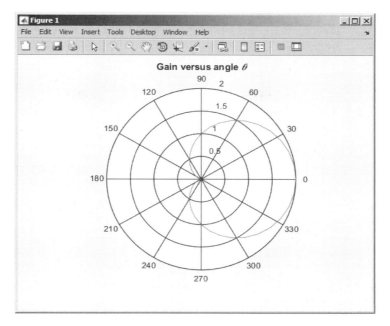

Figure 3.10 Gain of a cardioid microphone. [See color insert.]

3.3 Annotating and Saving Plots

Once a plot has been created by a MATLAB program, a user can edit and annotate the plot using the GUI-based tools available from the plot toolbar. Figure 3.11 shows the tools available, which allow the user to edit the properties of any objects on the plot or to add annotations to the plot. When the editing button ( ) is selected from the toolbar, the editing tools become available for use. When the button is depressed, clicking any line or text on the figure will cause it to be selected for editing, and double-clicking the line or text will open a Property Editor window that allows you to modify any or all of the characteristics of that object. Figure 3.12 shows Figure 3.10 after a user has clicked on the red line to change it to a 3-pixel-wide solid blue line.

The figure toolbar also includes a Plot Browser button ( ). When this button is depressed, the Plot Browser is displayed. This tool gives the user complete control over the figure. He or she can add axes, edit object properties, modify data values, and add annotations such as lines and text boxes.

If it is not otherwise displayed, the user can enable a Plot Edit Toolbar by selecting the View/Plot Edit Toolbar menu item. This toolbar allows a user to add lines, arrows, text, rectangles, and ellipses to annotate and explain a plot. Figure 3.13 shows a Figure Window with the Plot Edit Toolbar enabled.

Figure 3.14 shows the plot in Figure 3.10 after the Plot Browser and the Plot Edit Toolbar have been enabled. In this figure, the user has used the controls on the Plot Edit Toolbar to add an arrow and a comment to the plot.

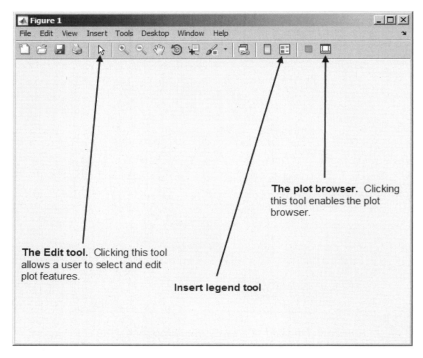

Figure 3.11 The editing tools on the figure toolbar.

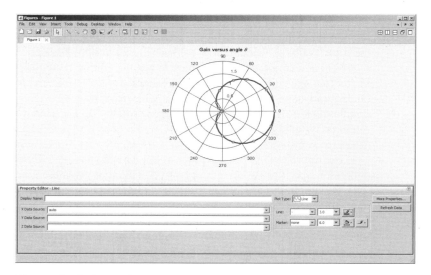

Figure 3.12 Figure 3.10 after the line has been modified using the editing tools built into the figure toolbar. [See color insert.]

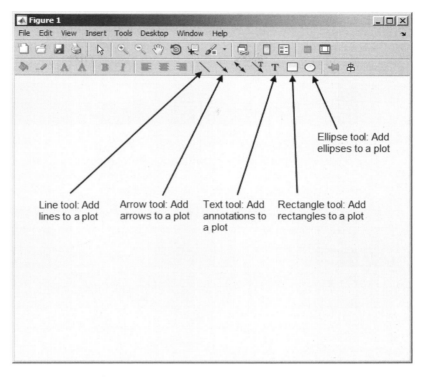

Figure 3.13 A figure window showing the Plot Edit Toolbar.

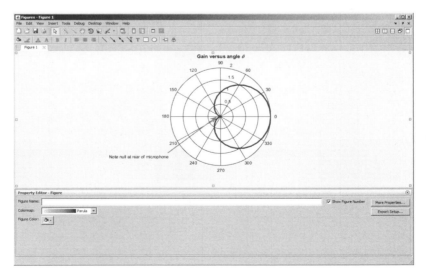

Figure 3.14 Figure 3.10 after the Plot Browser has been used to add an arrow and annotation.

When the plot has been edited and annotated, you can save the entire plot in a modifiable form using the File/Save As menu item from the Figure Window. The resulting figure file (`*.fig`) contains all the information required to re-create the figure plus annotations at any time in the future.

Quiz 3.1

This quiz provides a quick check to see if you have understood the concepts introduced in Section 3.5. If you have trouble with the quiz, reread the section, ask your instructor, or discuss the material with a fellow student. The answers to this quiz are found in the back of the book.

1. Write the MATLAB statements required to plot $\sin x$ versus $\cos 2x$ from 0 to 2π in steps of $\pi/10$. The points should be connected by a 2-pixel-wide red line, and each point should be marked with a 6-pixel-wide blue circular marker.
2. Use the Figure editing tools to change the markers on the previous plot into black squares. Add an arrow and annotation pointing to the location $x = \pi$ on the plot.

Write the MATLAB text string that will produce the following expressions:

3. $f(x) = \sin\theta \cos 2\phi$

4. **Plot of $\sum x^2$ versus x**

Write the expression produced by the following text strings:

5. `'\tau\it_{m}'`
6. `'\bf\itx_{1}^{ 2} + x_{2}^{ 2} \rm(units: \bfm^{2}\rm)'`
7. Plot the function $r = 10* \cos(3\theta)$ for $0 \le \theta \le 2\pi$ is steps of 0.01 π using a polar plot.
8. Plot the function $y(x) = \dfrac{1}{2x^2}$ for $0.01 \le x \le 100$ on a linear and a loglog plot. Take advantage of `linspace` and `logspace` when creating the plots. What is the shape of this function on a loglog plot?

3.4 Additional Types of Two-Dimensional Plots

In addition to the two-dimensional plots that we have already seen, MATLAB supports *many* other more specialized plots. In fact, the MATLAB help system lists more than 20 types of two-dimensional plots! Examples include **stem plots**, **stair plots**, **bar plots**, **pie plots**, and **compass plots**. A *stem plot* is a plot in which each data value is represented by a marker and a line connecting the marker vertically to the *x*-axis. A *stair plot* is a plot in which each data point is represented by a horizontal line, and successive points are connected by vertical lines, producing a stair-step effect. A *bar plot* is a plot in which each point is represented by a vertical bar or horizontal bar. A *pie plot* is a plot represented by "pie slices" of various sizes. [See color insert] Finally, a *compass plot* is a type of polar plot in which each value is represented by an arrow whose length is proportional to its value. These types of plots are summarized in Table 3.3, and examples of all of the plots are shown in Figure 3.15.

Stair, stem, vertical bar, horizontal bar, and compass plots are all similar to `plot`, and they are used in the same manner. For example, the following code produces the stem plot shown in Figure 3.15*a*.

```
x = [ 1 2 3 4 5 6];
y = [ 2 6 8 7 8 5];
stem(x,y);
title('\bfExample of a Stem Plot');
xlabel('\bf\itx');
ylabel('\bf\ity');
axis([0 7 0 10]);
```

Stair, bar, and compass plots can be created by substituting `stairs`, `bar`, `barh`, or `compass` for `stem` in the above code. The details of all of these plots, including any optional parameters, can be found in the MATLAB online help system.

Function `pie` behaves differently than the other plots described above. To create a pie plot, an engineer passes an array x containing the data to be plotted, and function `pie` determines the *percentage of the total pie* that each element of x represents. For example, if the array x is [1 2 3 4], then `pie` will calculate

Table 3.3: Additional Two-Dimensional Plotting Functions

Function	Description
`bar(x,y)`	This function creates a *vertical* bar plot, with the values in x used to label each bar and the values in y used to determine the height of the bar.
`barh(x,y)`	This function creates a *horizontal* bar plot, with the values in x used to label each bar and the values in y used to determine the horizontal length of the bar.
`compass(x,y)`	This function creates a polar plot, with an arrow drawn from the origin to the location of each (x, y) point. Note that the locations of the points to plot are specified in Cartesian coordinates, not polar coordinates.
`pie(x)` `pie(x,explode)`	This function creates a pie plot. This function determines the percentage of the total pie corresponding to each value of x, and plots pie slices of that size. The optional array `explode` controls whether or not individual pie slices are separated from the remainder of the pie.
`stairs(x,y)`	This function creates a stair plot, with each stair step centered on an (x, y) point.
`stem(x,y)`	This function creates a stem plot, with a marker at each (x, y) point and a stem drawn vertically from that point to the *x*-axis.

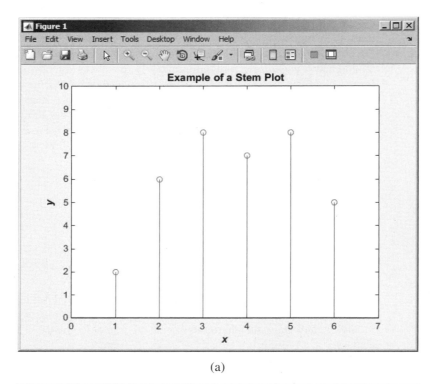

(a)

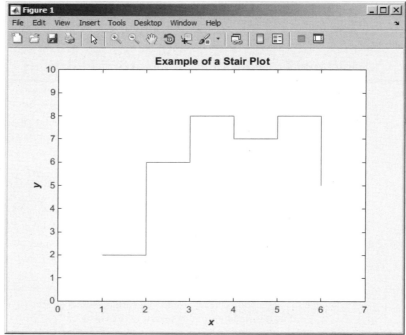

(b)

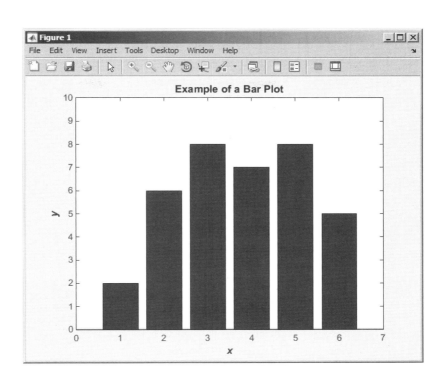

(c)

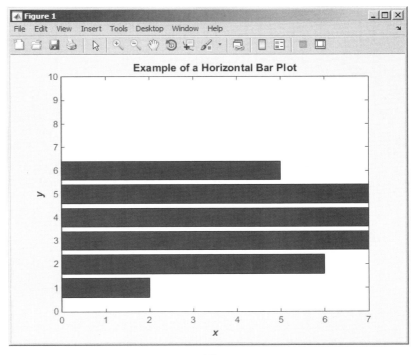

(d)

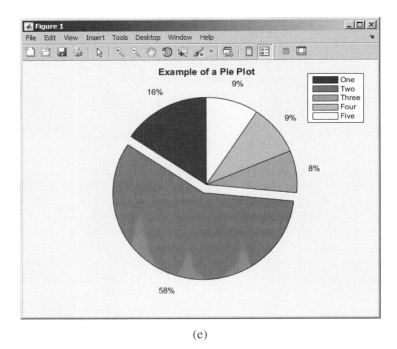

(e)

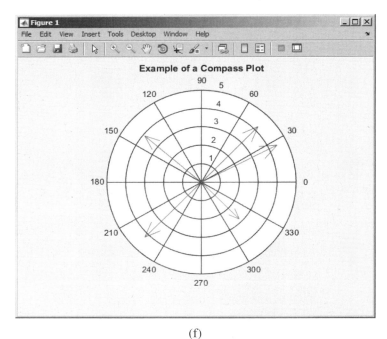

(f)

Figure 3.15 Additional types of 2D plots: (a) stem plot; (b) stair plot; (c) vertical bar plot; (d) horizontal bar plot; (e) pie plot; (f) compass plot. [See color insert for (e).]

that the first element x(1) is 1/10 or 10% of the pie, the second element x(2) is 2/10 or 20% of the pie, and so forth. The function then plots those percentages as pie slices.

Function pie also supports an optional parameter, explode. If present, explode is a logical array of 1's and 0's, with an element for each element in array x. If a value in explode is 1, then the corresponding pie slice is drawn slightly separated from the pie. For example, the code shown below produces the pie plot in Figure 3.15e. Note that the second slice of the pie is "exploded."

```
data = [10 37 5 6 6];
explode = [0 1 0 0 0];
pie(data,explode);
title('\bfExample of a Pie Plot');
legend('One','Two','Three','Four','Five');
```

3.5 Using the plot Function with Two-Dimensional Arrays

In all of the previous examples in this book, we have plotted data one vector at a time. What would happen if, instead of a vector of data, we had a two-dimensional array of data? The answer is that MATLAB treats each *column* of the 2D array as a separate line, and it plots as many lines as there are columns in the data set. For example, suppose that we create an array containing the function $f(x) = \sin x$ in column 1, $f(x) = \cos x$ in column 2, $f(x) = \sin^2 x$ in column 3, and $f(x) = \cos^2 x$ in column 4, each for $x = 0$ to 10 in steps of 0.1. This array can be created using the following statements

```
x = 0:0.1:10;
y = zeros(length(x),4);
y(:,1) = sin(x);
y(:,2) = cos(x);
y(:,3) = sin(x).^2;
y(:,4) = cos(x).^2;
```

If this array is plotted using the plot(x,y) command, the results are as shown in Figure 3.16. Note that each column of array y has become a separate line on the plot.

The bar and barh plots can also take two-dimensional array arguments. If an array argument is supplied to these plots, the program will display each column as a separately colored bar on the plot. For example, the following code produces the bar plot shown in Figure 3.17.

```
x = 1:5;
y = zeros(5,3);
y(1,:) = [1 2 3];
y(2,:) = [2 3 4];
y(3,:) = [3 4 5];
y(4,:) = [4 5 4];
y(5,:) = [5 4 3];
```

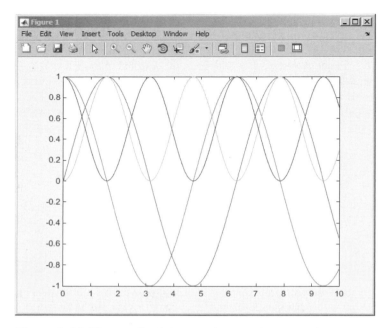

Figure 3.16 The result of plotting the two-dimensional array y. Note that each column is a separate line on the plot.

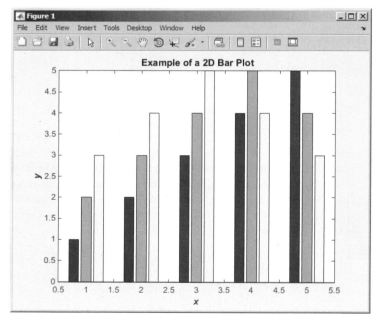

Figure 3.17 A bar plot created from a two-dimensional array y. Note that each column is a separate colored bar on the plot.

```
bar(x,y);
title('\bfExample of a 2D Bar Plot');
xlabel('\bf\itx');
ylabel('\bf\ity');
```

3.6 Summary

Chapter 3 extended our knowledge of two-dimensional plots, which were introduced in Chapter 2. Two-dimensional plots can take many different forms, as summarized in Table 3.4.

The `axis` command allows an engineer to select the specific range of x and y data to be plotted. The `hold` command allows later plots to be plotted on top of earlier ones, so that elements can be added to a graph a piece at a time. The `figure` command allows an engineer to create and select among multiple Figure Windows,

Table 3.4: Summary of Two-Dimensional Plots

Function	Description
plot(x,y)	This function plots points or lines with a linear scale on the x- and y-axes.
semilogx(x,y)	This function plots points or lines with a logarithmic scale on the x-axis and a linear scale on the y-axis.
semilogy(x,y)	This function plots points or lines with a linear scale on the x-axis and a logarithmic scale on the y-axis.
loglog(x,y)	This function plots points or lines with a logarithmic scale on the x-axis and a logarithmic scale on the y-axis.
polar(theta,r)	This function plots points or lines on a polar plot, where `theta` is the angle (in radians) of a point counterclockwise from the right-hand horizontal axis, and `r` is distance from the center of the plot to the point.
barh(x,y)	This function creates a *horizontal* bar plot, with the values in x used to label each bar and the values in y used to determine the horizontal length of the bar.
bar(x,y)	This function creates a *vertical* bar plot, with the values in x used to label each bar and the values in y used to determine the height of the bar.
compass(x,y)	This function creates a polar plot, with an arrow drawn from the origin to the location of each (x, y) point. Note that the locations of the points to plot are specified in Cartesian coordinates, not polar coordinates.
pie(x) pie(x,explode)	This function creates a pie plot. This function determines the percentage of the total pie corresponding to each value of x, and plots pie slices of that size. The optional array `explode` controls whether or not individual pie slices are separated from the remainder of the pie.
stairs(x,y)	This function creates a stair plot, with each stair step centered on an (x, y) point.
stem(x,y)	This function creates a stem plot, with a marker at each (x, y) point and a stem drawn vertically from that point to the x-axis.

so that a program can create multiple plots in separate windows. The `subplot` command allows an engineer to create and select among multiple plots within a single Figure Window.

We also learned how to control additional characteristics of our plots, such as the line width and marker color. These properties may be controlled by specifying `'PropertyName', value` pairs in the plot command after the data to be plotted.

Text strings in plots may be enhanced with stream modifiers and escape sequences. Stream modifiers allow an engineer to specify features like bold face, italic, superscripts, subscripts, font size, and font name. Escape sequences allow the engineer to include special characters such as Greek and mathematical symbols in the text string.

3.6.1 Summary of Good Programming Practice

The following guidelines should be adhered to when working with MATLAB functions.

1. Consider the type of data you are working with when determining how to best plot it. If the range of the data to plot covers many orders of magnitude, use a logarithmic scale to represent the data properly. If the range of the data to plot is an order of magnitude or less, then use a linear scale.
2. Use stream modifiers to create effects such as bold, italics, superscripts, subscripts, and special characters in your plot titles and labels.

3.6.2 MATLAB Summary

The following summary lists all of the MATLAB commands and functions described in this chapter, along with a brief description of each one.

Commands and Functions

`axis`	(a) Set the x and y limits of the data to be plotted.
	(b) Get the x and y limits of the data to be plotted.
	(c) Set other axis-related properties.
`bar(x,y)`	Create a vertical bar plot.
`barh(x,y)`	Create a horizontal bar plot.
`compass(x,y)`	Create a compass plot.
`figure`	Select a Figure Window to be the current Figure Window. If the selected Figure Window does not exist, it is automatically created.
`hold`	Allows multiple plot commands to write on top of each other.
`linspace`	Create an array of samples with equal spacing on a linear scale.
`loglog(x,y)`	Create a log/log plot.
`logspace`	Create an array of samples with equal spacing on a logarithmic scale.
`pie(x)`	Create a pie plot.

(continued)

`polar(theta,r)`	Create a polar plot.
`semilogx(x,y)`	Create a log/linear plot.
`semilogy(x,y)`	Create a linear/log plot.
`stairs(x,y)`	Create a stair plot.
`stem(x,y)`	Create a stem plot.
`subplot`	Select a subplot in the current Figure Window. If the selected subplot does not exist, it is automatically created. If the new subplot conflicts with a previously existing set of axes, they are automatically deleted.

3.7 Exercises

3.1 Plot the function $y(x) = e^{-0.5x} \sin 2x$ for 100 values of x between 0 and 10. Use a 2-point-wide solid blue line for this function. Then plot the function $y(x) = e^{-0.5x} \cos 2x$ on the same axes. Use a 3-point-wide dashed red line for this function. Be sure to include a legend, title, axis labels, and grid on the plots.

3.2 Use the MATLAB plot editing tools to modify the plot in Exercise 3.1. Change the line representing the function $y(x) = e^{-0.5x} \sin 2x$ to be a black dashed line that is 1-point-wide.

3.3 Plot the functions in Exercise 3.1 on a log/linear plot. Be sure to include a legend, title, axis labels, and grid on the plots.

3.4 Plot the function $y(x) = e^{-0.5x} \sin 2x$ on a bar plot. Use 100 values of x between 0 and 10 in the plot. Be sure to include a legend, title, axis labels, and grid on the plots.

3.5 Create a polar plot of the function $r(\theta) = \sin(2\theta) \cos \theta$ for $0 \le \theta \le 2\pi$.

3.6 Plot the function $f(x) = x^4 - 3x^3 + 10x^2 - x - 2$ for $-6 \le x \le 6$. Draw the function as a solid black 2-point-wide line, and turn on the grid. Be sure to include a title and axis labels, and include the equation for the function being plotted in the title string. (Note that you will need steam modifiers to get the italics and the superscripts in the title string.)

3.7 Plot the function $f(x) = \dfrac{x^2 - 6x + 5}{x - 3}$ using 200 points over the range $-2 \le x \le 8$. Note that there is an asymptote at $x = 3$, so the function will tent to infinity near that point. In order to see the rest of the plot properly, you will need to limit the y-axis to a reasonable size, so use the `axis` command to limit the y-axis to the range -10 to 10.

3.8 Suppose that George, Sam, Betty, Charlie, and Suzie contributed $15, $5, $10, $5, and $15 respectively to a colleague's going-away present. Create a pie chart of their contributions. What percentage of the cost was paid by Sam?

3.9 Plot the function $y(x) = e^{-x} \sin x$ for x between 0 and 4 in steps of 0.1. Create the following plot types: (*a*) linear plot; (*b*) log/linear plot; (*c*) stem plot; (*d*) stair plot; (*e*) bar plot; (*f*) horizontal bar plot; (*g*) compass plot. Be sure to include titles and axis labels on all plots.

3.10 Why does it not make sense to plot the function $y(x) = e^{-x} \sin x$ from the previous exercise on a linear/log or a log/log plot?

3.11 Assume that the complex function $f(t)$ is defined by the equation

$$f(t) = (1 + 0.25i)\,t - 2.0 \tag{3.4}$$

Plot the amplitude and phase of function f for $0 \leq t \leq 4$ on two separate subplots within a single figure. Be sure to provide appropriate titles and axis labels. [*NOTE:* You can calculate the amplitude of the function using the MATLAB function `abs` and the phase of the function using the MATLAB function `phase`.]

3.12 Create an array of 100 input samples in the range 1 to 100 using the `linspace` function, and plot the equation

$$y(x) = 20 \log_{10}(2x) \tag{3.5}$$

on a `semilogx` plot. Draw a solid blue line of width 2, and label each point with a red circle. Now create an array of 100 input samples in the range 1 to 100 using the `logspace` function, and plot Equation (3.5) on a `semilogx` plot. Draw a solid red line of width 2, and label each point with a black star. How does the spacing of the points on the plot compare when using `linspace` and `logspace`?

3.13 Error Bars When plots are made from real measurements recorded in the laboratory, the data that we plot is often the *average* of many separate measurements. This kind of data has two important pieces of information: the average value of the measurement and the amount of variation in the measurements that went into the calculation.

It is possible to convey both pieces of information on the same plot by adding *error bars* to the data. An error bar is a small vertical line that shows the amount of variation that went into the measurement at each point. The MATLAB function `errorbar` supplies this capability for MATLAB plots.

Look up `errorbar` in the MATLAB documentation, and learn how to use it. Note that there are two versions of this call, one that shows a single error that is applied equally on either side of the average point, and one that allows you to specify upper limits and lower limits separately.

Suppose that you wanted to use this capability to plot the mean high temperature at a location by month, as well as the minimum and maximum extremes. The data might take the form of the following table:

Temperatures at Location (°F)

Month	Average Daily High	Extreme High	Extreme Low
January	66	88	16
February	70	92	24

(continued)

March	75	100	25
April	84	105	35
May	93	114	39
June	103	122	50
July	105	121	63
August	103	116	61
September	99	116	47
October	88	107	34
November	75	96	27
December	66	87	22

Create a plot of the mean high temperature by month at this location, showing the extremes as error bars. Be sure to label your plot properly.

3.14 **The Spiral of Archimedes** The spiral of Archimedes is a curve described in polar coordinates by the equation

$$r = k\theta \tag{3.6}$$

where r is the distance of a point from the origin, and θ is the angle of that point in radians with respect to the origin. Plot the spiral of Archimedes for $0 \leq \theta \leq 6\pi$ when $k = 0.5$. Be sure to label you plot properly.

3.15 **Output Power from a Motor** The output power produced by a rotating motor is given by the equation

$$P = \tau_{IND} \, \omega_m \tag{3.7}$$

where τ_{IND} is the induced torque on the shaft in newton-meters, ω_m is the rotational speed of the shaft in radians per second, and P is in watts. Assume that the rotational speed of a particular motor shaft is given by the equation

$$\omega_m = 188.5(1 - e^{-0.2t}) \text{ rad/s} \tag{3.8}$$

and the induced torque on the shaft is given by

$$\tau_{IND} = 10e^{-0.2t} \text{ N} \cdot \text{m} \tag{3.9}$$

Plot the torque, speed, and power supplied by this shaft versus time in three subplots aligned vertically within a single figure for $0 \leq t \leq 10$ s. Be sure to label your plots properly with the symbols τ_{IND} and ω_m where appropriate. Create two separate plots, one with the power and torque displayed on a linear scale and one with the output power displayed on a logarithmic scale. Time should always be displayed on a linear scale.

3.16 **Plotting Orbits** When a satellite orbits the Earth, the satellite's orbit will form an ellipse with the Earth located at one of the focal points of the ellipse. The satellite's orbit can be expressed in polar coordinates as

$$r = \frac{p}{1 - \varepsilon \cos \theta} \tag{3.10}$$

where r and θ are the distance and angle of the satellite from the center of the Earth, p is a parameter specifying the size of the orbit, and ε is a parameter representing the eccentricity of the orbit. A circular orbit has an eccentricity ε of 0. An elliptical orbit has an eccentricity of $0 \le \varepsilon < 1$. If $\varepsilon > 1$, the satellite follows a hyperbolic path and escapes from the Earth's gravitational field.

Consider a satellite with a size parameter $p = 1000$ km. Plot the orbit of this satellite if (*a*) $\varepsilon = 0$; (*b*) $\varepsilon = 0.25$; (*c*) $\varepsilon = 0.5$. How close does each orbit come to the Earth? How far away does each orbit get from the Earth? Compare the three plots you created. Can you determine what the parameter p means from looking at the plots?

Chapter 4

Branching Statements and Program Design

In Chapter 2, we developed several complete working MATLAB programs. However, all of the programs were very simple, consisting of a series of MATLAB statements that were executed one after another in a fixed order. Such programs are called *sequential* programs. They read input data, process it to produce a desired answer, print out the answer, and quit. There is no way to repeat sections of the program more than once, and there is no way to selectively execute only certain portions of the program depending on values of the input data.

In the next two chapters, we will introduce a number of MATLAB statements that allow us to control the order in which statements are executed in a program. There are two broad categories of control statements: **branches**, which select specific sections of the code to execute; and **loops**, which cause specific sections of the code to be repeated. Branches will be discussed in this chapter, and loops will be discussed in Chapter 5.

With the introduction of branches and loops, our programs are going to become more complex, and it will get easier to make mistakes. To help avoid programming errors, we will introduce a formal program design procedure based upon the technique known as top-down design. We will also introduce a common algorithm development tool known as pseudocode.

We will also study the MATLAB logical data type before discussing branches, because branches are controlled by logical values and expressions.

4.1 Introduction to Top-Down Design Techniques

Suppose that you are an engineer working in industry and need to write a program to solve some problem. How do you begin?

When given a new problem, there is a natural tendency to sit down at a keyboard and start programming without "wasting" a lot of time thinking about the problem first. It is often possible to get away with this "on the fly" approach to programming for very small problems, such as many of the examples in this book. In the real world, however, problems are larger, and an engineer attempting this approach will become hopelessly bogged down. For larger problems, it pays to completely think out the problem and the approach you are going to take to it before writing a single line of code.

We will introduce a formal program design process in this section and then apply that process to every major application developed in the remainder of the book. For some of the simple examples that we will be doing, the design process will seem like overkill. However, as the problems that we solve get larger and larger, the process becomes more and more essential to successful programming.

When I was an undergraduate, one of my professors was fond of saying, "Programming is easy. It's knowing what to program that's hard." His point was forcefully driven home to me after I left university and began working in industry on larger-scale software projects. I found that the most difficult part of my job was to *understand the problem* I was trying to solve. Once I really understood the problem, it became easy to break it apart into smaller, more easily manageable pieces with well-defined functions and then to tackle those pieces one at a time.

Top-down design is the process of starting with a large task and breaking it down into smaller, more easily understandable pieces (sub-tasks), which perform a portion of the desired task. Each sub-task may in turn be subdivided into smaller sub-tasks if necessary. Once the program is divided into small pieces, each piece can be coded and tested independently. We do not attempt to combine the sub-tasks into a complete task until each of the sub-tasks has been verified to work properly by itself.

The concept of top-down design is the basis of our formal program design process. We will now introduce the details of the process, which is illustrated in Figure 4.1. The steps involved are:

1. *Clearly state the problem that you are trying to solve.*

    Programs are usually written to fill some perceived need but that need may not be articulated clearly by the person requesting the program. For example, a user may ask for a program to solve a system of simultaneous linear equations. This request is not clear enough to allow an engineer to design a program to meet the need; he or she must first know much more about the problem to be solved. Is the system of equations to be solved real or complex? What is the maximum number of equations and unknowns that the program must handle? Are there any symmetries in the equations which might be exploited to make the task easier? The program designer will have to talk with the user requesting the program, and the two of them will have to come up with a clear statement of exactly what they are trying to accomplish. A clear statement of the problem will prevent misunderstandings, and it will also help the program designer to properly organize his or her thoughts. In the example we were describing, a proper statement of the problem might have been:

    Design and create a program to solve a system of simultaneous linear equations having real coefficients and with up to 20 equations in 20 unknowns.

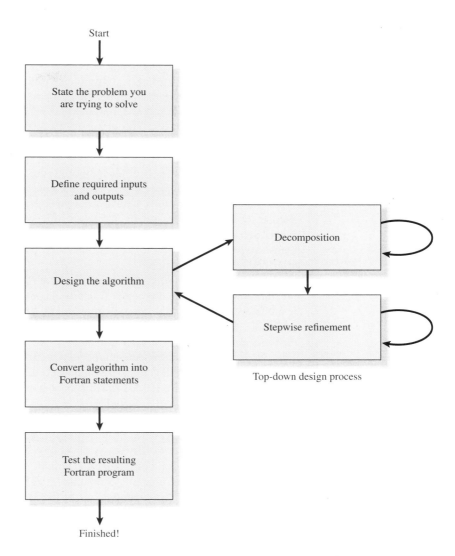

Figure 4.1 The program design process used in this book.

2. *Define the inputs required by the program and the outputs to be produced by the program.*

   The inputs to the program and the outputs produced by the program must be specified so that the new program will properly fit into the overall processing scheme. In the above example, the coefficients of the equations to be solved are probably in some pre-existing order, and our new program needs to be able to read them in that order. Similarly, it needs to produce the answers required by the programs that may follow it in the overall processing scheme and to write out those answers in the format needed by those programs.

3. *Design the algorithm that you intend to implement in the program.*

An **algorithm** is a step-by-step procedure for finding the solution to a problem. It is at this stage in the process that top-down design techniques come into play. The designer looks for logical divisions within the problem and divides it up into sub-tasks along those lines. This process is called *decomposition*. If the sub-tasks are themselves large, the designer can break them up into even smaller sub-sub-tasks. This process continues until the problem has been divided into many small pieces, each of which does a simple, clearly understandable job.

After the problem has been decomposed into small pieces, each piece is further refined through a process called *stepwise refinement*. In stepwise refinement, a designer starts with a general description of what the piece of code should do and then defines the functions of the piece in greater and greater detail until they are specific enough to be turned into MATLAB statements. Stepwise refinement is usually done with **pseudocode**, which will be described in the next section.

It is often helpful to solve a simple example of the problem by hand during the algorithm development process. If the designer understands the steps that he or she went through in solving the problem by hand, then he or she will be better able to apply decomposition and stepwise refinement to the problem.

4. *Turn the algorithm into MATLAB statements.*

If the decomposition and refinement process was carried out properly, this step will be very simple. All the engineer will have to do is to replace pseudocode with the corresponding MATLAB statements on a one-for-one basis.

5. *Test the resulting MATLAB program.*

This step is the real killer. The components of the program must first be tested individually, if possible, and then the program as a whole must be tested. When testing a program, we must verify that it works correctly for *all legal input data sets*. It is very common for a program to be written, tested with some standard data set, and released for use, only to find that it produces the wrong answers (or crashes) with a different input data set. If the algorithm implemented in a program includes different branches, we must test all of the possible branches to confirm that the program operates correctly under every possible circumstance. This exhaustive testing can be almost impossible in really large programs, so bugs can be discovered after the program has been in regular use for years.

Because the programs in this book are fairly small, we will not go through the sort of extensive testing described above. However, we will follow the basic principles in testing all of our programs.

Good Programming Practice

Follow the steps of the program design process to produce reliable, understandable MATLAB programs.

In a large programming project, the time actually spent programming is surprisingly small. In his book *The Mythical Man-Month*[1], Frederick P. Brooks Jr. suggests that in a typical large software project, 1/3 of the time is spent planning what to do (steps 1 through 3), 1/6 of the time is spent actually writing the program (step 4), and fully 1/2 of the time is spent in testing and debugging the program! Clearly, anything that we can do to reduce the testing and debugging time will be very helpful. We can best reduce the testing and debugging time by doing a very careful job in the planning phase and by using good programming practices. Good programming practices will reduce the number of bugs in the program and will make the ones that do creep in easier to find.

4.2 Use of Pseudocode

As a part of the design process, it is necessary to describe the algorithm that you intend to implement. The description of the algorithm should be in a standard form that is easy for both you and other people to understand, and the description should aid you in turning your concept into MATLAB code. The standard forms that we use to describe algorithms are called **constructs** (or sometimes structures), and an algorithm described using these constructs is called a structured algorithm. When the algorithm is implemented in a MATLAB program, the resulting program is called a **structured program**.

The constructs used to build algorithms can be described in a special way called pseudocode. **Pseudocode** is a hybrid mixture of MATLAB and English. It is structured like MATLAB, with a separate line for each distinct idea or segment of code, but the descriptions on each line are in English. Each line of the pseudocode should describe its idea in plain, easily understandable English. Pseudocode is very useful for developing algorithms, since it is flexible and easy to modify. It is especially useful since pseudocode can be written and modified with the same editor or word processor used to write the MATLAB program—no special graphical capabilities are required.

For example, the pseudocode for the algorithm in Example 2-3 is:

```
Prompt user to enter temperature in degrees Fahrenheit
Read temperature in degrees Fahrenheit (temp_f)
temp_k in kelvins <- (5/9) * (temp_f - 32) + 273.15
Write temperature in kelvins
```

Notice that a left arrow (<-) is used instead of an equal sign (=) to indicate that a value is stored in a variable, since this avoids any confusion between assignment and equality. Pseudocode is intended to aid you in organizing your thoughts before converting them into MATLAB code.

4.3 The Logical Data Type

The `logical` data type is a special type of data that can have one of only two possible values: `true` or `false`. These values are produced by the two special functions `true` and `false`. They are also produced by two types of MATLAB operators: relational operators and logic operators.

[1]*The Mythical Man-Month, Anniversary Edition*, by Frederick P. Brooks Jr., Addison-Wesley, 1995.

Logical values are stored in a single byte of memory, so they take up much less space than numbers, which usually occupy 8 bytes.

The operation of many MATLAB branching constructs is controlled by logical variables or expressions. If the result of a variable or expression is true, then one section of code is executed. If not, then a different section of code is executed.

To create a `logical` variable, just assign a logical value to it in an assignment statement. For example, the statement

```
a1 = true;
```

creates a logical variable `a1` containing the logical value `true`. If this variable is examined with the `whos` command, we can see that it has the logical data type:

```
» whos a1
Name        Size        Bytes        Class
a1          1x1             1        logical array
```

Unlike programming languages such as Java, C++, and Fortran, it is legal in MATLAB to mix numerical and logical data in expressions. If a logical value is used in a place where a numerical value is expected, `true` values are converted to 1 and `false` values are converted to 0, and then used as numbers. If a numerical value is used in a place where a logical value is expected, non-zero values are converted to `true` and 0 values are converted to `false`, and then used as logical values.

It is also possible to explicitly convert numerical values to logical values, and vice versa. The `logical` function converts numerical data to logical data, and the `real` function converts logical data to numerical data.

4.3.1 Relational and Logic Operators

Relational and logic operators are operators that produce a `true` or `false` result. These operators are very important, because they control which code gets executed in some MATLAB branching structures.

Relational operators are operators that compare two numbers and produce a true or false result. For example, a `>` b is a relational operator that compares the numbers in variables a and b. If the value in a is greater than the value in b, then this operator returns a true result. Otherwise, the operator returns a false result.

Logic operators are operators that compare one or two logical values, and produce a true or false result. For example, `&&` is a logical AND operator. The operation a `&&` b compares the logical values stored in variables a and b. If both a and b are true (nonzero), then the operator returns a true result. Otherwise, the operator returns a false result.

4.3.2 Relational Operators

Relational operators are operators with two numerical or string operands that return true (1) or false (0), depending on the relationship between the two operands. The general form of a relational operator is

$$a_1 \text{ op } a_2$$

where a_1 and a_2 are arithmetic expressions, variables, or strings, and op is one of the following relational operators:

Table 4.1: Relational Operators

Operator	Operation
==	Equal to
~=	Not equal to
>	Greater than
>=	Greater than or equal to
<	Less than
<=	Less than or equal to

If the relationship between a_1 and a_2 expressed by the operator is true, then the operation returns a true value; otherwise, the operation returns false.

Some relational operations and their results are given below:

Operation	Result
3 < 4	true (1)
3 <= 4	true (1)
3 == 4	false (0)
3 > 4	false (0)
4 <= 4	true (1)
'A' < 'B'	true (1)

The last relational operation is true because characters are evaluated in alphabetical order.

Relational operators may be used to compare a scalar value with an array. For example, if a $= \begin{bmatrix} 1 & 0 \\ -2 & 1 \end{bmatrix}$ and b $= 0$, then the expression a > b will yield the array $\begin{bmatrix} 1 & 0 \\ 0 & 1 \end{bmatrix}$. Relational operators may also be used to compare two arrays, as long as both arrays have the same size. For example, if a $= \begin{bmatrix} 1 & 0 \\ -2 & 1 \end{bmatrix}$ and b $= \begin{bmatrix} 0 & 2 \\ -2 & -1 \end{bmatrix}$, then the expression a >= b will yield the array $\begin{bmatrix} 1 & 0 \\ 1 & 1 \end{bmatrix}$. If the arrays have different sizes, a runtime error will result.

Note that since strings are really arrays of characters, *relational operators can only compare two strings if they are of equal lengths*. If they are of unequal lengths, the comparison operation will produce an error. We will learn of a more general way to compare strings in Chapter 9.

The equivalence relational operator is written with two equal signs, while the assignment operator is written with a single equal sign. These are very different operators that beginning engineers often confuse. The == symbol is a *comparison* operation that returns a logical (0 or 1) result, while the = symbol *assigns* the value of the expression to the right of the equal sign to the variable on the left of the equal sign. It is a very common mistake for beginning engineers to use a single equal sign when trying to do a comparison.

▣ Programming Pitfalls

Be careful not to confuse the equivalence relational operator (==) with the assignment operator (=).

In the hierarchy of operations, relational operators are evaluated after all arithmetic operators have been evaluated. Therefore, the following two expressions are equivalent (both are true).

```
7 + 3 < 2 + 11
(7 + 3) < (2 + 11)
```

4.3.3 A Caution About the == and ~= Operators

The equivalence operator (==) returns a `true` value (1) when the two values being compared are equal, and a `false` (0) when the two values being compared are different. Similarly, non-equivalence operator (~=) returns a `false` (0) when the two values being compared are equal, and a `true` (1) when the two values being compared are different. These operators are generally safe to use for comparing strings, but they can sometimes produce surprising results when two numeric values are compared. Due to **roundoff errors** during computer calculations, two theoretically equal numbers can differ slightly, causing an equality or inequality test to fail.

For example, consider the following two numbers, both of which should be equal to 0.0.

```
a = 0;
b = sin(pi);
```

Since these numbers are theoretically the same, the relational operation a == b *should* produce a 1. In fact, the results of this MATLAB calculation are

```
» a = 0;
» b = sin(pi);
» a == b
ans =
     0
```

MATLAB reports that a and b are different because a slight roundoff error in the calculation of `sin(pi)` makes the result be 1.2246×10^{-16} instead of exactly zero. The two theoretically equal values differ slightly due to roundoff error!

Instead of comparing two numbers for *exact* equality, you should set up your tests to determine if the two numbers are *nearly* equal to each other within some accuracy that takes into account the roundoff error expected for the numbers being compared. The test

```
» abs(a - b) < 1.0E-14
ans =
     1
```

produces the correct answer, despite the roundoff error in calculating b.

Good Programming Practice

Be cautious about testing for equality with numeric values, since roundoff errors may cause two variables that should be equal to fail a test for equality. Instead, test to see if the variables are *nearly* equal within the roundoff error to be expected on the computer you are working with.

4.3.4 Logic Operators

Logic operators are operators with one or two logical operands that yield a logical result. There are five binary logic operators: AND (& and &&), inclusive OR (| and ||), and exclusive OR (xor), and one unary logic operator: NOT (~). The general form of a binary logic operation is

$$l_1 \text{ op } l_2$$

and the general form of a unary logic operation is

$$\text{op } l_1$$

where l_1 and l_2 are expressions or variables, and op is one of the following logic operators shown in Table 4.2.

Table 4.2: Logic Operators

Operator	Operation
&	Logical AND
&&	Logical AND with shortcut evaluation
|	Logical Inclusive OR
||	Logical Inclusive OR with shortcut evaluation
xor	Logical Exclusive OR
~	Logical NOT

If the relationship between l_1 and l_2 expressed by the operator is true, then the operation returns a true (1); otherwise, the operation returns a false (0). Note that logic operators treat any nonzero value as true, and any zero value as false.

The results of the operators are summarized in **truth tables**, which show the result of each operation for all possible combinations of l_1 and l_2. Table 4.3 shows the truth tables for all logic operators.

Logical ANDs

The result of an AND operator is true (1) if and only if both input operands are true. If either or both operands are false, the result is false (0), as shown in Table 4.3.

Note that there are two logical AND operators: && and &. Why are there two AND operators, and what is the difference between them? The basic difference between && and & is that && supports *short-circuit evaluations* (or *partial evaluations*), while & doesn't. That is, && will evaluate expression l_1 and immediately return a false (0) value if l_1 is false. If l_1 is false, the operator never evaluates l_2, because the result of the operator will be false regardless of the value of l_2. In contrast, the & operator always evaluates both l_1 and l_2 before returning an answer.

A second difference between && and & is that && only works between scalar values, while & works with either scalar or array values, as long as the sizes of the arrays are compatible.

When should you use && and when should you use & in a program? Most of the time, it doesn't matter which AND operation is used. If you are comparing scalars, and it is not necessary to always evaluate l_2, then use the && operator. The partial evaluation will make the operation faster in the cases where the first operand is `false`.

Sometimes it is important to use shortcut expressions. For example, suppose that we wanted to test for the situation where the ratio of two variables a and b is greater than 10. The code to perform this test is:

```
x = a / b > 10.0
```

This code normally works fine, but what about the case where b is zero? In that case, we would be dividing by zero, which produces an `Inf` instead of a number. The test could be modified to avoid this problem as follows:

```
x = (b ~= 0) && (a/b > 10.0)
```

This expression uses partial evaluation, so if b = 0, the expression `a/b > 10.0` will never be evaluated, and no `Inf` will occur.

Table 4.3: Truth Tables for Logic Operators

Inputs		and		or		xor	not
l_1	l_2	l_1 & l_2	l_1 && l_2	l_1 | l_2	l_1 || l_2	$\text{xor}(l_1, l_2)$	$\sim l_1$
false	false	false	false	false	false	false	true
false	true	false	false	true	true	true	true
true	false	false	false	true	true	true	false
true	true	true	true	true	true	false	false

▣ Good Programming Practice

Use the & AND operator if it is necessary to ensure that both operands are evaluated in an expression, or if the comparison is between arrays. Otherwise, use the && AND operator, since the partial evaluation will make the operation faster in the cases where the first operand is `false`. The & operator is preferred in most practical cases.

Logical Inclusive ORs

The result of an inclusive OR operator is true (1) if either or both of the input operands are true. If both operands are false, the result is false (0), as shown in Table 4.3.

Note that there are two inclusive OR operators: || and |. Why are there two inclusive OR operators, and what is the difference between them? The basic difference between || and | is that || supports partial evaluations, while | doesn't. That is, || will evaluate expression l_1 and immediately return a true value if l_1 is true. If l_1 is true, the operator never evaluates l_2, because the result of the operator will be true regardless of the value of l_2. In contrast, the | operator always evaluates both l_1 and l_2 before returning an answer.

A second difference between || and | is that || only works between scalar values, while | works with either scalar or array values, as long as the sizes of the arrays are compatible.

When should you use || and when should you use | in a program? Most of the time, it doesn't matter which OR operation is used. If you are comparing scalars, and it is not necessary to always evaluate l_2, use the || operator. The partial evaluation will make the operation faster in the cases where the first operand is true.

▣ Good Programming Practice

Use the | inclusive OR operator if it is necessary to ensure that both operands are evaluated in an expression, or if the comparison is between arrays. Otherwise, use the || operator, since the partial evaluation will make the operation faster in the cases where the first operand is `true`. The | operator is preferred in most practical cases.

Logical Exclusive OR

The result of an exclusive OR operator is true if and only if one operand is true and the other one is false. If both operands are true or both operands are false, then the result is false, as shown in Table 4.3. Note that both operands must always be evaluated in order to calculate the result of an exclusive OR.

The logical exclusive OR operation is implemented as a function. For example,

```
a = 10;
b = 0;
x = xor(a, b);
```

The value in a is nonzero, so it is treated as true. The value in b is zero, so it is treated as false. Since one value is true and the other is false, the result of the xor operation will be true, and it will return a value of 1.

Logical NOT

The NOT operator (~) is a unary operator, having only one operand. The result of a NOT operator is true (1) if its operand is zero, and false (0) if its operand is nonzero, as shown in Table 4.3.

Hierarchy of Operations

In the hierarchy of operations, logic operators are evaluated *after all arithmetic operations and all relational operators have been evaluated.* The order in which the operators in an expression are evaluated is:

1. All arithmetic operators are evaluated first in the order previously described.
2. All relational operators (==, ~=, >, >=, <, <=) are evaluated, working from left to right.
3. All ~ operators are evaluated.
4. All & and && operators are evaluated, working from left to right.
5. All |, ||, and xor operators are evaluated, working from left to right.

As with arithmetic operations, parentheses can be used to change the default order of evaluation. Examples of some logic operators and their results are given below.

►Example 4.1—Evaluating Logical Expressions:

Assume that the following variables are initialized with the values shown, and calculate the result of the specified expressions:

```
value1 = 1
value2 = 0
value3 = 1
value4 = -10
value5 = 0
value6 = [1 2; 0 1]
```

Expression	Result	Comment
(a) ~value1	false (0)	
(b) ~value3	false (0)	The number 1 is treated as true, and the not operations is applied
(c) value1 | value2	true (1)	
(d) value1 & value2	false (0)	
(e) value4 & value5	false (0)	−10 is treated as true and 0 is treated as false when the AND operation is applied

(f)	`~(value4 & value5)`	true (1)	-10 is treated as true and 0 is treated as false when the AND operation is applied, and then the NOT operation reverses the result
(g)	`value1 + value4`	-9	
(h)	`value1 + (~value4)`	1	The number `value4` is nonzero, and so considered true. When the NOT operation is performed, the result is false (0). Then `value1` is added to the 0, so the final result is $1 + 0 = 1$.
(i)	`value3 && value6`	Illegal	The `&&` operator must be used with scalar operands.
(j)	`value3 & value6`	$\begin{bmatrix} 1 & 1 \\ 0 & 1 \end{bmatrix}$	AND between a scalar and an array operand. The nonzero values of array `value6` are treated as true.

The ~ operator is evaluated before other logic operators. Therefore, the parentheses in part (f) of the above example were required. If they had been absent, the expression in part (f) would have been evaluated in the order (`~value4`) `& value5`.

4.3.5 Logical Functions

MATLAB includes a number of logical functions that return `true` whenever the condition they test for is true and return `false` whenever the condition they test for is false. These functions can be used with relational and logic operators to control the operation of branches and loops.

A few of the more important logical functions are given in Table 4.4.

Table 4.4: Selected MATLAB Logical Functions

Function	Purpose
`false`	Returns a `false` (0) value.
`ischar(a)`	Returns `truev` if a is a character array and `false` otherwise.
`isempty(a)`	Returns `true` if a is an empty array and `false` otherwise.
`isinf(a)`	Returns `true` if the value of a is infinite (`Inf`) and `false` otherwise.
`isnan(a)`	Returns `true` if the value of a is NaN (not a number) and `false` otherwise.

(continued)

Table 4.4: Selected MATLAB Logical Functions. (*Continued*)

Function	Purpose
isnumeric(a)	Returns true if a is a numeric array and false otherwise.
logical	Converts numerical values to logical values: if a value is nonzero, it is converted to true. If it is zero, it is converted to false.
true	Returns a true (1) value.

Quiz 4.1

This quiz provides a quick check to see if you have understood the concepts introduced in Section 4.3. If you have trouble with the quiz, reread the sections, ask your instructor, or discuss the material with a fellow student. The answers to this quiz are found in the back of the book.

Assume that a, b, c, and d are as defined, and evaluate the following expressions.

$$a = 20; \qquad b = -2;$$
$$c = 0; \qquad d = 1;$$

1. a > b
2. b > d
3. a > b && c > d
4. va == b
5. a && b > c
6. ~~b

Assume that a, b, c, and d are as defined, and evaluate the following expressions.

$$a = 2; \qquad b = \begin{bmatrix} 1 & -2 \\ 0 & 10 \end{bmatrix};$$

$$c = \begin{bmatrix} 0 & 1 \\ 2 & 0 \end{bmatrix}; \qquad d = \begin{bmatrix} -2 & 1 & 2 \\ 0 & 1 & 0 \end{bmatrix};$$

7. ~(a > b)
8. a > c && b > c
9. c <= d
10. logical(d)
11. a * b > c
12. a * (b > c)

Assume that a, b, c, and d are as defined. Explain the order in which each of the following expressions are evaluated, and specify the results in each case:

$$a = 2; \qquad b = 3;$$
$$c = 10; \qquad d = 0;$$

13. `a*b^2 > a*c`
14. `d || b > a`
15. `(d | b) > a`

Assume that a, b, c, and d are as defined, and evaluate the following expressions.

$$a = 20; \qquad b = -2;$$
$$c = 0; \qquad d = \text{'Test'};$$

16. `isinf(a/b)`
17. `isinf(a/c)`
18. `a > b && ischar(d)`
19. `isempty(c)`
20. `(~a) & b`
21. `(~a) + b`

4.4 Branches

Branches are MATLAB statements that permit us to select and execute specific sections of code (called *blocks*) while skipping other sections of code. They are variations of the `if` construct, the `switch` construct, and the `try/catch` construct.

4.4.1 The `if` Construct

The `if` construct has the form

```
if control_expr_1
    Statement 1
    Statement 2           Block 1
    ...
elseif control_expr_2
    Statement 1
    Statement 2           Block 2
    ...
else
    Statement 1
    Statement 2           Block 3
    ...
end
```

where the control expressions are logical expressions that control the operation of the `if` construct. If *control_expr_1* is true (nonzero), then the program executes the statements in Block 1, and skips to the first executable statement following the `end`. Otherwise, the program checks for the status of *control_expr_2*. If *control_expr_2* is true (nonzero), then the program executes the statements in Block 2, and skips to the first executable statement following the `end`. If all control expressions are zero, then the program executes the statements in the block associated with the `else` clause.

There can be any number of `elseif` clauses (0 or more) in an `if` construct, but there can be at most one `else` clause. The control expression in each clause will be tested only if the control expressions in every clause above are false (0). Once one of the expressions proves to be true and the corresponding code block is executed, the program skips to the first executable statement following the `end`. If all control expressions are false, then the program executes the statements in the block associated with the `else` clause. If there is no `else` clause, then execution continues after the `end` statement without executing any part of the `if` construct.

Note that the MATLAB keyword `end` in this construct is *completely different* from the MATLAB function `end` that we used in Chapter 2 to return the highest value of a given subscript. MATLAB tells the difference between these two uses of `end` from the context in which the word appears within an M-file.

In most circumstances, *the control expressions will be some combination of relational and logic operators.* As we learned earlier in this chapter, relational and logic operators produce a true (1) when the corresponding condition is true and a false (0) when the corresponding condition is false. When an operator is true, its result is nonzero, and the corresponding block of code will be executed.

As an example of an `if` construct, consider the solution of a quadratic equation of the form

$$ax^2 + bx + c = 0 \tag{4.1}$$

The solution to this equation is

$$x = \frac{-b \pm \sqrt{b^2 - 4ac}}{2a} \tag{4.2}$$

The term $b^2 - 4ac$ is known as the *discriminant* of the equation. If $b^2 - 4ac > 0$, then there are two distinct real roots to the quadratic equation. If $b^2 - 4ac = 0$, then there is a single repeated root to the equation, and if $b^2 - 4ac < 0$, then there are two complex roots to the quadratic equation.

Suppose that we wanted to examine the discriminant of a quadratic equation and to tell a user whether the equation has two complex roots, two identical real roots, or two distinct real roots. In pseudocode, this construct would take the form

```
if (b^2 - 4*a*c) < 0
   Write msg that equation has two complex roots.
elseif (b**2 - 4.*a*c) == 0
   Write msg that equation has two identical real roots.
else
   Write msg that equation has two distinct real roots.
end
```

The MATLAB statements to do this are

```
if (b^2 - 4*a*c) < 0
   disp('This equation has two complex roots.');
elseif (b^2 - 4*a*c) == 0
   disp('This equation has two identical real roots.');
```

```
else
   disp('This equation has two distinct real roots.');
end
```

For readability, the blocks of code within an `if` construct are usually indented by 3 or 4 spaces, but this is not actually required.

Good Programming Practice

Always indent the body of an `if` construct by 3 or more spaces to improve the readability of the code. Note that indentation is automatic if you use the MATLAB editor to write your programs.

It is possible to write a complete `if` construct on a single line by separating the parts of the construct by commas or semicolons. Thus the following two constructs are identical:

```
if x < 0
   y = abs(x);
end
```

and

```
if x < 0; y = abs(x); end
```

However, this should only be done for very simple constructs.

4.4.2. Examples Using `if` Constructs

We will now look at two examples that illustrate the use of `if` constructs.

▶ Example 4.2—The Quadratic Equation

Write a program to solve for the roots of a quadratic equation, regardless of type.

Solution We will follow the design steps outlined earlier in the chapter.

1. **State the problem**
   The problem statement for this example is very simple. We want to write a program that will solve for the roots of a quadratic equation, whether they are distinct real roots, repeated real roots, or complex roots.

2. **Define the inputs and outputs**
   The inputs required by this program are the coefficients a, b, and c of the quadratic equation

$$ax^2 + bx + c = 0 \qquad (4.1)$$

The output from the program will be the roots of the quadratic equation, whether they are distinct real roots, repeated real roots, or complex roots.

3. **Design the algorithm**

This task can be broken down into three major sections, whose functions are input, processing, and output:

```
Read the input data
Calculate the roots
Write out the roots
```

We will now break each of the above major sections into smaller, more detailed pieces. There are three possible ways to calculate the roots, depending on the value of the discriminant, so it is logical to implement this algorithm with a three-branched if construct. The resulting pseudocode is:

```
Prompt the user for the coefficients a, b, and c.
Read a, b, and c
discriminant ← b^2 - 4 * a * c
if discriminant > 0
   x1 ← ( -b + sqrt(discriminant) ) / ( 2 * a )
   x2 ← ( -b - sqrt(discriminant) ) / ( 2 * a )
   Write msg that equation has two distinct real
   roots.
   Write out the two roots.
elseif discriminant == 0
   x1 ← -b / ( 2 * a )
   Write msg that equation has two identical real
   roots.
   Write out the repeated root.
else
   real_part ← -b / ( 2 * a )
   imag_part ← sqrt ( abs ( discriminant ) ) / ( 2 * a )
   Write msg that equation has two complex roots.
   Write out the two roots.
end
```

4. **Turn the algorithm into MATLAB statements.**

The final MATLAB code is shown below:

```
% Script file: calc_roots.m
%
% Purpose:
%     This program solves for the roots of a quadratic equation
%     of the form a*x^2 + b*x + c = 0. It calculates the answers
%     regardless of the type of roots that the equation possesses.
%
```

```
% Record of revisions:
%     Date          Programmer          Description of change
%     ====          ==========          =====================
%   01/02/14      S. J. Chapman       Original code
%
% Define variables:
%     a               -- Coefficient of x^2 term of equation
%     b               -- Coefficient of x term of equation
%     c               -- Constant term of equation
%     discriminant    -- Discriminant of the equation
%     imag_part       -- Imag part of equation (for complex roots)
%     real_part       -- Real part of equation (for complex roots)
%     x1              -- First solution of equation (for real roots)
%     x2              -- Second solution of equation (for real roots)

% Prompt the user for the coefficients of the equation
disp ('This program solves for the roots of a quadratic');
disp ('equation of the form A*X^2 + B*X + C = 0.');
a = input ('Enter the coefficient A:');
b = input ('Enter the coefficient B:');
c = input ('Enter the coefficient C:');
% Calculate discriminant
discriminant = b^2 - 4 * a * c;

% Solve for the roots, depending on the value of the discriminant
if discriminant > 0 % there are two real roots, so...

      x1 = (-b + sqrt(discriminant) ) / (2 * a);
      x2 = (-b - sqrt(discriminant) ) / (2 * a);
      disp ('This equation has two real roots:');
      fprintf ('x1 = %f\n', x1);
      fprintf ('x2 = %f\n', x2);

elseif discriminant == 0 % there is one repeated root, so...

      x1 = (-b) / (2 * a);
      disp ('This equation has two identical real roots:');
      fprintf ('x1 = x2 = %f\n',x1);

else % there are complex roots, so ...

      real_part = ( -b ) / ( 2 * a );
      imag_part = sqrt ( abs ( discriminant ) ) / (2 * a);
      disp ('This equation has complex roots:');
      fprintf('x1 = %f +i %f\n', real_part, imag_part);
      fprintf('x1 = %f -i %f\n', real_part, imag_part);

end
```

5. **Test the program.**

Next, we must test the program using real input data. Since there are three possible paths through the program, we must test all three paths before we can be certain that the program is working properly. From Equation (4.2), it is possible to verify the solutions to the equations given below:

$$x^2 + 5x + 6 = 0 \qquad\qquad x = -2 \text{ and } x = -3$$

$$x^2 + 4x + 4 = 0 \qquad\qquad x = -2$$

$$x^2 + 2x + 5 = 0 \qquad\qquad x = -1 \pm i2$$

If this program is executed three times with the above coefficients, the results are as shown below (user inputs are shown in bold face):

```
» calc_roots
This program solves for the roots of a quadratic
equation of the form A*X^2 + B*X + C = 0.
Enter the coefficient A: 1
Enter the coefficient B: 5
Enter the coefficient C: 6
This equation has two real roots:
x1 = -2.000000
x2 = -3.000000
» calc_roots
This program solves for the roots of a quadratic
equation of the form A*X^2 + B*X + C = 0.
Enter the coefficient A: 1
Enter the coefficient B: 4
Enter the coefficient C: 4
This equation has two identical real roots:
x1 = x2 = -2.000000
» calc_roots
This program solves for the roots of a quadratic
equation of the form A*X^2 + B*X + C = 0.
Enter the coefficient A: 1
Enter the coefficient B: 2
Enter the coefficient C: 5
This equation has complex roots:
x1 = -1.000000 +i 2.000000
x1 = -1.000000 -i 2.000000
```

The program gives the correct answers for our test data in all three possible cases.

◀

►Example 4.3—Evaluating a Function of Two Variables

Write a MATLAB program to evaluate a function $f(x, y)$ for any two user-specified values x and y. The function $f(x, y)$ is defined as follows.

$$f(x, y) = \begin{cases} x + y & x \geq 0 \text{ and } y \geq 0 \\ x + y^2 & x \geq 0 \text{ and } y < 0 \\ x^2 + y & x < 0 \text{ and } y \geq 0 \\ x^2 + y^2 & x < 0 \text{ and } y < 0 \end{cases}$$

Solution The function $f(x, y)$ is evaluated differently depending on the signs of the two independent variables x and y. To determine the proper equation to apply, it will be necessary to check for the signs of the x and y values supplied by the user.

1. **State the problem**
   This problem statement is very simple: Evaluate the function $f(x, y)$ for any user-supplied values of x and y.

2. **Define the inputs and outputs**
   The inputs required by this program are the values of the independent variables x and y. The output from the program will be the value of the function $f(x, y)$.

3. **Design the algorithm**
   This task can be broken down into three major sections, whose functions are input, processing, and output:

   ```
   Read the input values x and y
   Calculate f(x,y)
   Write out f(x,y)
   ```

   We will now break each of the above major sections into smaller, more detailed pieces. There are four possible ways to calculate the function $f(x, y)$, depending upon the values of x and y, so it is logical to implement this algorithm with a four-branched if statement. The resulting pseudocode is:

   ```
   Prompt the user for the values x and y.
   Read x and y
   if x ≥ 0 and y ≥ 0
       fun ← x + y
   elseif x ≥ 0 and y < 0
       fun ← x + y^2
   elseif x < 0 and y ≥ 0
       fun ← x^2 + y
   else
       fun ← x^2 + y^2
   end
   Write out f(x,y)
   ```

4. **Turn the algorithm into MATLAB statements.**
   The final MATLAB code is shown below.

```
% Script file: funxy.m
%
% Purpose:
%     This program solves the function f(x,y) for a
%     user-specified x and y, where f(x,y) is defined as:
%
%                   [x + y              x >= 0 and y >= 0
%                   |x + y^2            x >= 0 and y < 0
%     f(x, y) =     |x^2 + y            x < 0  and y >= 0
%                   [x^2 + y^2          x < 0  and y < 0
%
% Record of revisions:
%     Date           Programmer          Description of change
%     ====           ==========          =====================
%     01/03/14       S. J. Chapman       Original code
%
% Define variables:
%     x   -- First independent variable
%     y   -- Second independent variable
%     fun -- Resulting function

% Prompt the user for the values x and y
x = input ('Enter the x value: ');
y = input ('Enter the y value: ');

% Calculate the function f(x,y) based upon
% the signs of x and y.
if x >= 0 && y >= 0
    fun = x + y;
elseif x >= 0 && y < 0
    fun = x + y^2;
elseif x < 0 && y >= 0
    fun = x^2 + y;
else % x < 0 and y < 0, so
    fun = x^2 + y^2;
end

% Write the value of the function.
disp (['The value of the function is ' num2str(fun)]);
```

5. **Test the program.**
   Next, we must test the program using real input data. Since there are four possible paths through the program, we must test all four paths before we can be certain that the program is working properly. To test all four possible paths, we will execute the program with the four sets of input values $(x, y) = (2, 3)$, $(2, -3)$, $(-2, 3)$, and $(-2, -3)$. Calculating by hand, we see that

$$f(2,3) = 2 + 3 = 5$$
$$f(2,-3) = 2 + (-3)^2 = 11$$
$$f(-2,3) = (-2)^2 + 3 = 7$$
$$f(-2,-3) = (-2)^2 + (-3)^2 = 13$$

If this program is compiled, and then run four times with the above values, the results are:

```
» funxy
Enter the x coefficient: 2
Enter the y coefficient: 3
The value of the function is 5
» funxy
Enter the x coefficient: 2
Enter the y coefficient: -3
The value of the function is 11
» funxy
Enter the x coefficient: -2
Enter the y coefficient: 3
The value of the function is 7
» funxy
Enter the x coefficient: -2
Enter the y coefficient: -3
The value of the function is 13
```

The program gives the correct answers for our test values in all four possible cases.

◀

4.4.3 Notes Concerning the Use of `if` Constructs

The `if` construct is very flexible. It must have one `if` statement and one `end` statement. In between, it can have any number of `elseif` clauses, and may also have one `else` clause. With this combination of features, it is possible to implement any desired branching construct.

In addition, `if` constructs may be **nested**. Two `if` constructs are said to be nested if one of them lies entirely within a single code block of the other one. The following two `if` constructs are properly nested.

```
if x > 0
   ...
   if y < 0
      ...
   end
   ...
end
```

The MATLAB interpreter always associates a given `end` statement with the most recent `if` statement, so the first `end` above closes the `if y < 0` statement, while the second `end` closes the `if x > 0` statement. This works well for a properly written program, but can cause the interpreter to produce confusing error messages in cases where the programmer makes a coding error. For example, suppose that we have a large program containing a construct like the one shown below.

```
...
if (test1)
   ...
   if (test2)
      ...
      if (test3)
         ...
      end
      ...
   end
   ...
end
```

This program contains three nested `if` constructs that may span hundreds of lines of code. Now suppose that the first `end` statement is accidentally deleted during an editing session. When that happens, the MATLAB interpreter will automatically associate the second `end` with the innermost `if (test3)` construct, and the third `end` with the middle `if (test2)`. When the interpreter reaches the end of the file, it will notice that the first `if (test1)` construct was never ended, and it will generate an error message saying that there is a missing `end`. Unfortunately, it can't tell *where* the problem occurred, so we will have to go back and manually search the entire program to locate the problem.

It is sometimes possible to implement an algorithm using either multiple `elseif` clauses or nested `if` statements. In that case, the program designer may choose whichever style he or she prefers.

▶ Example 4.4—Assigning Letter Grades

Suppose that we are writing a program which reads in a numerical grade and assigns a letter grade to it according to the following table:

```
95 < grade           A
86 < grade ≤ 95      B
76 < grade ≤ 86      C
66 < grade ≤ 76      D
 0 < grade ≤ 66      F
```

Write an `if` construct that will assign the grades as described above using (*a*) multiple `elseif` clauses and (*b*) nested `if` constructs.

Solution (a) One possible structure using `elseif` clauses is

```
if grade > 95.0
   disp('The grade is A.');
elseif grade > 86.0
   disp('The grade is B.');
elseif grade > 76.0
   disp('The grade is C.');
elseif grade > 66.0
   disp('The grade is D.');
else
   disp('The grade is F.');
end
```

(b) One possible structure using nested `if` constructs is

```
if grade > 95.0
   disp('The grade is A.');
else
   if grade > 86.0
      disp('The grade is B.');
   else
      if grade > 76.0
         disp('The grade is C.');
      else
         if grade > 66.0
            disp('The grade is D.');
         else
            disp('The grade is F.');
         end
      end
   end
end
```

It should be clear from the above example that if there are a lot of mutually exclusive options, a single `if` construct with multiple `elseif` clauses will be simpler than a nested `if` construct.

Good Programming Practice

For branches in which there are many mutually exclusive options, use a single `if` construct with multiple `elseif` clauses in preference to nested `if` constructs.

4.4.4 The `switch` Construct

The `switch` construct is another form of branching construct. It permits an engineer to select a particular code block to execute based on the value of a single integer, character, or logical expression. The general form of a `switch` construct is:

```
switch (switch_expr)
case case_expr_1
    Statement 1
    Statement 2          ⎫
                          ⎬  Block 1
    ...                  ⎭

case case_expr_2
    Statement 1
    Statement 2          ⎫
                          ⎬  Block 2
    ...                  ⎭

...
otherwise
    Statement 1
    Statement 2          ⎫
                          ⎬  Block n
    ...                  ⎭

end
```

If the value of *switch_expr* is equal to *case_expr_1*, then the first code block will be executed, and the program will jump to the first statement following the end of the `switch` construct. Similarly, if the value of *switch_expr* is equal to *case_expr_2*, then the second code block will be executed, and the program will jump to the first statement following the end of the `switch` construct. The same idea applies for any other cases in the construct. The `otherwise` code block is optional. If it is present, it will be executed whenever the value of `switch_expr` is outside the range of all of the case selectors. If it is not present and the value of `switch_expr` is outside the range of all of the case selectors, then none of the code blocks will be executed. The pseudocode for the case construct looks just like its MATLAB implementation.

If many values of the `switch_expr` should cause the same code to execute, all of those values may be included in a single block by enclosing them in brackets, as shown below. If the switch expression matches any of the case expressions in the list, then the block will be executed.

```
switch (switch_expr)
case {case_expr_1, case_expr_2, case_expr_3}
    Statement 1
    Statement 2          ⎫
                          ⎬  Block 1
    ...                  ⎭

otherwise
    Statement 1
    Statement 2          ⎫
                          ⎬  Block n
    ...                  ⎭

end
```

The *switch_expr* and each *case_expr* may be either numerical or string values.

Note that at most one code block can be executed. After a code block is executed, execution skips to the first executable statement after the end statement. Thus if the switch expression matches more than one case expression, *only the first one of them will be executed.*

Let's look at a simple example of a switch construct. The following statements determine whether an integer between 1 and 10 is even or odd, and print out an appropriate message. It illustrates the use of a list of values as case selectors, and also the use of the otherwise block.

```
switch (value)
case {1,3,5,7,9}
    disp('The value is odd.');
case {2,4,6,8,10}
    disp('The value is even.');
otherwise
    disp('The value is out of range.');
end
```

4.4.5 The try/catch Construct

The try/catch construct is a special form branching construct designed to trap errors. Ordinarily, when a MATLAB program encounters an error while running, the program aborts. The try/catch construct modifies this default behavior. If an error occurs in a statement in the try block of this construct, then instead of aborting, the code in the catch block is executed and the program keeps running. This allows a programmer to handle errors within the program without causing the program to stop.

The general form of a try/catch construct is:

```
try
    Statement 1  ⎫
    Statement 2  ⎬ Try Block
    ...          ⎭

catch
    Statement 1  ⎫
    Statement 2  ⎬ Catch Block
    ...          ⎭
end
```

When a try/catch construct is reached, the statements in the try block will be executed. If no error occurs, the statements in the catch block will be skipped, and execution will continue at the first statement following the end of the construct. On the other hand, if an error *does* occur in the try block, the program will stop executing the statements in the try block, and immediately execute the statements in the catch block.

A `catch` statement can take an optional `ME` argument, where `ME` stands for a `MException` (MATLAB exception) object. The `ME` object is created when a failure occurs during the execution of statements in the `try` block. The `ME` object contains details about the type of exception (`ME.identifier`), the error message (`ME.message`), the cause of the error (`ME.cause`), and the stack (`ME.stack`), which specifies exactly where the error occurred. This information can be displayed to the user, or the programmer can use this information to try to recover from the error and let the program proceed[2].

An example program containing a `try/catch` construct follows. This program creates an array, and asks the user to specify an element of the array to display. The user will supply a subscript number, and the program displays the corresponding array element. The statements in the `try` block will always be executed in this program, while the statements in the `catch` block will only be executed of an error occurs in the `try` block. If the user specifies an illegal subscript, execution will transfer to the catch block, and the `ME` object will contain data explaining what went wrong. In this simple program, this information is just echoed to the Command Window. In more complicated programs, it could be used to recover from the error.

```
% Test try/catch

% Initialize array
a = [ 1 -3 2 5];

try

    % Try to display an element
    index = input('Enter subscript of element to display:');
    disp(['a(' int2str(index) ')=' num2str(a(index))]);

catch ME

    % If we get here, an error occurred. Display the error.
    ME
    stack = ME.stack

end
```

When this program is executed with a legal subscript, the results are:

```
» test_try_catch
Enter subscript of element to display: 3
a(3) = 2
```

---

[2]We will learn more about exceptions when we study object-oriented programming in Chapter 12.

When this program is executed with an illegal subscript, the results are:

```
» test_try_catch
Enter subscript of element to display: 9
ME =
    MException with properties:

        identifier: 'MATLAB:badsubscript'
           message: 'Attempted to access a(9); index out of
           bounds because numel(a)=4.'
             cause: {}
             stack: [1x1 struct]
stack =
        file: 'C:\Data\book\matlab\5e\chap4\test_try_catch.m'
        name: 'test_try_catch'
        line: 10
```

Quiz 4.2

This quiz provides a quick check to see if you have understood the concepts introduced in Section 4.4. If you have trouble with the quiz, reread the section, ask your instructor, or discuss the material with a fellow student. The answers to this quiz are found in the back of the book.

Write MATLAB statements that perform the functions described below.

1. If x is greater than or equal to zero, then assign the square root of x to variable sqrt_x and print out the result. Otherwise, print out an error message about the argument of the square root function, and set sqrt_x to zero.
2. A variable fun is calculated as numerator/denominator. If the absolute value of denominator is less than 1.0E-300, write "Divide by 0 error." Otherwise, calculate and print out fun.
3. The cost per mile for a rented vehicle is $1.00 for the first 100 miles, $0.80 for the next 200 miles, and $0.70 for all miles in excess of 300 miles. Write MATLAB statements that determine the total cost and the average cost per mile for a given number of miles (stored in variable distance).

Examine the following MATLAB statements. Are they correct or incorrect? If they are correct, what do they output? If they are incorrect, what is wrong with them?

```
4. if volts > 125
       disp('WARNING: High voltage on line.');

   if volts < 105
       disp('WARNING: Low voltage on line.');
   else
       disp('Line voltage is within tolerances.');
   end
```

5. ```
   color = 'yellow';
   switch (color)
   case 'red',
      disp('Stop now!');
   case 'yellow',
      disp('Prepare to stop.');
   case 'green',
      disp('Proceed through intersection.');
   otherwise,
      disp('Illegal color encountered.');
   end
   ```
6. ```
   if temperature > 37
      disp('Human body temperature exceeded.');
   elseif temperature > 100
      disp('Boiling point of water exceeded.');
   end
   ```

▶Example 4.5—Electrical Engineering: Frequency Response of a Low-Pass Filter

A simple low-pass filter circuit is shown in Figure 4.2. This circuit consists of a resistor and capacitor in series, and the ratio of the output voltage Vo to the input voltage V_i is given by the equation

$$\frac{V_0}{V_i} = \frac{1}{1 + j2\pi fRC} \tag{4.3}$$

where V_i is a sinusoidal input voltage of frequency f, R is the resistance in ohms, C is the capacitance in farads, and j is $\sqrt{-1}$ (electrical engineers use j instead of i for $\sqrt{-1}$, because the letter i is traditionally reserved for the current in a circuit).

Assume that the resistance $R = 16$ kΩ, and capacitance $C = 1$ μF, and plot the amplitude and frequency response of this filter over the frequency range $0 <= f <= 1000$ Hz.

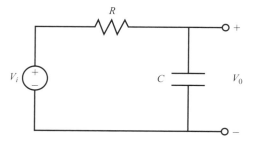

Figure 4.2 A simple low-pass filter circuit.

Solution The amplitude response of a filter is the ratio of the amplitude of the output voltage to the amplitude of the input voltage, and the phase response of the filter is the difference between the phase of the output voltage and the phase of the input voltage. The simplest way to calculate the amplitude and phase response of the filter is to evaluate Equation (4.3) at many different frequencies. The plot of the magnitude of Equation (4.3) versus frequency is the amplitude response of the filter, and the plot of the angle of Equation (4.3) versus frequency is the phase response of the filter.

Because the frequency and amplitude response of a filter can vary over a wide range, it is customary to plot both of these values on logarithmic scales. On the other hand, the phase varies over a very limited range, so it is customary to plot the phase of the filter on a linear scale. Therefore, we will use a `loglog` plot for the amplitude response, and a `semilogx` plot for the phase response of the filter. We will display both responses as two sub-plots within a figure.

We will also use stream modifiers to make the title and axis labels appear in bold face, as that improves the appearance of the plots.

The MATLAB code required to create and plot the responses is shown below.

```
% Script file: plot_filter.m
%
% Purpose:
%     This program plots the amplitude and phase responses
%     of a low-pass RC filter.
%
% Record of revisions:
%       Date            Programmer          Description of change
%       ====            ==========          =====================
%       01/05/14        S. J. Chapman       Original code
%
% Define variables:
%     amp      -- Amplitude response
%     C        -- Capacitance (farads)
%     f        -- Frequency of input signal (Hz)
%     phase    -- Phase response
%     R        -- Resistance (ohms)
%     res      -- Vo/Vi

% Initialize R & C
R = 16000;          % 16 k ohms
C = 1.0E-6;         % 1 uF

% Create array of input frequencies
f = 1:2:1000;

% Calculate response
res = 1 ./ (1 + j*2*pi*f*R*C);

% Calculate amplitude response
amp = abs(res);
```

```
% Calculate phase response
phase = angle(res);

% Create plots
subplot(2,1,1);
loglog(f, amp);
title('\bfAmplitude Response');
xlabel('\bfFrequency (Hz)');
ylabel('\bfOutput/Input Ratio');
grid on;

subplot(2,1,2);
semilogx(f, phase);
title('\bfPhase Response');
xlabel('\bfFrequency (Hz)');
ylabel('\bfOutput-Input Phase (rad)');
grid on;
```

The resulting amplitude and phase responses are shown in Figure 4.3. Note that this circuit is called a low-pass filter because low frequencies are passed through with little attenuation, while high frequencies are strongly attenuated.

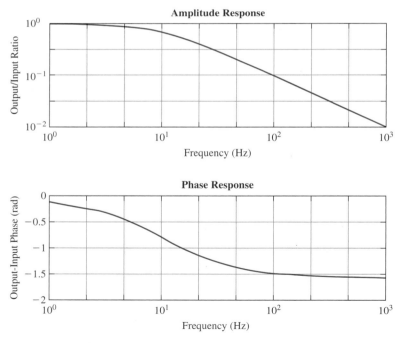

Figure 4.3 The amplitude and phase response of the low-pass filter circuit.

▶Example 4.6—Thermodynamics: The Ideal Gas Law

An ideal gas is one in which all collisions between molecules are perfectly elastic. It is possible to think of the molecules in an ideal gas as perfectly hard billiard balls that collide and bounce off of each other without losing kinetic energy.

Such a gas can be characterized by three quantities: absolute pressure (P), volume (V), and absolute temperature (T). The relationship among these quantities in an ideal gas is known as the Ideal Gas Law:

$$PV = nRT \qquad\qquad (4.4)$$

where P is the pressure of the gas in kilopascals (kPa), V is the volume of the gas in liters (L), n is the number of molecules of the gas in units of moles (mol), R is the universal gas constant (8.314 L·kPa/mol·K), and T is the absolute temperature in kelvins (K). (Note: 1 mol = 6.02×10^{23} molecules.)

Assume that a sample of an ideal gas contains 1 mole of molecules at a temperature of 273 K, and answer the following questions.

(a) How does the volume of this gas vary as its pressure varies from 1 to 1000 kPa? Plot pressure versus volume for this gas on an appropriate set of axes. Use a solid red line, with a width of 2 pixels.

(b) Suppose that the temperature of the gas is increased to 373 K. How does the volume of this gas vary with pressure now? Plot pressure versus volume for this gas on the same set of axes as part (a). Use a dashed blue line, with a width of 2 pixels.

Include a bold face title and x- and y-axis labels on the plot, as well as legends for each line.

Solution The values that we wish to plot both vary by a factor of 1000, so an ordinary linear plot will not produce a particularly useful result. Therefore, we will plot the data on a log-log scale.

Note that we must plot two curves on the same set of axes, so we must issue the command `hold on` after the first one is plotted, and `hold off` after the plot is complete. It will also be necessary to specify the color, style, and width of each line and to specify that labels be in bold face.

A program that calculates the volume of the gas as a function of pressure and creates the appropriate plot is shown below. Note that the special features controlling the style of the plot are shown in bold face.

```
% Script file: ideal_gas.m
%
% Purpose:
%  This program plots the pressure versus volume of an
%  ideal gas.
%
% Record of revisions:
%    Date          Programmer        Description of change
%    ====          ==========        =====================
%  01/16/14    S. J. Chapman      Original code
%
```

```
% Define variables:
%   n       -- Number of molecules (mol)
%   P       -- Pressure (kPa)
%   R       -- Ideal gas constant (L kPa/mol K)
%   T       -- Temperature (K)
%   V       -- volume (L)

% Initialize nRT
n = 1;              % Moles of atoms
R = 8.314;          % Ideal gas constant
T = 273;            % Temperature (K)

% Create array of input pressures. Note that this
% array must be quite dense to catch the major
% changes in volume at low pressures.
P = 1:0.1:1000;

% Calculate volumes
V = (n * R * T) ./ P;

% Create first plot
figure(1);
loglog(P,V,'r-','LineWidth',2);
title('\bfVolume vs Pressure in an Ideal Gas);
xlabel('\bfPressure (kPa)');
ylabel('\bfVolume (L)');
grid on;
hold on;

% Now increase temperature
T = 373;            % Temperature (K)

% Calculate volumes
V = (n * R * T) ./ P;

% Add second line to plot
figure(1);
loglog(P,V,'b--','LineWidth',2);
hold off;

% Add legend
legend('T = 273 K','T = 373 k');
```

The resulting volume versus pressure plot is shown in Figure 4.4.

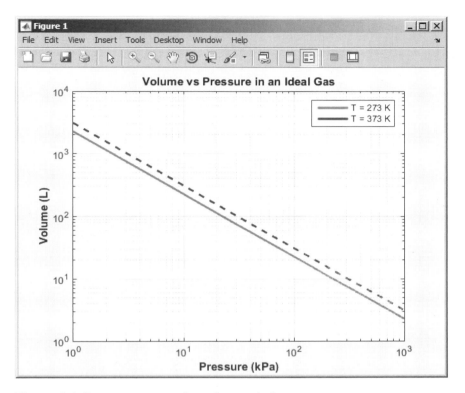

Figure 4.4 Pressure versus volume for an ideal gas.

4.5 More on Debugging MATLAB Programs

It is much easier to make a mistake when writing a program containing branches and loops than it is when writing simple sequential programs. Even after going through the full design process, a program of any size is almost guaranteed not to be completely correct the first time it is used. Suppose that we have built the program and tested it, only to find that the output values are in error. How do we go about finding the bugs and fixing them?

Once programs start to include loops and branches, the best way to locate an error is to use the symbolic debugger supplied with MATLAB. This debugger is integrated with the MATLAB editor.

To use the debugger, first open the file that you would like to debug using the File/Open menu selection in the MATLAB Command Window. When the file is opened, it is loaded into the editor and the syntax is automatically color-coded. Comments in the file appear in green, variables and numbers appear in black, character strings appear in red, and language keywords appear in blue. Figure 4.5 shows an example Edit/Debug Window containing the file `calc_roots.m`. [See color insert.]

Let's say that we would like to determine what happens when the program is executed. To do this, we can set one or more **breakpoints** by clicking time mouse on the horizontal dash mark at the left of the line(s) of interest. When a breakpoint is set, a red dot appears to the left of that line containing the breakpoint, as shown in Figure 4.6.

Once the breakpoints have been set, execute the program as usual by typing `calc_roots` in the Command Window. The program will run until it reaches the

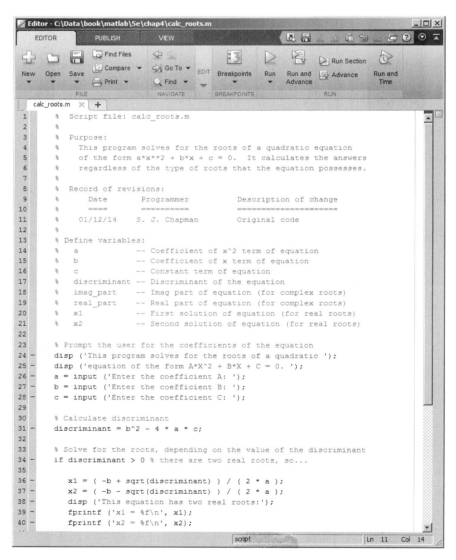

Figure 4.5 An Edit/Debug window with a MATLAB program loaded. [See color insert.]

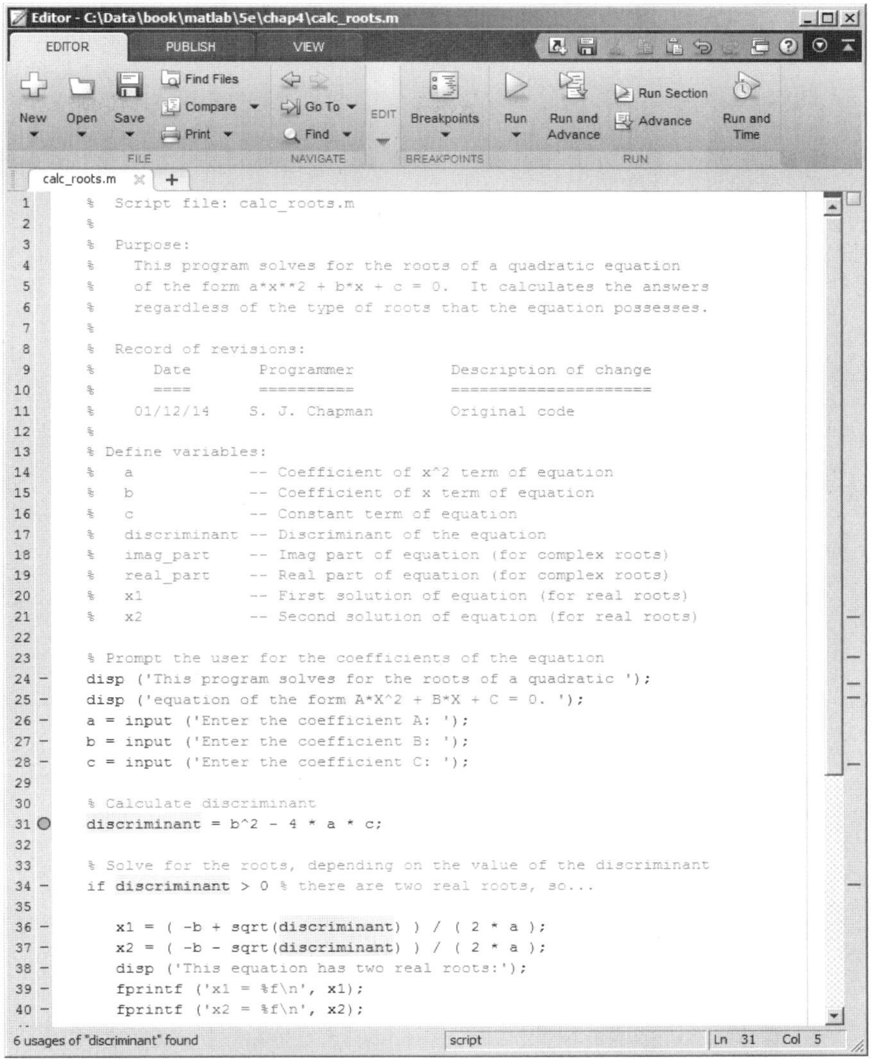

Figure 4.6 The window after a breakpoint has been set. Note the red dot to the left of the line with the breakpoint. [See color insert.]

first breakpoint and stop there. A green arrow will appear by the current line during the debugging process, as shown in Figure 4.7. When the breakpoint is reached, the programmer can examine and/or modify any variable in the workspace by typing its name in the Command Window or by examining the values in the Workspace Browser. When the programmer is satisfied with the program at that point, he or she can step through the program a line at a time by repeatedly pressing the F10 key or by clicking the Step tool ( ) on the Toolstrip. Alternatively, the programmer can run

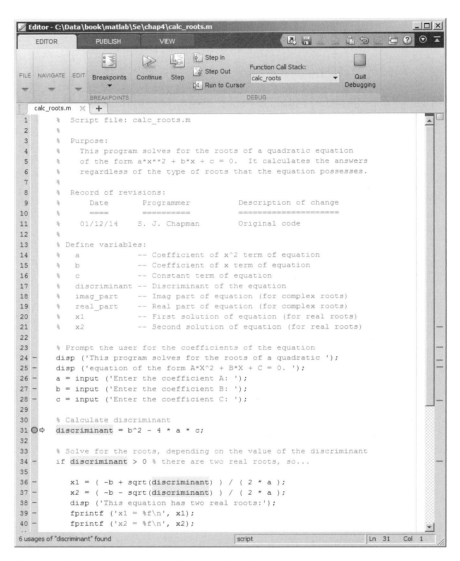

Figure 4.7 A green arrow will appear by the current line during the debugging process. [See color insert.]

to the next breakpoint by pressing the F5 key or by clicking the Continue tool (⬇). It is always possible to examine the values of any variable at any point in the program.

When a bug is found, the programmer can use the Editor to correct the MATLAB program and save the modified version to disk. Note that all breakpoints may be lost when the program is saved to disk with a new file name, so they may have to be set again before debugging can continue. This process is repeated until the program appears to be bug-free.

Two other very important features of the debugger are found in the Breakpoints group on the Toolstrip (see Figure 4.8a). The first feature is Set Condition, which sets or modifies a conditional breakpoint. A **conditional breakpoint** is a breakpoint where the code stops only if some condition is true. For example, a conditional breakpoint can be used to stop execution inside a `for` loop on its 200th execution. This can be very important if a bug only appears after a loop has been executed many

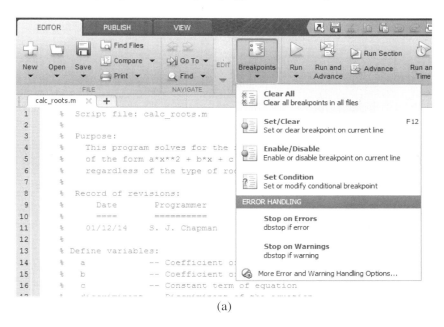

(a)

(b)

Figure 4.8 (a) Options in the Breakpoint group of the toolstrip. (b) Selecting the "Always stop if error" debugging option.

times. The condition that causes the breakpoint to stop execution can be modified, and the breakpoint can be enabled or disabled during debugging.

The second feature is Stop if Errors/Warnings, which appears if the user selects the More Error and Warning Handling option (see Figure 4.8b). If an error is occurring in a program that causes it to crash or generate warning messages, the program developer can select the "Always stop if error" or "Always stop if warning" radio button and execute the program. It will run to the point of the error or warning and stop there, allowing the developer to examine the values of variables and see exactly what is causing the problem.

A final critical feature is a tool called the Code Analyzer (previously called M-Lint). The Code Analyzer examines a MATLAB file and looks for potential problems. If it finds a problem, it shades that part of the code in the editor (see Figure 4.9).

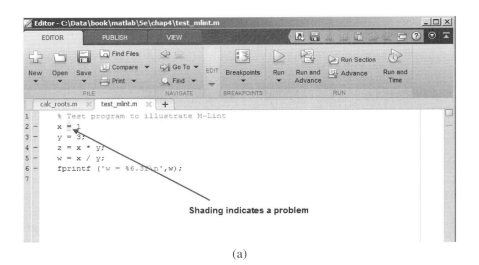

(a)

(b)

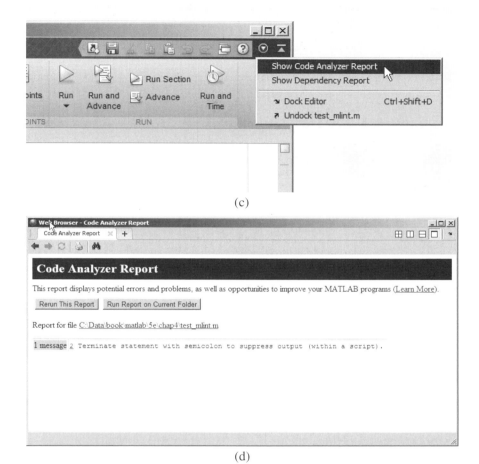

(c)

(d)

Figure 4.9 Using the Code Analyzer: (a) A shaded area in the Editor indicates a problem. (b) Placing the mouse over the shaded area produces a popup describing the problem. (c) A full report can also be generated using the "Tools > Show Code Analyzer Report" menu option. (d) A sample Code Analyzer Report.

If the developer places the mouse cursor over the shaded area, a popup will appear describing the problem, so that it can be fixed. It is also possible to display a complete list of all problems in a MATLAB file by clicking on the down arrow in the upper right-hand corner of the editor and selecting the Tools > Show Code Analyzer Report option.

The Code Analyzer is a *great* tool for locating errors, poor usage, or obsolete features in MATLAB code, including such things as variables that are defined but never used. The Code Analyzer is run automatically over any script loaded into the Edit/Debug Window, and the problem spots are shaded. Pay attention to its output and fix any problems that it reports.

4.6   Summary

In Chapter 4 we have presented the basic types of MATLAB branches and the relational and logic operations used to control them. The principal type of branch is the `if` construct. This construct is very flexible. It can have as many `elseif` clauses as needed to construct any desired test. Furthermore, `if` constructs can be nested to produce more complex tests. A second type of branch is the `switch` construct. It may be used to select among mutually exclusive alternatives specified by a control expression.

A third type of branch is the `try/catch` construct. It is used to trap errors that might occur during execution. The `catch` clause can have an optional exception object ME that provides information about the error that occurred.

The MATLAB symbolic debugger and related tools such as the Code Analyzer make debugging MATLAB code much easier. You should invest some time to become familiar with these tools.

4.6.1   Summary of Good Programming Practice

The following guidelines should be adhered to when programming with branch or loop constructs. By following them consistently, your code will contain fewer bugs, will be easier to debug, and will be more understandable to others who may need to work with it in the future.

1. Follow the steps of the program design process to produce reliable, understandable MATLAB programs.
2. Be cautious about testing for equality with numeric values, since roundoff errors may cause two variables that should be equal to fail a test for equality. Instead, test to see if the variables are *nearly* equal within the roundoff error to be expected on the computer you are working with.
3. Use the `&` AND operator if it is necessary to ensure that both operands are evaluated in an expression, or if the comparison is between arrays. Otherwise, use the `&&` AND operator, since the partial evaluation will make the operation faster in the cases where the first operand is `false`. The `&` operator is preferred in most practical cases.
4. Use the `|` inclusive OR operator if it is necessary to ensure that both operands are evaluated in an expression, or if the comparison is between arrays. Otherwise, use the `||` operator, since the partial evaluation will make the operation faster in the cases where the first operand is `true`. The `|` operator is preferred in most practical cases.
5. Always indent code blocks in `if`, `switch`, and `try/catch` constructs to make them more readable.
6. For branches in which there are many mutually exclusive options, use a single `if` construct with multiple `elseif` clauses in preference to nested `if` constructs.

4.6.2 MATLAB Summary

The following summary lists all of the MATLAB commands and functions described in this chapter, along with a brief description of each one.

Commands and Functions

`if` construct	Selects a block of statements to execute if a specified condition is satisfied.
`ischar(a)`	Returns a 1 if a is a character array and a 0 otherwise.
`isempty(a)`	Returns a 1 if a is an empty array and a 0 otherwise.
`isinf(a)`	Returns a 1 if the value of a is infinite (`Inf`) and a 0 otherwise.
`isnan(a)`	Returns a 1 if the value of a is NaN (not a number) and a 0 otherwise.
`isnumeric(a)`	Returns a 1 if a is a numeric array and a 0 otherwise.
`logical`	Converts numeric data to logical data, with nonzero values becoming `true` and zero values becoming `false`.
`poly`	Convert a list of roots of a polynomial into the polynomial coefficients.
`root`	Calculate the roots of a polynomial expressed as a series of coefficients.
`switch` construct	Selects a block of statements to execute from a set of mutually exclusive choices based on the result of a single expression.
`try/catch` construct	A special construct used to trap errors. It executes construct the code in the `try` block. If an error occurs, execution stops immediately and transfers to the code in the `catch` construct.

4.7 Exercises

4.1 Evaluate the following MATLAB expressions.

(a) `5 >= 5.5`
(b) `20 > 20`
(c) `xor(17 - pi < 15, pi < 3)`
(d) `true > false`
(e) `~~(35 / 17) == (35 / 17)`
(f) `(7 <= 8) == (3 / 2 == 1)`
(g) `17.5 && (3.3 > 2.)`

4.2 The tangent function is defined as $\tan\theta = \sin\theta\,/\cos\theta$. This expression can be evaluated to solve for the tangent as long as the magnitude of $\cos\theta$ is not too near to 0. (If $\cos\theta$ is 0, evaluating the equation for $\tan\theta$ will produce the non-numerical value `Inf`.) Assume that θ is given in *degrees*, and write the MATLAB statements to evaluate $\tan\theta$ as long as the magnitude of $\cos\theta$ is greater than or equal to 10^{-2}. If the magnitude of $\cos\theta$ is less than 10^{-2}, write out an error message instead.

4.3 The following statements are intended to alert a user to dangerously high oral thermometer readings (values are in degrees Fahrenheit). Are they correct or incorrect? If they are incorrect, explain why and correct them.

```
if temp < 97.5
    disp('Temperature below normal');
elseif temp > 97.5
    disp('Temperature normal');
elseif temp > 99.5
    disp('Temperature slightly high');
elseif temp > 103.0
    disp('Temperature dangerously high');
end
```

4.4 The cost of sending a package by an express delivery service is $15.00 for the first two pounds and $5.00 for each pound or fraction thereof over two pounds. If the package weighs more than 70 pounds, a $15.00 excess weight surcharge is added to the cost. No package over 100 pounds will be accepted. Write a program that accepts the weight of a package in pounds and computes the cost of mailing the package. Be sure to handle the case of overweight packages.

4.5 In Example 4.3, we wrote a program to evaluate the function $f(x, y)$ for any two user-specified values x and y, where the function $f(x, y)$ was defined as follows.

$$f(x, y) = \begin{cases} x + y & x \geq 0 \text{ and } y \geq 0 \\ x + y^2 & x \geq 0 \text{ and } y < 0 \\ x^2 + y & x < 0 \text{ and } y \geq 0 \\ x^2 + y^2 & x < 0 \text{ and } y < 0 \end{cases}$$

The problem was solved by using a single `if` construct with four code blocks to calculate $f(x, y)$ for all possible combinations of x and y. Rewrite program `funxy` to use nested `if` constructs, where the outer construct evaluates the value of x and the inner constructs evaluate the value of y.

4.6 Write a MATLAB program to evaluate the function

$$y(x) = \ln\frac{1}{1 - x}$$

for any user-specified value of x, where x is a number < 1.0 (note that ln is the natural logarithm, the logarithm to the base e). Use an `if` structure to verify that the value passed to the program is legal. If the value of x is legal, calculate $y(x)$. If not, write a suitable error message and quit.

4.7 Write a program that allows a user to enter a string containing a day of the week ('Sunday', 'Monday', 'Tuesday', etc.), and uses a `switch` construct to convert the day to its corresponding number, where Sunday is considered the first day of the week, and Saturday is considered the last day of the week. Print out the resulting day number. Also, be sure to handle the case of an illegal day name with an `otherwise` statement! (*Note:* Be sure to use the `'s'` option on function `input` so that the input is treated as a string.)

4.8 Suppose that a student has the option of enrolling for a single elective during a term. The student must select a course from a limited list of options: English, History, Astronomy, or Literature. Construct a fragment of MATLAB code that will prompt the student for his or her choice, read in the choice, and use the answer as the case expression for a `switch` construct. Be sure to include a default case to handle invalid inputs.

4.9 **Ideal Gas Law** The Ideal Gas Law was defined in Example 4.6. Assume that the volume of 1 mole of this gas is 10 L, and plot the pressure of the gas as a function of temperature as the temperature is changed from 250 to 400 kelvins. What sort of plot (linear, semilogx, and so forth) is most appropriate for this data?

4.10 **Ideal Gas Law** A tank holds an amount of gas pressurized at 200 kPa in the winter when the temperature of the tank is $0°$ C. What would the pressure in the tank be if it holds the same amount of gas when the temperature is $100°$ C? Create a plot showing the expected pressure as the temperature in the tank increases from $0°$ C to $200°$ C.

4.11 **van der Waals Equation** The Ideal Gas Law describes the temperature, pressure, and volume of an ideal gas. It is

$$PV = nRT \tag{4.4}$$

where P is the pressure of the gas in kilopascals (kPa), V is the volume of the gas in liters (L), n is the number of molecules of the gas in units of moles (mol), R is the universal gas constant (8.314 L·kPa/mol·K), and T is the absolute temperature in kelvins (K). (Note: 1 mol $= 6.02 \times 10^{23}$ molecules.)

Real gasses are not ideal because the molecules of the gas are not perfectly elastic—they tend to cling together a bit. The relationship between the temperature, pressure, and volume of a real gas can be represented by a modification of the ideal gas law called *van der Waals Equation*. It is

$$\left(P + \frac{n^2 a}{V^2}\right)(V - nb) = nRT \tag{4.5}$$

where P is the pressure of the gas in kilopascals (kPa), V is the volume of the gas in liters (L), a is a measure of attraction between the particles, n is the number of molecules of the gas in units of moles (mol), and b is the volume of one mole of the particles, R is the universal gas constant (8.314 L·kPa/mol·K), and T is the absolute temperature in kelvins (K).

This equation can be solved for P to give pressure as a function of temperature and volume.

$$P = \frac{nRT}{V - nb} - \frac{n^2 a}{V^2} \tag{4.6}$$

For carbon dioxide, the value of $a = 0.396$ kPa $\cdot$ L and the value of $b = 0.0427$ L/mol. Assume that a sample of carbon dioxide gas contains 1 mole of molecules at a temperature of $0°$ C (273 K) and occupies 30 L of volume. Answer the following questions.

(a) What is the pressure of the gas according to the Ideal Gas Law?
(b) What is the pressure of the gas according to the van der Waals equation?
(c) Plot the pressure versus volume at this temperature according to the Ideal Gas Law and according to van der Waals equation on the same axes. Is the pressure of a real gas higher or lower than the pressure of an ideal gas under the same temperature conditions?

4.12 Suppose that a polynomial equation has the following 6 roots: $-6, -2, 1 + i\sqrt{2}, 1 - i\sqrt{2}, 2,$ and 6. Find the coefficients of the polynomial.

4.13 Find the roots of the polynomial equation

$$y(x) = x^6 - x^5 - 6x^4 + 14x^3 - 12x^2$$

Plot the resulting function, and compare the observed roots to the calculated roots. Also, plot the location of the roots on a complex plane.

4.14 **Antenna Gain Pattern** The gain G of a certain microwave dish antenna can be expressed as a function of angle by the equation

$$G(\theta) = \left| \text{sinc } 4\theta \right| \quad \text{for} \quad -\frac{\pi}{2} \leq \theta \leq \frac{\pi}{2} \tag{4.7}$$

where θ is measured in radians from the boresight of the dish, and sinc $x = \sin x / x$. Plot this gain function on a polar plot with the title "**Antenna Gain vs θ**" in bold face.

4.15 The author of this book now lives in Australia. In 2009, individual citizens and residents of Australia paid the following income taxes:

Taxable Income (in A$)	Tax on This Income
$0–$6,000	None
$6,001–$34,000	15¢ for each $1 over $6,000
$34,001–$80,000	$4,200 plus 30¢ for each $1 over $34,000
$80,001–$180,000	$18,000 plus 40¢ for each $1 over $80,000
Over $180,000	$58,000 plus 45¢ for each $1 over $180,000

In addition, a flat 1.5% Medicare levy is charged on all income. Write a program to calculate how much income tax a person will owe based on this information. The program should accept a total income figure from the user, and calculate the income tax, Medicare levy, and total tax payable by the individual.

4.16 In 2002, individual citizens and residents of Australia paid the following income taxes:

Taxable Income (in A$)	Tax on This Income
$0–$6,000	None
$6,001–$20,000	17¢ for each $1 over $6,000
$20,001–$50,000	$2,380 plus 30¢ for each $1 over $20,000
$50,001–$60,000	$11,380 plus 42¢ for each $1 over $50,000
Over $60,000	$15,580 plus 47¢ for each $1 over $60,000

In addition, a flat 1.5% Medicare levy was charged on all income. Write a program to calculate how much *less* income tax a person paid on a given amount of income in 2009 than he or she would have paid in 2002.

4.17 Refraction When a ray of light passes from a region with an index of refraction n_1 into a region with a different index of refraction n_2, the light ray is bent (see Figure 4.10). The angle at which the light is bent is given by *Snell's Law*

$$n_1 \sin \theta_1 = n_2 \sin \theta_2 \qquad (4.8)$$

where θ_1 is the angle of incidence of the light in the first region, and θ_2 is the angle of incidence of the light in the second region. Using Snell's Law, it is possible to predict the angle of incidence of a light ray in Region 2 if the angle

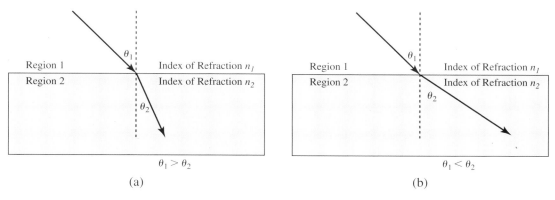

Figure 4.10 A ray of light bends as it passes from one medium into another one. (a) If the ray of light passes from a region with a low index of refraction into a region with a higher index of refraction, the ray of light bends more towards the normal. (b) If the ray of light passes from a region with a high index of refraction into a region with a lower index of refraction, the ray of light bends away from the normal.

of incidence θ_1 in Region 1 and the indices of refraction n_1 and n_2 are known. The equation to perform this calculation is

$$\theta_2 = \sin^{-1}\left[\frac{n_1}{n_2}\sin\theta_1\right] \tag{4.9}$$

Write a program to calculate the angle of incidence (in degrees) of a light ray in Region 2 given the angle of incidence θ_1 in Region 1 and the indices of refraction n_1 and n_2. (*Note:* If $n_1 > n_2$, then for some angles θ_1, Equation (4.9) will have no real solution because the absolute value of the quantity $\left[\frac{n_2}{n_1}\sin\theta_1\right]$ will be greater than 1.0. When this occurs, all light is reflected back into Region 1, and no light passes into Region 2 at all. Your program must be able to recognize and properly handle this condition.)

The program should also create a plot showing the incident ray, the boundary between the two regions, and the refracted ray on the other side of the boundary.

Test your program by running it for the following two cases: (*a*) $n_1 = 1.0$, $n_2 = 1.7$, and $\theta_1 = 45°$. (*b*) $n_1 = 1.7$, $n_2 = 1.0$, and $\theta_1 = 45°$.

4.18 High-Pass Filter Figure 4.11 shows a simple high-pass filter consisting of a resistor and a capacitor. The ratio of the output voltage V_o to the input voltage V_i is given by the equation

$$\frac{V_o}{V_i} = \frac{j2\pi fRC}{1 + j2\pi fRC} \tag{4.10}$$

Assume that $R = 16$ kΩ and $C = 1$ μF. Calculate and plot the amplitude and phase response of this filter as a function of frequency.

4.19 As we saw in Chapter 2, the `load` command can be used to load data from a MAT file into the MATLAB workspace. Write a script that prompts a user for the name of a file to load, and then loads the data from that file. The script should be in a `try/catch` construct to catch and display errors if the specified file cannot be opened. Test your script file for loading both valid and invalid MAT files.

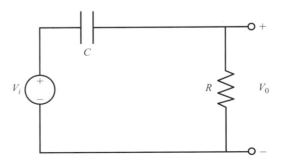

Figure 4.11 A simple high-pass filter circuit.

Loops and Vectorization

Loops are MATLAB constructs that permit us to execute a sequence of statements more than once. There are two basic forms of loop constructs: **while loops** and **for loops**. The major difference between these two types of loops is in how the repetition is controlled. The code in a `while` loop is repeated an indefinite number of times until some user-specified condition is satisfied. By contrast, the code in a `for` loop is repeated a specified number of times, and the number of repetitions is known before the loops starts.

Vectorization is an alternate and faster way to perform the same function as many MATLAB `for` loops. After introducing loops, this chapter will show how to replace many loops with vectorized code for increased speed.

MATLAB programs that use loops often process very large amounts of data, and the programs need an efficient way to read that data in for processing. This chapter introduces the `textread` function to make it simple to read large data sets in from disk files.

5.1 The `while` Loop

A **while loop** is a block of statements that are repeated indefinitely as long as some condition is satisfied. The general form of a `while` loop is

```
while expression
    . . .
    . . .              } Code block
    . . .
end
```

The controlling expression produces a logical value. If the *expression* is true, the code block will be executed, and then control will return to the `while` statement. If the *expression* is still true, the statements will be executed again. This process will be repeated until the *expression* becomes false. When control returns to the while statement and the expression is false, the program will execute the first statement after the end.

The pseudocode corresponding to a while loop is

```
while expr

    . . .
    . . .
    . . .

end
```

We will now show an example statistical analysis program that is implemented using a while loop.

►Example 5.1—Statistical Analysis

It is very common in science and engineering to work with large sets of numbers, each of which is a measurement of some particular property that we are interested in. A simple example would be the grades on the first test in this course. Each grade would be a measurement of how much a particular student has learned in the course to date.

Much of the time, we are not interested in looking closely at every single measurement that we make. Instead, we want to summarize the results of a set of measurements with a few numbers that tell us a lot about the overall data set. Two such numbers are the *average* (or *arithmetic mean*) and the *standard deviation* of the set of measurements. The average, or arithmetic mean, of a set of numbers is defined as

$$\bar{x} = \frac{1}{N}\sum_{i=1}^{N} x_i \tag{5.1}$$

where x_i is sample i out of N samples. If all of the input values are available in an array, the average of a set of numbers may be calculated by MATLAB function mean. The standard deviation of a set of numbers is defined as

$$s = \sqrt{\frac{N\sum_{i=1}^{N} x_i^2 - \left(\sum_{i=1}^{N} x_i\right)^2}{N(N-1)}} \tag{5.2}$$

Standard deviation is a measure of the amount of scatter on the measurements; the greater the standard deviation, the more scattered the points in the data set are.

Implement an algorithm that reads in a set of measurements and calculates the mean and the standard deviation of the input data set.

Solution This program must be able to read in an arbitrary number of measurements and then calculate the mean and standard deviation of those measurements. We will use a while loop to accumulate the input measurements before performing the calculations.

When all of the measurements have been read, we must have some way of telling the program that there is no more data to enter. For now, we will assume that all the input measurements are either positive or zero, and we will use a negative input value as a *flag* to indicate that there is no more data to read. If a negative value is entered, then the program will stop reading input values and will calculate the mean and standard deviation of the data set.

1. **State the problem.**
   Since we assume that the input numbers must be positive or zero, a proper statement of this problem would be: *calculate the average and the standard deviation of a set of measurements, assuming that all of the measurements are either positive or zero, and assuming that we do not know in advance how many measurements are included in the data set. A negative input value will mark the end of the set of measurements.*

2. **Define the inputs and outputs.**
   The inputs required by this program are an unknown number of positive or zero numbers. The outputs from this program are a printout of the mean and the standard deviation of the input data set. In addition, we will print out the number of data points input to the program, since this is a useful check that the input data was read correctly.

3. **Design the algorithm.**
   This program can be broken down into three major steps:

   ```
   Accumulate the input data
   Calculate the mean and standard deviation
   Write out the mean, standard deviation, and number
     of points
   ```

   The first major step of the program is to accumulate the input data. To do this, we will have to prompt the user to enter the desired numbers. When the numbers are entered, we will have to keep track of the number of values entered, their sum, and the sum of the squares of those values. The pseudocode for these steps is:

   ```
   Initialize n, sum_x, and sum_x2 to 0
   Prompt user for first number
   Read in first x
   while x > = 0
      n ← n + 1
      sum_x ← sum_x + x
      sum_x2 ← sum_x2 + x^2
      Prompt user for next number
      Read in next x
   end
   ```

   Note that we have to read in the first value before the while loop starts so that the while loop can have a value to test the first time it executes.

Next, we must calculate the mean and standard deviation. The pseudocode for this step is just the MATLAB versions of Equations (5.1) and (5.2).

```
x_bar ← sum_x / n
std_dev ← sqrt((n*sum_x2 - sum_x^2) / (n*(n-1)))
```

Finally, we must write out the results.

```
Write out the mean value x_bar
Write out the standard deviation std_dev
Write out the number of input data points n
```

4. **Turn the algorithm into MATLAB statements.**
   The final MATLAB program is shown below:

```
%   Script file: stats_1.m
%
%   Purpose:
%     To calculate mean and the standard deviation of
%     an input data set containing an arbitrary number
%     of input values.
%
%   Record of revisions:
%       Date            Engineer            Description of change
%       ====            ========            =====================
%     01/24/14        S. J. Chapman         Original code
%
% Define variables:
%     n        -- The number of input samples
%     std_dev  -- The standard deviation of the input samples
%     sum_x    -- The sum of the input values
%     sum_x2   -- The sum of the squares of the input values
%     x        -- An input data value
%     xbar     -- The average of the input samples

% Initialize sums.
n = 0; sum_x = 0; sum_x2 = 0;

% Read in first value
x = input('Enter first value:');

% While Loop to read input values.
while x >= 0

   % Accumulate sums:
   n       = n + 1;
   sum_x   = sum_x + x;
   sum_x2  = sum_x2 + x^2;
```

```
    % Read in next value
    x = input('Enter next value:');

end

% Calculate the mean and standard deviation
x_bar = sum_x / n;
std_dev = sqrt( (n * sum_x2 - sum_x^2) / (n * (n-1)) );

% Tell user.
fprintf('The mean of this data set is: %f\n', x_bar);
fprintf('The standard deviation is:    %f\n', std_dev);
fprintf('The number of data points is: %f\n', n);
```

5. **Test the program.**

To test this program, we will calculate the answers by hand for a simple data set, and then compare the answers to the results of the program. If we used three input values: 3, 4, and 5, then the mean and standard deviation would be

$$\bar{x} = \frac{1}{N}\sum_{i=1}^{N}x_i = \frac{1}{3}(12) = 4$$

$$s = \sqrt{\frac{N\sum_{i=1}^{N}x_i^2 - \left(\sum_{i=1}^{N}x_i\right)^2}{N(N-1)}} = 1$$

When the above values are fed into the program, the results are

```
» stats_1
Enter first value: 3
Enter next value:  4
Enter next value:  5
Enter next value:  -1
The mean of this data set is: 4.000000
The standard deviation is:    1.000000
The number of data points is: 3.000000
```

The program gives the correct answers for our test data set.

◄

In the example above, we failed to follow the design process completely. This failure has left the program with a fatal flaw! Did you spot it?

We have failed because *we did not completely test the program for all possible types of inputs*. Look at the example once again. If we enter either no numbers or only one number, then we will be dividing by zero in the above equations! The division-by-zero error will cause divide-by-zero warnings to be printed, and the output values will be NaN. We need to modify the program to detect this problem, tell the user what the problem is, and stop gracefully.

A modified version of the program called stats_2 is shown below. Here, we check to see if there are enough input values before performing the calculations. If not, the program will print out an intelligent error message and quit. Test the modified program for yourself.

```
%  Script file: stats_2.m
%
%  Purpose:
%    To calculate mean and the standard deviation of
%    an input data set containing an arbitrary number
%    of input values.
%
%  Record of revisions:
%     Date          Engineer         Description of change
%     ====          ==========       =====================
%    01/24/14    S. J. Chapman       Original code
% 1. 01/24/14    S. J. Chapman       Correct divide-by-0 error if
%                                    0 or 1 input values given.
%
% Define variables:
%    n        -- The number of input samples
%    std_dev  -- The standard deviation of the input samples
%    sum_x    -- The sum of the input values
%    sum_x2   -- The sum of the squares of the input values
%    x        -- An input data value
%    xbar     -- The average of the input samples

% Initialize sums.
n = 0; sum_x = 0; sum_x2 = 0;

% Read in first value
x = input('Enter first value: ');

% While Loop to read input values.
while x >= 0

   % Accumulate sums.
   n      = n + 1;
   sum_x  = sum_x + x;
   sum_x2 = sum_x2 + x^2;

   % Read in next value
   x = input('Enter next value:');
end
```

```
% Check to see if we have enough input data.
if n < 2    % Insufficient information

   disp('At least 2 values must be entered!');

else % There is enough information, so
     % calculate the mean and standard deviation

     x_bar = sum_x / n;
     std_dev = sqrt((n * sum_x2 - sum_x^2)/(n * (n-1)));

     % Tell user.
     fprintf('The mean of this data set is: %f\n', x_bar);
     fprintf('The standard deviation is:    %f\n', std_dev);
     fprintf('The number of data points is: %f\n', n);

end
```

Note that the average and standard deviation could have been calculated with the built-in MATLAB functions mean and std if all of the input values are saved in a vector, and that vector is passed to these functions. You will be asked to create a version of the program that uses the standard MATLAB functions in an exercise at the end of this chapter.

5.2 The for Loop

The **for loop** is a loop that executes a block of statements a specified number of times. The for loop has the form

```
for index = expr
   . . .
   . . .                    } Body
   . . .
end
```

where index is the loop variable (also known as the **loop index**) and *expr* is the loop control expression, whose result is an array. The columns in the array produced by *expr* are stored one at a time in the variable index and then the loop body is executed, so that the loop is executed once for each column in the array produced by *expr*. The expression usually takes the form of a vector in shortcut notation first:incr:last.

The statements between the for statement and the end statement are known as the *body* of the loop. They are executed repeatedly during each pass of the for loop. The for loop construct functions as follows:

1. At the beginning of the loop, MATLAB generates an array by evaluating the control expression.
2. The first time through the loop, the program assigns the first column of the array to the loop variable index, and the program executes the statements within the body of the loop.

3. After the statements in the body of the loop have been executed, the program assigns the next column of the array to the loop variable index, and the program executes the statements within the body of the loop again.
4. Step 3 is repeated over and over as long as there are additional columns in the array.

Let's look at a number of specific examples to make the operation of the for loop clearer. First, consider the following example:

```
for ii = 1:10
   Statement 1
   ...
   Statement n
end
```

In this loop, the control index is the variable ii[1]. In this case, the control expression generates a 1×10 array, so statements 1 through n will be executed 10 times. The loop index ii will be 1 the first time, 2 the second time, and so on. The loop index will be 10 on the last pass through the statements. When control is returned to the for statement after the tenth pass, there are no more columns in the control expression, so execution transfers to the first statement after the end statement. Note that the loop index ii is still set to 10 after the loop finishes executing.

Second, consider the following example:

```
for ii = 1:2:10
   Statement 1
   ...
   Statement n
end
```

In this case, the control expression generates a 1×5 array, so statements 1 through n will be executed 5 times. The loop index ii will be 1 the first time, 3 the second time, and so on. The loop index will be 9 on the fifth and last pass through the statements. When control is returned to the for statement after the fifth pass, there are no more columns in the control expression, so execution transfers to the first statement after the end statement. Note that the loop index ii is still set to 9 after the loop finishes executing.

Third, consider the following example:

```
for ii = [5 9 7]
   Statement 1
   ...
   Statement n
end
```

Here, the control expression is an explicitly written 1×3 array, so statements 1 through n will be executed 3 times with the loop index set to 5 the first time, 9 the second time, and 7 the final time. The loop index ii is still set to 7 after the loop finishes executing.

---

[1]By habit, programmers working in most programming languages use simple variable names like i and j as loop indices. However, MATLAB pre-defines the variables i and j to be the value $\sqrt{-1}$. Because of this definition, the examples in the book use ii and jj as example loop indices.

Finally, consider the example:

```
for ii = [1 2 3;4 5 6]
    Statement 1
    ...
    Statement n
end
```

In this case, the control expression is a 2×3 array, so statements 1 through n will be executed 3 times. The loop index `ii` will be the column vector $\begin{bmatrix} 1 \\ 4 \end{bmatrix}$ the first time, $\begin{bmatrix} 2 \\ 5 \end{bmatrix}$ the second time, and $\begin{bmatrix} 3 \\ 6 \end{bmatrix}$ the third time. The loop index `ii` is still set to $\begin{bmatrix} 3 \\ 6 \end{bmatrix}$ after the loop finishes executing. This example illustrates the fact that a loop index can be a vector.

The pseudocode corresponding to a `for` loop looks like the loop itself:

```
for index = expression
    Statement 1
    ...
    Statement n
end
```

►Example 5.2—The Factorial Function

To illustrate the operation of a `for` loop, we will use a `for` loop to calculate the factorial function. The factorial function is defined for any integer ≥ 0 as

$$n! = \begin{cases} 1 & n = 0 \\ n \times (n-1) \times (n-2) \times \ldots \times 2 \times 1 & n > 0 \end{cases} \qquad (5.3)$$

The MATLAB code to calculate N factorial for positive value of N would be

```
n_factorial = 1
for ii = 1:n
    n_factorial = n_factorial * ii;
end
```

Suppose that we wish to calculate the value of 5!. If n is 5, the `for` loop control expression would be the row vector `[1  2  3  4  5]`. This loop will be executed 5 times, with the variable `ii` taking on values of 1, 2, 3, 4, and 5 in the successive loops. The resulting value of `n_factorial` will be $1 \times 2 \times 3 \times 4 \times 5 = 120$.

◄

►Example 5.3—Calculating the Day of Year

The *day of year* is the number of days (including the current day) that have elapsed since the beginning of a given year. It is a number in the range 1 to 365 for ordinary years and 1 to 366 for leap years. Write a MATLAB program that accepts a day, month, and year and calculates the day of year corresponding to that date.

Solution To determine the day of year, this program will need to sum up the number of days in each month preceding the current month plus the number of elapsed days in the current month. A `for` loop will be used to perform this sum. Since the number of days in each month varies, it is necessary to determine the correct number of days to add for each month. A `switch` construct will be used to determine the proper number of days to add for each month.

During a leap year, an extra day must be added to the day of year for any month after February. This extra day accounts for the presence of February 29 in the leap year. Therefore, to perform the day of year calculation correctly, we must determine which years are leap years. In the Gregorian calendar, leap years are determined by the following rules:

1. Years evenly divisible by 400 are leap years.
2. Years evenly divisible by 100 but *not* by 400 are not leap years.
3. All years divisible by 4 but *not* by 100 are leap years.
4. All other years are not leap years.

We will use the `mod` (for modulus) function to determine whether or not a year is evenly divisible by a given number. The `mod` function returns the remainder after the division of two numbers. For example, the remainder of 9/4 is 1, since 4 goes into 9 twice with a remainder of 1. If the result of the function `mod(year,4)` is zero, then we know that the year is evenly divisible by 4. Similarly, if the result of the function `mod(year,400)` is zero, then we know that the year is evenly divisible by 400.

A program to calculate the day of year is shown below. Note that the program sums up the number of days in each month before the current month and that it uses a `switch` construct to determine the number of days in each month.

```
%   Script file: doy.m
%
%   Purpose:
%     This program calculates the day of year corresponding
%     to a specified date. It illustrates the use of switch and
%     for constructs.
%
%   Record of revisions:
%       Date              Engineer            Description of change
%       ====              ==========          =====================
%     01/27/14          S. J. Chapman         Original code
%
% Define variables:
%     day               -- Day (dd)
%     day_of_year       -- Day of year
%     ii                -- Loop index
%     leap_day          -- Extra day for leap year
%     month             -- Month (mm)
%     year              -- Year (yyyy)
```

```
% Get day, month, and year to convert
disp('This program calculates the day of year given the');
disp('specified date.');
month = input('Enter specified month (1-12):');
day   = input('Enter specified day(1-31):   ');
year  = input('Enter specified year(yyyy):   ');

% Check for leap year, and add extra day if necessary
if mod(year,400) == 0
   leap_day = 1;        % Years divisible by 400 are leap years
elseif mod(year,100) == 0
   leap_day = 0;        % Other centuries are not leap years
elseif mod(year,4) == 0
   leap_day = 1;        % Otherwise every 4th year is a leap year
else
   leap_day = 0;        % Other years are not leap years
end

% Calculate day of year by adding current day to the
% days in previous months.
day_of_year = day;
for ii = 1:month-1

   % Add days in months from January to last month
   switch (ii)
   case {1,3,5,7,8,10,12},
      day_of_year = day_of_year + 31;
   case {4,6,9,11},
      day_of_year = day_of_year + 30;
   case 2,
      day_of_year = day_of_year + 28 + leap_day;
   end

end

% Tell user
fprintf('The date %2d/%2d/%4d is day of year %d.\n', ...
        month, day, year, day_of_year);
```

We will use the following known results to test the program:

1. Year 1999 is not a leap year. January 1 must be day of year 1, and December 31 must be day of year 365.
2. Year 2000 is a leap year. January 1 must be day of year 1, and December 31 must be day of year 366.
3. Year 2001 is not a leap year. March 1 must be day of year 60, since January has 31 days, February has 28 days, and this is the first day of March.

If this program is executed five times with the above dates, the results are

```
» doy
This program calculates the day of year given the
specified date.
Enter specified month (1-12): 1
Enter specified day(1-31):    1
Enter specified year(yyyy):    1999
The date  1/ 1/1999 is day of year 1.
» doy
This program calculates the day of year given the
specified date.
Enter specified month (1-12): 12
Enter specified day(1-31):    31
Enter specified year(yyyy):    1999
The date 12/31/1999 is day of year 365.
» doy
This program calculates the day of year given the
specified date.
Enter specified month (1-12): 1
Enter specified day(1-31):    1
Enter specified year(yyyy):    2000
The date  1/ 1/2000 is day of year 1.
» doy
This program calculates the day of year given the
specified date.
Enter specified month (1-12): 12
Enter specified day(1-31):    31
Enter specified year(yyyy):    2000
The date 12/31/2000 is day of year 366.
» doy
This program calculates the day of year given the
specified date.
Enter specified month (1-12): 3
Enter specified day(1-31):    1
Enter specified year(yyyy):    2001
The date  3/ 1/2001 is day of year 60.
```

The program gives the correct answers for our test dates in all five test cases.

◀

▶ Example 5.4—Statistical Analysis

Implement an algorithm that reads in a set of measurements and calculates the mean and the standard deviation of the input data set, when any value in the data set can be positive, negative, or zero.

Solution This program must be able to read in an arbitrary number of measurements and then calculate the mean and standard deviation of those measurements. Each measurement can be positive, negative, or zero.

Since we cannot use a data value as a flag this time, we will ask the user for the number of input values, and then use a `for` loop to read in those values. The modified program that permits the use of any input value is shown below. Verify its operation for yourself by finding the mean and standard deviation of the following 5 input values: 3, –1, 0, 1, and –2.

```
%  Script file: stats_3.m
%
%  Purpose:
%    To calculate mean and the standard deviation of
%    an input data set, where each input value can be
%    positive, negative, or zero.
%
%  Record of revisions:
%      Date            Engineer          Description of change
%      ====            ==========        =====================
%    01/27/14      S. J. Chapman        Original code
%
% Define variables:
%   ii       -- Loop index
%   n        -- The number of input samples
%   std_dev  -- The standard deviation of the input samples
%   sum_x    -- The sum of the input values
%   sum_x2   -- The sum of the squares of the input values
%   x        -- An input data value
%   xbar     -- The average of the input samples

% Initialize sums.
sum_x = 0; sum_x2 = 0;

% Get the number of points to input.
n = input('Enter number of points:');

% Check to see if we have enough input data.
if n < 2   % Insufficient data

   disp ('At least 2 values must be entered.');

else % we will have enough data, so let's get it.

   % Loop to read input values.
   for ii = 1:n

      % Read in next value
      x = input('Enter value: ');
```

```
   % Accumulate sums.
   sum_x  = sum_x + x;
   sum_x2 = sum_x2 + x^2;

end

% Now calculate statistics.
x_bar = sum_x / n;
std_dev = sqrt((n * sum_x2 - sum_x^2) / (n * (n-1)));

% Tell user.
fprintf('The mean of this data set is: %f\n', x_bar);
fprintf('The standard deviation is:    %f\n', std_dev);
fprintf('The number of data points is: %f\n', n);

end
```

5.2.1 Details of Operation

Now that we have seen examples of a `for` loop in operation, we must examine some important details required to use `for` loops properly.

1. **Indent the bodies of loops.** It is not necessary to indent the body of a `for` loop as we have shown above. MATLAB will recognize the loop even if every statement in it starts in column 1. However, the code is much more readable if the body of the `for` loop is indented, so you should always indent the bodies of loops.

🄰 Good Programming Practice

Always indent the body of a `for` loop by 3 or more spaces to improve the readability of the code.

2. **Don't modify the loop index within the body of a loop.** The loop index of a `for` loop *should not be modified anywhere within the body of the loop.* The index variable is often used as a counter within the loop, and modifying its value can cause strange and hard-to-find errors. The example shown below is intended to initialize the elements of an array, but the statement "ii = 5" has been accidentally inserted into the body of the loop. As a result, only a (5) is initialized, and it gets the values that should have gone into a (1), a (2), and so forth.

```
for ii = 1:10
   ...
   ii = 5;     % Error!
   ...
   a(ii) = <calculation>
end
```

 Good Programming Practice

Never modify the value of a loop index within the body of the loop.

3. **Preallocating Arrays.** We learned in Chapter 2 that it is possible to extend an existing array simply by assigning a value to a higher array element. For example, the statement

```
arr = 1:4;
```

defines a 4-element array containing the values [1 2 3 4]. If the statement

```
arr(8) = 6;
```

is executed, the array will be automatically extended to 8 elements, and will contain the values [1 2 3 4 0 0 0 6]. Unfortunately, each time that an array is extended, MATLAB has to (1) create a new array, (2) copy the contents of the old array to the new longer array, (3) add the new value to the array, and then (4) delete the old array. This process is very time-consuming for long arrays.

When a for loop stores values in a previously undefined array, the loop forces MATLAB to go through this process each time the loop is executed. On the other hand, if the array is **preallocated** to its maximum size before the loop starts executing, no copying is required, and the code executes much faster. The code fragment shown below shows how to preallocate an array before starting the loop.

```
square = zeros(1,100);
for ii = 1:100
    square(ii) = ii^2;
end
```

Good Programming Practice

Always preallocate all arrays used in a loop before executing the loop. This practice greatly increases the execution speed of the loop.

5.2.2 Vectorization: A Faster Alternative to Loops

Many loops are used to apply the same calculations over and over to the elements of an array. For example, the following code fragment calculates the squares, square roots, and cube roots of all integers between 1 and 100 using a for loop.

```
for ii = 1:100
   square(ii) = ii^2;
   square_root(ii) = ii^(1/2);
   cube_root(ii) = ii^(1/3);
end
```

Here, the loop is executed 100 times, and one value of each output array is calculated during each cycle of the loop.

MATLAB offers a faster alternative for calculations of this sort: **vectorization**. Instead of executing each statement 100 times, MATLAB can do the calculation for all the elements in an array in a *single* statement. Because of the way MATLAB is designed, this single statement can be much faster than the loop, and perform exactly the same calculation.

For example, the following code fragment uses vectors to perform the same calculation as the loop shown above. We first calculate a vector of the indices into the arrays and then perform each calculation only once, doing all 100 elements in the single statement.

```
ii = 1:100;
square = ii.^2;
square_root = ii.^(1/2);
cube_root = ii.^(1/3);
```

Even though these two calculations produce the same answers, they are *not* equivalent. The version with the `for` loop can be *more than 15 times slower* than the vectorized version! This happens because the statements in the `for` loop must be interpreted[2] and executed a line at a time by MATLAB during each pass of the loop. In effect, MATLAB must interpret and execute 300 separate lines of code. In contrast, MATLAB only has to interpret and execute 4 lines in the vectorized case. Since MATLAB is designed to implement vectorized statements in a very efficient fashion, it is much faster in that mode.

In MATLAB, the process of replacing loops by vectorized statements is known as vectorization. Vectorization can yield dramatic improvements in performance for many MATLAB programs.

Good Programming Practice

If it is possible to implement a calculation either with a `for` loop or using vectors, implement the calculation with vectors. Your program will be much faster.

//

5.2.3 The MATLAB Just-In-Time (JIT) Compiler

A just-in-time (JIT) compiler was added to MATLAB 6.5 and later versions. The JIT compiler examines MATLAB code before it is executed and, where possible, compiles the code before executing it. Since the MATLAB code is compiled instead

---

[2]But see the next item about the MATLAB Just-In-Time compiler.

of being interpreted, it runs almost as fast as vectorized code. The JIT compiler can often dramatically speed up the execution of for loops.

The JIT compiler is a very nice tool when it works, since it speeds up the loops without any action by the engineer. However, the JIT compiler has some limitations that prevent it from speeding up all loops. The JIT compiler limitations vary with MATLAB version, with fewer limitations in later versions of the program[3].

 Good Programming Practice

Do not rely on the JIT compiler to speed up your code. It has limitations that vary with the version of MATLAB you are using, and an engineer can typically do a better job with manual vectorization.

▶ **Example 5.5—Comparing Loops and Vectors**

To compare the execution speeds of loops and vectors, perform and time the following four sets of calculations.

1. Calculate the squares of every integer from 1 to 10000 in a for loop without initializing the array of squares first.
2. Calculate the squares of every integer from 1 to 10000 in a for loop, using the zeros function to preallocate the array of squares first and calculating the square of the number in-line. (This will allow the JIT compiler to function.)
3. Calculate the squares of every integer from 1 to 10000 with vectors.

Solution This program must calculate the squares of the integers from 1 to 10000 in each of the three ways described above, timing the executions in each case. The timing can be accomplished using the MATLAB functions tic and toc. Function tic resets the built-in elapsed time counter, and function toc returns the elapsed time in seconds since the last call to function tic.

Since the real-time clocks in many computers have a fairly coarse granularity, it may be necessary to execute each set of instructions multiple times to get a valid average time.

A MATLAB program to compare the speeds of the three approaches is shown below:

```
%   Script file: timings.m
%
%   Purpose:
%       This program calculates the time required to
```

---

[3]Mathworks refuses to release a list of situations in which the JIT compiler works and situations in which it doesn't work, saying that it is complicated and that it varies between different versions of MATLAB. They suggest that you write your loops and then time them to see if they are fast or slow! The good news is that the JIT compiler works properly in more and more situations with each release, but you never know….

```
%       calculate the squares of all integers from 1 to
%       10,000 in three different ways:
%       1.  Using a for loop with an uninitialized output
%           array.
%       2.  Using a for loop with a preallocated output
%           array and the JIT compiler.
%       3.  Using vectors.
%
% Record of revisions:
%     Date          Engineer           Description of change
%     ====          ==========         =====================
%   01/29/14   S. J. Chapman      Original code
%
% Define variables:
%   ii, jj        -- Loop index
%   average1      -- Average time for calculation 1
%   average2      -- Average time for calculation 2
%   average3      -- Average time for calculation 3
%   maxcount      -- Number of times to loop calculation
%   square        -- Array of squares

% Perform calculation with an uninitialized array
% "square".  This calculation is done only 10 times
% because it is so slow.
maxcount = 10;                  % Number of repetitions
tic;                            % Start timer
for jj = 1:maxcount
   clear square                 % Clear output array
    for ii = 1:10000
       square(ii) = ii^2;       % Calculate square
    end
end
average1 = (toc)/maxcount;  % Calculate average time

% Perform calculation with a preallocated array
% "square". This calculation is averaged over 1000
% loops.
maxcount = 1000;                % Number of repetitions
tic;                            % Start timer
for jj = 1:maxcount
   clear square                 % Clear output array
   square = zeros(1,10000); % Pre-initialize array
    for ii = 1:10000
       square(ii) = ii^2;       % Calculate square
    end
end
average2 = (toc)/maxcount;  % Calculate average time
```

```
% Perform calculation with vectors. This calculation
% averaged over 1000 executions.
maxcount = 1000;              % Number of repetitions
tic;                         % Start timer
for jj = 1:maxcount
   clear square              % Clear output array
   ii = 1:10000;            % Set up vector
   square = ii.^2;          % Calculate square
end
average3 = (toc)/maxcount;   % Calculate average time

% Display results
fprintf('Loop / uninitialized array     = %8.5f\n', average1);
fprintf('Loop / initialized array / JIT = %8.5f\n', average2);
fprintf('Vectorized                     = %8.5f\n', average3);
```

When this program is executed using MATLAB 2014B on my computer, the results are:

```
» timings
Loop / uninitialized array     =   0.00275
Loop / initialized array / JIT =   0.00012
Vectorized                     =   0.00003
```

The loop with the uninitialized array was very slow compared the loop executed with the JIT compiler or the vectorized loop. The vectorized loop was fastest way to perform the calculation, but if the JIT compiler works for your loop, you get most of the acceleration without having to do anything! As you can see, designing loops to allow the JIT compiler to function or replacing the loops with vectorized calculations can make an incredible difference in the speed of your MATLAB code. ◀

The Code Analyzer code checking tool can help you identify problems with uninitialized arrays that can slow the execution of a MATLAB program. For example, if we run the Code Analyzer on program timings.m, the code checker will identify the uninitialized array and write out a warning message (see Figure 5.1).

5.2.4 The break and continue Statements

There are two additional statements that can be used to control the operation of while loops and for loops: the break and continue statements. The break statement terminates the execution of a loop and passes control to the next statement after the end of the loop, and the continue statement terminates the current pass through the loop and returns control to the top of the loop.

If a break statement is executed in the body of a loop, the execution of the body will stop and control will be transferred to the first executable statement after the loop. An example of the break statement in a for loop is shown below.

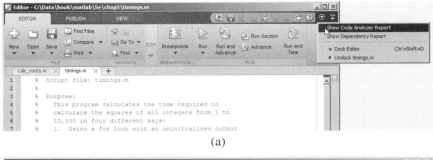

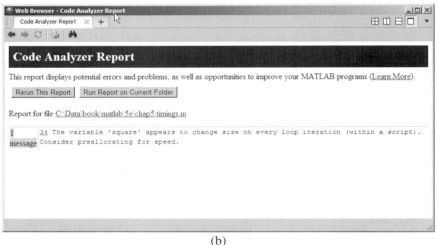

Figure 5.1 The Code Analyzer can identify some problems that will slow down the execution of MATLAB loops: (a) Running the Code Analyzer on programs timings.m. (b) The Code Analyzer report identifies the uninitalized array in the program.

```
for ii = 1:5
   if ii == 3
      break;
   end
   fprintf('ii = %d\n',ii);
end
disp(['End of loop!']);
```

When this program is executed, the output is:

```
» test_break
ii = 1
ii = 2
End of loop!
```

Note that the `break` statement was executed on the iteration when `ii` was 3, and control transferred to the first executable statement after the loop without executing the `fprintf` statement.

If a continue statement is executed in the body of a loop, the execution of the current pass through the loop will stop and control will return to the top of the loop. The controlling variable in the for loop will take on its next value, and the loop will be executed again. An example of the continue statement in a for loop is shown below.

```
for ii = 1:5
   if ii == 3
      continue;
   end
   fprintf('ii = %d\n',ii);
end
disp(['End of loop!']);
```

When this program is executed, the output is:

```
» test_continue
ii = 1
ii = 2
ii = 4
ii = 5
End of loop!
```

Note that the continue statement was executed on the iteration when ii was 3, and control transferred to the top of the loop without executing the fprintf statement.

The break and continue statements work with both while loops and for loops.

5.2.5 Nesting Loops

It is possible for one loop to be completely inside another loop. If one loop is completely inside another one, the two loops are called **nested loops**. The following example shows two nested for loops used to calculate and write out the product of two integers.

```
for ii = 1:3

   for jj = 1:3
      product = ii * jj;
      fprintf('%d * %d = %d\n',ii,jj,product);
   end
end
```

In this example, the outer for loop will assign a value of 1 to index variable ii and then the inner for loop will be executed. The inner for loop will be executed 3 times with index variable jj having values 1, 2, and 3. When the entire inner for loop has been completed, the outer for loop will assign a value of 2 to index variable ii, and the inner for loop will be executed again. This process repeats until the outer for loop has executed 3 times, and the resulting output is

```
1 * 1 = 1
1 * 2 = 2
1 * 3 = 3
2 * 1 = 2
```

```
2 * 2 = 4
2 * 3 = 6
3 * 1 = 3
3 * 2 = 6
3 * 3 = 9
```

Note that the inner `for` loop executes completely before the index variable of the outer `for` loop is incremented.

When MATLAB encounters an `end` *statement, it associates that statement with the innermost currently open construct.* Therefore, the first `end` statement above closes the "`for jj = 1:3`" loop, and the second `end` statement above closes the "`for ii = 1:3`" loop. This fact can produce hard-to-find errors if an `end` statement is accidentally deleted somewhere within a nested loop construct.

If `for` *loops are nested, they should have independent loop index variables.* If they have the same index variable, then the inner loop will change the value of the loop index that the outer loop just set.

If a `break` or `continue` statement appears inside a set of nested loops, then that statement refers to the *innermost* of the loops containing it. For example, consider the following program

```
for ii = 1:3
    for jj = 1:3
        if jj == 3
            break;
        end
        product = ii * jj;
        fprintf('%d * %d = %d\n',ii,jj,product);
    end
    fprintf('End of inner loop\n');
end
fprintf('End of outer loop\n');
```

If the inner loop counter `jj` is equal to 3, then the `break` statement will be executed. This will cause the program to exit the innermost loop. The program will print out "End of inner loop", the index of the outer loop will be increased by 1, and execution of the innermost loop will start over. The resulting output values are

```
1 * 1 = 1
1 * 2 = 2
End of inner loop
2 * 1 = 2
2 * 2 = 4
End of inner loop
3 * 1 = 3
3 * 2 = 6
End of inner loop
End of outer loop
```

5.3   Logical Arrays and Vectorization

We learned about logical data in Chapter 4. Logical data can have one of two possible values: `true` (1) or `false` (0). Scalars and arrays of logical data are created as the output of relational and logic operators.

For example, consider the following statements:

```
a = [1 2 3; 4 5 6; 7 8 9];
b = a > 5;
```

These statements produced two arrays a and b. Array a is a `double` array containing the values $\begin{bmatrix} 1 & 2 & 3 \\ 4 & 5 & 6 \\ 7 & 8 & 9 \end{bmatrix}$, while array b is a `logical` array containing the values $\begin{bmatrix} 0 & 0 & 0 \\ 0 & 0 & 1 \\ 1 & 1 & 1 \end{bmatrix}$. When the `whos` command is executed, the results are as shown below.

```
» whos
    Name        Size        Bytes        Class

     a          3x3           72         double array
     b          3x3            9         logical array

Grand total is 18 elements using 81 bytes
```

Logical arrays have a very important special property—*they can serve as a **mask** for arithmetic operations*. A mask is an array that selects the elements of another array for use in an operation. The specified operation will be applied to the selected elements, and *not* to the remaining elements.

For example, suppose that arrays a and b are as defined above. Then the statement a(b) = sqrt(a(b)) will take the square root of all elements for which the logical array b is `true` and leave all the other elements in the array unchanged.

```
» a(b) = sqrt(a(b))
a =
    1.0000    2.0000    3.0000
    4.0000    5.0000    2.4495
    2.6458    2.8284    3.0000
```

This is a very fast and very clever way of performing an operation on a subset of an array without needing loops and branches.

The following two code fragments both take the square root of all elements in array a whose value is greater than 5, but the vectorized approach is more compact, elegant, and faster than the loop approach.

```
for ii = 1:size(a,1)
    for jj = 1:size(a,2)
        if a(ii,jj) > 5
            a(ii,jj) = sqrt(a(ii,jj));
        end
    end
end
```

```
b = a > 5;
a(b) = sqrt(a(b));
```

5.3.1 Creating the Equivalent of `if/else` Constructs with Logical Arrays

Logical arrays can also be used to implement the equivalent of an `if/else` construct inside a set of `for` loops. As we saw in the last section, it is possible to apply an operation to selected elements of an array using a logical array as a mask. It is also possible to apply a different set of operations to the *unselected* elements of the array by simply adding the not operator (~) to the logical mask. For example, suppose that we wanted to take the square root of any elements in a two-dimensional array whose value is greater than 5, and to square the remaining elements in the array. The code for this operation using loops and branches is

```
for ii = 1:size(a,1)
    for jj = 1:size(a,2)
        if a(ii,jj) > 5
            a(ii,jj) = sqrt(a(ii,jj));
        else
            a(ii,jj) = a(ii,jj)^2;
        end
    end
end
```

The vectorized code for this operation is

```
b = a > 5;
a(b) = sqrt(a(b));
a(~b) = a(~b).^2;
```

The vectorized code is significantly faster than the loops-and-branches version.

Quiz 5.1

This quiz provides a quick check to see if you have understood the concepts introduced in Sections 5.1 through 5.3. If you have trouble with the quiz, reread the section, ask your instructor, or discuss the material with a fellow student. The answers to this quiz are found in the back of the book.

Examine the following `for` loops and determine how many times each loop will be executed.

1. `for index = 7:10`
2. `for jj = 7:-1:10`
3. `for index = 1:10:10`
4. `for ii = -10:3:-7`
5. `for kk = [0 5 ; 3 3]`

Examine the following loops and determine the value in `ires` at the end of each of the loops.

6.
```
ires = 0;
for index = 1:10
    ires = ires + 1;
end
```

7.
```
ires = 0;
for index = 1:10
    ires = ires + index;
end
```

8.
```
ires = 0;
for index1 = 1:10
    for index2 = index1:10
        if index2 == 6
            break;
        end
        ires = ires + 1;
    end
end
```

9.
```
ires = 0;
for index1 = 1:10
    for index2 = index1:10
        if index2 == 6
            continue;
        end
        ires = ires + 1;
    end
end
```

10. Write the MATLAB statements to calculate the values of the function

$$f(t) = \begin{cases} \sin t & \text{for all } t \text{ where } \sin t > 0 \\ 0 & \text{elsewhere} \end{cases}$$

for $-6\pi \le t \le 6\pi$ at intervals of $\pi/10$. Do this twice, once using loops and branches and once using vectorized code.

5.4 The MATLAB Profiler

MATLAB includes a profiler, which can be used to identify the parts of a program that consume the most execution time. The profiler can identify "hot spots," where optimizing the code will result in major increases in speed.

Figure 5.2 (a) The MATLAB Profiler is opened using the "Desktop/Profiler" menu option on the MATLAB Desktop. (b) The profiler has a box in which to type the name of the program to execute and a pushbutton to start profiling.

The MATLAB profiler is started by selecting the Run and Time tool (Run and Time) from the Code section of the Home Tab. A Profiler Window opens, with a field in which to enter the name of the program to profile and a pushbutton to start the profile process running[4] (see Figure 5.2).

After the profiler runs, a Profile Summary is displayed, showing how much time is spent in each function being profiled (see Figure 5.3a). Clicking on any profiled function brings up a more detailed display showing exactly how much time was spent on each line when that function was executed (see Figure 5.3b). With this

---

[4]There is also a "Run and Time" tool on the Editor tab. Clicking that tool automatically profiles the current displayed M-file.

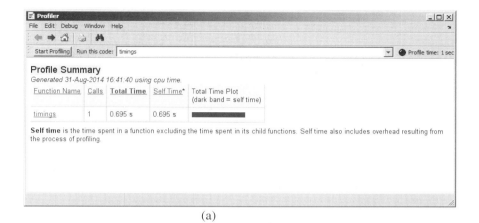

(a)

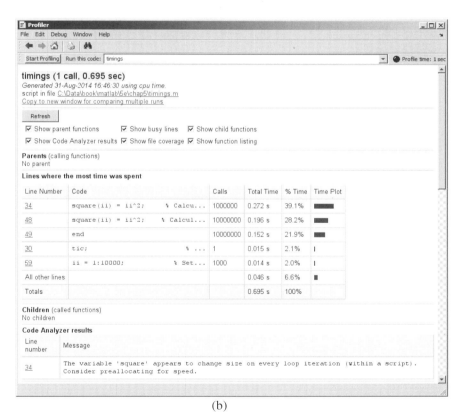

(b)

Figure 5.3 (a) The MATLAB Profiler is opened using the "Run and Time" tool in the Code section of the Home tab on the toolstrip. (b) The profiler has a box in which to type the name of the program to execute, and a pushbutton to start profiling.

information, the engineer can identify the slow portions of the code and work to speed them up with vectorization and similar techniques. For example, the profiler will highlight loops that run slowly because they can't be handled by the JIT compiler.

Normally, the profiler should be run *after a program is working properly*. **It is a waste of time to profile a program before it is working.**

 Good Programming Practice

Use the MATLAB Profiler to identify the parts of programs that consume the most CPU time. Optimizing those parts of the program will speed up the overall execution of the program.

5.5 Additional Examples

►Example 5.6—Fitting a Line to a Set of Noisy Measurements

The velocity of a falling object in the presence of a constant gravitational field is given by the equation

$$v(t) = at + v_0 \qquad (5.4)$$

where $v(t)$ is the velocity at any time t, a is the acceleration due to gravity, and v_0 is the velocity at time 0. This equation is derived from elementary physics—it is known to every freshman physics student. If we plot velocity versus time for the falling object, our (v, t) measurement points should fall along a straight line. However, the same freshman physics student also knows that if we go out into the laboratory and attempt to *measure* the velocity versus time of an object, our measurements will *not* fall along a straight line. They may come close, but they will never line up perfectly. Why not? This happens because we can never make perfect measurements. The measurements always include some *noise*, which distorts them.

There are many cases in science and engineering where there are noisy sets of data such as this, and we wish to estimate the straight line that "best fits" the data. This problem is called the *linear regression* problem. Given a noisy set of measurements (x, y) that appear to fall along a straight line, how can we find the equation of the line

$$y = mx + b \qquad (5.5)$$

that "best fits" the measurements? If we can determine the regression coefficients m and b, then we can use this equation to predict the value of y at any given x by evaluating Equation (5.5) for that value of x.

A standard method for finding the regression coefficients m and b is the *method of least squares*. This method is named "least squares" because it produces the line $y = mx + b$ for which the sum of the squares of the differences between the observed y values and the predicted y values is as small as possible. The slope of the least squares line is given by

$$m = \frac{\left(\sum xy\right) - \left(\sum x\right)\bar{y}}{\left(\sum x^2\right) - \left(\sum x\right)\bar{x}} \tag{5.6}$$

and the intercept of the least squares line is given by

$$b = \bar{y} - m\bar{x} \tag{5.7}$$

where

$\sum x$ is the sum of the x values

$\sum x^2$ is the sum of the squares of the x values

$\sum xy$ is the sum of the products of the corresponding x and y values

$\bar{x}$ is the mean (average) of the x values

$\bar{y}$ is the mean (average) of the y values

Write a program that will calculate the least-squares slope m and y-axis intercept b for a given set of noisy measured data points (x, y). The data points should be read from the keyboard, and both the individual data points and the resulting least-squares fitted line should be plotted.

Solution

1. **State the problem.**

   Calculate the slope m and intercept b of a least-squares line that best fits an input data set consisting of an arbitrary number of (x, y) pairs. The input (x, y) data is read from the keyboard. Plot both the input data points and the fitted line on a single plot.

2. **Define the inputs and outputs**.

   The inputs required by this program are the number of points to read, plus the pairs of points (x, y).

   The outputs from this program are the slope and intercept of the least-squares fitted line, the number of points going into the fit, and a plot of the input data and the fitted line.

3. **Describe the algorithm**.

   This program can be broken down into six major steps

   ```
   Get the number of input data points
   Read the input statistics
   Calculate the required statistics
   Calculate the slope and intercept
   Write out the slope and intercept
   Plot the input points and the fitted line
   ```

   The first major step of the program is to get the number of points to read in. To do this, we will prompt the user and read his or her answer with an `input` function. Next, we will read the input (x,y) pairs one pair at a time using an `input` function in a `for` loop. Each pair of input value will be placed in an array (`[x  y]`), and the function will return that array to the calling program. Note that a `for` loop is appropriate because we know in advance how many times the loop will be executed.

The pseudocode for these steps is shown below.

```
Print message describing purpose of the program
n_points ← input('Enter number of [x y] pairs:');
for ii = 1:n_points
    temp ← input('Enter [x y] pair:');
    x(ii) ← temp(1)
    y(ii) ← temp(2)
end
```

Next, we must accumulate the statistics required for the calculation. These statistics are the sums Σx, Σy, Σx^2, and Σxy. The pseudocode for these steps is:

```
Clear the variables sum_x, sum_y, sum_x2, and sum_y2
for ii = 1:n_points
    sum_x ← sum_x + x(ii)
    sum_y ← sum_y + y(ii)
    sum_x2 ← sum_x2 + x(ii)^2
    sum_xy ← sum_xy + x(ii)*y(ii)
end
```

Next, we must calculate the slope and intercept of the least-squares line. The pseudocode for this step is just the MATLAB versions of Equations (5.6) and (5.7).

```
x_bar ← sum_x / n_points
y_bar ← sum_y / n_points
slope ← (sum_xy-sum_x * y_bar)/(sum_x2 - sum_x * x_bar)
y_int ← y_bar - slope * x_bar
```

Finally, we must write out and plot the results. The input data points should be plotted with circular markers and without a connecting line, while the fitted line should be plotted as a solid 2-pixel-wide line. To do this, we will need to plot the points first, set `hold on`, plot the fitted line, and set `hold off`. We will add titles and a legend to the plot for completeness.

4. **Turn the algorithm into MATLAB statements.**

The final MATLAB program is shown below:

```
%
% Purpose:
%   To perform a least-squares fit of an input data set
%   to a straight line and print out the resulting slope
%   and intercept values. The input data for this fit
%   comes from a user-specified input data file.
%
```

```
% Record of revisions:
%     Date          Engineer          Description of change
%     ====          ==========        =====================
%   01/30/14      S. J. Chapman       Original code
%
% Define variables:
%   ii            -- Loop index
%   n_points      -- Number in input [x y] points
%   slope         -- Slope of the line
%   sum_x         -- Sum of all input x values
%   sum_x2        -- Sum of all input x values squared
%   sum_xy        -- Sum of all input x*y values
%   sum_y         -- Sum of all input y values
%   temp          -- Variable to read user input
%   x             -- Array of x values
%   x_bar         -- Average x value
%   y             -- Array of y values
%   y_bar         -- Average y value
%   y_int         -- y-axis intercept of the line
disp('This program performs a least-squares fit of an');
disp('input data set to a straight line.');
n_points = input('Enter the number of input [x y] points:');

% Read the input data
for ii = 1:n_points
   temp = input('Enter [x y] pair:');
   x(ii) = temp(1);
   y(ii) = temp(2);
end

% Accumulate statistics
sum_x = 0;
sum_y = 0;
sum_x2 = 0;
sum_xy = 0;
for ii = 1:n_points
   sum_x  = sum_x + x(ii);
   sum_y  = sum_y + y(ii);
   sum_x2 = sum_x2 + x(ii)^2;
   sum_xy = sum_xy + x(ii) * y(ii);
end

% Now calculate the slope and intercept.
x_bar = sum_x / n_points;
y_bar = sum_y / n_points;
slope = (sum_xy - sum_x * y_bar) / (sum_x2 - sum_x * x_bar);
y_int = y_bar - slope * x_bar;
```

```
% Tell user.
disp('Regression coefficients for the least-squares line:');
fprintf('Slope (m)      = %8.3f\n', slope);
fprintf('Intercept (b) = %8.3f\n', y_int);
fprintf('No. of points = %d\n', n_points);

% Plot the data points as blue circles with no
% connecting lines.
plot(x,y,'bo');
hold on;

% Create the fitted line
xmin = min(x);
xmax = max(x);
ymin = slope * xmin + y_int;
ymax = slope * xmax + y_int;

% Plot a solid red line with no markers
plot([xmin xmax],[ymin ymax],'r-','LineWidth',2);
hold off;

% Add a title and legend
title ('\bfLeast-Squares Fit');
xlabel('\bf\itx');
ylabel('\bf\ity');
legend('Input data','Fitted line');
grid on
```

5. **Test the program.**

To test this program, we will try a simple data set. For example, if every point in the input data set falls exactly along a line, then the resulting slope and intercept should be exactly the slope and intercept of that line. Thus the data set

```
[1.1 1.1]
[2.2 2.2]
[3.3 3.3]
[4.4 4.4]
[5.5 5.5]
[6.6 6.6]
[7.7 7.7]
```

should produce a slope of 1.0 and an intercept of 0.0. If we run the program with these values, the results are:

```
» lsqfit
```
This program performs a least-squares fit of an input data set to a straight line.

```
Enter the number of input [x y] points: 7
Enter [x y] pair: [1.1 1.1]
Enter [x y] pair: [2.2 2.2]
Enter [x y] pair: [3.3 3.3]
Enter [x y] pair: [4.4 4.4]
Enter [x y] pair: [5.5 5.5]
Enter [x y] pair: [6.6 6.6]
Enter [x y] pair: [7.7 7.7]
Regression coefficients for the least-squares line:
  Slope (m)     =     1.000
  Intercept (b) =     0.000
  No. of points =         7
```

Now let's add some noise to the measurements. The data set becomes

```
[1.1 1.01]
[2.2 2.30]
[3.3 3.05]
[4.4 4.28]
[5.5 5.75]
[6.6 6.48]
[7.7 7.84]
```

If we run the program with these values, the results are:

```
» lsqfit
This program performs a least-squares fit of an
input data set to a straight line.
Enter the number of input [x y] points: 7
Enter [x y] pair: [1.1 1.01]
Enter [x y] pair: [2.2 2.30]
Enter [x y] pair: [3.3 3.05]
Enter [x y] pair: [4.4 4.28]
Enter [x y] pair: [5.5 5.75]
Enter [x y] pair: [6.6 6.48]
Enter [x y] pair: [7.7 7.84]
Regression coefficients for the least-squares line:
  Slope (m)     =     1.024
  Intercept (b) =    -0.120
  No. of points =         7
```

If we calculate the answer by hand, it is easy to show that the program gives the correct answers for our two test data sets. The noisy input data set and the resulting least-squares fitted line are shown in Figure 5.4.

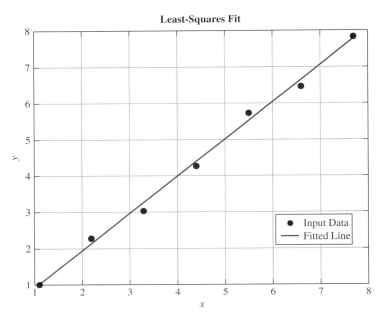

Figure 5.4 A noisy data set with a least-squares fitted line.

This example uses several of the plotting capabilities that we introduced in Chapter 3. It uses the `hold` command to allow multiple plots to be placed on the same axes, the `LineWidth` property to set the width of the least-squares fitted line, and escape sequences to make the title bold face and the axis labels bold italic.

▶Example 5.7—Physics—The Flight of a Ball

If we assume negligible air friction and ignore the curvature of the Earth, a ball that is thrown into the air from any point on the Earth's surface will follow a parabolic flight path (see Figure 5.5a). The height of the ball at any time t after it is thrown is given by Equation (5.8)

$$y(t) = y_0 + v_{y0}t + \frac{1}{2}gt^2 \tag{5.8}$$

where y_0 is the initial height of the object above the ground, v_{y0} is the initial vertical velocity of the object, and g is the acceleration due to the Earth's gravity. The horizontal distance (range) traveled by the ball as a function of time after it is thrown is given by Equation (5.9)

$$x(t) = x_0 + v_{x0}t \tag{5.9}$$

where x_0 is the initial horizontal position of the ball on the ground, and v_{x0} is the initial horizontal velocity of the ball.

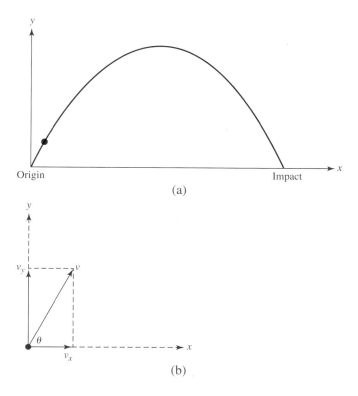

Figure 5.5 (a) When a ball is thrown upwards, it follows a parabolic trajectory. (b) The horizontal and vertical components of a velocity vector v at an angle θ with respect to the horizontal.

If the ball is thrown with some initial velocity v_0 at an angle of θ degrees with respect to the Earth's surface, then the initial horizontal and vertical components of velocity will be

$$v_{x0} = v_0 \cos \theta \qquad (5.10)$$

$$v_{y0} = v_0 \sin \theta \qquad (5.11)$$

Assume that the ball is initially thrown from position $(x_0, y_0) = (0, 0)$ with an initial velocity v_0 of 20 meters per second at an initial angle of θ degrees. Write a program that will plot the trajectory of the ball and also determine the horizontal distance traveled before it touches the ground again. The program should plot the trajectories of the ball for all angles θ from 5 to 85° in 10° steps, and should determine the horizontal distance traveled for all angles θ from 0 to 90° in 1° steps. Finally, it should determine the angle θ that maximizes the range of the ball and plot that particular trajectory in a different color with a thicker line.

Solution To solve this problem, we must determine an equation for the time that the ball returns to the ground. Then, we can calculate the (x, y) position of the ball using

Equations (5.8) through (5.11). If we do this for many times between 0 and the time that the ball returns to the ground, we can use those points to plot the ball's trajectory.

The time that the ball will remain in the air after it is thrown may be calculated from Equation (5.8). The ball will touch the ground at the time t for which $y(t) = 0$. Remembering that the ball will start from ground level ($y(0) = 0$), and solving for t, we get:

$$y(t) = y_0 + v_{y0}t + \frac{1}{2}gt^2 \tag{5.8}$$

$$0 = 0 + v_{y0}t + \frac{1}{2}gt^2$$

$$0 = \left(v_{y0} + \frac{1}{2}gt\right)t$$

so the ball will be at ground level at time $t_1 = 0$ (when we threw it) and at time

$$t_2 = -\frac{2v_{y0}}{g} \tag{5.12}$$

From the problem statement, we know that the initial velocity v_0 is 20 meters per second, and that the ball will be thrown at all angles from $0°$ to $90°$ in $1°$ steps. Finally, any elementary physics textbook will tell us that the acceleration due to the Earth's gravity is -9.81 meters per second squared.

Now let's apply our design technique to this problem.

1. **State the problem.**
   A proper statement of this problem would be: *Calculate the range that a ball would travel when it is thrown with an initial velocity of v_0 of 20 m/s at an initial angle θ. Calculate this range for all angles between $0°$ and $90°$, in $1°$ steps. Determine the angle θ that will result in the maximum range for the ball. Plot the trajectory of the ball for angles between $5°$ and $85°$, in $10°$ increments. Plot the maximum-range trajectory in a different color and with a thicker line. Assume that there is no air friction.*

2. **Define the inputs and outputs.**
   As the problem is defined above, no inputs are required. We know from the problem statement what v_0 and θ will be, so there is no need to input them. The outputs from this program will be a table showing the range of the ball for each angle θ, the angle θ for which the range is maximum, and a plot of the specified trajectories.

3. **Design the algorithm.**
   This program can be broken down into the following major steps

```
Calculate the range of the ball for θ between 0 and 90°
Write a table of ranges
Determine the maximum range and write it out
Plot the trajectories for θ between 5 and 85°
Plot the maximum-range trajectory
```

Since we know the exact number of times that the loops will be repeated, for loops are appropriate for this algorithm. We will now refine the pseudocode for each of the major steps above.

To calculate the maximum range of the ball for each angle, we will first calculate the initial horizontal and vertical velocity from Equations (5.10) and (5.11). Then we will determine the time when the ball returns to Earth from Equation (5.12). Finally, we will calculate the range at that time from Equation (5.8). The detailed pseudocode for these steps is shown below. Note that we must convert all angles to radians before using the trig functions!

```
Create and initialize an array to hold ranges
for ii = 1:91
    theta ← ii - 1
    vxo ← vo * cos(theta*conv)
    vyo ← vo * sin(theta*conv)
    max_time ← -2 * vyo / g
    range(ii) ← vxo * max_time
end
```

Next, we must write a table of ranges. The pseudocode for this step is:

```
Write heading
for ii = 1:91
    theta ← ii - 1
    print theta and range
end
```

The maximum range can be found with the max function. Recall that this function returns both the maximum value and its location. The pseudocode for this step is:

```
[maxrange index] ← max(range)
Print out maximum range and angle (=index-1)
```

We will use nested for loops to calculate and plot the trajectories. To get all of the plots to appear on the screen, we must plot the first trajectory and then set hold on before plotting any other trajectories. After plotting the last trajectory, we must set hold off. To perform this calculation, we will divide each trajectory into 21 time steps, and find the x and y positions of the ball for each time step. Then, we will plot those (x, y) positions. The pseudocode for this step is:

```
for ii = 5:10:85

    % Get velocities and max time for this angle
    theta ← ii - 1
    vxo ← vo * cos(theta*conv)
    vyo ← vo * sin(theta*conv)
    max_time ← -2 * vyo / g

    Initialize x and y arrays
    for jj = 1:21
```

```
                time ← (jj-1) * max_time/20
                x(time) ← vxo * time
                y(time) ← vyo * time + 0.5 * g * time^2
            end
            plot(x,y) with thin green lines
            Set "hold on" after first plot
        end
        Add titles and axis labels
```

Finally, we must plot the maximum range trajectory in a different color and with a thicker line.

```
        vxo ← vo * cos(max_angle*conv)
        vyo ← vo * sin(max_angle*conv)
        max_time ← -2 * vyo / g

        Initialize x and y arrays
        for jj = 1:21
            time ← (jj-1) * max_time/20
            x(jj) ← vxo * time
            y(jj) ← vyo * time + 0.5 * g * time^2
        end
        plot(x,y) with a thick red line
        hold off
```

4. **Turn the algorithm into MATLAB statements.**
   The final MATLAB program is shown below.

```
%   Script file: ball.m
%
%   Purpose:
%     This program calculates the distance traveled by a ball
%     thrown at a specified angle "theta" and a specified
%     velocity "vo" from a point on the surface of the Earth,
%     ignoring air friction and the Earth's curvature. It
%     calculates the angle yielding maximum range and also
%     plots selected trajectories.
%
%   Record of revisions:
%       Date              Engineer           Description of change
%       ====              ==========          ======================
%     01/30/14          S. J. Chapman         Original code
%
% Define variables:
%   conv          -- Degrees-to-radians conv factor
%   g             -- Accel. due to gravity (m/s^2)
%   ii, jj        -- Loop index
%   index         -- Location of maximum range in array
%   maxangle      -- Angle that gives maximum range (deg)
```

```
%    maxrange       -- Maximum range (m)
%    range          -- Range for a particular angle (m)
%    time           -- Time (s)
%    theta          -- Initial angle (deg)
%    traj_time      -- Total trajectory time (s)
%    vo             -- Initial velocity (m/s)
%    vxo            -- X-component of initial velocity (m/s)
%    vyo            -- Y-component of initial velocity (m/s)
%    x              -- X-position of ball (m)
%    y              -- Y-position of ball (m)

%   Constants
conv = pi / 180;   % Degrees-to-radians conversion factor
g = -9.81;         % Accel. due to gravity
vo = 20;           % Initial velocity

%Create an array to hold ranges
range = zeros(1,91);

% Calculate maximum ranges
for ii = 1:91
   theta = ii -1;
   vxo = vo * cos(theta*conv);
   vyo = vo * sin(theta*conv);
   max_time = -2 * vyo / g;
   range(ii) = vxo * max_time;
end

% Write out table of ranges
fprintf ('Range versus angle theta:\n');
for ii = 1:91
   theta = ii -1;
   fprintf(' %2d    %8.4f\n',theta, range(ii));
end

% Calculate the maximum range and angle
[maxrange index] = max(range);
maxangle = index - 1;
fprintf ('\nMax range is %8.4f at %2d degrees.\n',...
        maxrange, maxangle);

% Now plot the trajectories
for ii = 5:10:85

   % Get velocities and max time for this angle
   theta = ii;
```

```
      vxo = vo * cos(theta*conv);
      vyo = vo * sin(theta*conv);
      max_time = -2 * vyo / g;

      % Calculate the (x,y) positions
      x = zeros(1,21);
      y = zeros(1,21);
      for jj = 1:21
          time = (jj-1) * max_time/20;
          x(jj) = vxo * time;
          y(jj) = vyo * time + 0.5 * g * time^2;
      end
      plot(x,y,'b');
      if ii == 5
          hold on;
      end
   end

   % Add titles and axis labels
   title ('\bfTrajectory of Ball vs Initial Angle \theta');
   xlabel ('\bf\itx \rm\bf(meters)');
   ylabel ('\bf\ity \rm\bf(meters)');
   axis ([0 45 0 25]);
   grid on;

   % Now plot the max range trajectory
   vxo = vo * cos(maxangle*conv);
   vyo = vo * sin(maxangle*conv);
   max_time = -2 * vyo / g;

   % Calculate the (x,y) positions
   x = zeros(1,21);
   y = zeros(1,21);
   for jj = 1:21
       time = (jj-1) * max_time/20;
       x(jj) = vxo * time;
       y(jj) = vyo * time + 0.5 * g * time^2;
   end
   plot(x,y,'r','LineWidth',3.0);
   hold off
```

The acceleration due to gravity at sea level can be found in any physics text. It is about 9.81 m/s^2, directed downward.

5. **Test the program.**

   To test this program, we will calculate the answers by hand for a few of the angles, and compare the results with the output of the program.

θ	$v_{x0} = v_0 \cos \theta$	$v_{y0} = v_0 \sin \theta$	$t_2 = -\dfrac{2v_{y0}}{g}$	$x = v_{x0}t_2$
0°	20 m/s	0 m/s	0 s	0 m
5°	19.92 m/s	1.74 m/s	0.355 s	7.08 m
40°	15.32 m/s	12.86 m/s	2.621 s	40.15 m
45°	14.14 m/s	14.14 m/s	2.883 s	40.77 m

When program `ball` is executed, a 91-line table of angles and ranges is produced. To save space, only a portion of the table is reproduced below.

```
» ball
Range versus angle theta:
    0      0.0000
    1      1.4230
    2      2.8443
    3      4.2621
    4      5.6747
    5      7.0805
...
   40     40.1553
   41     40.3779
   42     40.5514
   43     40.6754
   44     40.7499
   45     40.7747
   46     40.7499
   47     40.6754
   48     40.5514
   49     40.3779
   50     40.1553
...
   85      7.0805
   86      5.6747
   87      4.2621
   88      2.8443
   89      1.4230
   90      0.0000

Max range is  40.7747 at 45 degrees.
```

The resulting plot is shown in Figure 5.6. The program output matches our hand calculation for the angles calculated above to the 4-digit accuracy of the hand calculation. Note that the maximum range occurred at an angle of 45°.

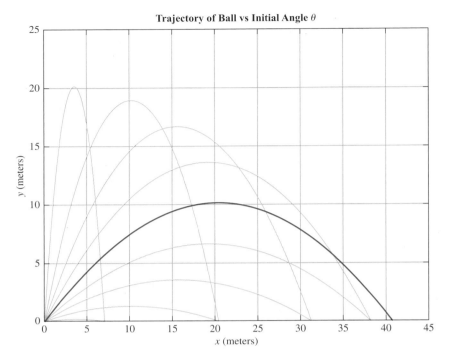

Figure 5.6 Possible trajectories for the ball.

This example uses several of the plotting capabilities that we introduced in Chapter 3. It uses the `axis` command to set the range of data to display, the `hold` command to allow multiple plots to be placed on the same axes, the `LineWidth` property to set the width of the line corresponding to the maximum-range trajectory, and escape sequences to create the desired title and *x*- and *y*-axis labels.

However, this program is not written in the most efficient manner, since there are a number of loops that could have been better replaced by vectorized statements. You will be asked to re-write and improve `ball.m` in Exercise 5.11 at the end of this chapter.

5.6 The `textread` Function

In the least-squares fit problem in Example 5.6, we had to enter each (*x*,*y*) pair of data point from the keyboard and include them in an array constructor ([ ]). This would be a *very* tedious process if we wanted to enter large amounts of data into a program, so we need a better way to load data into our programs. Large data sets are almost always stored in files, not typed at the command line, so what we really need is an easy way to read data from a file and use it in a MATLAB program. The `textread` function serves that purpose.

The textread function reads ASCII files that are formatted into columns of data, where each column can be of a different type, and stores the contents of each column in a separate output array. This function is *very* useful for importing large amounts of data printed out by other applications.

The form of the textread function is

```
[a,b,c,...] = textread(filename,format,n)
```

where filename is the name of the file to open, format is a string containing a description of the type of data in each column, and n is the number of lines to read. (If n is missing, the function reads to the end of the file.) The format string contains the same types of format descriptors as function fprintf. Note that the number of output arguments must match the number of columns that you are reading.

For example, suppose that file test_input.dat contains the following data:

```
James    Jones   O+    3.51    22    Yes
Sally    Smith   A+    3.28    23    No
```

The first three columns in this file contain character data, the next two contain numbers, and the last column contains character data. This data could be read into a series of arrays with the following function:

```
[first,last,blood,gpa,age,answer] = ...
textread('test_input.dat','%s %s %s %f %d %s')
```

Note the string descriptors %s for the columns where there is string data, and the numeric descriptors %f and %d for the columns where there is floating point and integer data. String data is returned in a cell array (which we will learn about in Chapter 10), and numeric data is always returned in a double array.

When this command is executed, the results are:

```
» [first,last,blood,gpa,age,answer] = ...
textread('test_input.dat','%s %s %s %f %d %s')

first =
    'James'
    'Sally'
last =
    'Jones'
    'Smith'
blood =
    'O+'
    'A+'
gpa =
    3.5100
    3.2800
age =
    42
    28
```

```
answer=
    'Yes'
    'No'
```

This function can also skip selected columns by adding an asterisk to the corresponding format descriptor (for example, `%*s`). The following statement reads only the `first`, `last`, and `gpa` from the file:

```
» [first,last,gpa] = ...
            textread('test_input.dat','%s %s %*s %f
            %*d %*s')

first =
    'James'
    'Sally'
last =
    'Jones'
    'Smith'
gpa =
    3.5100
    3.2800
```

Function `textread` is much more useful and flexible than the `load` command. The `load` command assumes that all of the data in the input file is of a single type—it cannot support different types of data in different columns. In addition, it stores all of the data into a single array. In contrast, the `textread` function allows each column to go into a separate variable, which is *much* more convenient when working with columns of mixed data.

Function `textread` has a number of additional options that increase its flexibility. Consult the MATLAB online help system for details of these options.

5.7 Summary

There are two basic types of loops in MATLAB, the `while` loop and the `for` loop. The `while` loop is used to repeat a section of code in cases where we do not know in advance how many times the loop must be repeated. The `for` loop is used to repeat a section of code in cases where we know in advance how many times the loop should be repeated. It is possible to exit from either type of loop at any time using the `break` statement.

A `for` loop can often be replaced by vectorized code, which performs the same calculations in single statements instead of in a loop. Because of the way MATLAB is designed, vectorized code is faster than loops, so it pays to replace loops with vectorized code whenever possible.

The MATLAB Just-in-Time (JIT) compiler also speeds up loop execution in some cases, but the exact cases that it works for vary in different versions of MATLAB. If it works, the JIT compiler will produce code that is almost as fast as vectorized statements.

The textread function can be used to read selected columns of an ASCII data file into a MATLAB program for processing. This function is quite flexible, making it easy to read output files created by other programs.

5.7.1 Summary of Good Programming Practice

The following guidelines should be adhered to when programming with loop constructs. By following them consistently, your code will contain fewer bugs, will be easier to debug, and will be more understandable to others who may need to work with it in the future.

1. Always indent code blocks in while and for constructs to make them more readable.
2. Use a while loop to repeat sections of code when you don't know in advance how often the loop will be executed.
3. Use a for loop to repeat sections of code when you know in advance how often the loop will be executed.
4. Never modify the values of a for loop index while inside the loop.
5. Always preallocate all arrays used in a loop before executing the loop. This practice greatly increases the execution speed of the loop.
6. If it is possible to implement a calculation either with a for loop or using vectors, implement the calculation with vectors. Your program will be much faster.
7. Do not rely on the JIT compiler to speed up your code. It has many limitations, and an engineer can typically do a better job with manual vectorization.
8. Use the MATLAB Profiler to identify the parts of programs that consume the most CPU time. Optimizing those parts of the program will speed up the overall execution of the program.

5.7.2 MATLAB Summary

The following summary lists all of the MATLAB commands and functions described in this chapter, along with a brief description of each one.

Commands and Functions

break	Stop the execution of a loop and transfer control to the first statement after the end of the loop.
continue	Stop the execution of a loop and transfer control to the top of the loop for the next iteration.
factorial	Calculate the factorial function.
for loop	Loops over a block of statements a specified number of times.
tic	Resets elapsed time counter.

(*continued*)

Commands and Functions (*Continued*)

textread	Reads text data from a file into one or more input variables.
toc	Returns elapsed time since last call to tic.
while loop	Loops over a block of statements until a test condition becomes 0 (false).

5.8 Exercises

5.1 Write the MATLAB statements required to calculate $y(t)$ from the equation

$$y(t) = \begin{cases} -3t^2 + 5 & t \geq 0 \\ 3t^2 + 5 & t < 0 \end{cases}$$

for values of t between -9 and 9 in steps of 0.5. Use loops and branches to perform this calculation.

5.2 Rewrite the statements required to solve Exercise 5.1 using vectorization.

5.3 Write the MATLAB statements required to calculate and print out the squares of all the even integers between 0 and 50. Create a table consisting of each integer and its square, with appropriate labels over each column.

5.4 Write an M-file to evaluate the equation $y(x) = x^2 - 3x + 2$ for all values of x between -1 and 3, in steps of 0.1. Do this twice, once with a for loop and once with vectors. Plot the resulting function using a 3-point-thick dashed red line.

5.5 Write an M-file to calculate the factorial function N!, as defined in Example 5.2. Be sure to handle the special case of 0! Also, be sure to report an error if N is negative or not an integer.

5.6 Examine the following for statements and determine how many times each loop will be executed.

(a) for ii = -32768:32767
(b) for ii = 32768:32767
(c) for kk = 2:4:3
(d) for jj = ones(5,5)

5.7 Examine the following for loops and determine the value of ires at the end of each of the loops and also the number of times each loop executes.

```
(a) ires = 0;
    for index = -10:10
      ires = ires + 1;
    end

(b) ires = 0;
    for index = 10:-2:4
      if index == 6
          continue;
       end
      ires = ires + index;
    end
```

```
(c) ires = 0;
    for index = 10:-2:4
        if index == 6
            break;
        end
        ires = ires + index;
    end

(d) ires = 0;
    for index1 = 10:-2:4
        for index2 = 2:2:index1
            if index2 == 6
                break
            end
            ires = ires + index2;
        end
    end
```

5.8 Examine the following `while` loops and determine the value of `ires` at the end of each of the loops, and the number of times each loop executes.

```
(a) ires = 1;
    while mod(ires,10) ~= 0
        ires = ires + 1;
    end

(b) ires = 2;
    while ires <= 200
        ires = ires^2;
    end

(c) ires = 2;
    while ires > 200
        ires = ires^2;
    end
```

5.9 What is contained in array `arr1` after each of the following sets of statements is executed?

```
(a) arr1 = [1 2 3 4; 5 6 7 8; 9 10 11 12];
    mask = mod(arr1,2) == 0;
    arr1(mask) = -arr1(mask);
(b) arr1 = [1 2 3 4; 5 6 7 8; 9 10 11 12];
    arr2 = arr1 <= 5;
    arr1(arr2) = 0;
    arr1(~arr2) = arr1(~arr2).^2;
```

5.10 How can a logical array be made to behave as a logical mask for vector operations?

5.11 Modify program `ball` from Example 5.7 by replacing the inner `for` loops with vectorized calculations.

5.12 Modify program `ball` from Example 5.7 to read in the acceleration due to gravity at a particular location, and to calculate the maximum range of the ball for that acceleration. After modifying the program, run it with accelerations of -9.8 m/s^2, -9.7 m/s^2, and -9.6 m/s^2. What effect does the reduction in gravitational attraction have on the range of the ball? What effect does the reduction in gravitational attraction have on the best angle θ at which to throw the ball?

5.13 Modify program `ball` from Example 5.7 to read in the initial velocity with which the ball is thrown. After modifying the program, run it with initial velocities of 10 m/s, 20 m/s, and 30 m/s. What effect does changing the initial velocity v_0 have on the range of the ball? What effect does it have on the best angle θ at which to throw the ball?

5.14 Program `lsqfit` from Example 5.6 required the user to specify the number of input data points before entering the values. Modify the program so that it reads an arbitrary number of data values using a `while` loop and stops reading input values when the user presses the Enter key without typing any values. Test your program using the same two data sets that were used in Example 5.6. (*Hint:* The `input` function returns an empty array (`[]`) if a user presses Enter without supplying any data. You can use function `isempty` to test for an empty array, and stop reading data when one is detected.)

5.15 Modify program `lsqfit` from Example 5.6 to read its input values from an ASCII file named `input1.dat`. The data in the file will be organized in rows, with one pair of (x, y) values on each row, as shown below:

```
1.1    2.2
2.2    3.3
. . .
```

Use the `load` function to read the input data. Test your program using the same two data sets that were used in Example 5.6.

5.16 Modify program `lsqfit` from Example 5.6 to read its input values from a user-specified ASCII file named `input1.dat`. The data in the file will be organized in rows, with one pair of (x, y) values on each row, as shown below:

```
1.1    2.2
2.2    3.3
. . .
```

Use the `textread` function to read the input data. Test your program using the same two data sets that were used in Example 5.6.

5.17 **Factorial Function** MATLAB includes a standard function called `factorial` to calculate the factorial function. Use the MATLAB help system to look up this function, and then calculate 5!, 10!, and 15! using both the program in Example 5.2 and the `factorial` function. How do the results compare?

5.18 **Running Average Filter** Another way of smoothing a noisy data set is with a *running average filter*. For each data sample in a running average filter, the program examines a subset of n samples centered on the sample under test, and it replaces that sample with the average value from the n samples. (*Note:* For points near the beginning and the end of the data set, use a smaller number of samples in the running average, but be sure to keep an equal number of samples on either side of the sample under test.)

Write a program that allows the user to specify the name of an input data set and the number of samples to average in the filter and then performs a running average filter on the data. The program should plot both the original data and the smoothed curve after the running average filter.

Test your program using the data in the file `input3.dat`, which is available from the book's website.

5.19 Median Filter Another way of smoothing a noisy data set is with a *median filter*. For each data sample in a median filter, the program examines a subset of n samples centered on the sample under test, and it replaces that sample with the median value from the n samples. (*Note:* For points near the beginning and the end of the data set, use a smaller number of samples in the median calculation, but be sure to keep an equal number of samples on either side of the sample under test.) This type of filter is very effective against data sets containing isolated "wild" points that are very far away from the other nearby points.

Write a program that allows the user to specify the name of an input data set and the number of samples to use in the filter and then performs a median filter on the data. The program should plot both the original data and the smoothed curve after the median filter.

Test your program using the data in the file `input3.dat`, which is available from the book's website. Is the median filter better or worse than the running average filter for smoothing this data set? Why?

5.20 Fourier Series A Fourier series is an infinite series representation of a periodic function in terms of sines and cosines at a fundamental frequency (matching the period of the waveform) and multiples of that frequency. For example, consider a square wave function of period L, whose amplitude is 1 for [0 $L/2$), -1 for [$L/2$ L), 1 for [L $3L/2$), and so forth. This function is plotted in Figure 5.7. This function can be represented by the Fourier series

$$f(x) = \sum_{n=1,3,5,\dots}^{\infty} \frac{1}{n} \sin\left(\frac{n\pi x}{L}\right) \tag{5.13}$$

Plot the original function assuming $L = 1$, and calculate and plot Fourier series approximations to that function containing 3, 5, and 10 terms.

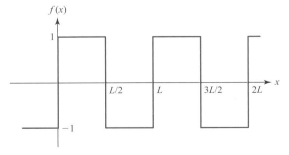

Figure 5.7 A square wave waveform.

5.21 Program doy in Example 5.3 calculates the day of year associated with any given month, day, and year. As written, this program does not check to see if the data entered by the user is valid. It will accept nonsense values for months and days and do calculations with them to produce meaningless results. Modify the program so that it checks the input values for validity before using them. If the inputs are invalid, the program should tell the user what is wrong and quit. The year should be a number greater than zero, the month should be a number between 1 and 12, and the day should be a number between 1 and a maximum that depends on the month. Use a switch construct to implement the bounds checking performed on the day.

5.22 Write a MATLAB program to evaluate the function

$$y(x) = \ln \frac{1}{1 - x} \tag{5.14}$$

for any user-specified value of x, where ln is the natural logarithm (logarithm to the base e). Write the program with a while loop, so that the program repeats the calculation for each legal value of x entered into the program. When an illegal value of x is entered, terminate the program. (Any $x \geq 1$ is considered an illegal value.)

5.23 Fibonacci Numbers The nth Fibonacci number is defined by the following recursive equations:

$$f(1) = 1$$
$$f(2) = 2$$
$$f(n) = f(n - 1) + f(n - 2) \quad n > 2$$

Therefore, $f(3) = f(2) + f(1) = 2 + 1 = 3$, and so forth for higher numbers. Write an M-file to calculate and write out the nth Fibonacci number for $n > 2$, where n is input by the user. Use a while loop to perform the calculation.

5.24 Current through a Diode The current flowing through the semiconductor diode shown in Figure 5.8 is given by the equation

$$i_D = I_0\left(e^{\frac{qv_D}{kT}} - 1\right) \tag{5.15}$$

where i_D = the voltage across the diode, in volts
  v_D = the current flow through the diode, in amps
  I_0 = the leakage current of the diode, in amps
  q = the charge on an electron, 1.602×10^{-19} coulombs
  k = Boltzmann's constant, 1.38×10^{-23} joule/K
  T = temperature, in kelvins (K)

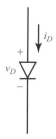

Figure 5.8 A semiconductor diode.

The leakage current I_0 of the diode is 2.0 μA. Write a program to calculate the current flowing through this diode for all voltages from -1.0 V to $+0.6$ V, in 0.1 V steps. Repeat this process for the following temperatures: 75 °F, 100 °F, and 125 °F. Create a plot of the current as a function of applied voltage, with the curves for the three different temperatures appearing as different colors.

5.25 **Tension on a Cable** A 100-kg object is to be hung from the end of a rigid 2-meter horizontal pole of negligible weight, as shown in Figure 5.9. The pole is attached to a wall by a pivot and is supported by a 2-meter cable that is attached to the wall at a higher point. The tension on this cable is given by the equation

$$T = \frac{W \cdot lc \cdot lp}{d\sqrt{lp^2 - d^2}} \tag{5.16}$$

where T is the tension on the cable, W is the weight of the object, lc is the length of the cable, lp is the length of the pole, and d is the distance along the pole at which the cable is attached. Write a program to determine the distance d at which to attach the cable to the pole in order to minimize the tension on the cable. To do this, the program should calculate the tension on the cable at regular 0.1 m intervals from $d = 0.3$ m to $d = 1.8$ m, and should locate the position d that produces the minimum tension. Also, the program should plot the tension on the cable as a function of d, with appropriate titles and axis labels.

5.26 Modify the program created in Exercise 5.25 to determine how sensitive the tension on the cable is to the precise location d at which the cable is attached. Specifically, determine the range of d values that will keep the tension on the cable within 10% of its minimum value.

5.27 **Area of a Parallelogram** The area of a parallelogram with two adjacent sides defined by vectors **A** and **B** can be found from Equation (5.17) (see Figure 5.10).

$$area = \left| \mathbf{A} \times \mathbf{B} \right| \tag{5.17}$$

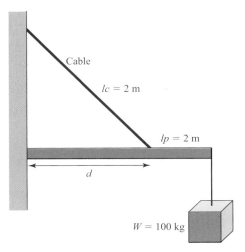

Figure 5.9 A 100 kg weight suspended from a rigid bar supported by a cable.

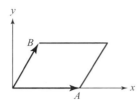

Figure 5.10 A parallelogram.

Write a program to read vectors **A** and **B** from the user, and calculate the resulting area of the parallelogram. Test your program by calculating the area of a parallelogram bordered by vectors $\mathbf{A} = 10\,\hat{\mathbf{i}}$ and $\mathbf{B} = 5\,\hat{\mathbf{i}} + 8.66\,\hat{\mathbf{j}}$.

5.28 Area of a Rectangle The area of the rectangle in Figure 5.11 is given by Equation (5.18) and the perimeter of the rectangle is given by Equation (5.19).

$$area = W \times H \tag{5.18}$$

$$perimeter = 2W + 2H \tag{5.19}$$

Assume that the total perimeter of a rectangle is limited to 10, and write a program that calculates and plots the area of the rectangle as its width is varied from the smallest possible value to the largest possible value. At what width is the area of the rectangle maximized?

5.29 Bacterial Growth Suppose that a biologist performs an experiment in which he or she measures the rate at which a specific type of bacterium reproduces asexually in different culture media. The experiment shows that in medium A the bacteria reproduce once every 60 minutes, and in medium B the bacteria reproduce once every 90 minutes. Assume that a single bacterium is placed on each culture medium at the beginning of the experiment. Write a program that calculates and plots the number of bacteria present in each culture at intervals of three hours from the beginning of the experiment until 24 hours have elapsed. Make two plots, one a linear *xy* plot and the other a linear-log (`semilogy`) plot. How do the numbers of bacteria compare on the two media after 24 hours?

5.30 Decibels Engineers often measure the ratio of two power measurements in *decibels*, or dB. The equation for the ratio of two power measurements in decibels is

$$dB = 10\log_{10}\frac{P_2}{P_1} \tag{5.20}$$

where P_2 is the power level being measured, and P_1 is some reference power level. Assume that the reference power level P_1 is 1 watt, and write a program

Figure 5.11 A rectangle.

that calculates the decibel level corresponding to power levels between 1 and 20 watts, in 0.5 W steps. Plot the dB-versus-power curve on a log-linear scale.

5.31 Geometric Mean The *geometric mean* of a set of positive numbers x_1 through x_n is defined as the *n*th root of the product of the numbers:

$$\text{geometric mean} = \sqrt[n]{x_1 x_2 x_3 \cdots x_n} \tag{5.21}$$

Write a MATLAB program that will accept an arbitrary number of positive input values and calculate both the arithmetic mean (*i.e.*, the average) and the geometric mean of the numbers. Use a `while` loop to get the input values, and terminate the inputs when a user enters a negative number. Test your program by calculating the average and geometric mean of the four numbers 10, 5, 2, and 5.

5.32 RMS Average The *root-mean-square* (rms) *average* is another way of calculating a mean for a set of numbers. The rms average of a series of numbers is the square root of the arithmetic mean of the squares of the numbers:

$$\text{rms average} = \sqrt{\frac{1}{N}\sum_{i=1}^{N} x_i^2} \tag{5.22}$$

Write a MATLAB program that will accept an arbitrary number of positive input values and calculate the rms average of the numbers. Prompt the user for the number of values to be entered, and use a `for` loop to read in the numbers. Test your program by calculating the rms average of the four numbers 10, 5, 2, and 5.

5.33 Harmonic Mean The *harmonic mean* is yet another way of calculating a mean for a set of numbers. The harmonic mean of a set of numbers is given by the equation:

$$\text{harmonic mean} = \frac{N}{\dfrac{1}{x_1} + \dfrac{1}{x_2} + \cdots + \dfrac{1}{x_n}} \tag{5.23}$$

Write a MATLAB program that will read in an arbitrary number of positive input values and calculate the harmonic mean of the numbers. Use any method that you desire to read in the input values. Test your program by calculating the harmonic mean of the four numbers 10, 5, 2, and 5.

5.34 Write a single program that calculates the arithmetic mean (average), rms average, geometric mean, and harmonic mean for a set of positive numbers. Use any method that you desire to read in the input values. Compare these values for each of the following sets of numbers:

(a) 4, 4, 4, 4, 4, 4, 4
(b) 4, 3, 4, 5, 4, 3, 5
(c) 4, 1, 4, 7, 4, 1, 7
(d) 1, 2, 3, 4, 5, 6, 7

5.35 Mean Time Between Failure Calculations The reliability of a piece of electronic equipment is usually measured in terms of Mean Time Between Failures (MTBF), where MTBF is the average time that the piece of equipment can operate before a failure occurs in it. For large systems containing many pieces

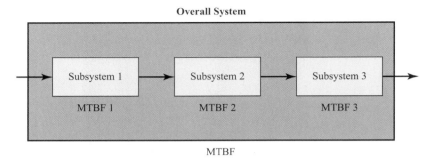

Figure 5.12 An electronic system containing three subsystems with known MTBFs.

of electronic equipment, it is customary to determine the MTBFs of each component and to calculate the overall MTBF of the system from the failure rates of the individual components. If the system is structured like the one shown in Figure 5.12, every component must work in order for the whole system to work, and the overall system MTBF can be calculated as

$$
\text{MTBF}_{sys} = \cfrac{1}{\cfrac{1}{\text{MTBF}_1} + \cfrac{1}{\text{MTBF}_2} + \cdots + \cfrac{1}{\text{MTBF}_n}} \tag{5.24}
$$

Write a program that reads in the number of series components in a system and the MTBFs for each component and then calculates the overall MTBF for the system. To test your program, determine the MTBF for a radar system consisting of an antenna subsystem with an MTBF of 2000 hours, a transmitter with an MTBF of 800 hours, a receiver with an MTBF of 3000 hours, and a computer with an MTBF of 5000 hours.

Chapter 6

Basic User-Defined Functions

In Chapter 4, we learned the importance of good program design. The basic technique that we employed was **top-down design**. In top-down design, the engineer starts with a statement of the problem to be solved and the required inputs and outputs. Next, he or she describes the algorithm to be implemented by the program in broad outline and applies *decomposition* to break the algorithm down into logical subdivisions called sub-tasks. Then, the engineer breaks down each sub-task until he or she winds up with many small pieces, each of which does a simple, clearly understandable job. Finally, the individual pieces are turned into MATLAB code.

Although we have followed this design process in our examples, the results have been somewhat restricted, because we have had to combine the final MATLAB code generated for each sub-task into a single large program. There has been no way to code, verify, and test each sub-task independently before combining them into the final program.

Fortunately, MATLAB has a special mechanism designed to make sub-tasks easy to develop and debug independently before building the final program. It is possible to code each sub-task as a separate **function**, and each function can be tested and debugged independently of all of the other sub-tasks in the program.

Well-designed functions enormously reduce the effort required on a large programming project. Their benefits include:

1. **Independent testing of sub-tasks**. Each sub-task can be written as an independent unit. The sub-task can be tested separately to ensure that it performs properly by itself before combining it into the larger program. This step is known as **unit testing**. It eliminates a major source of problems before the final program is even built.

2. **Reusable code**. In many cases, the same basic sub-task is needed in many parts of a program. For example, it may be necessary to sort a list of values into ascending order many different times within a program, or even in other programs. It is possible to design, code, test, and debug a

229

single function to do the sorting and then to reuse that function whenever sorting is required. This reusable code has two major advantages: it reduces the total programming effort required, and it simplifies debugging, since the sorting function only needs to be debugged once.

3. **Isolation from unintended side effects**. Functions receive input data from the program that invokes them through a list of variables called an **input argument list**, and return results to the program through an **output argument list**. Each function has its own workspace with its own variables, independent of all other functions and of the calling program. *The only variables in the calling program that can be seen by the function are those in the input argument list, and the only variables in the function that can be seen by the calling program are those in the output argument list.* This is very important, since accidental programming mistakes within a function can only affect the variables within the function in which the mistake occurred.

Once a large program is written and released, it has to be *maintained*. Program maintenance involves fixing bugs and modifying the program to handle new and unforeseen circumstances. The engineer who modifies a program during maintenance is often not the person who originally wrote it. In poorly written programs, it is common for the engineer modifying the program to make a change in one region of the code and to have that change cause unintended side effects in a totally different part of the program. This happens because variable names are re-used in different portions of the program. When the engineer changes the values left behind in some of the variables, those values are accidentally picked up and used in other portions of the code.

The use of well-designed functions minimizes this problem by **data hiding**. The variables in the main program are not visible to the function (except for those in the input argument list), and the variables in the main program cannot be accidentally modified by anything occurring in the function. Therefore, mistakes or changes in the function's variables cannot accidentally cause unintended side effects in the other parts of the program.

 Good Programming Practice

Break large program tasks into functions whenever practical to achieve the important benefits of independent component testing, reusability, and isolation from undesired side effects.

6.1 Introduction to MATLAB Functions

All of the M-files that we have seen so far have been **script files**. Script files are just collections of MATLAB statements that are stored in a file. When a script file is executed, the result is the same as it would be if all of the commands had been typed directly into the Command Window. Script files share the Command Window's workspace, so any variables that were defined before the script file starts are visible to the script file, and any variables created by the script file remain in the workspace after the script file finishes executing. A script file has no input arguments and returns no results, but script files can communicate with other script files through the data left behind in the workspace.

In contrast, a **MATLAB function** is a special type of M-file that runs in its own independent workspace. It receives input data through an **input argument list** and returns results to the caller through an **output argument list**. The general form of a MATLAB function is

```
function [outarg1, outarg2, ...] = fname(inarg1, inarg2, ...)
% H1 comment line
% Other comment lines
...
(Executable code)
...
(return)
end
```

The `function` statement marks the beginning of the function. It specifies the name of the function and the input and output argument lists. The input argument list appears in parentheses after the function name, and the output argument list appears in brackets to the left of the equal sign. (If there is only one output argument, the brackets can be dropped.)

Each ordinary MATLAB function should be placed in a file with the same name (including capitalization) as the function, and the file extension ".m". For example, if a function is named `My_fun`, then that function should be placed in a file named `My_fun.m`.

The input argument list is a list of names representing values that will be passed from the caller to the function. These names are called **dummy arguments**. They are just placeholders for actual values that are passed from the caller when the function is invoked. Similarly, the output argument list contains a list of dummy arguments that are placeholders for the values returned to the caller when the function finishes executing.

A function is invoked by naming it in an expression together with a list of **actual arguments**. A function can be invoked by typing its name directly in the Command Window or by including it in a script file or another function. The name in the calling program must *exactly match* the function name (including capitalization)[1]. When the function is invoked, the value of the first actual argument is used in place of the first dummy argument, and so forth for each other actual argument/dummy argument pair.

Execution begins at the top of the function and ends when a `return` statement, an `end` statement, or the end of the function is reached. Because execution stops at the end of a function anyway, the `return` statement is not actually required in most

---

[1]For example, suppose that a function has been declared with the name My_Fun, and placed in file My_Fun.m. Then this function should be called with the name My_Fun, not my_fun or MY_FUN. If the capitalization fails to match, this will produce an error on Linux and Macintosh computers, and a warning on Windows-based computers.

functions and is rarely used. Each item in the output argument list must appear on the left side of a least one assignment statement in the function. When the function returns, the values stored in the output argument list are returned to the caller and may be used in further calculations.

The use of an `end` statement to terminate a function is a new feature as of MATLAB 7.0. It is optional unless a file includes nested functions, which are described in Chapter 7. We will not use the `end` statement to terminate a function unless it is actually needed, so you will rarely see it used in this book.

The initial comment lines in a function serve a special purpose. The first comment line after the function statement is called the **H1 comment line**. It should always contain a one-line summary of the purpose of the function. The special significance of this line is that it is searched and displayed by the `lookfor` command. The remaining comment lines from the H1 line until the first blank line or the first executable statement are displayed by the `help` command. They should contain a brief summary of how to use the function.

A simple example of a user-defined function is shown below. Function `dist2` calculates the distance between points (x_1, y_1) and (x_2, y_2) in a Cartesian coordinate system.

```
function distance = dist2 (x1, y1, x2, y2)
%DIST2 Calculate the distance between two points
% Function DIST2 calculates the distance between
% two points (x1,y1) and (x2,y2) in a Cartesian
% coordinate system.
%
% Calling sequence:
%     distance = dist2(x1, y1, x2, y2)

% Define variables:
%     x1         -- x-position of point 1
%     y1         -- y-position of point 1
%     x2         -- x-position of point 2
%     y2         -- y-position of point 2
%     distance -- Distance between points

% Record of revisions:
%     Date          Programmer          Description of change
%     ====          ==========          =====================
%     02/01/14      S. J. Chapman       Original code
% Calculate distance.
distance = sqrt((x2-x1).^2 + (y2-y1).^2);
```

This function has four input arguments and one output argument. A simple script file using this function is shown below.

```
% Script file: test_dist2.m
%
% Purpose:
```

```
%       This program tests function dist2.
%
%  Record of revisions:
%      Date            Programmer          Description of change
%      ====            ==========          =====================
%      02/01/14        S. J. Chapman       Original code
%
%  Define variables:
%      ax              -- x-position of point a
%      ay              -- y-position of point a
%      bx              -- x-position of point b
%      by              -- y-position of point b
%      result          -- Distance between the points

%  Get input data.
disp('Calculate the distance between two points:');
ax = input('Enter x value of point a:');
ay = input('Enter y value of point a:');
bx = input('Enter x value of point b:');
by = input('Enter y value of point b:');

%  Evaluate function
result = dist2 (ax, ay, bx, by);

%  Write out result.
fprintf('The distance between points a and b is %f\n',result);
```

When this script file is executed, the results are:

```
» test_dist2
Calculate the distance between two points:
Enter x value of point a:  1
Enter y value of point a:  1
Enter x value of point b:  4
Enter y value of point b:  5
The distance between points a and b is 5.000000
```

These results are correct, as we can verify from simple hand calculations.

Function dist2 also supports the MATLAB help subsystem. If we type "help dist2", the results are:

```
» help dist2
DIST2 Calculate the distance between two points
    Function DIST2 calculates the distance between
    two points (x1,y1) and (x2,y2) in a Cartesian
    coordinate system.

    Calling sequence:
      res = dist2(x1, y1, x2, y2)
```

Similarly, "lookfor distance" produces the result

```
» lookfor distance
DIST2 Calculate the distance between two points
MAHAL Mahalanobis distance.
DIST Distances between vectors.
NBDIST Neighborhood matrix using vector distance.
NBGRID Neighborhood matrix using grid distance.
NBMAN Neighborhood matrix using Manhattan-distance.
```

To observe the behavior of the MATLAB workspace before, during, and after the function is executed, we will load function dist2 and the script file test_dist2 into the MATLAB debugger, and set breakpoints before, during, and after the function call (see Figure 6.1). When the program stops at the breakpoint *before* the function call, the workspace is as shown in Figure 6.2 (*a*). Note that

Figure 6.1 M-file test_dist2 and function dist2 are loaded into the debugger, with breakpoints set before, during, and after the function call.

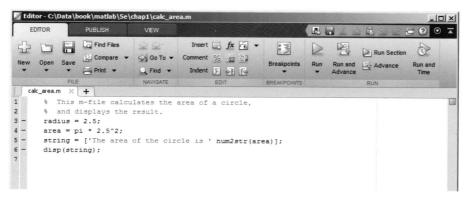

Figure 1.5 (d) The MATLAB Editor, displayed as an independent window.

Comments in an M-file file appear in green, variables and numbers appear in black, complete character strings appear in magenta, incomplete character strings appear in red, and language keywords appear in blue.

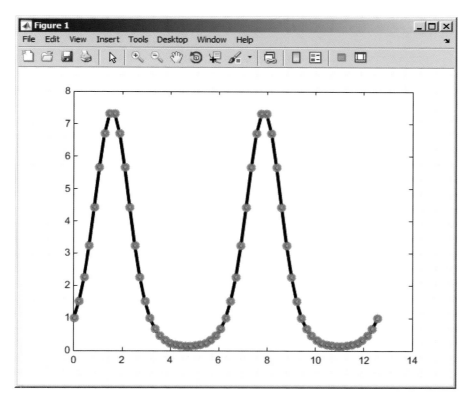

Figure 3.8 A plot illustrating the use of the LineWidth and Marker properties.

```
plot(x,y,'-ko','LineWidth',3.0,'MarkerSize',6,...'MarkerEdgeColor',
'r','MarkerFaceColor','g')
```

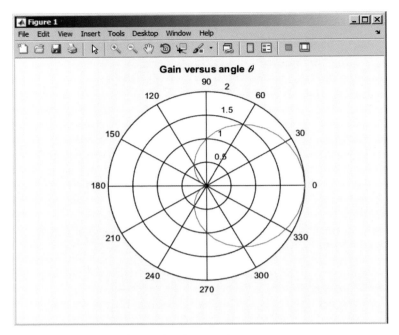

Figure 3.10 Gain of a cardioid microphone.

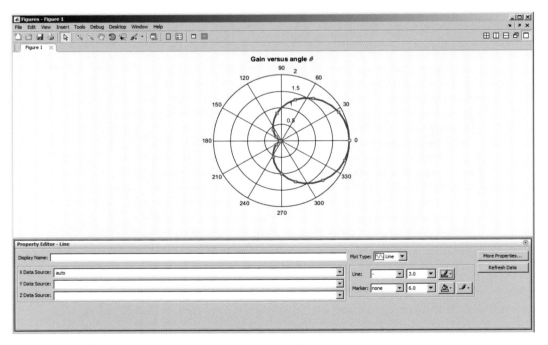

Figure 3.12 Figure 3.10 after the line has been modified using the editing tools built into the figure toolbar.

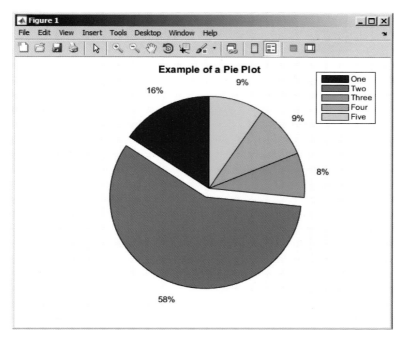

Figure 3.15 (e) Pie plot.

```
13      % Define variables:
14      %   a            -- Coefficient of x^2 term of equation
15      %   b            -- Coefficient of x term of equation
16      %   c            -- Constant term of equation
17      %   discriminant -- Discriminant of the equation
18      %   imag_part    -- Imag part of equation (for complex roots)
19      %   real_part    -- Real part of equation (for complex roots)
20      %   x1           -- First solution of equation (for real roots)
21      %   x2           -- Second solution of equation (for real roots)
22
23      % Prompt the user for the coefficients of the equation
24 -    disp ('This program solves for the roots of a quadratic ');
25 -    disp ('equation of the form A*X^2 + B*X + C = 0. ');
26 -    a = input ('Enter the coefficient A: ');
27 -    b = input ('Enter the coefficient B: ');
28 -    c = input ('Enter the coefficient C: ');
29
30      % Calculate discriminant
31 -    discriminant = b^2 - 4 * a * c;
32
33      % Solve for the roots, depending on the value of the discriminant
34 -    if discriminant > 0 % there are two real roots, so...
35
36 -       x1 = ( -b + sqrt(discriminant) ) / ( 2 * a );
37 -       x2 = ( -b - sqrt(discriminant) ) / ( 2 * a );
38 -       disp ('This equation has two real roots:');
39 -       fprintf ('x1 = %f\n', x1);
40 -       fprintf ('x2 = %f\n', x2);
```

script   Ln 11   Col 14

Figure 4.5 An Edit/Debug window with a MATLAB program loaded.

```
23      % Prompt the user for the coefficients of the equation
24  -   disp ('This program solves for the roots of a quadratic ');
25  -   disp ('equation of the form A*X^2 + B*X + C = 0. ');
26  -   a = input ('Enter the coefficient A: ');
27  -   b = input ('Enter the coefficient B: ');
28  -   c = input ('Enter the coefficient C: ');
29
30      % Calculate discriminant
31 O    discriminant = b^2 - 4 * a * c;
32
33      % Solve for the roots, depending on the value of the discriminant
34  -   if discriminant > 0 % there are two real roots, so...
35
36  -       x1 = ( -b + sqrt(discriminant) ) / ( 2 * a );
37  -       x2 = ( -b - sqrt(discriminant) ) / ( 2 * a );
38  -       disp ('This equation has two real roots:');
39  -       fprintf ('x1 = %f\n', x1);
40  -       fprintf ('x2 = %f\n', x2);
```

6 usages of "discriminant" found   script   Ln 31   Col 5

Figure 4.6 The window after a breakpoint has been set. Note the red dot to the left of the line with the breakpoint.

```
22
23      % Prompt the user for the coefficients of the equation
24  -   disp ('This program solves for the roots of a quadratic ');
25  -   disp ('equation of the form A*X^2 + B*X + C = 0. ');
26  -   a = input ('Enter the coefficient A: ');
27  -   b = input ('Enter the coefficient B: ');
28  -   c = input ('Enter the coefficient C: ');
29
30      % Calculate discriminant
31 O⇨   discriminant = b^2 - 4 * a * c;
32
33      % Solve for the roots, depending on the value of the discriminant
34  -   if discriminant > 0 % there are two real roots, so...
35
36  -       x1 = ( -b + sqrt(discriminant) ) / ( 2 * a );
37  -       x2 = ( -b - sqrt(discriminant) ) / ( 2 * a );
38  -       disp ('This equation has two real roots:');
39  -       fprintf ('x1 = %f\n', x1);
40  -       fprintf ('x2 = %f\n', x2);
```

6 usages of "discriminant" found   script   Ln 31   Col 1

Figure 4.7 A green arrow will appear by the current line during the debugging process.

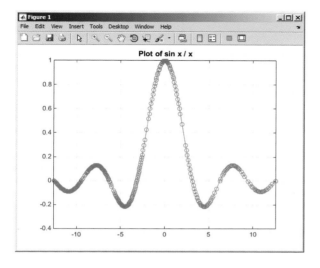

5

Figure 7.8 The function $f(x) = \sin x/x$, plotted with function `fplot`.

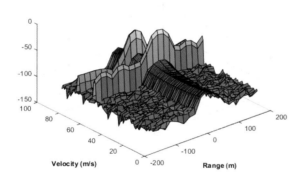

Figure 7.10 A radar range-velocity space containing two targets and background noise.

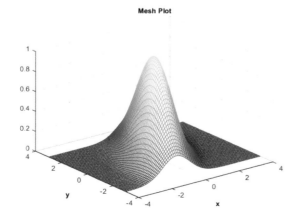

Figure 8.9 (a) A mesh plot of the function $z(x,y) = e^{-0.5[x^2 + 0.5(x-y)^2]}$.

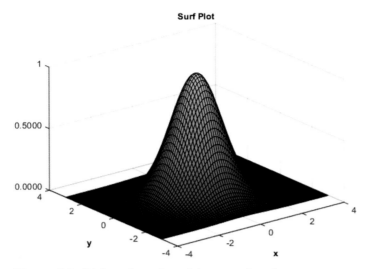

Figure 8.9 (b) A surface plot of the same function.

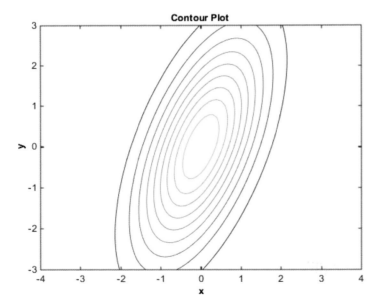

Figure 8.9 (c) A contour plot of the same function.

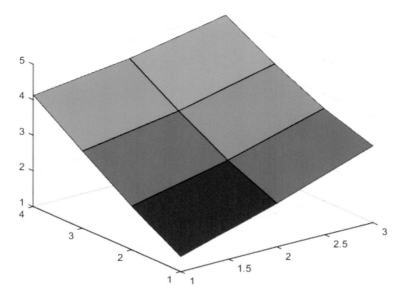

Figure 8.10 A surface plot of the function $z(x,y) = \sqrt{x^2 + y^2}$ for $x = 0, 1$, and 2, and for $y = 0, 1, 2$, and 3.

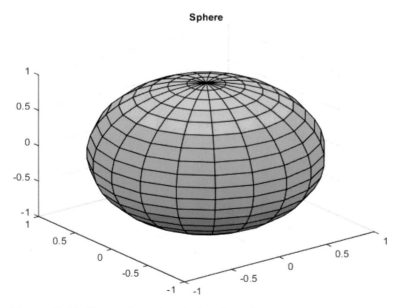

Figure 8.11 Three-dimensional plot of a sphere.

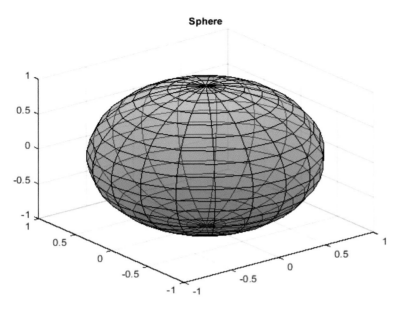

Figure 8.12 A partially transparent sphere, created with an alpha value of 0.5.

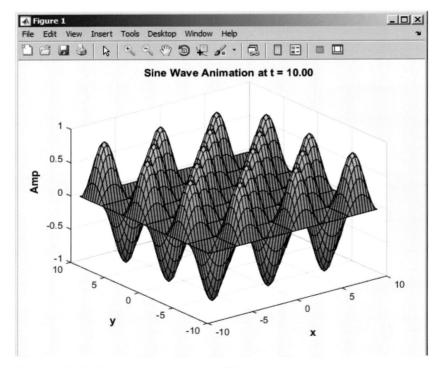

Figure 13.8 One snapshot from the 3D sine wave animation.

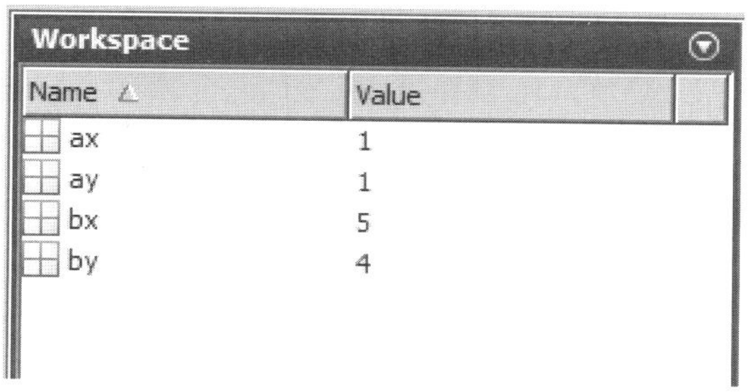

(a)

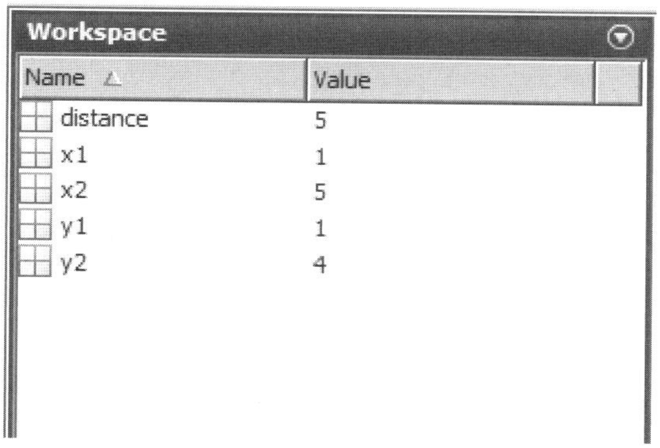

(b)

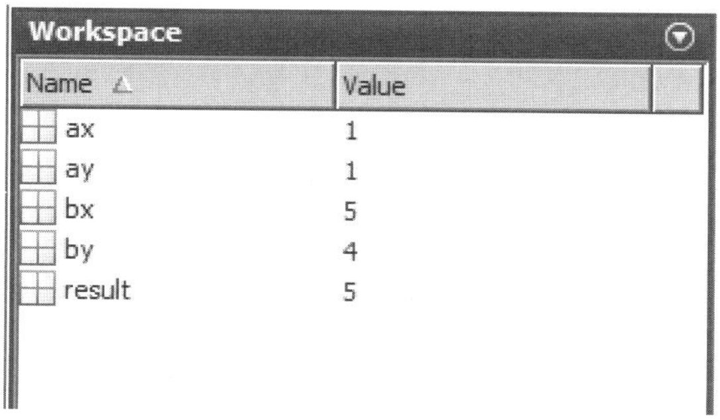

(c)

Figure 6.2 (a) The workspace before the function call.
(b) The workspace during the function call.
(c) The workspace after the function call.

variables `ax`, `ay`, `bx`, and `by` are defined in the workspace, with the values that we have entered. When the program stops at the breakpoint *within* the function call, the function's workspace is active. It is as shown in Figure 6.2 (*b*). Note that variables `x1`, `x2`, `y1`, `y2`, and `distance` are defined in the function's workspace, and the variables defined in the calling M-file are not present. When the program stops in the calling program at the breakpoint *after* the function call, the workspace is as shown in Figure 6.2 (*c*). Now the original variables are back, with the variable `result` added to contain the value returned by the function. These figures show that the workspace of the function is different than the workspace of the calling M-file.

6.2 Variable Passing in MATLAB: The Pass-By-Value Scheme

MATLAB programs communicate with their functions using a **pass-by-value** scheme. When a function call occurs, MATLAB makes a *copy* of the actual arguments and passes them to the function. This copying is very significant, because it means that even if the function modifies the input arguments, it won't affect the original data in the caller. This feature helps to prevent unintended side effects, in which an error in the function might unintentionally modify variables in the calling program.

This behavior is illustrated in the function shown below. This function has two input arguments: `a` and `b`. During its calculations, it modifies both input arguments.

```
function out = sample(a, b)
fprintf('In      sample: a = %f, b = %f %f\n',a,b);
a = b(1) + 2*a;
b = a .* b;
out = a + b(1);
fprintf('In      sample: a = %f, b = %f %f\n',a,b);
```

A simple test program to call this function is shown below.

```
a = 2; b = [6 4];
fprintf('Before sample: a = %f, b = %f %f\n',a,b);
out = sample(a,b);
fprintf('After  sample: a = %f, b = %f %f\n',a,b);
fprintf('After  sample: out = %f\n',out);
```

When this program is executed, the results are:

```
» test_sample
Before sample: a = 2.000000, b = 6.000000 4.000000
In      sample: a = 2.000000, b = 6.000000 4.000000
In      sample: a = 10.000000, b = 60.000000 40.000000
After  sample: a = 2.000000, b = 6.000000 4.000000
After  sample: out = 70.000000
```

Note that `a` and `b` were both changed inside function `sample`, but those changes had *no effect on the values in the calling program*.

Users of the C language will be familiar with the pass-by-value scheme, since C uses it for scalar values passed to functions. However C does *not* use the pass-by-value scheme when passing arrays, so an unintended modification to a dummy array in a C function can cause side effects in the calling program. MATLAB improves on this by using the pass-by-value scheme for both scalars and arrays[2].

▶ Example 6.1—Rectangular-to-Polar Conversion

The location of a point in a Cartesian plane can be expressed in either the rectangular coordinates (x, y) or the polar coordinates (r, θ), as shown in Figure 6.3. The relationships among these two sets of coordinates are given by the following equations:

$$x = r \cos \theta \tag{6.1}$$

$$x = r \sin \theta \tag{6.2}$$

$$r = \sqrt{x^2 + y^2} \tag{6.3}$$

$$\theta = \tan^{-1} \frac{y}{x} \tag{6.4}$$

Write two functions `rect2polar` and `polar2rect` that convert coordinates from rectangular to polar form, and vice versa, where the angle θ is expressed in degrees.

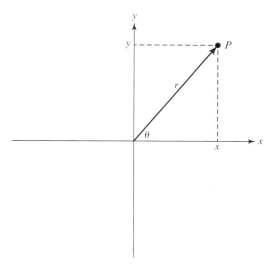

Figure 6.3 A point *P* in a Cartesian plane can be located by either the rectangular coordinates (x, y) or the polar coordinates (r, θ).

[2]The implementation of argument passing in MATLAB is actually more sophisticated than this discussion indicates. As pointed out above, the copying associated with pass-by-value takes up a lot of time, but it provides protection against unintended side effects. MATLAB actually uses the best of both approaches: it analyzes each argument of each function and determines whether or not the function modifies that argument. If the function modifies the argument, then MATLAB makes a copy of it. If it does not modify the argument, then MATLAB simply points to the existing value in the calling program. This practice increases speed while still providing protection against side effects!

Solution We will apply our standard problem-solving approach to creating these functions. Note that MATLAB's trigonometric functions work in radians, so we must convert from degrees to radians and vice versa when solving this problem. The basic relationship between degrees and radians is

$$180° = \pi \text{ radians} \qquad (6.5)$$

1. **State the problem**

   A succinct statement of the problem is:

   > Write a function that converts a location on a Cartesian plane expressed in rectangular coordinates into the corresponding polar coordinates, with the angle θ expressed in degrees. Also, write a function that converts a location on a Cartesian plane expressed in polar coordinates with the angle θ expressed in degrees into the corresponding rectangular coordinates.

2. **Define the inputs and outputs**

   The inputs to function `rect2polar` are the rectangular (x, y) location of a point. The outputs of the function are the polar (r, θ) location of the point. The inputs to function `polar2rect` are the polar (r, θ) location of a point. The outputs of the function are the rectangular (x, y) location of the point.

3. **Describe the algorithm**

   These functions are very simple, so we can directly write the final pseudocode for them. The pseudocode for function `polar2rect` is:

   ```
   x ← r * cos(theta * pi/180)
   y ← r * sin(theta * pi/180)
   ```

   The pseudocode for function `rect2polar` will use the function `atan2`, because that function works over all four quadrants of the Cartesian plane. (Look that function up in the MATLAB Help Browser!)

   ```
   r ← sqrt(x.^2 + y.^2)
   theta ← 18/pi * atan2(y,x)
   ```

4. **Turn the algorithm into MATLAB statements**

   The MATLAB code for the selection `polar2rect` function is shown below.

```
function [x, y] = polar2rect(r,theta)
%POLAR2RECT Convert rectangular to polar coordinates
% Function POLAR2RECT accepts the polar coordinates
% (r,theta), where theta is expressed in degrees,
% and converts them into the rectangular coordinates
% (x,y).
%
% Calling sequence:
%   [x, y] = polar2rect(r,theta)

% Define variables:
%   r          -- Length of polar vector
```

```
%   theta       -- Angle of vector in degrees
%   x           -- x-position of point
%   y           -- y-position of point

%   Record of revisions:
%       Date        Programmer          Description of change
%       ====        ==========          =====================
%       02/01/14    S. J. Chapman       Original code

x = r * cos(theta * pi/180);
y = r * sin(theta * pi/180);
```

The MATLAB code for the selection rect2polar function is shown below.

```
function [r, theta] = rect2polar(x,y)
%RECT2POLAR Convert rectangular to polar coordinates
% Function RECT2POLAR accepts the rectangular coordinates
% (x,y) and converts them into the polar coordinates
% (r,theta), where theta is expressed in degrees.
%
% Calling sequence:
%   [r, theta] = rect2polar(x,y)

% Define variables:
%   r           -- Length of polar vector
%   theta       -- Angle of vector in degrees
%   x           -- x-position of point
%   y           -- y-position of point

%   Record of revisions:
%       Date        Programmer          Description of change
%       ====        ==========          =====================
%       02/01/14    S. J. Chapman       Original code

r = sqrt (x.^2 + y .^2);
theta = 180/pi * atan2(y,x);
```

Note that these functions both include help information, so they will work properly with MATLAB's help subsystem and with the lookfor command.

5. **Test the program.**

To test these functions, we will execute them directly in the MATLAB Command Window. We will test the functions using the 3-4-5 triangle, which is familiar to most people from secondary school. The smaller angle within a 3-4-5 triangle is approximately 36.87°. We will also test the function in all four quadrants of the Cartesian plane to ensure that the conversions are correct everywhere.

```
» [r, theta] = rect2polar(4,3)
r =
      5
theta =
   36.8699
» [r, theta] = rect2polar(-4,3)
r =
      5
theta =
   143.1301
» [r, theta] = rect2polar(-4,-3)
r =
      5
theta =
  -143.1301
» [r, theta] = rect2polar(4,-3)
r =
      5
theta =
   -36.8699
» [x, y] = polar2rect(5,36.8699)
x =
    4.0000
y =
    3.0000
» [x, y] = polar2rect(5,143.1301)
x =
   -4.0000
y =
    3.0000
» [x, y] = polar2rect(5,-143.1301)
x =
   -4.0000
y =
   -3.0000
» [x, y] = polar2rect(5,-36.8699)
x =
    4.0000
y =
   -3.0000
»
```

These functions appear to be working correctly in all quadrants of the
Cartesian plane.

▶Example 6.2—Sorting Data

In many scientific and engineering applications, it is necessary to take a random input data set and to sort it so that the numbers in the data set are either all in *ascending order* (lowest-to-highest) or all in *descending order* (highest-to-lowest). For example, suppose that you were a zoologist studying a large population of animals, and that you wanted to identify the largest 5% of the animals in the population. The most straightforward way to approach this problem would be to sort the sizes of all of the animals in the population into ascending order and take the top 5% of the values.

Sorting data into ascending or descending order seems to be an easy job. After all, we do it all the time. It is a simple matter for us to sort the data (10, 3, 6, 4, 9) into the order (3, 4, 6, 9, 10). How do we do it? We first scan the input data list (10, 3, 6, 4, 9) to find the smallest value in the list (3) and then scan the remaining input data (10, 6, 4, 9) to find the next smallest value (4), and so forth, until the complete list is sorted.

In fact, sorting can be a very difficult job. As the number of values to be sorted increases, the time required to perform the simple sort described above increases rapidly, since we must scan the input data set once for each value sorted. For very large data sets, this technique just takes too long to be practical. Even worse, how would we sort the data if there were too many numbers to fit into the main memory of the computer? The development of efficient sorting techniques for large data sets is an active area of research and is the subject of whole courses all by itself.

In this example, we will confine ourselves to the simplest possible algorithm to illustrate the concept of sorting. This simplest algorithm is called the **selection sort**. It is just a computer implementation of the mental math described above. The basic algorithm for the selection sort is:

1. Scan the list of numbers to be sorted to locate the smallest value in the list. Place that value at the front of the list by swapping it with the value currently at the front of the list. If the value at the front of the list is already the smallest value, then do nothing.
2. Scan the list of numbers from position 2 to the end to locate the next smallest value in the list. Place that value in position 2 of the list by swapping it with the value currently at that position. If the value in position 2 is already the next smallest value, then do nothing.
3. Scan the list of numbers from position 3 to the end to locate the third smallest value in the list. Place that value in position 3 of the list by swapping it with the value currently at that position. If the value in position 3 is already the third smallest value, then do nothing.
4. Repeat this process until the next-to-last position in the list is reached. After the next-to-last position in the list has been processed, the sort is complete.

Note that if we are sorting N values, this sorting algorithm requires N − 1 scans through the data to accomplish the sort.

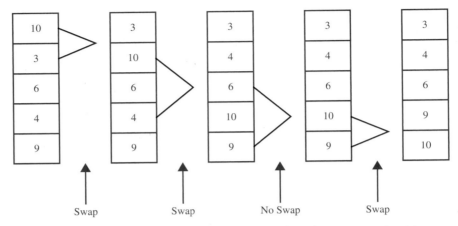

Figure 6.4 An example problem demonstrating the selection sort algorithm.

This process is illustrated in Figure 6.4. Since there are five values in the data set to be sorted, we will make four scans through the data. During the first pass through the entire data set, the minimum value is 3, so the 3 is swapped with the 10 which was in position 1. Pass 2 searches for the minimum value in positions 2 through 5. That minimum is 4, so the 4 is swapped with the 10 in position 2. Pass 3 searches for the minimum value in positions 3 through 5. That minimum is 6, which is already in position 3, so no swapping is required. Finally, pass 4 searches for the minimum value in positions 4 through 5. That minimum is 9, so the 9 is swapped with the 10 in position 4, and the sort is completed.

☒ Programming Pitfalls

The selection-sort algorithm is the easiest sorting algorithm to understand, but it is computationally inefficient. *It should never be applied to sort large data sets* (say, sets with more than 1000 elements). Over the years, computer scientists have developed much more efficient sorting algorithms. The `sort` and `sortrows` functions built into MATLAB are extremely efficient and should be used for all real work.

We will now develop a program to read in a data set from the Command Window, sort it into ascending order, and display the sorted data set. The sorting will be done by a separate user-defined function.

Solution This program must be able to ask the user for the input data, sort the data, and write out the sorted data. The design process for this problem is given below.

1. **State the problem**

   We have not yet specified the type of data to be sorted. If the data is numeric, then the problem may be stated as follows:

   > Develop a program to read an arbitrary number of numeric input values from the Command Window, sort the data into ascending order using a separate sorting function, and write the sorted data to the Command Window.

2. **Define the inputs and outputs**

   The inputs to this program are the numeric values typed in the Command Window by the user. The outputs from this program are the sorted data values written to the Command Window.

3. **Describe the algorithm**

   This program can be broken down into three major steps

   ```
   Read the input data into an array
   Sort the data in ascending order
   Write the sorted data
   ```

   The first major step is to read in the data. We must prompt the user for the number of input data values and then read in the data. Since we will know how many input values there are to read, a `for` loop is appropriate for reading in the data. The detailed pseudocode is shown below:

   ```
   Prompt user for the number of data values
   Read the number of data values (nvals)
   Preallocate an input array
   for ii = 1:nvals
      Prompt for next value
      Read value
   end
   ```

   Next, we have to sort the data in a separate function. We will need to make **nvals**-1 passes through the data, finding the smallest remaining value each time. We will use a pointer to locate the smallest value in each pass. Once the smallest value is found, it will be swapped to the top of the list if it is not already there. The detailed pseudocode is shown below:

```
for ii = 1:nvals-1

   % Find the minimum value in a(ii) through a(nvals)
   iptr ← ii
   for jj = ii+1 to nvals
      if a(jj) < a(iptr)
         iptr ← jj
      end
   end
```

```
      % iptr now points to the min value, so swap a(iptr)
      % with a(ii) if iptr ~= ii.
      if ii ~= iptr
         temp ← a(ii)
         a(ii) ← a(iptr)
         a(iptr) ← temp
      end
   end
```

The final step is writing out the sorted values. No refinement of the pseudo-code is required for that step. The final pseudocode is the combination of the reading, sorting and writing steps.

4. **Turn the algorithm into MATLAB statements.**

The MATLAB code for the selection sort function is shown below.

```
function out = ssort(a)
%SSORT Selection sort data in ascending order
% Function SSORT sorts a numeric data set into
% ascending order. Note that the selection sort
% is relatively inefficient. DO NOT USE THIS
% FUNCTION FOR LARGE DATA SETS. Use MATLAB's
% "sort" function instead.

% Define variables:
%   a          -- Input array to sort
%   ii         -- Index variable
%   iptr       -- Pointer to min value
%   jj         -- Index variable
%   nvals      -- Number of values in "a"
%   out        -- Sorted output array
%   temp       -- Temp variable for swapping

%   Record of revisions:
%      Date          Programmer        Description of change
%      ====          ==========        =====================
%      02/02/14      S. J. Chapman     Original code

% Get the length of the array to sort
nvals = length(a);

% Sort the input array
for ii = 1:nvals-1
   % Find the minimum value in a(ii) through a(n)
   iptr = ii;
   for jj = ii+1:nvals
      if a(jj) < a(iptr)
```

```
            iptr = jj;
         end
      end

      % iptr now points to the minimum value, so swap a(iptr)
      % with a(ii) if ii ~= iptr.
      if ii ~= iptr
         temp    = a(ii);
         a(ii)   = a(iptr);
         a(iptr) = temp;
      end
   end

% Pass data back to caller
out = a;
```

The program to invoke the selection sort function is shown below.

```
% Script file: test_ssort.m
%
% Purpose:
%   To read in an input data set, sort it into ascending
%   order using the selection sort algorithm, and to
%   write the sorted data to the Command Window. This
%   program calls function "ssort" to do the actual
%   sorting.
%
% Record of revisions:
%     Date          Programmer          Description of change
%     ====          ==========          =====================
%   02/02/14      S. J. Chapman       Original code
%
% Define variables:
%   array  -- Input data array
%   ii     -- Index variable
%   nvals  -- Number of input values
%   sorted -- Sorted data array

% Prompt for the number of values in the data set
nvals = input('Enter number of values to sort: ');

% Preallocate array
array = zeros(1,nvals);

% Get input values
for ii = 1:nvals
```

```
    % Prompt for next value
    string = ['Enter value ' int2str(ii) ':   '];
    array(ii) = input(string);

end

% Now sort the data
sorted = ssort(array);

% Display the sorted result.
fprintf('\nSorted data:\n');
for ii = 1:nvals
    fprintf('  %8.4f\n',sorted(ii));
end
```

5. **Test the program.**

To test this program, we will create an input data set and run the program with it. The data set should contain a mixture of positive and negative numbers as well as at least one duplicated value to see if the program works properly under those conditions.

```
» test_ssort
Enter number of values to sort:  6
Enter value 1:  -5
Enter value 2:  4
Enter value 3:  -2
Enter value 4:  3
Enter value 5:  -2
Enter value 6:  0

Sorted data:
  -5.0000
  -2.0000
  -2.0000
   0.0000
   3.0000
   4.0000
```

The program gives the correct answers for our test data set. Note that it works for both positive and negative numbers as well as for repeated numbers.

◄

6.3   Optional Arguments

Many MATLAB functions support optional input arguments and output arguments. For example, we have seen calls to the plot function with as few as two or as many as seven input arguments. On the other hand, the function max supports either one or two output arguments. If there is only one output argument, max returns the maximum value of an array. If there are two output arguments, `max` returns both the maximum value and the location of the maximum value in an array. How do MATLAB functions know how many input and output arguments are present, and how do they adjust their behavior accordingly?

There are eight special functions that can be used by MATLAB functions to get information about their optional arguments and to report errors in those arguments. Six of these functions are introduced here, and the remaining two will be introduced in Chapter 10 after we learn about the cell array data type. The functions introduced now are:

- `nargin`—This function returns the number of actual input arguments that were used to call the function.
- `nargout`—This function returns the number of actual output arguments that were used to call the function.
- `nargchk`—This function returns a standard error message if a function is called with too few or too many arguments.
- `error`—Display error message and abort the function producing the error. This function is used if the argument errors are fatal.
- `warning`—Display warning message and continue function execution. This function is used if the argument errors are not fatal, and execution can continue.
- `inputname`—This function returns the actual name of the variable that corresponds to a particular argument number.

When functions `nargin` and `nargout` are called within a user-defined function, these functions return the number of actual input arguments and the number of actual output arguments that were used when the user-defined function was called.

Function `nargchk` generates a string containing a standard error message if a function is called with too few or too many arguments. The syntax of this function is

```
message = nargchk(min_args,max_args,num_args);
```

where `min_args` is the minimum number of arguments, `max_args` is the maximum number of arguments, and `num_args` is the actual number of arguments. If the number of arguments is outside the acceptable limits, a standard error message is produced. If the number of arguments is within acceptable limits, then an empty string is returned.

Function `error` is a standard way to display an error message and abort the user-defined function causing the error. The syntax of this function is `error ('msg')`, where `msg` is a character string containing an error message. When `error` is executed, it halts the current function and returns control to the keyboard, displaying the error message in the Command Window. If the message string is empty, `error` does nothing, and execution continues. This function works well with `nargchk`, which produces a message string when an error occurs and an empty string when there is no error.

Function `warning` is a standard way to display a warning message that includes the function and line number where the problem occurred but let execution continue.

The syntax of this function is `warning('msg')`, where `msg` is a character string containing a warning message. When `warning` is executed, it displays the warning message in the Command Window, and lists the function name and line number where the warning came from. If the message string is empty, `warning` does nothing. In either case, execution of the function continues.

Function `inputname` returns the name of the actual argument used when a function is called. The syntax of this function is

```
name = inputname(argno);
```

where `argno` is the number of the argument. If argument is a variable, then its name is returned. If the argument is an expression, then this function will return an empty string. For example, consider the function

```
function myfun(x,y,z)
name = inputname(2);
disp(['The second argument is named' name]);
```

When this function is called, the results are

```
» myfun(dog,cat)
The second argument is named cat
» myfun(1,2+cat)
The second argument is named
```

Function `inputname` is useful for displaying argument names in warning and error messages.

►Example 6.3—Using Optional Arguments

We will illustrate the use of optional arguments by creating a function that accepts an (*x*, *y*) value in rectangular coordinates and produces the equivalent polar representation, consisting of a magnitude and an angle in degrees. The function will be designed to support two input arguments, *x* and *y*. However, if only one argument is supplied, the function will assume that the *y* value is zero and proceed with the calculation. The function will normally return both the magnitude and the angle in degrees, but if only one output argument is present, it will return only the magnitude. This function is shown below:

```
function [mag, angle] = polar_value(x,y)
%POLAR_VALUE Converts (x,y) to (r,theta)
% Function POLAR_VALUE converts an input (x,y)
% value into (r,theta), with theta in degrees.
% It illustrates the use of optional arguments.

% Define variables:
%   angle    -- Angle in degrees
%   msg      -- Error message
```

```
%    mag        -- Magnitude
%    x          -- Input x value
%    y          -- Input y value (optional)

% Record of revisions:
%    Date        Programmer         Description of change
%    ====        ==========         =====================
%    02/03/14    S. J. Chapman      Original code

% Check for a legal number of input arguments.
msg = nargchk(1,2,nargin);
error(msg);

% If the y argument is missing, set it to 0.
if nargin < 2
   y = 0;
end

% Check for (0,0) input arguments, and print out
% a warning message.
if x == 0 & y == 0
   msg = 'Both x any y are zero: angle is meaningless!';
   warning(msg);
end

% Now calculate the magnitude.
mag = sqrt(x.^2 + y.^2);

% If the second output argument is present, calculate
% angle in degrees.
if nargout == 2
   angle = atan2(y,x) * 180/pi;
end
```

We will test this function by calling it repeatedly from the Command Window. First, we will try to call the function with too few or too many arguments.

```
» [mag angle] = polar_value
??? Error using ==> polar_value
Not enough input arguments.

» [mag angle] = polar_value(1,-1,1)
??? Error using ==> polar_value
Too many input arguments.
```

The function provides proper error messages in both cases. Next, we will try to call the function with one or two input arguments.

```
» [mag angle] = polar_value(1)
mag =
       1
angle =
       0
» [mag angle] = polar_value(1,-1)
mag =
      1.4142
angle =
     -45
```

The function provides the correct answer in both cases. Next, we will try to call the function with one or two output arguments.

```
» mag = polar_value(1,-1)
mag =
      1.4142
» [mag angle] = polar_value(1,-1)
mag =
      1.4142
angle =
       -45
```

The function provides the correct answer in both cases. Finally, we will try to call the function with both *x* and *y* equal to zero.

```
» [mag angle] = polar_value(0,0)

Warning: Both x any y are zero: angle is meaningless!
> In d:\book\matlab\chap6\polar_value.m at line 32
mag =
       0
angle =
       0
```

In this case, the function displays the warning message, but execution continues. ◄

Note that a MATLAB function may be declared to have more output arguments than are actually used, and this is *not* an error. The function does not actually have to check `nargout` to determine if an output argument is present. For example, consider the following function:

```
function [z1, z2] = junk(x,y)
z1 = x + y;
z2 = x - y;
end % function junk
```

This function can be called successfully with one or two output arguments.

```
» a = junk(2,1)
a =
    3
» [a b] = junk(2,1)
a =
    3
b =
    1
```

The reason for checking `nargout` in a function is to prevent useless work. If a result is going to be thrown away anyway, why bother to calculate it in the first place? An engineer can speed up the operation of a program by not bothering with useless calculations.

Quiz 6.1

This quiz provides a quick check to see if you have understood the concepts introduced in Sections 6.1 through 6.3. If you have trouble with the quiz, reread the section, ask your instructor, or discuss the material with a fellow student. The answers to this quiz are found in the back of the book.

1. What are the differences between a script file and a function?
2. How does the `help` command work with user-defined functions?
3. What is the significance of the H1 comment line in a function?
4. What is the pass-by-value scheme? How does it contribute to good program design?
5. How can a MATLAB function be designed to have optional arguments?

For questions 6 and 7, determine whether the function calls are correct or not. If they are in error, specify what is wrong with them.

```
6. out = test1(6);

   function res = test1(x,y)
   res = sqrt(x.^2 + y.^2);


7. out = test2(12);

   function res = test2(x,y)
   error(nargchk(1,2,nargin));
   if nargin == 2
      res = sqrt(x.^2 + y.^2);
   else
      res = x;
   end
```

6.4 Sharing Data Using Global Memory

We have seen that programs exchange data with the functions they call through their argument lists. When a function is called, each actual argument is copied, and the copy is used by the function.

In addition to the argument list, MATLAB functions can exchange data with each other and with the base workspace through global memory. **Global memory** is a special type of memory that can be accessed from any workspace. If a variable is declared to be global in a function, then it will be placed in the global memory instead of the local workspace. If the same variable is declared to be global in another function, then that variable will refer to the *same memory location* as the variable in the first function. Each script file or function that declares the global variable will have access to the same data values, so *global memory provides a way to share data between functions*.

A global variable is declared with the **global statement**. The form of a `global` statement is

```
global var1 var2 var3 ...
```

where `var1`, `var2`, `var3`, and so forth are the variables to be placed in global memory. By convention, global variables are declared in all capital letters, but this is not actually a requirement.

Good Programming Practice

Declare global variables in all capital letters to make them easy to distinguish from local variables.

///

Each global variable must be declared to be global before it is used for the first time in a function—it is an error to declare a variable to be global after it has already been created in the local workspace[3]. To avoid this error, it is customary to declare global variables immediately after the initial comments and before the first executable statement in a function.

Good Programming Practice

Declare global variables immediately after the initial comments and before the first executable statement in each function that uses them.

///

---

[3]If a variable is declared `global` after it has already been defined in a function, MATLAB will issue a warning message and then change the local value to match the global value. You should never rely on this capability, though, because future versions of MATLAB will not allow it.

Global variables are especially useful for sharing very large volumes of data among many functions, because the entire data set does not have to be copied each time that a function is called. The downside of using global memory to exchange data among functions is that the functions will only work for that specific data set. A function that exchanges data through input arguments can be reused by simply calling it with different arguments, but a function that exchanges data through global memory must actually be modified to allow it to work with a different data set.

Global variables are also useful for sharing hidden data among a group of related functions while keeping it invisible from the invoking program unit.

Good Programming Practice

You may use global memory to pass large amounts of data among functions within a program.

▶Example 6.4—Random Number Generator

It is impossible to make perfect measurements in the real world. There will always be some *measurement noise* associated with each measurement. This fact is an important consideration in the design of systems to control the operation of such real-world devices as airplanes, refineries, and nuclear reactors. A good engineering design must take these measurement errors into account, so that the noise in the measurements will not lead to unstable behavior (no plane crashes, refinery explosions, or meltdowns!).

Most engineering designs are tested by running *simulations* of the operation of the system before it is ever built. These simulations involve creating mathematical models of the behavior of the system and feeding the models a realistic string of input data. If the models respond correctly to the simulated input data, then we can have reasonable confidence that the real-world system will respond correctly to the real-world input data.

The simulated input data supplied to the models must be corrupted by a simulated measurement noise, which is just a string of random numbers added to the ideal input data. The simulated noise is usually produced by a *random number generator*.

A random number generator is a function that will return a different and apparently random number each time it is called. Since the numbers are in fact generated by a deterministic algorithm, they only appear to be random[4]. However, if the algorithm used to generate them is complex enough, the numbers will be random enough to use in the simulation.

[4]For this reason, some people refer to these functions as *pseudorandom number generators*.

One simple random number generator algorithm is described below[5]. It relies on the unpredictability of the modulo function when applied to large numbers. Recall from Chapter 2 that the modulus function mod returns the remainder after the division of two numbers. Consider the following equation:

$$n_{i+1} = \text{mod}(8121n_i + 28411, 134456) \tag{6.6}$$

Assume that n_i is a non-negative integer. Then because of the modulo function, n_{i+1} will be a number between 0 and 134455 inclusive. Next, n_{i+1} can be fed into the equation to produce a number n_{i+2} that is also between 0 and 134455. This process can be repeated forever to produce a series of numbers in the range [0,134455]. If we didn't know the numbers 8121, 28411, and 134456 in advance, it would be impossible to guess the order in which the values of n would be produced. Furthermore, it turns out that there is an equal (or uniform) probability that any given number will appear in the sequence. Because of these properties, Equation (6.6) can serve as the basis for a simple random number generator with a uniform distribution.

We will now use Equation (6.6) to design a random number generator whose output is a real number in the range [0.0, 1.0)[6].

Solution We will write a function that generates one random number in the range $0 \le ran < 1.0$ each time that it is called. The random number will be based on the equation

$$ran_i = \frac{n_i}{134456} \tag{6.7}$$

where n_i is a number in the range 0 to 134455 produced by Equation (6.7).

The particular sequence produced by Equations (6.6) and (6.7) will depend on the initial value of n_0 (called the *seed*) of the sequence. We must provide a way for the user to specify n_0 so that the sequence may be varied from run to run.

1. **State the problem**

   Write a function random0 that will generate and return an array ran containing one or more numbers with a uniform probability distribution in the range $0 \le$ ran < 1.0, based on the sequence specified by Equations (6.6) and (6.7). The function should have one or two input arguments (m and n) specifying the size of the array to return. If there is one argument, the function should generate a square array of size m × m. If there are two arguments, the function should generate an array of size m × n. The initial value of the seed n_0 will be specified by a call to a function called seed.

2. **Define the inputs and outputs**

   There are two functions in this problem: seed and random0. The input to function seed is an integer to serve as the starting point of the sequence.

---

[5]This algorithm is adapted from the discussion found in Chapter 7 of *Numerical Recipes: The Art of Scientific Programming*, by Press, Flannery, Teukolsky, and Vetterling, Cambridge University Press, 1986.
[6]The notation [0.0,1.0) implies that the range of the random numbers is between 0.0 and 1.0, including the number 0.0, but excluding the number 1.0.

There is no output from this function. The input to function random0 is one or two integers specifying the size of the array of random numbers to be generated. If only argument m is supplied, the function should generate a square array of size m × m. If both arguments m and n are supplied, the function should generate an array of size n × m. The output from the function is the array of random values in the range [0.0, 1.0].

3. **Describe the algorithm**

The pseudocode for function random0 is:

```
function ran = random0 ( m, n )
Check for valid arguments
Set n ← m if not supplied
Create output array with "zeros" function
for ii = 1:number of rows
    for jj = 1:number of columns
        ISEED ← mod (8121 * ISEED + 28411, 134456)
        ran(ii,jj) ← iseed / 134456
    end
end
```

where the value of ISEED is placed in global memory so that it is saved between calls to the function. The pseudocode for function seed is trivial:

```
function seed (new_seed)
new_seed ← round(new_seed)
ISEED ← abs(new_seed)
```

The round function is used in case the user fails to supply an integer, and the absolute value function is used in case the user supplies a negative seed. The user will not have to know in advance that only positive integers are legal seeds.

The variable ISEED will be placed in global memory so that it may be accessed by both functions.

4. **Turn the algorithm into MATLAB statements.**

Function random0 is shown below.

```
function ran = random0 (m,n)
%RANDOM0 Generate uniform random numbers in [0,1]
% Function RANDOM0 generates an array of uniform
% random numbers in the range [0,1]. The usage
% is:
%
% random0(m)     -- Generate an m x m array
% random0(m,n)   -- Generate an m x n array

% Define variables:
%    ii         -- Index variable
%    ISEED      -- Random number seed (global)
```

```
%    jj        -- Index variable
%    m         -- Number of columns
%    msg       -- Error message
%    n         -- Number of rows
%    ran       -- Output array
%
%   Record of revisions:
%      Date         Programmer           Description of change
%      ====         ==========           =====================
%    02/04/14     S. J. Chapman       Original code

% Declare global values
global ISEED              % Seed for random number generator

% Check for a legal number of input arguments.
msg = nargchk(1,2,nargin);
error(msg);

% If the n argument is missing, set it to m.
if nargin < 2
   n = m;
end

% Initialize the output array
ran = zeros(m,n);

% Now calculate random values
for ii = 1:m
   for jj = 1:n
      ISEED = mod(8121*ISEED + 28411, 134456);
      ran(ii,jj) = ISEED / 134456;
   end
end
```

Function seed is shown below.

```
function seed(new_seed)
%SEED Set new seed for function RANDOM0
% Function SEED sets a new seed for function
% RANDOM0. The new seed should be a positive
% integer.

% Define variables:
%   ISEED      -- Random number seed (global)
%   new_seed   -- New seed
```

```
%  Record of revisions:
%       Date         Programmer        Description of change
%       ====         ==========        =====================
%     02/04/14    S. J. Chapman        Original code
%
% Declare global values
global ISEED              % Seed for random number generator

% Check for a legal number of input arguments.
msg = nargchk(1,1,nargin);
error(msg);

% Save seed
new_seed = round(new_seed);
ISEED = abs(new_seed);
```

5. **Test the resulting MATLAB programs.**
   If the numbers generated by these functions are truly uniformly distributed random numbers in the range $0 \leq \text{ran} < 1.0$, then the average of many numbers should be close to 0.5 and the standard deviation of the numbers should be close to $\frac{1}{\sqrt{12}}$.

   Furthermore, if the range between 0 and 1 is divided into a number of equal-size bins, the number of random values falling in each bin should be about the same. A **histogram** is a plot of the number of values falling in each bin. MATLAB function hist will create and plot a histogram from an input data set, so we will use it to verify the distribution of random numbers generated by random0.

   To test the results of these functions, we will perform the following tests:

   1. Call seed with new_seed set to 1024.
   2. Call random0(4) to see that the results appear random.
   3. Call random0(4) to verify that the results differ from call to call.
   4. Call seed again with new_seed set to 1024.
   5. Call random0(4) to see that the results are the same as in (2) above. This verifies that the seed is properly being reset.
   6. Call random0(2,3) to verify that both input arguments are being used correctly.
   7. Call random0(1,100000) and calculate the average and standard deviation of the resulting data set using MATLAB functions mean and std. Compare the results to 0.5 and $\frac{1}{\sqrt{12}}$.
   8. Create a histogram of the data from (7) to see if approximately equal numbers of values fall in each bin.

We will perform these tests interactively, checking the results as we go.

```
» seed(1024)
» random0(4)
ans =
      0.0598      1.0000      0.0905      0.2060
      0.2620      0.6432      0.6325      0.8392
      0.6278      0.5463      0.7551      0.4554
      0.3177      0.9105      0.1289      0.6230
» random0(4)
ans =
      0.2266      0.3858      0.5876      0.7880
      0.8415      0.9287      0.9855      0.1314
      0.0982      0.6585      0.0543      0.4256
      0.2387      0.7153      0.2606      0.8922
» seed(1024)
» random0(4)
ans =
      0.0598      1.0000      0.0905      0.2060
      0.2620      0.6432      0.6325      0.8392
      0.6278      0.5463      0.7551      0.4554
      0.3177      0.9105      0.1289      0.6230
» random0(2,3)
ans =
      0.2266      0.3858      0.5876
      0.7880      0.8415      0.9287
» arr = random0(1,100000);
» mean(arr)
ans =
      0.5001
» std(arr)
ans =
      0.2887
» hist(arr,10)
» title('\bfHistogram of the Output of random0');
» xlabel('Bin');
» ylabel('Count');
```

The results of these tests look reasonable, so the function appears to be working. The average of the data set was 0.5001, which is quite close to the theoretical value of 0.5000, and the standard deviation of the data set was 0.2887, which is equal to the theoretical value of 0.2887 to the accuracy displayed. The histogram is shown in Figure 6.5, and the distribution of the random values is roughly even across all of the bins.

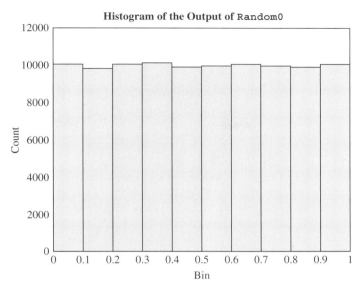

Figure 6.5 Histogram of the output of function `random0`.

6.5 Preserving Data Between Calls to a Function

When a function finishes executing, the special workspace created for that function is destroyed, so the contents of all local variables within the function will disappear. The next time that the function is called, a new workspace will be created, and all of the local variables will be returned to their default values. This behavior is usually desirable, since it ensures that MATLAB functions behave in a repeatable fashion every time they are called.

However, it is sometimes useful to preserve some local information within a function between calls to the function. For example, we might which to create a counter to count the number of times that the function has been called. If such a counter were destroyed every time the function exited, the count would never exceed 1!

MATLAB includes a special mechanism to allow local variables to be preserved between calls to a function. **Persistent memory** is a special type of memory that can only be accessed from within the function, but is preserved unchanged between calls to the function.

A persistent variable is declared with the **persistent statement**. The form of a global statement is

```
persistent var1 var2 var3 ...
```

where *var1*, *var2*, *var3*, and so forth are the variables to be placed in persistent memory.

Good Programming Practice

Use persistent memory to preserve the values of local variables within a function between calls to the function.

▶Example 6.5—Running Averages

It is sometimes desirable to calculate running statistics on a data set on-the-fly as the values are being entered. The built-in MATLAB functions `mean` and `std` could perform this function, but we would have to pass the entire data set to them for recalculation after each new data value is entered. A better result can be achieved by writing a special function that keeps track of the appropriate running sums between calls and only needs the latest value to calculate the current average and standard deviation.

The average, or arithmetic mean, of a set of numbers is defined as

$$\bar{x} = \frac{1}{N} \sum_{i=1}^{N} x_i \tag{6.8}$$

where x_i is sample i out of N samples. The standard deviation of a set of numbers is defined as

$$s = \sqrt{\frac{N \sum_{i=1}^{N} x_i^2 - \left(\sum_{i=1}^{N} x_i \right)^2}{N(N-1)}} \tag{6.9}$$

Standard deviation is a measure of the amount of scatter on the measurements—the greater the standard deviation, the more scattered the points in the data set are. If we can keep track of the number of values N, the sum of the values Σx, and the sum of the squares of the values Σx^2, then we can calculate the average and standard deviation at any time from Equations (6.8) and (6.9).

Write a function to calculate the running average and standard deviation of a data set as it is being entered.

Solution This function must be able to accept input values one at a time and keep running sums of N, Σx, and Σx^2, which will be used to calculate the current average and standard deviation. It must store the running sums in global memory so that they are preserved between calls. Finally, there must be a mechanism to reset the running sums.

1. **State the problem**
   Create a function to calculate the running average and standard deviation of a data set as new values are entered. The function must also include a feature to reset the running sums when desired.

2. **Define the inputs and outputs**

There are two types of inputs required by this function:

1. The character string `'reset'` to reset running sums to zero.
2. The numeric values from the input data set, present one value per function call.

The outputs from this function are the mean and standard deviation of the data supplied to the function so far.

3. **Design the algorithm**

This function can be broken down into four major steps

```
Check for a legal number of arguments
Check for a 'reset', and reset sums if present
Otherwise, add current value to running sums
Calculate and return running average and std dev
     if enough data is available. Return zeros if
     not enough data is available.
```

The detailed pseudocode for these steps is:

```
Check for a legal number of arguments
if x == 'reset'
    n ← 0
    sum_x ← 0
    sum_x2 ← 0
else
    n ← n + 1
    sum_x ← sum_x + x
    sum_x2 ← sum_x2 + x^2
end

% Calculate ave and sd
if n == 0
    ave ← 0
    std ← 0
elseif n == 1
    ave ← sum_x
    std ← 0
else
    ave ← sum_x / n
    std ← sqrt((n*sum_x2 - sum_x^2) / (n*(n-1)))
end
```

4. **Turn the algorithm into MATLAB statements.**

The final MATLAB function is shown below.

```
function [ave, std] = runstats(x)
%RUNSTATS Generate running ave & std deviation
```

```
% Function RUNSTATS generates a running average
% and standard deviation of a data set. The
% values x must be passed to this function one
% at a time. A call to RUNSTATS with the argument
% 'reset' will reset the running sums.

% Define variables:
%   ave         -- Running average
%   msg         -- Error message
%   n           -- Number of data values
%   std         -- Running standard deviation
%   sum_x       -- Running sum of data values
%   sum_x2      -- Running sum of data values squared
%   x           -- Input value
%
%   Record of revisions:
%       Date          Programmer          Description of change
%       ====          ==========          =====================
%       02/05/14      S. J. Chapman       Original code

% Declare persistent values
persistent n                % Number of input values
persistent sum_x            % Running sum of values
persistent sum_x2           % Running sum of values squared

% Check for a legal number of input arguments.
msg = nargchk(1,1,nargin);
error(msg);

% If the argument is 'reset', reset the running sums.
if x == 'reset'
   n = 0;
   sum_x = 0;
   sum_x2 = 0;
else
   n = n + 1;
   sum_x = sum_x + x;
   sum_x2 = sum_x2 + x^2;
end

% Calculate ave and sd
if n == 0
   ave = 0;
   std = 0;
elseif n == 1
   ave = sum_x;
```

```
   std = 0;
else
   ave = sum_x / n;
   std = sqrt((n*sum_x2 - sum_x^2) / (n*(n-1)));
end
```

5. **Test the program.**

> To test this function, we must create a script file that resets `runstats`, reads input values, calls `runstats`, and displays the running statistics. An appropriate script file is shown below:

```
%  Script file: test_runstats.m
%
%  Purpose:
%    To read in an input data set and calculate the
%    running statistics on the data set as the values
%    are read in. The running stats will be written
%    to the Command Window.
%
%  Record of revisions:
%     Date          Programmer        Description of change
%     ====          ==========        =====================
%    02/05/14     S. J. Chapman       Original code
%
% Define variables:
%    array   -- Input data array
%    ave     -- Running average
%    std     -- Running standard deviation
%    ii      -- Index variable
%    nvals   -- Number of input values
%    std     -- Running standard deviation

% First reset running sums
[ave std] = runstats('reset');

% Prompt for the number of values in the data set
nvals = input('Enter number of values in data set:  ');

% Get input values
for ii = 1:nvals

   % Prompt for next value
   string = ['Enter value ' int2str(ii) ':  '];
   x = input(string);

   % Get running statistics
   [ave std] = runstats(x);
```

```
        % Display running statistics
        fprintf('Average = %8.4f; Std dev = %8.4f\n',ave,std);

end
```

To test this function, we will calculate running statistics by hand for a set of 5 numbers and compare the hand calculations to the results from the program. If a data set is created with the following 5 input values

$$3., \quad 2., \quad 3., \quad 4., \quad 2.8$$

then the running statistics calculated by hand would be:

Value	n	Σx	Σx^2	Average	Std_dev
3.0	1	3.0	9.0	3.00	0.000
2.0	2	5.0	13.0	2.50	0.707
3.0	3	8.0	22.0	2.67	0.577
4.0	4	12.0	38.0	3.00	0.816
2.8	5	14.8	45.84	2.96	0.713

The output of the test program for the same data set is:

```
» test_runstats
Enter number of values in data set:   5
Enter value 1:   3
Average =   3.0000; Std dev =    0.0000
Enter value 2:   2
Average =   2.5000; Std dev =    0.7071
Enter value 3:   3
Average =   2.6667; Std dev =    0.5774
Enter value 4:   4
Average =   3.0000; Std dev =    0.8165
Enter value 5:   2.8
Average =   2.9600; Std dev =    0.7127
```

so the results check to the accuracy shown in the hand calculations.

◀

6.6 Built-in MATLAB Functions: Sorting Functions

MATLAB includes two built-in sorting functions that are extremely efficient and should be used instead of the simple sort function we created in Example 6.2. These functions are enormously faster than the sort we created in Example 6.2, and the speed difference increases rapidly as the size of the data set to sort increases.

Function `sort` sorts a data set into ascending or descending order. If the data is a column or row vector, the entire data set is sorted. If the data is a two-dimensional matrix, the columns of the matrix are sorted separately.

The most common forms of the `sort` function are

```
res = sort(a);               % Sort in ascending order
res = sort(a,'ascend');      % Sort in ascending order
res = sort(a,'descend');     % Sort in descending order
```

If `a` is a vector, the data set is sorted in the specified order. For example,

```
» a = [1 4 5 2 8];
» sort(a)
ans =
       1      2      4      5      8
» sort(a,'ascend')
ans =
       1      2      4      5      8
» sort(a,'descend')
ans =
       8      5      4      2      1
```

If `b` is a matrix, the data set is sorted independently by column. For example,

```
» b = [1 5 2; 9 7 3; 8 4 6]
b =
       1      5      2
       9      7      3
       8      4      6
» sort(b)
ans =
       1      4      2
       8      5      3
       9      7      6
```

Function `sortrows` sorts a matrix of data into ascending or descending order *according to one or more specified columns*.

The most common forms of the `sortrows` function are

```
res = sortrows(a);           % Ascending sort of col 1
res = sortrows(a,n);         % Ascending sort of col n
res = sortrows(a,-n);        % Descending order of col n
```

It is also possible to sort by more than one column. For example, the statement

```
res = sortrows(a, [m n]);
```

would sort the rows by column m and if two or more rows have the same value in column m, further sort those rows by column n.

For example, suppose b is a matrix as defined below. Then `sortrows(b)` will sort the rows in ascending order of column 1, and `sortrows(b,[2 3])` will sort the row in ascending order of columns 2 and 3.

```
» b = [1 7 2; 9 7 3; 8 4 6]
b =
      1      7      2
      9      7      3
      8      4      6
» sortrows(b)
ans =
      1      7      2
      8      4      6
      9      7      3
» sortrows(b,[2 3])
ans =
      8      4      6
      1      7      2
      9      7      3
```

6.7 Built-in MATLAB Functions: Random Number Functions

MATLAB includes two standard functions that generate random values from different distributions. They are

- rand—Generates random values from a uniform distribution on the range [0,1)
- randn—Generates random values from a standard normal distribution

Both of them are much faster and much more "random" than the simple function that we have created. If you really need random numbers in your programs, use one of these functions.

In a uniform distribution, every number in the range [0,1) has an equal probability of appearing. In contrast, the normal distribution is a classic "bell shaped curve" with the most likely number being 0.0 and a standard deviation of 1.0.

Functions rand and randn have the following calling sequences:

- rand() or randn()—Generates a single random value
- rand(n) or randn(n)—Generates an $n \times n$ array of random values
- rand(m,n) or randn(m,n)—Generates an $m \times n$ array of random values

6.8 Summary

In Chapter 6, we presented an introduction to user-defined functions. Functions are special types of M-files that receive data through input arguments and return results through output arguments. Each function has its own independent workspace. Each function should appear in a separate file with the same name as the function, *including capitalization*.

Functions are called by naming them in the Command Window or another M-file. The names used should match the function name exactly, including capitalization.

Arguments are passed to functions using a pass-by-value scheme, meaning that MATLAB copies each argument and passes the copy to the function. This copying is important, because the function can freely modify its input arguments without affecting the actual arguments in the calling program.

MATLAB functions can support varying numbers of input and output arguments. Function `nargin` reports the number of actual input arguments used in a function call, and function `nargout` reports the number of actual output arguments used in a function call.

Data can also be shared between MATLAB functions by placing the data in global memory. Global variables are declared using the `global` statement. Global variables may be shared by all functions that declare them. By convention, global variable names are written in all capital letters.

Internal data within a function can be preserved between calls to that function by placing the data in persistent memory. Persistent variables are declared using the `persistent` statement.

6.8.1 Summary of Good Programming Practice

The following guidelines should be adhered to when working with MATLAB functions.

1. Break large program tasks into smaller, more understandable functions whenever possible.
2. Declare global variables in all capital letters to make them easy to distinguish from local variables.
3. Declare global variables immediately after the initial comments and before the first executable statement in each function that uses them.
4. You may use global memory to pass large amounts of data among functions within a program.
5. Use persistent memory to preserve the values of local variables within a function between calls to the function.

6.8.2 MATLAB Summary

The following summary lists all of the MATLAB commands and functions described in this chapter, along with a brief description of each one.

Commands and Functions

error	Displays error message and aborts the function producing the error. This function is used if the argument errors are fatal.
global	Declares global variables.
nargchk	Returns a standard error message if a function is called with too few or too many arguments.
nargin	Returns the number of actual input arguments that were used to call the function.
nargout	Returns the number of actual output arguments that were used to call the function.

(continued)

Commands and Functions (*Continued*)

persistent	Declares persistent variables.
rand	Generates random values from a uniform distribution.
randn	Generates random values from a standard normal distribution.
return	Stop executing a function and return to caller.
sort	Sort data in ascending or descending order.
sortrows	Sort rows of a matrix in ascending or descending order based on a specified column.
warning	Displays a warning message and continues function execution. This function is used if the argument errors are not fatal, and execution can continue.

6.9 Exercises

6.1 What is the difference between a script file and a function?

6.2 When a function is called, how is data passed from the caller to the function, and how are the results of the function returned to the caller?

6.3 What are the advantages and disadvantages of the pass-by-value scheme used in MATLAB?

6.4 Modify the selection sort function developed in this chapter so that it accepts a second optional argument that may be either 'up' or 'down'. If the argument is 'up', sort the data in ascending order. If the argument is 'down', sort the data in descending order. If the argument is missing, the default case is to sort the data in ascending order. (Be sure to handle the case of invalid arguments, and be sure to include the proper help information in your function.)

6.5 The inputs to MATLAB functions sin, cos, and tan are in radians, and the output of functions asin, acos, atan, and atan2 are in radians. Create a new set of functions sin_d, cos_d, and so forth, whose inputs and outputs are in degrees. Be sure to test your functions. (*Note*: Recent versions of MATLAB have built-in functions sind, cosd, and so forth, which work with inputs in degrees instead of radians. You can evaluate your functions and the corresponding built-in functions with the same input values to verify the proper operation of your functions.)

6.6 Write a function f_to_c that accepts a temperature in degrees Fahrenheit and returns the temperature in degrees Celsius. The equation is

$$T \text{ (in °C)} = \frac{5}{9}\Big[T(\text{in °F}) - 32.0\Big] \qquad (6.10)$$

6.7 Write a function c_to_f that accepts a temperature in degrees Celsius and returns the temperature in degrees Fahrenheit. The equation is

$$T \text{ (in °F)} = \frac{9}{5}T \text{ (in °C)} + 32 \qquad (6.11)$$

Demonstrate that this function is the inverse of the one in Exercise 6.6. In other words, demonstrate that the expression c_to_f(f_to_c(temp)) is just the original temperature temp.

6.8 The area of a triangle whose three vertices are points (x_1, y_1), (x_2, y_2), and (x_3, y_3) (see Figure 6.6) can be found from the equation

$$A = \frac{1}{2} \begin{vmatrix} x_1 & x_2 & x_3 \\ y_1 & y_2 & y_3 \\ 1 & 1 & 1 \end{vmatrix} \tag{6.12}$$

where $||$ is the determinant operation. The area returned will be positive if the points are taken in counterclockwise order, and negative if the points are taken in clockwise order. This determinant can be evaluated by hand to produce the following equation

$$A = \frac{1}{2} \left[x_1(y_2 - y_3) - x_2(y_1 - y_3) + x_3(y_1 - y_2) \right] \tag{6.13}$$

Write a function `area2d` that calculates the area of a triangle given the three bounding points (x_1, y_1), (x_2, y_2), and (x_3, y_3) using Equation (6.13). Then test your function by calculating the area of a triangle bounded by the points (0, 0), (10, 0), and (15, 5).

6.9 The area inside any polygon can be broken down into a series of triangles, as shown in Figure 6.7. If this is an n-sided polygon, then it can be divided into $n - 2$ triangles. Create a function that calculates the perimeter of the polygon and the area enclosed by the polygon. Use function `area2d` from the previous exercise to calculate the area of the polygon. Write a program that accepts an ordered list of points bounding a polygon and calls your function to return the perimeter and area of the polygon. Then test your function by calculating the perimeter and area of a polygon bounded by the points (0, 0), (10, 0), (8, 8), (2, 10), and (−4, 5).

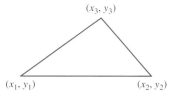

Figure 6.6 A triangle bounded by points (x_1, y_1), (x_2, y_2), and (x_3, y_3).

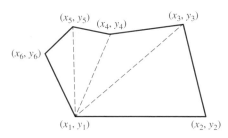

Figure 6.7 An arbitrary polygon can be divided up into a series of triangles. If there are n sides to the polygon, then it can be divided into $n - 2$ triangles.

6.10 Inductance of a Transmission Line   The inductance per meter of a single-phase two-wire transmission line is given by the equation

$$L = \frac{\mu_0}{\pi}\left[\frac{1}{4} + \ln\left(\frac{D}{r}\right)\right] \tag{6.14}$$

where L is the inductance in henrys per meter of line, $\mu_0 = 4\pi \times 10^{-7}$ H/m is the permeability of free space, D is the distance between the two conductors, and r is the radius of each conductor. Write a function that calculates the total inductance of a transmission line as a function of its length in kilometers, the spacing between the two conductors, and the diameter of each conductor. Use this function to calculate the inductance of a 100 km transmission line with conductors of radius $r = 2$ cm and distance $D = 1.5$ m.

6.11 Based on Equation (6.14), would the inductance of a transmission line increase or decrease if the diameter of its conductors increases? How much would the inductance of the line change if the diameter of each conductor were doubled?

6.12 Capacitance of a Transmission Line   The capacitance per meter of a single-phase two-wire transmission line is given by the equation

$$C = \frac{\pi\varepsilon}{\ln\left(\dfrac{D-r}{r}\right)} \tag{6.15}$$

where C is the capacitance in farads per meter of line, $\varepsilon_0 = 4\pi \times 10^{-7}$ F/m is the permittivity of free space, D is the distance between the two conductors, and r is the radius of each conductor. Write a function that calculates the total capacitance of a transmission line as a function of its length in kilometers, the spacing between the two conductors, and the diameter of each conductor. Use this function to calculate the capacitance of a 100 km transmission line with conductors of radius $r = 2$ cm and distance $D = 1.5$ m.

6.13 What happens to the inductance and capacitance of a transmission line as the distance between the two conductors increases?

6.14 Use function random0 to generate a set of 100,000 random values. Sort this data set twice, once with the ssort function of Example 6.2, and once with MATLAB's built-in sort function. Use tic and toc to time the two sort functions. How do the sort times compare? (*Note:* Be sure to copy the original array and present the same data to each sort function. To have a fair comparison, both functions must get the same input data set.)

6.15 Try the sort functions in Exercise 6.14 for array sizes of 10,000, 100,000, and 200,000. How does the sorting time increase with data set size for the sort function of Example 6.2? How does the sorting time increase with data set size for the built-in sort function? Which function is more efficient?

6.16 Modify function random0 so that it can accept 0, 1, or 2 calling arguments. If it has no calling arguments, it should return a single random value. If it has 1 or 2 calling arguments, it should behave as it currently does.

6.17 As function random0 is currently written, it will fail if function seed is not called first. Modify function random0 so that it will function properly with some default seed even if function seed is never called.

6.18 Dice Simulation It is often useful to be able to simulate the throw of a fair die. Write a MATLAB function `dice` that simulates the throw of a fair die by returning some random integer between 1 and 6 every time that it is called. (*Hint:* Call `random0` to generate a random number. Divide the possible values out of `random0` into six equal intervals, and return the number of the interval that a given random value falls into.)

6.19 Road Traffic Density Function `random0` produces a number with a *uniform* probability distribution in the range [0.0, 1.0). This function is suitable for simulating random events if each outcome has an equal probability of occurring. However, in many events, the probability of occurrence is *not* equal for every event, and a uniform probability distribution is not suitable for simulating such events.

For example, when traffic engineers studied the number of cars passing a given location in a time interval of length t, they discovered that the probability of k cars passing during the interval is given by the equation

$$P(k, t) = e^{-\lambda t} \frac{(\lambda t)^k}{k!} \text{ for } t \geq 0, \lambda > 0, \text{ and } k = 0, 1, 2, \ldots \quad (6.16)$$

This probability distribution is known as the *Poisson distribution*; it occurs in many applications in science and engineering. For example, the number of calls k to a telephone switchboard in time interval t, the number of bacteria k in a specified volume t of liquid, and the number of failures k of a complicated system in time interval t all have Poisson distributions.

Write a function to evaluate the Poisson distribution for any k, t, and λ. Test your function by calculating the probability of 0, 1, 2, ..., 5 cars passing a particular point on a highway in 1 minute, given that λ is 1.6 per minute for that highway. Plot the Poisson distribution for $t = 1$ and $\lambda = 1.6$.

6.20 Write three MATLAB functions to calculate the hyperbolic sine, cosine, and tangent functions:

$$\sinh(x) = \frac{e^x - e^{-x}}{2} \qquad \cosh(x) = \frac{e^x + e^{-x}}{2} \qquad \tanh(x) = \frac{e^x - e^{-x}}{e^x + e^{-x}}$$

Use your functions to plot the shapes of the hyperbolic sine, cosine, and tangent functions.

6.21 Write a MATLAB function to perform a running average filter on a data set, as described in Exercise 5.18. Test your function using the same data set used in Exercise 5.18.

6.22 Write a MATLAB function to perform a median filter on a data set, as described in Exercise 5.19. Test your function using the same data set used in Exercise 5.19.

6.23 Sort with Carry It is often useful to sort an array `arr1` into ascending order, while simultaneously carrying along a second array `arr2`. In such a sort, each time an element of array `arr1` is exchanged with another element of `arr1`, the corresponding elements of array `arr2` are also swapped. When the sort is over, the elements of array `arr1` are in ascending order, while the elements of array `arr2` that were associated with particular elements of array `arr1` are still associated with them. For example, suppose we have the following two arrays:

Element	arr1	arr2
1.	6.	1.
2.	1.	0.
3.	2.	10.

After sorting array `arr1` while carrying along array `arr2`, the contents of the two arrays will be:

Element	arr1	arr2
1.	1.	0.
2.	2.	10.
3.	6.	1.

Write a function to sort one real array into ascending order while carrying along a second one. Test the function with the following two 9-element arrays:

```
a = [1, 11, -6, 17, -23, 0, 5, 1, -1];
b = [31, 101, 36, -17, 0, 10, -8, -1, -1];
```

6.24 The sort-with-carry function of Exercise 6.23 is a special case of the built-in function `sortrows`, where the number of columns is two. Create a single matrix c with two columns consisting of the data in vectors a and b in the previous exercise, and sort the data using `sortrows`. How does the sorted data compare to the results of Exercise 6.23?

6.25 Compare the performance of `sortrows` with the sort-with-carry function created in Exercise 6.23. To do this, create two copies of a 10,000 × 2 element array containing random values, and sort column 1 of each array while carrying along column 2 using both functions. Determine the execution times of each sort function using `tic` and `toc`. How does the speed of your function compare with the speed of the standard function `sortrows`?

6.26 Figure 6.8 shows two ships steaming on the ocean. Ship 1 is at position (x_1, y_1) and steaming on heading θ_1. Ship 2 is at position (x_2, y_2) and steaming on

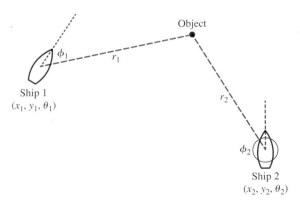

Figure 6.8 Two ships at positions (x_1, y_1) and (x_2, y_2) respectively. Ship 1 is traveling at heading θ_1, and Ship 2 is traveling at heading θ_2.

heading θ_2. Suppose that Ship 1 makes radar contact with an object at range r_1 and bearing ϕ_1. Write a MATLAB function that will calculate the range r_2 and bearing ϕ_2 at which Ship 2 should see the object.

6.27 Linear Least Squares Fit Develop a function that will calculate slope m and intercept b of the least-squares line that best fits an input data set. The input data points (x, y) will be passed to the function in two input arrays, x and y. (The equations describing the slope and intercept of the least-squares line are in Example 5.6 in the previous chapter.) Test your function using a test program and the following 20-point input data set:

Sample Data to Test Least Squares Fit Routine

No.	x	y	No.	x	y
1	−4.91	−8.18	11	−0.94	0.21
2	−3.84	−7.49	12	0.59	1.73
3	−2.41	−7.11	13	0.69	3.96
4	−2.62	−6.15	14	3.04	4.26
5	−3.78	−6.62	15	1.01	6.75
6	−0.52	−3.30	16	3.60	6.67
7	−1.83	−2.05	17	4.53	7.70
8	−2.01	−2.83	18	6.13	7.31
9	0.28	−1.16	19	4.43	9.05
10	1.08	0.52	20	4.12	10.95

6.28 Correlation Coefficient of Least Squares Fit Develop a function that will calculate both the slope m and intercept b of the least-squares line that best fits an input data set and also the correlation coefficient of the fit. The input data points (x, y) will be passed to the function in two input arrays, x and y. The equations describing the slope and intercept of the least-squares line are given in Example 5.1, and the equation for the correlation coefficient is

$$r = \frac{n\left(\sum xy\right) - \left(\sum x\right)\left(\sum y\right)}{\sqrt{\left[\left(n\sum x^2\right) - \left(\sum x\right)^2\right]\left[n\sum y^2 - \left(\sum y\right)^2\right]}} \tag{6.17}$$

where

$\sum x$ is the sum of the x values
$\sum y$ is the sum of the y values
$\sum x^2$ is the sum of the squares of the x values
$\sum y^2$ is the sum of the squares of the y values
$\sum xy$ is the sum of the products of the corresponding x and y values
n is the number of points included in the fit

Test your function using a test driver program and the 20-point input data set given in the previous problem.

6.29 Create a function `random1` that uses function `random0` to generate uniform random values in the range $[-1, 1)$. Test your function by calculating and displaying 20 random samples.

6.30 Gaussian (Normal) Distribution Function `random0` returns a uniformly-distributed random variable in the range $[0, 1)$, which means that there is an equal probability of any given number in the range occurring on a given call to the function. Another type of random distribution is the Gaussian distribution, in which the random value takes on the classic bell-shaped curve shown in Figure 6.9. A Gaussian distribution with an average of 0.0 and a standard deviation of 1.0 is called a *standard normal distribution*, and the probability of any given value occurring in the standardized normal distribution is given by the equation

$$p(x) = \frac{1}{\sqrt{2\pi}} e^{-x^2/2} \tag{6.18}$$

It is possible to generate a random variable with a standard normal distribution starting from a random variable with a uniform distribution in the range $[-1, 1)$ as follows:

1. Select two uniform random variables x_1 and x_2 from the range $[-1, 1)$ such that $x_1^2 + x_2^2 < 1$. To do this, generate two uniform random variables in the range $[-1, 1)$, and see if the sum of their squares happens to be less than 1. If so, use them. If not, try again.

2. Then, each of the values y_1 and y_2 in the equations below will be a normally distributed random variable.

$$y_1 = \sqrt{\frac{-2\ln r}{r}} x_1 \tag{6.19}$$

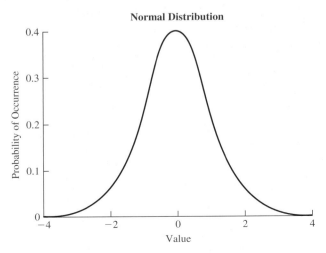

Figure 6.9 A normal probability distribution.

$$y_2 = \sqrt{\frac{-2\ln r}{r}}\, x_2 \tag{6.20}$$

where
$$p(x) = \frac{1}{\sqrt{2\pi}}\, e^{-x^2/2} \tag{6.21}$$

Write a function that returns a normally distributed random value each time that it is called. Test your function by getting 1000 random values, calculating the standard deviation, and plotting a histogram of the distribution. How close to 1.0 was the standard deviation?

6.31 Gravitational Force   The gravitational force F between two bodies of masses m_1 and m_2 is given by the equation

$$F = \frac{Gm_1m_2}{r^2} \tag{6.22}$$

where G is the gravitational constant ($6.672 \times 10^{-11}\,\mathrm{N\,m^2/kg^2}$), m_1 and m_2 are the masses of the bodies in kilograms, and r is the distance between the two bodies. Write a function to calculate the gravitational force between two bodies given their masses and the distance between them. Test you function by determining the force on an 800 kg satellite in orbit 38,000 km above the Earth. (The mass of the Earth is 6.98×10^{24} kg and the radius of the Earth is 6.371×10^6 m.)

6.32 Rayleigh Distribution   The Rayleigh distribution is another random number distribution that appears in many practical problems. A Rayleigh-distributed random value can be created by taking the square root of the sum of the squares of two normally distributed random values. In other words, to generate a Rayleigh-distributed random value r, get two normally distributed random values (n_1 and n_2), and perform the following calculation:

$$r = \sqrt{n_1^2 + n_2^2} \tag{6.23}$$

(a) Create a function `rayleigh(n,m)` that returns an n x m array of Rayleigh-distributed random numbers. If only one argument is supplied [`rayleigh(n)`], the function should return an n x n array of Rayleigh-distributed random numbers. Be sure to design your function with input argument checking and with proper documentation for the MATLAB help system.

(b) Test your function by creating an array of 20,000 Rayleigh-distributed random values and plotting a histogram of the distribution. What does the distribution look like?

(c) Determine the mean and standard deviation of the Rayleigh distribution.

Chapter **7**

Advanced Features of User-Defined Functions

In Chapter 6, we introduced the basic features of user-defined functions. This chapter continues the discussion with a selection of more advanced features.

7.1 Function Functions

"Function function" is the rather awkward name that MATLAB gives to a function whose input arguments include the names or handles of other functions. The functions that are passed to the "function function" are normally used during that function's execution.

For example, MATLAB contains a function function called `fzero`. This function locates a zero of the function that is passed to it. For example, the statement `fzero('cos',[0 pi])` locates a zero of the function `cos` between 0 and π, and `fzero('exp(x)-2',[0 1])` locates a zero of the function "`exp(x)-2`" between 0 and 1. When these statements are executed, the result is:

```
» fzero('cos',[0 pi])
ans =

    1.5708
» fzero('exp(x)-2',[0 1])
ans =
    0.6931
```

The keys to the operation of function functions are two special MATLAB functions, `eval` and `feval`. Function `eval` *evaluates a character string* as though it had been typed in the Command Window, while function `feval` *evaluates a named function* at a specific input value.

Function `eval` evaluates a character string as though it has been typed in the Command Window. This function gives MATLAB functions a chance to construct executable statements during execution. The form of the `eval` function is

```
eval(string)
```

For example, the statement `x = eval('sin(pi/4)')` produces the result

```
» x = eval('sin(pi/4)')
x =
    0.7071
```

An example in which a character string is constructed and evaluated using the `eval` function is shown below:

```
x = 1;
str = ['exp(' num2str(x) ') -1'];
res = eval(str);
```

In this case, `str` contains the character string `'exp(1) -1'`, which `eval` evaluates to get the result 1.7183.

Function `feval` evaluates a *named function* defined by an M-file at a specified input value. The general form of the `feval` function is

```
feval(fun,value)
```

For example, the statement `x = feval('sin',pi/4)` produces the result

```
» x = feval('sin',pi/4)
x =
    0.7071
```

Some of the more common MATLAB function functions are listed in Table 7.1. Type `help fun_name` to learn how to use each of these functions.

Table 7.1: Common MATLAB Function Functions

Function Name	Description
`fminbnd`	Minimize a function of one variable.
`fzero`	Find a zero of a function of one variable.
`quad`	Numerically integrate a function.
`ezplot`	Easy to use function plotter.
`fplot`	Plot a function by name.

►Example 7.1—Creating a Function Function

Create a function function that will plot any MATLAB function of a single variable between specified starting and ending values.

Solution This function has two input arguments, the first one containing the name of the function to plot and the second one containing a two-element vector with the range of values to plot.

1. **State the problem**

   Create a function to plot any MATLAB function of a single variable between two user-specified limits.

2. **Define the inputs and outputs**

   There are two inputs required by this function:

   (a) A character string containing the name of a function.

   (b) A two-element vector containing the first and last values to plot.

   The output from this function is a plot of the function specified in the first input argument.

3. **Design the algorithm**

   This function can be broken down into four major steps

   ```
   Check for a legal number of arguments
   Check that the second argument has two elements
   Calculate the value of the function between the
       start and stop points
   Plot and label the function
   ```

   The detailed pseudocode for the evaluation and plotting steps is:

   ```
   n_steps ← 100
   step_size ← (xlim(2) – xlim(1)) / n_steps
   x ← xlim(1):step_size:xlim(2)
   y ← feval(fun,x)
   plot(x,y)
   title(['\bfPlot of function ' fun '(x)'])
   xlabel('\bfx')
   ylabel(['\bf' fun '(x)'])
   ```

4. **Turn the algorithm into MATLAB statements.**

   The final MATLAB function is shown below.

```
function quickplot(fun,xlim)
%QUICKPLOT Generate quick plot of a function
% Function QUICKPLOT generates a quick plot
% of a function contained in a external m-file,
% between user-specified x limits.
```

```
% Define variables:
%   fun         -- Name of function to plot in a char string
%   msg         -- Error message
%   n_steps     -- Number of steps to plot
%   step_size   -- Step size
%   x           -- X-values to plot
%   y           -- Y-values to plot
%   xlim        -- Plot x limits
%
% Record of revisions:
%      Date         Programmer        Description of change
%      ====         ==========        =====================
%   02/07/14      S. J. Chapman       Original code
% Check for a legal number of input arguments.
msg = nargchk(2,2,nargin);
error(msg);

% Check the second argument to see if it has two
% elements. Note that this double test allows the
% argument to be either a row or a column vector.
if ( size(xlim,1) == 1 && size(xlim,2) == 2 ) | ...
   ( size(xlim,1) == 2 && size(xlim,2) == 1 )

   % Ok--continue processing.
   n_steps = 100;
   step_size = (xlim(2) - xlim(1)) / n_steps;
   x = xlim(1):step_size:xlim(2);
   y = zeros(size(x));
   h = str2func(fun)

   for ii = 1:length(x)
      y(ii) = feval(h,x(ii));
   end

   plot(x,y);
   title(['\bfPlot of function ' fun '(x)']);
   xlabel('\bfx');
   ylabel(['\bf' fun '(x)']);

else
   % Else wrong number of elements in xlim.
   error('Incorrect number of elements in xlim.');
end
```

5. **Test the program.**

   To test this function, we must call it with correct and incorrect input arguments, verifying that it handles both correct inputs and errors properly. The
   results are shown below:

   ```
   » quickplot('sin')
   ??? Error using ==> quickplot
   Not enough input arguments.

   » quickplot('sin',[-2*pi 2*pi],3)
   ??? Error using ==> quickplot
   Too many input arguments.

   » quickplot('sin',-2*pi)
   ??? Error using ==> quickplot
   Incorrect number of elements in xlim.

   » quickplot('sin',[-2*pi 2*pi])
   ```

   The last call was correct, and it produced the plot shown in Figure 7.1.

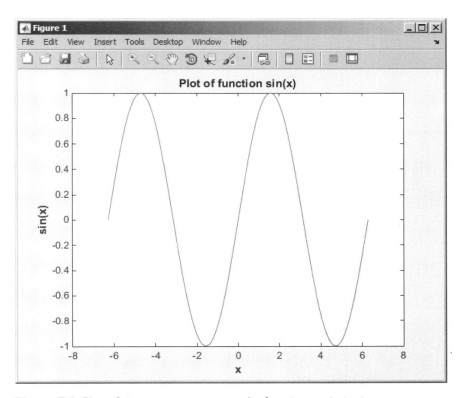

Figure 7.1 Plot of sin *x* versus *x* generate by function `quickplot`.

7.2 Local Functions, Private Functions, and Nested Functions

MATLAB includes several special types of functions that behave differently than the ordinary functions we have used so far. Ordinary functions can be called by any other function, as long as they are in the same directory or in any directory on the MATLAB path.

The **scope** of a function is defined as the locations within MATLAB from which the function can be accessed. The scope of an ordinary MATLAB function is the current working directory. If the function lies in a directory on the MATLAB path, then the scope extends to all MATLAB functions in a program, because they all check the path when trying to find a function with a given name.

In contrast, the scope of the other function types that we will discuss in the rest of this chapter is more limited in one way or another.

7.2.1 Local Functions

It is possible to place more than one function in a single file. If more than one function is present in a file, the top function is a normal or **primary function**, while the ones below it are called **local functions** or **subfunctions**. The primary function should have the same name as the file it appears in. Local functions look just like ordinary functions, but they are only accessible to the other functions within the same file. In other words, the scope of a local function is the other functions within the same file (see Figure 7.2).

Local functions are often used to implement "utility" calculations for a main function. For example, the file `mystats.m` shown below contains a primary

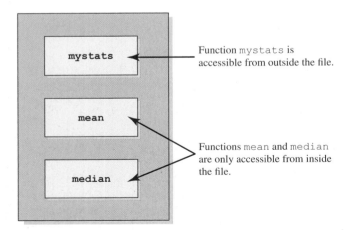

Figure 7.2 The first function in a file is called the primary function. It should have the same name as the file it appears in, and it is accessible from outside the file. The remaining functions in the file are subfunctions; they are only accessible from within the file.

function mystats and two local functions mean and median. Function mystats is a normal MATLAB function, so it can be called by any other MATLAB function in the same directory. If this file is in a directory included in the MATLAB search path, it can be called by any other MATLAB function, even if the other function is not in the same directory. By contrast, the scope of functions mean and median is restricted to other functions within the same file. Function mystats can call them and they can call each other, but a function outside of the file cannot. They are "utility" functions that perform a part of the job of the main function mystats.

```
function [avg, med] = mystats(u)
%MYSTATS Find mean and median with internal functions.
% Function MYSTATS calculates the average and median
% of a data set using local functions.

n = length(u);
avg = mean(u,n);
med = median(u,n);

function a = mean(v,n)
% Subfunction to calculate average.
a = sum(v)/n;

function m = median(v,n)
% Subfunction to calculate median.
w = sort(v);
if rem(n,2) == 1
    m = w((n+1)/2);
else
    m = (w(n/2)+w(n/2+1))/2;
end
```

7.2.2 Private Functions

Private functions are functions that reside in subdirectories with the special name private. They are only visible to other functions in the private directory or to functions in the parent directory. In other words, the scope of these functions is restricted to the private directory and to the parent directory that contains it.

For example, assume the directory testing is on the MATLAB search path. A subdirectory of testing called private can contain functions that only the functions in testing can call. Because private functions are invisible outside of the parent directory, they can use the same names as functions in other directories. This is useful if you want to create your own version of a particular function while retaining the original in another directory. Because MATLAB looks for private functions before standard M-file functions, it will find a private function named test.m before a non-private function named test.m.

You can create your own private directories simply by creating a subdirectory called private under the directory containing your functions. Do not place these private directories on your search path.

When a function is called from within an M-file, MATLAB first checks the file to see if the function is a local function defined in the same file. If not, it checks for a private function with that name. If it is not a private function, MATLAB checks current directory for the function name. If it is not in the current directory, MATLAB checks the standard search path for the function.

If you have special-purpose MATLAB functions that should only be used by other functions and never be called directly by the user, consider hiding them as local functions or private functions. Hiding the functions will prevent their accidental use and will also prevent conflicts with other public functions of the same name.

7.2.3 Nested Functions

Nested functions are functions that are defined *entirely within the body of another function*, called the **host function**. They are only visible to the host function in which they are embedded and to other nested functions embedded at the same level within the same host function.

A nested function has access to any variables defined with it, *plus any variables defined within the host function* (see Figure 7.3). In other words, the **scope** of the variables declared in the host function includes both the host function and any nested functions within it. The only exception occurs if a variable in the nested function has the same name as a variable within the host function. In that case, the variable within the host function is not accessible.

Note that if a file contains one or more nested functions, then *every function in the file* must be terminated with an end statement. This is the only time when the end statement is required at the end of a function—at all other times it is optional.

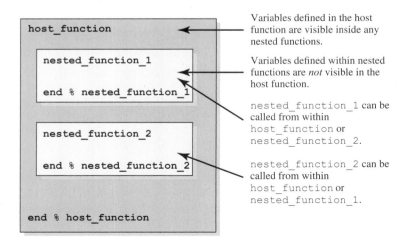

Figure 7.3 Nested functions are defined within a host function, and they inherit variables defined within the host function.

Programming Pitfalls

If a file contains one or more nested functions, then *every function in the file* must be terminated with an `end` statement. It is an error to omit `end` statements in this case.

The following program illustrates the use of variables in nested functions. It contains a host function `test_nested_1` and a nested function `fun1`. When the program starts, variables a, b, x, and y are initialized as shown in the host function, and their values are displayed. Then the program calls `fun1`. Since `fun1` is nested, it inherits a, b, and x from the host function. Note that it does *not* inherit y, because `fun1` defines a local variable with that name. When the values of the variables are displayed at the end of `fun1`, we see that a has been increased by 1 (due to the assignment statement) and that y is set to 5. When execution returns to the host function, a is still increased by 1, showing that the variable a in the host function and the variable a in the nested function are really the same. On the other hand, y is again 9, because the variable y in the host function is not the same as the variable y in the nested function.

```
function res = test_nested_1

% This is the top level function.
% Define some variables.
a = 1; b = 2; x = 0; y = 9;

% Display variables before call to fun1
fprintf('Before call to fun1:\n');
fprintf('a, b, x, y = %2d %2d %2d %2d\n', a, b, x, y);

% Call nested function fun1
x = fun1(x);

% Display variables after call to fun1
fprintf('\nAfter call to fun1:\n');
fprintf('a, b, x, y = %2d %2d %2d %2d\n', a, b, x, y);

   % Declare a nested function
   function res = fun1(y)

   % Display variables at start of call to fun1
   fprintf('\nAt start of call to fun1:\n');
   fprintf('a, b, x, y = %2d %2d %2d %2d\n', a, b, x, y);

   y = y + 5;
   a = a + 1;
   res = y;
```

```
        % Display variables at end of call to fun1
        fprintf('\nAt end of call to fun1:\n');
        fprintf('a, b, x, y = %2d %2d %2d %2d\n', a, b, x, y);

    end % function fun1
end % function test_nested_1
```

When this program is executed, the results are:

```
» test_nested_1
Before call to fun1:
a, b, x, y =  1   2   0   9

At start of call to fun1:
a, b, x, y =  1   2   0   0

At end of call to fun1:
a, b, x, y =  2   2   0   5

After call to fun1:
a, b, x, y =  2   2   5   9
```

Like local functions, nested functions can be used to perform special-purpose calculations within a host function.

Good Programming Practice

Use local functions, private functions, or nested functions to hide special-purpose calculations that should not be generally accessible to other functions. Hiding the functions will prevent their accidental use and will also prevent conflicts with other public functions of the same name.

7.2.4 Order of Function Evaluation

In a large program, there could possibly be multiple functions (local functions, private functions, nested functions, and public functions) with the same name. When a function with a given name is called, how do we know which copy of the function will be executed?

The answer to this question is that MATLAB locates functions in a specific order as follows:

1. MATLAB checks to see if there is a nested function within the current function with the specified name. If so, it is executed.
2. MATLAB checks to see if there is a local function within the current file with the specified name. If so, it is executed.

3. MATLAB checks for a private function with the specified name. If so, it is executed.
4. MATLAB checks for a function with the specified name in the current directory. If so, it is executed.
5. MATLAB checks for a function with the specified name on the MATLAB path. MATLAB will stop searching and execute the first function with the right name found on the path.

7.3 Function Handles

A **function handle** is a MATLAB data type that holds information to be used in referencing a function. When you create a function handle, MATLAB captures all the information about the function that it needs to execute it later on. Once the handle is created, it can be used to execute the function at any time.

Function handles are key to the operation of MATLAB functions that use other functions.

7.3.1 Creating and Using Function Handles

A function handle can be created either of two possible ways: the @ operator or the `str2func` function. To create a function handle with the @ operator, just place it in front of the function name. To create a function handle with the `str2func` function, call the function with the function name in a string. For example, suppose that function `my_func` is defined as follows:

```
function res = my_func(x)
res = x.^2 - 2*x + 1;
```

Then either of the following lines will create a function handle for function `my_func`:

```
hndl = @my_func
hndl = str2func('my_func');
```

Once a function handle has been created, the function can be executed by naming the function handle followed by any calling parameters. The result will be exactly the same as if the function itself were named.

```
» hndl = @my_func
hndl =
    @my_func
» hndl(4)
ans =
    9
» my_func(4)
ans =
    9
```

If a function has no calling parameters, then the function handle must be followed by empty parentheses when it is used to call the function:

```
» h1 = @randn;
» h1()
ans =
   -0.4326
```

After a function handle is created, it appears in the current workspace with the data type "function handle":

```
» whos
Name      Size      Bytes      Class               Attributes

ans       1x1           8      double
h1        1x1          16      function_handle
hndl      1x1          16      function_handle
```

A function handle can also be executed using the feval function. This provides a convenient way to execute function handles within a MATLAB program.

```
» feval(hndl,4)
ans =
     9
```

It is possible to recover the function name from a function handle using the func2str function.

```
» func2str(hndl)
ans =
my_func
```

This feature is very useful when we want to create descriptive messages, error messages, or labels inside a function that accepts and evaluates function handles. For example, the function shown below accepts a function handle in the first argument and plots the function at the points specified in the second argument. It also prints out a title containing the name of the function being plotted.

```
function plotfunc(fun,points)
% PLOTFUNC Plots a function between the specified points.
% Function PLOTFUNC accepts a function handle and
% plots the function at the points specified.

% Define variables:
%   fun        -- Function handle
%   msg        -- Error message
%
%  Record of revisions:
%    Date          Programmer        Description of change
%    ====          ==========        =====================
%   03/05/14       S. J. Chapman      Original code
```

```
% Check for a legal number of input arguments.
msg = nargchk(2,2,nargin);
error(msg);

% Get function name
fname = func2str(fun);

% Plot the data and label the plot
plot(points,fun(points));
title(['\bfPlot of ' fname '(x) vs x']);
xlabel('\bfx');
ylabel(['\bf' fname '(x)']);
grid on;
```

For example, this function can be used to plot the function $\sin x$ from -2π to 2π with the following statement:

```
plotfunc(@sin, [-2*pi:pi/10:2*pi])
```

The resulting function is shown in Figure 7.4.

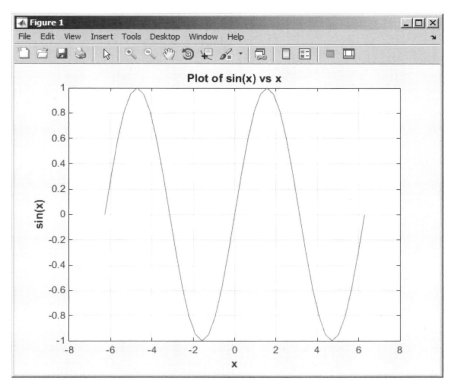

Figure 7.4 Plot of function $\sin x$ from -2π to 2π, created using function `plotfunc`.

Table 7.2: MATLAB Functions that manipulate Function Handles

Function	Description
@	Create a function handle.
feval	Evaluates a function using a function handle.
func2str	Recover the function name associated with a given function handle.
functions	Recover miscellaneous information from a function handle. The data is returned in a structure.
str2func	Create a function handle from a specified string.

Note that the function functions such as `feval` and `fzero` accept function handles as well as function names in their calling arguments. For example, the following two statements are equivalent and produce the same answer:

```
» res = feval('sin',3*pi/2)
res =
    -1
» res = feval(@sin,3*pi/2)
res =
    -1
```

Some common MATLAB functions used with function handles are summarized in Table 7.2.

7.3.2 The Significance of Function Handles

Either function names or function handles can be used to execute most functions. However, function handles have certain advantages over function names. These advantages include:

1. **Passing Function Access Information to Other Functions.** As we saw in the previous section, you can pass a function handle as an argument in a call to another function. The function handle enables the receiving function to call the function attached to the handle. You can execute a function handle from within another function *even if the handle's function is not in the scope of the evaluating function*. This is because the function handle has a complete description of the function to execute—the calling function does not have to search for it.
2. **Improved Performance in Repeated Operations.** MATLAB performs a search for a function at the time that you create a function handle and then stores this access information in the handle itself. Once defined, you can use this handle over and over, without having to look it up again. This makes function execution faster.
3. **Allow Wider Access to Local Functions (Subfunctions) and Private Functions.** All MATLAB functions have a certain scope. They are visible to

other MATLAB entities within that scope but not visible outside of it. You can call a function directly from another function that is within its scope but *not* from a function outside that scope. Local functions, private functions, and nested functions are limited in their visibility to other MATLAB functions. You can invoke a local function only from another function that is defined within the same M-file. You can invoke a private function only from a function in the directory immediately above the `private` subdirectory. You can invoke a nested function only from within the host function or another nested function at the same level. However, when you create a handle to a function that has limited scope, the function handle stores all the information MATLAB needs to evaluate the function from *any* location in the MATLAB environment. If you create a handle to a local function within the M-file that defines the local function, you can then pass the handle to code that resides outside of that M-file and evaluate the local function from beyond its usual scope. The same holds true for private functions and nested functions.

4. **Include More Functions per M-File for Easier File Management.** You can use function handles to help reduce the number of M-files required to contain your functions. The problem with grouping a number of functions in one M-file has been that this defines them as local functions and thus reduces their scope in MATLAB. Using function handles to access these local functions removes this limitation. This enables you to group functions as you want and reduce the number of files you have to manage.

7.3.3   Function Handles and Nested Functions

When MATLAB invokes an ordinary function, a special workspace is created to contain the function's variables. The function executes to completion, and then the workspace is destroyed. All the data in the function workspace is lost, except for any values labeled `persistent`. If the function is executed again, a completely new workspace is created for the new execution.

By contrast, when a host function creates a handle for a nested function and returns that handle to a calling program, the host function's workspace is created and *remains in existence for as long as the function handle remains in existence.* Since the nested function has access to the host function's variables, MATLAB has to preserve the host's function's data as long as there is any chance that the nested function will be used. This means that *we can save data in a function between uses.*

This idea is illustrated in the function shown below. When function `count_calls` is executed, it initializes a local variable `current_count` to a user-specified initial count and then creates and returns a handle to the nested function `increment_count`. When `increment_count` is called using that function handle, the count is increased by one and the new value is returned.

```
function fhandle = count_calls(initial_value)

% Save initial value in a local variable
% in the host function.
current_count = initial_value;
```

```
% Create and return a function handle to the
% nested function below.
fhandle = @increment_count;

    % Define a nested function to increment counter
    function count = increment_count
    current_count = current_count + 1;
    count = current_count;
    end % function increment_count

end % function count_calls
```

When this program is executed, the results are as shown below. Each call to the function handle increments the count by one.

```
» fh = count_calls(4);
» fh()
ans =
      5
» fh()
ans =
      6
» fh()
ans =
      7
```

Even more importantly, *each function handle created for a function has its own independent workspace*. If we create two different handles for this function, each one will have its own local data, and they will be independent of each other. As you can see, we can increment either counter independently by calling the function with the proper handle.

```
» fh1 = count_calls(4);
» fh2 = count_calls(20);
» fh1()
ans =
       5
» fh1()
ans =
       6
» fh2()
ans =
      21
» fh1()
ans =
       7
```

You can use this feature to run multiple counters and so forth within a program without them interfering with each other.

7.3.4 An Example Application: Solving Ordinary Differential Equations

One very important application of function handles occurs in the MATLAB functions designed to solve ordinary differential equations. MATLAB includes a plethora of functions to solve differential equations under various conditions, but the most all-round useful of them is `ode45`. This function solves ordinary differential equations of the form

$$y' = f(t,y) \tag{7.1}$$

using a Runge-Kutta (4,5) integration algorithm, and it works well for many types of equations with many different input conditions.

The calling sequence for this function is

```
[t,y] = ode45(odefun_handle,tspan,y0,options)
```

where the calling parameters are:

`odefun_handle`	A *handle* to a function $f(t,y)$ that calculates the derivative y' of the differential equation.
`tspan`	A vector containing the times to integrate. If this is a two-element array `[t0 tend]`, then the values are interpreted as the starting and ending times to integrate. The integrator applies the initial conditions at time `t0` and integrates the equation until time `tend`. If the array has more than two elements, then the integrator returns the values of the differential equation at exactly the specified times.
`y0`	The initial conditions for the variable at time `t0`
`options`	A structure of optional parameters that change the default integration properties. (We will not use this parameter in this book.)

and the results are:

`t`	A column vector of time points at which the differential equation was solved.
`y`	The solution array. Each row of `y` contains the solutions to all variables at the time specified in the same row of `t`.

This function also works well for systems of simultaneous first order differential equations, where there are vectors of dependent variables y_1, y_2, and so forth.

We will try a few example differential equations to get a better understanding of this function. First, consider the simple first order linear time-invariant differential equation

$$\frac{dy}{dt} + 2y = 0 \tag{7.2}$$

with the initial condition $y(0) = 1$. The function that would specify the derivative of the differential equation is

$$\frac{dy}{dt} = -2y \tag{7.3}$$

This function could be programmed in MATLAB as follows:

```
function yprime = fun1(t,y)
yprime = -2 * y;
```

Function ode45 could be used to solve Equation (7.2) for $y(t)$

```
%  Script file: ode45_test1.m
%
%  Purpose:
%    This program solves a differential equation of the
%    form dy/dt + 2 * y = 0, with the initial condition
%    y(0) = 1.
%
%  Record of revisions:
%      Date                Programmer           Description of change
%      ====                ==========           =====================
%    03/15/14            S. J. Chapman          Original code
%
% Define variables:
%  odefun_handle -- Handle to function that defines the derivative
%    tspan          -- Duration to solve equation for
%    yo             -- Initial condition for equation
%    t              -- Array of solution times
%    y              -- Array of solution values

% Get a handle to the function that defines the
% derivative.
odefun_handle = @fun1;

% Solve the equation over the period 0 to 5 seconds
tspan = [0 5];

% Set the initial conditions
y0 = 1;

% Call the differential equation solver.
[t,y] = ode45(odefun_handle,tspan,y0);
```

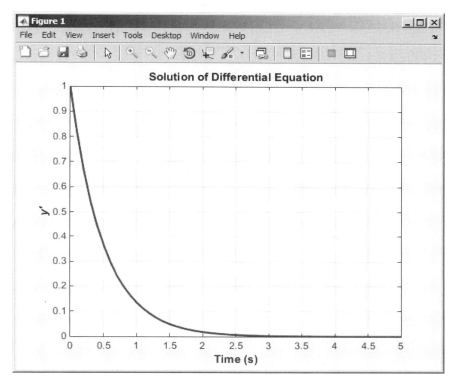

Figure 7.5 Solution to the differential equation $dy/dt + 2y = 0$ with the initial condition $y(0) = 1$.

```
% Plot the result
figure(1);
plot(t,y,'b-','LineWidth',2);
grid on;
title('\bfSolution of Differential Equation');
xlabel('\bfTime (s)');
ylabel('\bf\ity''');
```

When this script file is executed, the resulting output is shown in Figure 7.5. This sort of exponential decay is exactly what would be expected for a first-order linear differential equation.

► Example 7.2—Radioactive Decay Chains

The radioactive isotope thorium-227 decays into radium-223 with a half-life of 18.68 days, and radium-223 in turn decays into radon-219 with a half-life of 11.43 days. The radioactive decay constant for thorium 227 is $\lambda_{th} = 0.03710638$/day, and the radioactive decay constant for radon is $\lambda_{ra} = 0.0606428$/day. Assume that initially we have 1 million atoms of thorium-227 and calculate and plot the amount of thorium-227 and radium-223 that will be present as a function of time.

Solution The rate of decrease in thorium-227 is equal to the amount of thorium-227 present at a given moment times the decay constant for the material.

$$\frac{dn_{th}}{dt} = -\lambda_{th} n_{th} \tag{7.4}$$

where n_{th} is the amount of thorium-227 and λ_{th} is the decay rate per day. The rate of decrease in radium-223 is equal to the amount of radium-223 present at a given moment times the decay constant for the material. However, the amount of radium-223 is *increased* by the number of atoms of thorium-227 that have decayed, so the total change in the amount of radium-223 is

$$\frac{dn_{ra}}{dt} = -\lambda_{ra} n_{ra} - \frac{dn_{th}}{dt}$$

$$\frac{dn_{ra}}{dt} = -\lambda_{ra} n_{ra} + \lambda_{th} n_{th} \tag{7.5}$$

where n_{ra} is the amount of radon-219 and λ_{ra} is the decay rate per day. Equations (7.4) and (7.5) must be solved simultaneously to determine the amount of thorium-227 and radium-223 present at any given time.

1. **State the problem**
   Calculate and plot the amount of thorium-227 and radium-223 present as a function of time, given that there were initially 1,000,000 atoms of thorium-227 and no radium-223.

2. **Define the inputs and outputs**
   There are no inputs to this program. The outputs from this program are the plots of thorium-227 and radium-223 as a function of time.

3. **Describe the algorithm**
   This program can be broken down into three major steps

   ```
   Create a function to describe the derivatives of
   thorium-227 and radium-223
   Solve the differential equations using ode45
   Plot the resulting data
   ```

   The first major step is to create a function that calculates the rate of change of thorium-227 and radium-223. This is just a direct implementation of Equations (7.4) and (7.5). The detailed pseudocode is shown below:

   ```
   function yprime = decay1(t,y)
   yprime(1) = -lambda_th * y(1);
   yprime(2) = -lambda_ra * y(2) + lambda_th * y(1);
   ```

Next we have to solve the differential equation. To do this, we need to set the initial conditions and the duration, and then call `ode45`. The detailed pseudocode is shown below:

```
% Get a function handle.
odefun_handle = @decay1;

% Solve the equation over the period 0 to 100 days
tspan = [0 100];

% Set the initial conditions
y0(1) = 1000000;      % Atoms of thorium-227
y0(2) = 0;            % Atoms of radium-223

% Call the differential equation solver.
[t,y] = ode45(odefun_handle,tspan,y0);
```

The final step is writing and plotting the results. Each result appears in its own column, so `y(:,1)` will contain the amount of thorium-227 and `y(:,2)` will contain the amount radium-223.

4. **Turn the algorithm into MATLAB statements.**
   The MATLAB code for the selection sort function is shown below.

```
%   Script file: calc_decay.m
%
%   Purpose:
%     This program calculates the amount of thorium-227 and
%     radium-223 left as a function of time, given an initial
%     concentration of 1000000 atoms of thorium-227
%     and no atoms 0 radium-223.%
%   Record of revisions:
%      Date             Programmer         Description of change
%      ====             ==========         =====================
%    03/15/14         S. J. Chapman        Original code
%
% Define variables:
%  odefun_handle -- Handle to function that defines the derivative
%    tspan        -- Duration to solve equation for
%    yo           -- Initial condition for equation
%    t            -- Array of solution times
%    y            -- Array of solution values

% Get a handle to the function that defines the derivative.
odefun_handle = @decay1;

% Solve the equation over the period 0 to 100 days
tspan = [0 100];

% Set the initial conditions
y0(1) = 1000000;      % Atoms of thorium-227
y0(2) = 0;            % Atoms of radium-223
```

```
% Call the differential equation solver.
[t,y] = ode45(odefun_handle,tspan,y0);

% Plot the result
figure(1);
plot(t,y(:,1),'b-','LineWidth',2);
hold on;
plot(t,y(:,2),'k--','LineWidth',2);
title('\bfAmount of Thorium-227 and Radium-223 vs Time');
xlabel('\bfTime (days)');
ylabel('\bfNumber of Atoms');
legend('Thorium-227','Radium-223');
grid on;
hold off;
```

The function to calculate the derivatives is shown below.

```
function yprime = decay1(t,y)
%DECAY1 Calculates the decay rates of thorium-227 and radium-223.
% Function DECAY1 Calculates the rates of change of thorium-227
% and radium-223 (yprime) for a given current concentration y.

% Define variables:
%   t           -- Time (in days)
%   y           -- Vector of current concentrations
%
%   Record of revisions:
%       Date            Programmer          Description of change
%       ====            ==========          =====================
%       03/15/07        S. J. Chapman       Original code

% Set decay constants.
lambda_th = 0.03710636;
lambda_ra = 0.0606428;

% Calculate rates of decay
yprime = zeros(2,1);
yprime(1) = -lambda_th * y(1);
yprime(2) = -lambda_ra * y(2) + lambda_th * y(1);
```

5. **Test the program.**

When this program is executed, the results are as shown in Figure 7.6. These results look reasonable. The initial amount of thorium-227 starts high and decreases exponentially with a half-life of about 18 days. The initial amount of radium-223 starts at zero and rises rapidly due to the decay of thorium-227 and then starts decreasing as the amount of increase from the decay of thorium 227 becomes less than the rate of decay of radium-223.

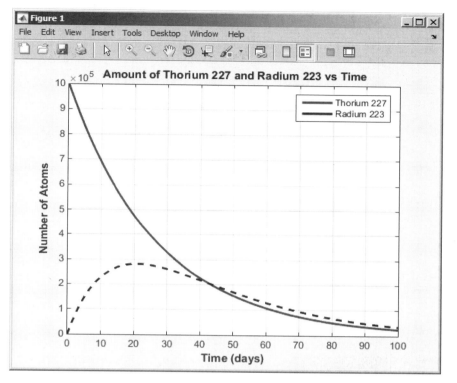

Figure 7.6 Plot of radioactive decay of thorium-227 and radium-223 versus time.

7.4 Anonymous Functions

An anonymous function is a function "without a name"[1]. It is a function that is declared in a single MATLAB statement that returns a function handle, which can then be used to execute the function. The form of an anonymous function is

```
fhandle = @ (arglist) expr
```

where `fhandle` is a function handle used to reference the function, `arglist` is a list of calling variables, and `expr` is an expression involving the argument list that evaluates the function. For example, we can create a function to evaluate the expression $f(x) = x^2 - 2x - 2$ as follows:

```
myfunc = @ (x) x.^2 - 2*x - 2
```

---
[1]This is the meaning of the word "anonymous"!

The function can then be invoked using the function handle. For example, we can evaluate $f(2)$ as follows:

```
» myfunc(2)
ans =
     -2
```

Anonymous functions are a quick way to write short functions that can then be used in function functions. For example, we can find a root of the function $f(x) = x^2 - 2x - 2$ by passing the anonymous function to `fzero` as follows:

```
» root = fzero(myfunc, [0 4])
root =
     2.7321
```

7.5 Recursive Functions

A function is said to be **recursive** if the function calls itself. The factorial function is a good example of a recursive function. In Chapter 5, we defined the factorial function as

$$n! = \begin{cases} 1 & n = 0 \\ n \times (n-1) \times (n-2) \times \ldots \times 2 \times 1 & n > 0 \end{cases} \qquad (7.6)$$

This definition can also be written as

$$n! = \begin{cases} 1 & n = 0 \\ n \times (n-1)! & n > 0 \end{cases} \qquad (7.7)$$

where the value of the factorial function $n!$ is defined using the factorial function itself. MATLAB functions are designed to be recursive, so Equation (7.7) can be implemented directly in MATLAB.

▶Example 7.3—The Factorial Function

To illustrate the operation of a recursive function, we will implement the factorial function using the definition in Equation (7.7). The MATLAB code to calculate n factorial for positive value of n would be

```
function result = fact(n)
%FACT Calculate the factorial function
% Function FACT calculates the factorial function
% by recursively calling itself.

% Define variables:
%   n             -- Non-negative integer input
%
```

```
%  Record of revisions:
%     Date         Programmer              Description of change
%     ====         ==========              =====================
%     07/07/14  S. J. Chapman              Original code
% Check for a legal number of input arguments.
msg = nargchk(1,1,nargin);
error(msg);

% Calculate function
if n == 0
   result = 1;
else
   result = n * fact(n-1);
end
```

When this program is executed, the results are as expected.

```
» fact(5)
ans =
    120
» fact(0)
ans =
     1
```

◄

7.6 Plotting Functions

In all previous plots, we have created arrays of data to plot and passed those arrays to the plotting function. MATLAB also includes two functions that will plot a function directly, without the necessity of creating intermediate data arrays. These functions are ezplot and fplot.

Function ezplot takes one of the following forms.

```
ezplot(fun);
ezplot(fun, [xmin xmax]);
ezplot(fun, [xmin xmax], figure);
```

The argument fun is either a function handle, the name of an M-file function, or a *character string* containing the functional expression to be evaluated. The optional parameter [xmin xmax] specifies the range of the function to plot. If it is absent, the function will be plotted between -2π and 2π. The optional parameter figure specifies the figure number to plot the function on.

For example, the following statements plot the function $f(x) = \sin x/x$ between -4π and 4π. The output of these statements is shown in Figure 7.7[2].

---

[2]Note that the value of the function at exactly zero will be 0/0, which is undefined, and returns the value NaN (not a number). MATLAB ignores NaNs when it plots a vector, so the resulting plot appears continuous.

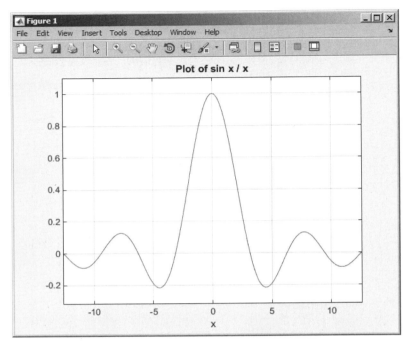

Figure 7.7 The function $f(x) = \sin x / x$, plotted with function `ezplot`.

```
ezplot('sin(x)/x',[-4*pi 4*pi]);
title('Plot of sin x / x');
grid on;
```

Function `fplot` is similar to but more sophisticated than `ezplot`. This function takes the following forms.

```
fplot(fun);
fplot(fun, [xmin xmax]);
fplot(fun, [xmin xmax], LineSpec);
[x, y] = fplot(fun, [xmin xmax], ...);
```

The argument `fun` is either a function handle, the name of an M-file function, or a *character string* containing the functional expression to be evaluated. The optional parameter [`xmin xmax`] specifies the range of the function to plot. If it is absent, the function will be plotted between -2π and 2π. The optional parameter `LineSpec` specifies the line color, line style, and marker style to use when displaying the function. The `LineSpec` values are the same as for the `plot` function. The final version of the `fplot` function returns the `x` and `y` values of the line without actually plotting the function.

Function `fplot` has the following advantages over `ezplot`:

1. Function `fplot` is *adaptive*, meaning that it calculates and displays more data points in the regions where the function being plotted is changing most

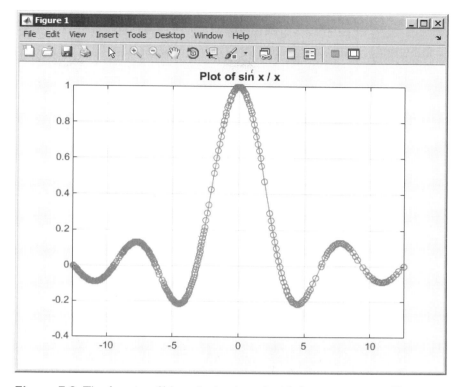

Figure 7.8 The function $f(x) = \sin x/x$, plotted with function `fplot`. [See color insert.]

rapidly. The resulting plot is more accurate at locations where a function's behavior changes suddenly.

2. Function `fplot` supports user-defined line specifications (color, line style, and marker style).

In general, you should use `fplot` in preference to `ezplot` whenever you plot functions.

The following statements plot the function $f(x) = \sin x/x$ between -4π and 4π using function `fplot`. Note that they specify a dashed red line with circular markers. The output of these statements is shown in Figure 7.8. [See color insert.]

```
fplot('sin(x)/x',[-4*pi 4*pi],'-or');
title('Plot of sin x / x');
grid on;
```

Good Programming Practice

Use function `fplot` to plot functions directly without having to create intermediate data arrays.

7.7 Histograms

A *histogram* is a plot showing the distribution of values within a data set. To create a histogram, the range of values within the data set is divided into evenly spaced bins, and the number of data values falling into each bin is determined. The resulting count can then be plotted as a function of bin number.

The standard MATLAB histogram function is `hist`. The forms of this function are shown below:

```
hist(y)
hist(y,nbins)
hist(y,x)
[n,xout] = hist(y,...)
```

The first form of the function creates and plots a histogram with 10 equally spaced bins, while the second form creates and plots a histogram with `nbins` equally spaced bins. The third form of the function allows the user to specify the bin centers to use in an array `x`; the function creates a bin centered on each element in the array. In all three of these cases, the function both creates and plots the histogram. The last form of the function creates a histogram and returns the bin centers in array `xout` and the count in each bin in array `n`, without actually creating a plot.

For example, the following statements create a data set containing 10,000 Gaussian random values, and generate a histogram of the data using 15 evenly spaced bins. The resulting histogram is shown in Figure 7.9.

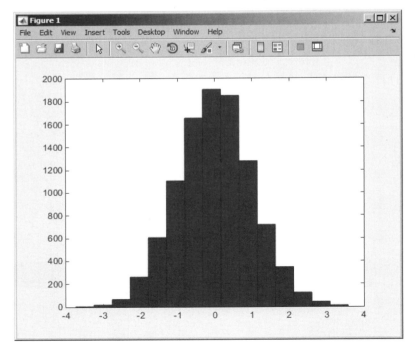

Figure 7.9 A histogram.

```
y = randn(10000,1);
hist(y,15);
```

MATLAB also includes a function `rose` to create and plot a histogram on radial axes. It is especially useful for distributions of angular data. You will be asked to use this function in an end-of-chapter exercise.

▶ Example 7.4—Radar Target Processing

Some modern radars use coherent integration, allowing them to determine both the range and the velocity of detected targets. Figure 7.10 shows the output of an integration interval from such a radar. This is a plot of amplitude (in dB milliwatts) versus relative range and velocity. Two targets are present in this data set—one at a relative range of about 0 meters and moving at about 80 meters per second, and a second one at a relative range of about 20 meters and moving at about 60 m/s. The remainder of the range and velocity space is filled with sidelobes and background noise.

To estimate the strength of the targets detected by this radar, we need to calculate the signal to noise ratio (SNR) of the targets. It is easy to find the amplitudes of each target, but how can we determine the noise level of the background? One common approach relies in recognizing that most of the range/velocity cells in the radar data contain only noise. If we can find the most common amplitude amongst the range/velocity cells, then that should correspond to the level of the noise. A

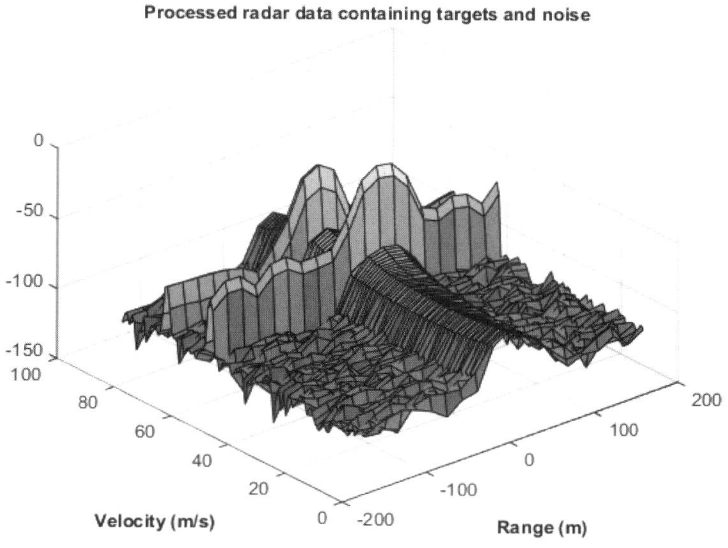

Figure 7.10 A radar range-velocity space containing two targets and background noise. [See color insert.]

good way to do this is to make a histogram of the amplitudes of all samples in the range/velocity space and then look for the amplitude bin containing the most samples.

Find the background noise level in this sample of processed radar data.

Solution

1. **State the problem**

   Determine the background noise level in a given sample of range/velocity radar data, and report that value to the user.

2. **Define the inputs and outputs**

   The input for this problem is a sample of radar data stored in file rd_space.mat. This MAT file contains a vector of range data called range, a vector of velocity data called velocity, and an array of amplitude values called amp. The output from this program is the amplitude of the largest bin in a histogram of data samples, which should correspond to the noise level.

3. **Describe the algorithm**

   This task can be broken down into four major sections:

   ```
   Read the input data set
   Calculate the histogram of the data
   Locate the peak bin in the data set
   Report the noise level to the user
   ```

   The first step is to read the data, which is trivial. The pseudocode for this step is:

   ```
   % Load the data
   load rd_space.mat
   ```

   Next, we must calculate the histogram of the data. Using the MATLAB help system, we can see that the histogram function requires a *vector* of input data, not a 2D array. We can convert the 2D array amp into a 1D vector of data using the form amp(:), as we described in Chapter 2. The form of the histogram function that specifies output parameters will return an array of bin counts and bin centers. The number of bins to use must also be chosen carefully. If there are too few bins, the estimate of the noise level will be coarse. If there are too many bins, there will not be enough samples in the range/velocity space to fill them properly. As a compromise, we will try 31 bins. The pseudocode for this step is:

   ```
   % Calculate histogram
   [nvals, amp_levels] = hist(amp(:), 31)
   ```

   where nvals is an array of the counts in each bin, and amp_levels is an array containing the central amplitude value for each bin.

   Now we must locate the peak bin in the output array nvals. The best way to do this is using the MATLAB function max, which returns the maximum value (and optionally the location of that maximum value) in an array. Use the MATLAB help system to look this function up. The form of this function that we need is:

```
[max_val, max_loc] = max(array)
```

where `max_val` is the maximum value in the array and `max_loc` is the array index of that maximum value. Once the location of the maximum amplitude is known, the signal strength of that bin can be found by looking at location `max_loc` in the `amp_levels` array. The pseudocode for this step is:

```
% Calculate histogram
[nvals, amp_levels] = hist(amp, 31)

% Get location of peak
[max_val, max_loc] = max(nvals)

% Get the power level of that bin
noise_power = amp_levels(max_loc)
```

The final step is to tell the user. This is trivial.

```
Tell user.
```

4. **Turn the algorithm into MATLAB statements.**
   The final MATLAB code is shown below.

```
% Script file: radar_noise_level.m
%
% Purpose:
%   This program calculates the background noise level
%   in a buffer of radar data.
%
% Record of revisions:
%     Date          Programmer              Description of change
%     ====          ==========              =====================
%   05/29/14      S. J. Chapman             Original code
%
% Define variables:
%   amp_levels    -- Amplitude level of each bin
%   noise_power   -- Power level of peak noise
%   nvals         -- Number of samples in each bin

% Load the data
load rd_space.mat

% Calculate histogram
[nvals, amp_levels] = hist(amp(:), 31);

% Get location of peak
[max_val, max_loc] = max(nvals);
```

```
% Get the power level of that bin
noise_power = amp_levels(max_loc);

% Tell user
fprintf('The noise level in the buffer is %6.2f dBm.\n', noise_power);
```

5. **Test the program.**

    Next, we must test the function using various strings.

    ```
    » radar_noise_level
    The noise level in the buffer is -104.92 dBm.
    ```

To verify this answer, we can plot the histogram of the data calling `hist` without output arguments.

```
hist(amp(:), 31);
xlabel('\bfAmplitude (dBm)');
ylabel('\bfCount');
title('\bfHistogram of Cell Amplitudes');
```

The resulting plot is shown in Figure 7.11. The target power appears to be about –20 dBm, and the noise power does appear to be about –105 dBm. This program appears to be working properly.

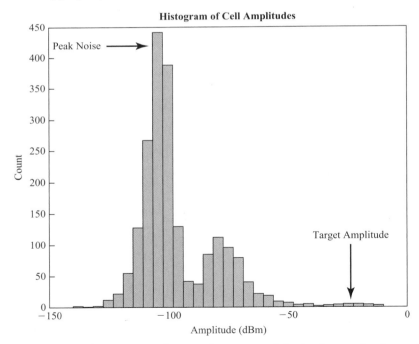

Figure 7.11 A histogram showing the power of the background noise and the power of the detected targets.

Quiz 7.1

This quiz provides a quick check to see if you have understood the concepts introduced in Sections 7.1 through 7.7. If you have trouble with the quiz, reread the section, ask your instructor, or discuss the material with a fellow student. The answers to this quiz are found in the back of the book.

1. What is a local function? How does it differ from an ordinary function?
2. What is meant by the term "scope"?
3. What is a private function? How does it differ from an ordinary function?
4. What are nested functions? What is the scope of a variable in the parent function of a nested function?
5. In what order does MATLAB decide search for a function to execute?
6. What is a function handle? How do you create a function handle? How do you call a function using a function handle?
7. What will be returned by the following function, if it is called with the expression `myfun(@cosh)`?

```
function res = myfun(x)
res = func2str(x);
end
```

7.8 Summary

In Chapter 7, we presented advanced features of user-defined functions.

Function functions are MATLAB functions whose input arguments include the names of other functions. The functions whose names are passed to the function function are normally used during that function's execution. Examples are some root-solving and plotting functions.

Local functions are additional functions placed within a single file. Local functions are only accessible from other functions within the same file. Private functions are functions placed in a special subdirectory called `private`. They are only accessible to functions in the parent directory. Local functions and private functions can be used to restrict access to MATLAB functions.

Function handles are a special data type containing all the information required to invoke a function. Function handles are created with the @ operator or the `str2func` function, and are used by naming the handle following by parentheses and the required calling arguments.

Anonymous functions are simple functions without a name, which are created in a single line and called by their function handles.

Functions `ezplot` and `fplot` are function functions that can directly plot a user-specified function without having to create output data first.

Histograms are plots of the number of samples from a data set that fall into each of a series of amplitude bins.

7.8.1 Summary of Good Programming Practice

The following guidelines should be adhered to when working with MATLAB functions.

1. Use local functions or private functions to hide special-purpose calculations that should not be generally accessible to other functions. Hiding the functions will prevent their accidental use and will also prevent conflicts with other public functions of the same name.
2. Use function `fplot` to plot functions directly without having to create intermediate data arrays.

7.8.2 MATLAB Summary

The following summary lists all of the MATLAB commands and functions described in this chapter, along with a brief description of each one.

Commands and Functions

@	Creates a function handle (or an anonymous function).
eval	Evaluates a character string as though it had been typed in the Command Window.
ezplot	Easy-to-use function plotter.
feval	Calculates the value of a function $f(x)$ defined by an M-file at a specific x.
fminbnd	Minimizes a function of one variable.
fplot	Plots a function by name.
functions	Recovers miscellaneous information from a function handle.
func2str	Recovers the function name associated with a given function handle.
fzero	Finds a zero of a function of one variable.
global	Declares global variables.
hist	Calculates and plots a histogram of a data set.
inputname	Returns the actual name of the variable that corresponds to a particular argument number.
nargchk	Returns a standard error message if a function is called with too few or too many arguments.
nargin	Returns the number of actual input arguments that were used to call the function.
nargout	Returns the number of actual output arguments that were used to call the function.
ode45	Function to solve ordinary differential equations using a Runge-Kutta (4,5) technique.
quad	Numerically integrate a function.
str2func	Creates a function handle from a specified string.

7.9 Exercises

7.1 Write a function that uses function `random0` from Chapter 6 to generate a random value in the range $[-1.0, 1.0)$. Make `random0` a local function of your new function.

7.2 Write a function that uses function `random0` to generate a random value in the range `[low, high)`, where `low` and `high` are passed as calling arguments. Make `random0` a private function called by your new function.

7.3 Write a single MATLAB function `hyperbolic` to calculate the hyperbolic sine, cosine, and tangent functions as defined in Exercise 6.20. The function should have two arguments. The first argument will be a string containing the function names `'sinh'`, `'cosh'`, or `'tanh'`, and the second argument will be the value of x at which to evaluate the function. The file should also contain three local functions `sinh1`, `cosh1`, and `tanh1` to perform the actual calculations, and the primary function should call the proper local function depending on the value in the string. (*Note:* Be sure to handle the case of an incorrect number of arguments and also the case of an invalid string. In either case, the function should generate an error.)

7.4 Write a program that creates three anonymous functions representing the functions $f(x) = 10 \cos x$, $g(x) = 5 \sin x$, and $h(a,b) = \sqrt{a^2 + b^2}$. Plot $h(f(x), g(x))$ over the range $-10 \leq x \leq 10$.

7.5 Plot the function $f(x) = 1/\sqrt{x}$ over the range $0.1 \leq x \leq 10.0$ using function `fplot`. Be sure to label your plot properly.

7.6 **Minimizing a Function of One Variable** Function `fminbnd` can be used to find the minimum of a function over a user-defined interval. Look up the details of this function in the MATLAB help, and find the minimum of the function $y(x) = x^4 - 3x^2 + 2x$ over the interval (0.5 1.5). Use an anonymous function for $y(x)$.

7.7 Plot the function $y(x) = x^4 - 3x^2 + 2x$ over the range $(-2, 2)$. Then use function `fminbnd` to find the minimum value over the interval $(-1.5, 0.5)$. Did the function actually find the minimum value over that region? What is going on here?

7.8 **Histogram** Create an array of 100,000 samples from function `randn`, the built-in MATLAB Gaussian random number generator. Plot a histogram of these samples over 21 bins.

7.9 **Rose Plot** Create an array of 100,000 samples from function `randn`, the built-in MATLAB Gaussian random number generator. Create a histogram of these samples over 21 bins, and plot them on a rose plot. (*Hint:* Look up rose plots in the MATLAB help subsystem.)

7.10 **Minima and Maxima of a Function** Write a function that attempts to locate the maximum and minimum values of an arbitrary function $f(x)$ over a certain range. The function handle of the function being evaluated should be passed to the function as a calling argument. The function should have the following input arguments:

```
first_value  --  The first value of x to search
last_value   --  The last value of x to search
num_steps    --  The number of steps to include in the search
func         --  The name of the function to search
```

The function should have the following output arguments:

```
xmin    -- The value of x at which the minimum was found
min_value  --  The minimum value of f(x) found
xmax    -- The value of x at which the maximum was found
max_value  --  The maximum value f(x) found
```

Be sure to check that there are a valid number of input arguments and that the MATLAB `help` and `lookfor` commands are properly supported.

7.11 Write a test program for the function generated in the previous exercise. The test program should pass to the function function the user-defined function $f(x) = x^3 - 5x^2 + 5x + 2$, and search for the minimum and maximum in 200 steps over the range $-1 \le x \le 3$. It should print out the resulting minimum and maximum values.

7.12 Write a program that locates the zeros of the function $f(x) = \cos^2 x - 0.25$ between 0 and 2π. Use the function `fzero` to actually locate the zeros of this function. Plot the function over that range and show that `fzero` has reported the correct values.

7.13 Write a program that evaluates the function $f(x) = \tan^2 x + x - 2$ between -2π and 2π in steps of $\pi/10$ and plots the results. Create a function handle for your function, and use function `feval` to evaluate your function at the specified points.

7.14 Write a program that locates and reports the positions of each radar target in the range-velocity space of Example 7.4. For each target, report range, velocity, amplitude, and signal-to-noise ratio (SNR).

7.15 **Derivative of a Function** The *derivative* of a continuous function $f(x)$ is defined by the equation

$$\frac{d}{dx} f(x) = \lim_{\Delta x \to 0} \frac{f(x + \Delta x) - f(x)}{\Delta x} \tag{7.8}$$

In a sampled function, this definition becomes

$$f'(x_i) = \frac{f(x_{i+1}) - f(x_i)}{\Delta x} \tag{7.9}$$

where $\Delta x = x_{i+1} - x_i$. Assume that a vector `vect` contains `nsamp` samples of a function taken at a spacing of `dx` per sample. Write a function that will calculate the derivative of this vector from Equation (7.9). The function should check to make sure that `dx` is greater than zero to prevent divide-by-zero errors in the function.

To check your function, you should generate a data set whose derivative is known, and compare the result of the function with the known correct answer. A good choice for a test function is $\sin x$. From elementary calculus, we know that $\frac{d}{dx}(\sin x) = \cos x$. Generate an input vector containing 100 values of the function $\sin x$ starting at $x = 0$ and using a step size Δx of 0.05. Take the derivative of the vector with your function, and then compare the resulting answers to the known correct answer. How close did your function come to calculating the correct value for the derivative?

7.16 Derivative in the Presence of Noise We will now explore the effects of input noise on the quality of a numerical derivative. First, generate an input vector containing 100 values of the function sin x starting at $x = 0$ and using a step size Δx of 0.05, just as you did in the previous problem. Next, use function `random0` to generate a small amount of random noise with a maximum amplitude of ± 0.02, and add that random noise to the samples in your input vector. Figure 7.12 shows an example of the sinusoid corrupted by noise. Note that the peak amplitude of the noise is only 2% of the peak amplitude of your signal, since the maximum value of sin x is 1. Now take the derivative of the function using the derivative function that you developed in the last problem. How close to the theoretical value of the derivative did you come?

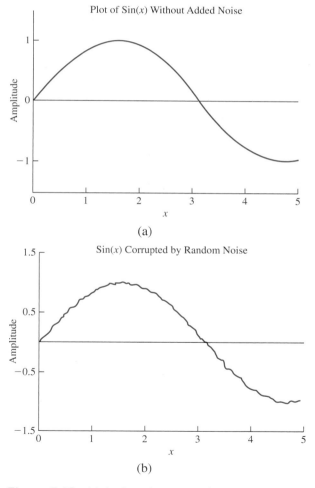

Figure 7.12 (a) A plot of sin x as a function of x with no noise added to the data. (b) A plot of sin x as a function of x with a 2% peak amplitude uniform random noise added to the data.

7.17 Create an anonymous function to evaluate the expression $y(x) = 2e^{-0.5x}\cos x - 0.2$, and find the roots of that function with `fzero` between 0 and 7.

7.18 The factorial function created in Example 7.4 does not check to ensure that the input values are non-negative integers. Modify the function to perform this check, and to write out an error if an illegal value is passed as a calling argument.

7.19 **Fibonacci Numbers** A function is said to be *recursive* if the function calls itself. MATLAB functions are designed to allow recursive operation. To test this feature, write a MATLAB function that derives the Fibonacci numbers. The nth Fibonacci number is defined by the equation:

$$F_n = \begin{cases} F_{n-1} + F_{n-2} & n > 1 \\ 1 & n = 1 \\ 0 & n = 0 \end{cases} \qquad (7.10)$$

where n is a non-negative integer. The function should check to make sure that there is a single argument n, and that n is a non-negative integer. If it is not, generate an error using the `error` function. If the input argument is a non-negative integer, the function should evaluate F_n using Equation (7.10). Test your function by calculating the Fibonacci numbers for $n = 1$, $n = 5$, and $n = 10$.

7.20 **The Birthday Problem** The Birthday Problem is: if there are a group of n people in a room, what is the probability that two or more of them have the same birthday (month and day, ignoring the year)? It is possible to determine the answer to this question by simulation. Write a function that calculates the probability that two or more of n people will have the same birthday, where n is a calling argument. (*Hint:* To do this, the function should create an array of size n and generate n birthdays in the range 1 to 365 randomly. It should then check to see if any of the n birthdays are identical. The function should perform this experiment at least 5000 times and calculate the fraction of those times in which two or more people had the same birthday.) Write a test program that calculates and prints out the probability that 2 or more of n people will have the same birthday for $n = 2, 3, \ldots, 40$.

7.21 **Constant False Alarm Rate (CFAR)** A simplified radar receiver chain is shown in Figure 7.13a. When a signal is received in this receiver, it contains both the desired information (returns from targets) and thermal noise. After the detection step in the receiver, we would like to be able to pick out received target returns from the thermal noise background. We can do this by setting a threshold level and then declaring that we see a target whenever the signal crosses that threshold. Unfortunately, it is occasionally possible for the receiver noise to cross the detection threshold even if no target is present. If that happens, we will declare the noise spike to be a target, creating a *false alarm*. The detection threshold needs to be set as low as possible so that we can detect weak targets, but it must not be set too low, or we get many false alarms.

After video detection, the thermal noise in the receiver has a Rayleigh distribution. Figure 7.13b shows 100 samples of a Rayleigh-distributed noise with a mean amplitude of 10 volts. Note that there would be one false alarm even if the detection threshold were as high as 26! The probability distribution of these noise samples is shown in Figure 7.13c.

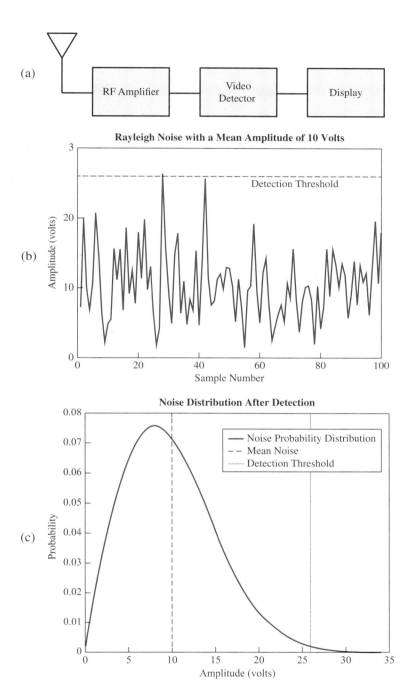

Figure 7.13 (a) A typical radar receiver. (b) Thermal noise with a mean of 10 volts output from the detector. The noise sometimes crosses the detection threshold. (c) Probability distribution of the noise out of the detector.

Detection thresholds are usually calculated as a multiple of the mean noise level so that if the noise level changes, the detection threshold will change with it to keep false alarms under control. This is known as *constant false alarm rate* (CFAR) detection. A detection threshold is typically quoted in decibels. The relationship between the threshold in dB and the threshold in volts is

$$\text{Threshold (volts)} = \text{Mean Noise Level (volts)} \times 10^{\frac{dB}{20}} \qquad (7.11)$$

or

$$dB = 20 \log_{10}\left(\frac{\text{Threshold (volts)}}{\text{Mean Noise Level (volts)}}\right) \qquad (7.12)$$

The false alarm rate for a given detection threshold is calculated as:

$$P_{fa} = \frac{\text{Number of False Alarms}}{\text{Total Number of Samples}} \qquad (7.13)$$

Write a program that generates 1,000,000 random noise samples with a mean amplitude of 10 volts and a Rayleigh noise distribution. Determine the false alarm rates when the detection threshold is set to 5, 6, 7, 8, 9, 10, 11, 12, and 13 dB above the mean noise level. At what level should the threshold be set to achieve a false alarm rate of 10^{-4}?

7.22 Function Generators   Write a nested function that evaluates a polynomial of the form $y = ax^2 + bx + c$. The host function gen_func should have three calling arguments a, b, and c to initialize the coefficients of the polynomial. It should also create and return a function handle for the nested function eval_func. The nested function eval_func(x) should calculate a value of y for a given value of x, using the values of a, b, and c stored in the host function. This is effectively a function generator, since each combination of a, b, and c values produces a function handle that evaluates a unique polynomial. Then perform the following steps:

(a) Call gen_func(1,2,1) and save the resulting function handle in variable h1. This handle now evaluates the function $y = x^2 + 2x + 1$.
(b) Call gen_func(1,4,3) and save the resulting function handle in variable h2. This handle now evaluates the function $y = x^2 + 4x + 3$.
(c) Write a function that accepts a function handle and plots the specified function between two specified limits.
(d) Use this function to plot the two polynomials generated in parts (a) and (b) above.

7.23 RC Circuits   Figure 7.14a shows a simple series *RC* circuit with the output voltage taken across the capacitor. Assume that there is no voltage or power in this circuit before time $t = 0$, and that the voltage $v_{in}(t)$ is applied at time $t \geq 0$. Calculate and plot the output voltage of this circuit for time $0 \leq t \leq 10$ s. (*Hint:* The output voltage from this circuit can be found by writing a Kirchoff's Current Law (KCL) equation at the output, and solving for $v_{out}(t)$. The KCL equation is

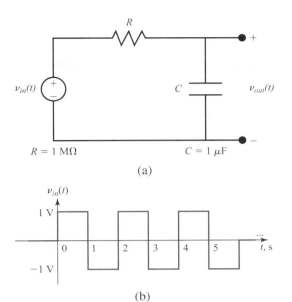

Figure 7.14 (a) A simple series RC circuit. (b) The input voltage to this circuit as a function of time. Note that the voltage is 0 for all times before zero and all times after $t = 6$ s.

$$\frac{v_{out}(t) - v_{in}(t)}{R} + C\frac{dv_{out}(t)}{dt} = 0 \tag{7.14}$$

Collecting terms in this equation produces the result

$$\frac{dv_{out}(t)}{dt} + \frac{1}{RC}v_{out}(t) = \frac{1}{RC}v_{in}(t) \tag{7.15}$$

Solve this equation for $v_{out}(t)$.

7.24 Calculate and plot the output v of the following differential equation:

$$\frac{dv(t)}{dt} + v(t) = \begin{cases} t & 0 \le t \le 5 \\ 0 & \text{elsewhere} \end{cases} \tag{7.16}$$

Chapter 8

Complex Numbers and 3D Plots

In this chapter, we will learn how to work with complex numbers and about the types of three-dimensional plots available in MATLAB.

8.1 Complex Data

Complex numbers are numbers with both a real and an imaginary component. Complex numbers occur in many problems in science and engineering. For example, complex numbers are used in electrical engineering to represent alternating current voltages, currents, and impedances. The differential equations that describe the behavior of most electrical and mechanical systems also give rise to complex numbers. Because they are so ubiquitous, it is impossible to work as an engineer without a good understanding of the use and manipulation of complex numbers.

A complex number has the general form

$$c = a + bi \tag{8.1}$$

where c is a complex number, a and b are both real numbers, and i is $\sqrt{-1}$. The number a is called the *real part* and b is called the *imaginary part* of the complex number c. Since a complex number has two components, it can be plotted as a point on a plane (see Figure 8.1). The horizontal axis of the plane is the real axis, and the vertical axis of the plane is the imaginary axis, so that any complex number $a + bi$ can be represented as a single point a units along the real axis and b units along the imaginary axis. A complex number represented this way is said to be in *rectangular coordinates*, since the real and imaginary axes define the sides of a rectangle.

A complex number can also be represented as a vector of length z and angle θ, where θ is the counterclockwise angle between the positive real (x) axis and the line

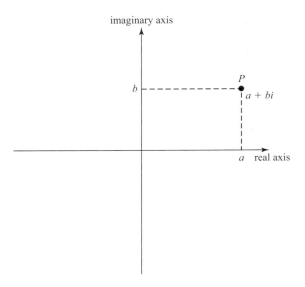

Figure 8.1 Representing a complex number in rectangular coordinates.

from the origin to the point c on the complex plane (see Figure 8.2). A complex number represented this way is said to be in *polar coordinates*.

$$c = a + bi = z\angle\theta \qquad (8.2)$$

The relationships among the rectangular and polar coordinate terms a, b, z, and θ are:

$$a = z\cos\theta \qquad (8.3)$$

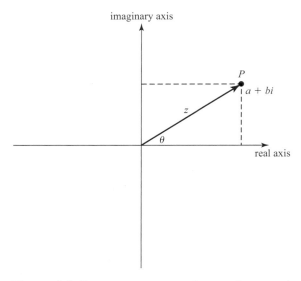

Figure 8.2 Representing a complex number in polar coordinates.

$$b = z \sin \theta \tag{8.4}$$

$$z = \sqrt{a^2 + b^2} \tag{8.5}$$

$$\theta = \tan^{-1} \frac{b}{a} \tag{8.6}$$

where $\tan^{-1}(\ )$ is the two-argument inverse tangent function `atan2(y, x)`, whose output is defined over the range $-\pi \leq \theta \leq \pi$.

MATLAB uses rectangular coordinates to represent complex numbers. Each complex number consists of a pair of real numbers (a,b). The first number (a) is the real part of the complex number, and the second number (b) is the imaginary part of the complex number.

If complex numbers c_1 and c_2 are defined as $c_1 = a_1 + b_1 i$ and $c_2 = a_2 + b_2 i$, then the addition, subtraction, multiplication, and division of c_1 and c_2 are defined as:

$$c_1 + c_2 = (a_1 + a_2) + (b_1 + b_2)i \tag{8.7}$$

$$c_1 - c_2 = (a_1 - a_2) + (b_1 - b_2)i \tag{8.8}$$

$$c_1 \times c_2 = (a_1 a_2 - b_1 b_2) + (a_1 b_2 - b_1 a_2)i \tag{8.9}$$

$$\frac{c_1}{c_2} = \frac{a_1 a_2 + b_1 b_2}{a_2^2 + b_2^2} + \frac{b_1 a_2 - a_1 b_2}{a_2^2 + b_2^2}i \tag{8.10}$$

Note that additions and subtractions are very simple in rectangular form, but multiplications and divisions are relatively complex. If complex numbers are expressed in polar form instead, multiplication and division are much simpler. In polar form, the multiplication of two complex numbers is performed by multiplying the magnitudes of the two numbers and adding the angles of the two numbers:

$$c_1 \times c_2 = z_1 z_2 \angle \theta_1 + \theta_2 \tag{8.11}$$

Similarly, division is performed by dividing the magnitudes of the two numbers and subtracting the angles of the two numbers:

$$\frac{c_1}{c_2} = \frac{z_1}{z_2} \angle \theta_1 - \theta_2 \tag{8.12}$$

When two complex numbers appear in a binary operation, MATLAB performs the required additions, subtractions, multiplications, or divisions between the two complex numbers using versions of the above formulas.

8.1.1 Complex Variables

A complex variable is created automatically when a complex value is assigned to a variable name. The easiest way to create a complex value is to use the intrinsic values `i` or `j`, both of which are pre-defined to be $\sqrt{-1}$. For example, the following statement stores the complex value $4 + i3$ into variable `c1`.

```
» c1 = 4 + i*3
c1 =
    4.0000 + 3.0000i
```

Alternately, the imaginary part can be specified by simply appending an `i` or `j` to the end of a number:

```
» c1 = 4 + 3i
c1 =
    4.0000 + 3.0000i
```

The function `isreal` can be used to determine whether a given array is real or complex. If any element of an array has an imaginary component, then the array is complex, and `isreal(array)` returns a 0.

8.1.2 Using Complex Numbers with Relational Operators

It is possible to compare two complex numbers with the `==` relational operator to see if they are equal to each other and to compare them with the `~=` operator to see if they are not equal to each other. Both of these operators produce the expected results. For example, if $c_1 = 4 + i3$ and $c_2 = 4 - i3$, then the relational operation $c_1 == c_2$ produces a 0 and the relational operation $c_1 ~= c_2$ produces a 1.

However, *comparisons with the* >, <, >=, *or* <= *operators do not produce the expected results.* When complex numbers are compared with these relational operators, only the *real parts* of the numbers are compared. For example, if $c_1 = 4 + i3$ and $c_2 = 3 + i8$, then the relational operation $c_1 > c_2$ produces a true (1) even though the magnitude of c_1 is really smaller than the magnitude of c_2.

If you ever need to compare two complex numbers with these operators, you will probably be more interested in the total magnitude of the number than in the magnitude of only its real part. The magnitude of a complex number can be calculated with the `abs` intrinsic function (see below), or directly from Equation (8.5).

$$|c| = \sqrt{a^2 + b^2} \tag{8.5}$$

If we compare the *magnitudes* of c_1 and c_2 above, the results are more reasonable: `abs(c_1) > abs(c_2)` produces a 0, since the magnitude of c_2 is greater than the magnitude of c_1.

///

⊠ Programming Pitfalls

Be careful when using the relational operators with complex numbers. The relational operators >, >=, <, and <= only compare the *real parts* of complex numbers, not their magnitudes. If you need these relational operators with complex number, it will probably be more sensible to compare the total magnitudes rather than only the real components.

///

8.1.3 Complex Functions

MATLAB includes many functions that support complex calculations. These functions fall into three general categories:

1. **Type conversion functions** These functions convert data from the complex data type to the real (double) data type. Function real returns the *real part* of a complex number as a double data type and throws away the imaginary part of the complex number. Function imag returns the *imaginary part* of a complex number as a double precision data type.

2. **Absolute value and angle functions** These functions convert a complex number to its polar representation. Function abs(c) calculates the absolute value of a complex number using the equation

$$abs(c) = \sqrt{a^2 + b^2}$$

where $c = a + bi$. Function angle(c) calculates the angle of a complex number using the equation

```
angle(c) = atan2(imag(c),real(c))
```

producing an answer in the range $-\pi \le \theta \le \pi$.

3. **Mathematical functions** Most elementary mathematical functions are defined for complex values. These functions include exponential functions, logarithms, trigonometric functions, and square roots. The functions sin, cos, log, sqrt, and so forth will work as well with complex data as they will with real data.

Some of the intrinsic functions that support complex numbers are listed in Table 8.1.

Table 8.1: Some Functions that Support Complex Numbers

Function	Description
conj(c)	Computes the complex conjugate of a number c. If $c = a + bi$, then conj(c) = $a - bi$.
real(c)	Returns the real part of the complex number c.
imag(c)	Returns the imaginary part of the complex number c.
isreal(c)	Returns true (1) if no element of array c has an imaginary component. Therefore, ~isreal(c) returns true (1) if any element of array c has an imaginary component.
abs(c)	Returns the magnitude of the complex number c.
angle(c)	Returns the angle of the complex number c in radians, computed from the expression atan2(imag(c),real(c)).

►Example 8.1—The Quadratic Equation (Revisited)

The availability of complex numbers often simplifies the calculations required to solve problems. For example, when we solved the quadratic equation in Example 4.2, it was necessary to take three separate branches through the program depending on the sign of the discriminant. With complex numbers available, the square root of a negative number presents no difficulties, so we can greatly simplify these calculations.

Write a general program to solve for the roots of a quadratic equation, regardless of type. Use complex variables so that no branches will be required based on the value of the discriminant.

Solution

1. **State the problem**

   Write a program that will solve for the roots of a quadratic equation, whether they are distinct real roots, repeated real roots, or complex roots, without requiring tests on the value of the discriminant.

2. **Define the inputs and outputs**

   The inputs required by this program are the coefficients a, b, and c of the quadratic equation

   $$ax^2 + bx + c = 0 \tag{8.13}$$

   The output from the program will be the roots of the quadratic equation, whether they are real, repeated, or complex.

3. **Describe the algorithm**

   This task can be broken down into three major sections, whose functions are input, processing, and output:

   ```
   Read the input data
   Calculate the roots
   Write out the roots
   ```

   We will now break each of the above major sections into smaller, more detailed pieces. In this algorithm, the value of the discriminant is unimportant in determining how to proceed. The resulting pseudocode is:

   ```
   Prompt the user for the coefficients a, b, and c.
   Read a, b, and c
   discriminant ← b^2 - 4 * a * c
       x1 ← ( -b + sqrt(discriminant) ) / ( 2 * a )
       x2 ← ( -b - sqrt(discriminant) ) / ( 2 * a )
       Print 'The roots of this equation are: '
       Print 'x1 = ', real(x1), ' +i ', imag(x1)
       Print 'x2 = ', real(x2), ' +i ', imag(x2)
   ```

4. **Turn the algorithm into MATLAB statements.**

   The final MATLAB code is shown below.

```
%   Script file: calc_roots2.m
%
%   Purpose:
%     This program solves for the roots of a quadratic equation
%     of the form a*x^2 + b*x + c = 0. It calculates the answers
%     regardless of the type of roots that the equation possesses.
%
%   Record of revisions:
%       Date          Engineer            Description of change
%       ====          ==========          =====================
%     02/24/14      S. J. Chapman          Original code
%
% Define variables:
%    a             -- Coefficient of x^2 term of equation
%    b             -- Coefficient of x term of equation
%    c             -- Constant term of equation
%    discriminant  -- Discriminant of the equation
%    x1            -- First solution of equation
%    x2            -- Second solution of equation

% Prompt the user for the coefficients of the equation
disp ('This program solves for the roots of a quadratic');
disp ('equation of the form A*X^2 + B*X + C = 0.');
a = input ('Enter the coefficient A:');
b = input ('Enter the coefficient B:');
c = input ('Enter the coefficient C:');

% Calculate discriminant
discriminant = b^2 - 4 * a * c;

% Solve for the roots
x1 = ( -b + sqrt(discriminant) ) / ( 2 * a );
x2 = ( -b - sqrt(discriminant) ) / ( 2 * a );

% Display results
disp ('The roots of this equation are:');
fprintf ('x1 = (%f) +i (%f)\n', real(x1), imag(x1));
fprintf ('x2 = (%f) +i (%f)\n', real(x2), imag(x2));
```

5. **Test the program.**

Next, we must test the program using real input data. We will test cases in which the discriminant is greater than, less than, and equal to 0 to be certain that the program is working properly under all circumstances. From Equation (4.2), it is possible to verify the solutions to the equations given below:

$$x^2 + 5x + 6 = 0 \qquad x = -2, \text{ and } x = -3$$
$$x^2 + 4x + 4 = 0 \qquad x = -2$$
$$x^2 + 2x + 5 = 0 \qquad x = -1 \pm 2i$$

When the above coefficients are fed into the program, the results are

```
» calc_roots2
This program solves for the roots of a quadratic
equation of the form A*X^2 + B*X + C = 0.
Enter the coefficient A: 1
Enter the coefficient B: 5
Enter the coefficient C: 6
The roots of this equation are:
x1 = (-2.000000) +i (0.000000)
x2 = (-3.000000) +i (0.000000)
» calc_roots2
This program solves for the roots of a quadratic
equation of the form A*X^2 + B*X + C = 0.
Enter the coefficient A: 1
Enter the coefficient B: 4
Enter the coefficient C: 4
The roots of this equation are:
x1 = (-2.000000) +i (0.000000)
x2 = (-2.000000) +i (0.000000)
» calc_roots2
This program solves for the roots of a quadratic
equation of the form A*X^2 + B*X + C = 0.
Enter the coefficient A: 1
Enter the coefficient B: 2
Enter the coefficient C: 5
The roots of this equation are:
x1 = (-1.000000) +i (2.000000)
x2 = (-1.000000) +i (-2.000000)
```

The program gives the correct answers for our test data in all three possible cases. Note how much simpler this program is compared to the quadratic root solver found in Example 4.2. The complex data type has greatly simplified our program.

◀

▶Example 8.2—Series RC Circuit

Figure 8.3 shows a resistor and capacitor connected in series and driven by a 100-volt AC power source. The output voltage of this circuit can be found from the *voltage divider rule*:

$$\mathbf{V}_{out} = \frac{Z_2}{Z_1 + Z_2}\mathbf{V}_{in} \qquad (8.14)$$

where $\mathbf{V}_{in}$ is the input voltage, $Z_1 = Z_R$ is the impedance of the resistor, and $Z_2 = Z_C$ is the impedance of the capacitor. If the input voltage is $\mathbf{V}_{in} = 100\angle 0°V$, the impedance of the resistor $Z_R = 100\ \Omega$, and the impedance of the capacitor $Z_C = -j100\ \Omega$, what is the output voltage of this circuit?

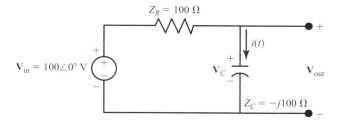

Figure 8.3 An AC voltage divider circuit.

Solution We will need to calculate the output voltage of this circuit in polar coordinates in order to get the magnitude output voltage. The output voltage in rectangular coordinates can be calculated from Equation (8.14), and then the magnitude of the output voltage can be found from Equation (8.5). The code to perform these calculations is

```
%   Script file: voltage_divider.m
%
%   Purpose:
%     This program calculates the output voltage across an
%     AC voltage divider circuit.
%
%   Record of revisions:
%       Date        Programmer          Description of change
%       ====        ==========          =====================
%     02/28/14    S. J. Chapman       Original code
%
% Define variables:
%   vin              -- Input voltage
%   vout             -- Output voltage across z2
%   z1               -- Impedance of first element
%   z2               -- Impedance of second element

% Prompt the user for the coefficients of the equation
disp ('This program calculates the output voltage across
a voltage divider.');
vin = input ('Enter input voltage:');
z1  = input ('Enter z1:');
z2  = input ('Enter z2:');

% Calculate the output voltage
vout = z2 / (z1 + z2) * vin;

% Display results
disp ('The output voltage is:');
fprintf ('vout  =  %f  at  an  angle  of  %f  degrees\n',
abs(vout), angle(vout)*180/pi);
```

When this program is executed, the results are

```
» This program calculates the output voltage across a
voltage divider.
Enter input voltage: 100
Enter z1: 100
Enter z2: -100j
The output voltage is:
vout = 70.710678 at an angle of -45.000000 degrees
```

The program uses complex numbers to calculate the output voltage from this circuit.

◀

8.1.4 Plotting Complex Data

Complex data has both real and imaginary components, and plotting complex data with MATLAB is a bit different than plotting real data. For example, consider the function

$$y(t) = e^{-0.2t}(\cos t + i \sin t) \tag{8.15}$$

If this function is plotted with the conventional `plot` function, only the real data will be plotted—the imaginary part will be ignored. The following statements produce the plot shown in Figure 8.4, together with a warning message that the imaginary part of the data is being ignored.

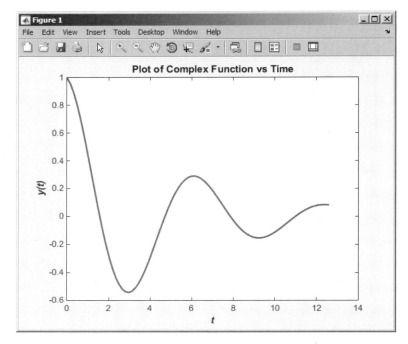

Figure 8.4 Plot of $y(t) = e^{-0.2t}(\cos t + i\sin t)$ using the command `plot(t,y)`.

```
t = 0:pi/20:4*pi;
y = exp(-0.2*t).*(cos(t)+i*sin(t));
plot(t,y,'LineWidth',2);
title('\bfPlot of Complex Function vs Time');
xlabel('\bf\itt');
ylabel('\bf\ity(t)');
```

If both the real and imaginary parts of the function are of interest, then the user has several choices. Both parts can be plotted as a function of time on the same axes using the statements shown below (see Figure 8.5).

```
t = 0:pi/20:4*pi;
y = exp(-0.2*t).*(cos(t)+i*sin(t));
plot(t,real(y),'b-','LineWidth',2);
hold on;
plot(t,imag(y),'r--','LineWidth',2);
title('\bfPlot of Complex Function vs Time');
xlabel('\bf\itt');
ylabel('\bf\ity(t)');
legend ('real','imaginary');
hold off;
```

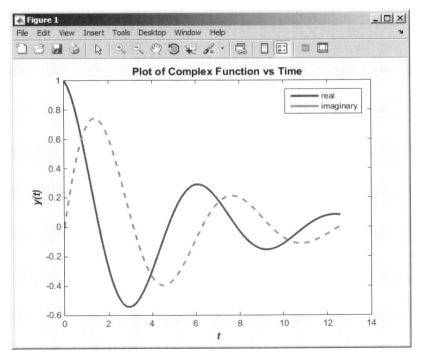

Figure 8.5 Plot of real and imaginary parts of y(t) versus time.

Alternatively, the real part of the function can be plotted versus the imaginary part. If a *single* complex argument is supplied to the `plot` function, it automatically generates a plot of the real part versus the imaginary part. The statements to generate this plot are shown below, and the result is shown in Figure 8.6.

```
t = 0:pi/20:4*pi;
y = exp(-0.2*t).*(cos(t)+i*sin(t));
plot(y,'b-','LineWidth',2);
title('\bfPlot of Complex Function');
xlabel('\bfReal Part');
ylabel('\bfImaginary Part');
```

Finally, the function can be plotted as a polar plot showing magnitude versus angle. The statements to generate this plot are shown below, and the result is shown in Figure 8.7.

```
t = 0:pi/20:4*pi;
y = exp(-0.2*t).*(cos(t)+i*sin(t));
polar(angle(y),abs(y));
title('\bfPlot of Complex Function');
```

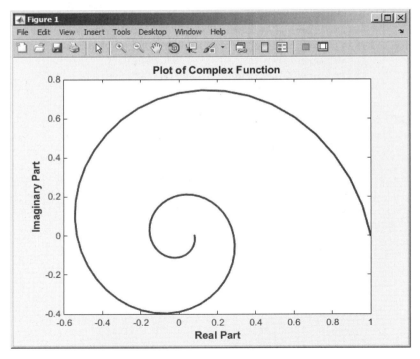

Figure 8.6 Plot of real versus imaginary parts of y(t).

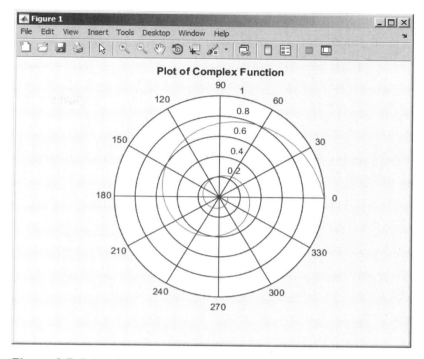

Figure 8.7 Polar plot of magnitude of y(t) versus angle.

This quiz provides a quick check to see if you have understood the concepts introduced in Section 8.1. If you have trouble with the quiz, reread the section, ask your instructor, or discuss the material with a fellow student. The answers to this quiz are found in the back of the book.

1. What is the value of `result` in the following statements?

   (a) ```
       x = 12 + i*5;
       y = 5 - i*13;
       result = x > y;
       ```
   (b) ```
       x = 12 + i*5;
       y = 5 - i*13;
       result = abs(x) > abs(y);
       ```
   (c) ```
       x = 12 + i*5;
       y = 5 - i*13;
       result = real(x) - imag(y);
       ```

2. If `array` is a complex array, what does the function `plot(array)` do?
       ```

8.2 Multidimensional Arrays

MATLAB also supports arrays with more than two dimensions. These **multi-dimensional arrays** are very useful for displaying data that intrinsically has more than two dimensions or for displaying multiple versions of 2-D data sets. For example, measurements of pressure and velocity throughout a three-dimensional volume are very important in such studies as aerodynamics and fluid dynamics. These sorts of areas naturally use multidimensional arrays.

Multidimensional arrays are a natural extension of two-dimensional arrays. Each additional dimension is represented by one additional subscript used to address the data.

It is very easy to create a multidimensional array. They can be created either by assigning values directly in assignment statements or by using the same functions that are used to create one- and two-dimensional arrays. For example, suppose that you have a two-dimensional array created by the assignment statement

```
» a = [ 1 2 3 4; 5 6 7 8]
a =
         1            2            3            4
         5            6            7            8
```

This is a 2 × 4 array, with each element addressed by two subscripts. The array can be extended to be a three-dimensional 2 × 4 × 3 array with the following assignment statements.

```
» a(:,:,2) = [  9 10 11 12; 13 14 15 16];
» a(:,:,3) = [ 17 18 19 20; 21 22 23 24]
a(:,:,1)  =
         1            2            3            4
         5            6            7            8
a(:,:,2)  =
         9           10           11           12
        13           14           15           16
a(:,:,3)  =
        17           18           19           20
        21           22           23           24
```

Individual elements in this multidimensional array can be addressed by the array name followed by three subscripts, and subsets of the data can be created using the colon operators. For example, the value of a(2,2,2) is

```
» a(2,2,2)
ans =
        14
```

and the vector a(1,1,:) is

```
» a(1,1,:)
ans(:,:,1)  =
         1
```

```
ans(:,:,2) =
          9
ans(:,:,3) =
         17
```

Multidimensional arrays can also be created using the same functions as other arrays, for example:

```
» b = ones(4,4,2)
b(:,:,1) =
          1       1       1       1
          1       1       1       1
          1       1       1       1
          1       1       1       1
b(:,:,2) =
          1       1       1       1
          1       1       1       1
          1       1       1       1
          1       1       1       1
» c = randn(2,2,3)
c(:,:,1) =
        -0.4326      0.1253
        -1.6656      0.2877
c(:,:,2) =
        -1.1465      1.1892
         1.1909     -0.0376
c(:,:,3) =
         0.3273     -0.1867
         0.1746      0.7258
```

The number of dimensions in a multidimensional array can be found using the ndims function, and the size of the array can be found using the size function.

```
» ndims(c)
ans =
          3
» size(c)
ans =
     2     2     3
```

If you are writing applications that need multidimensional arrays, see the MATLAB Users Guide for more details on the behavior of various MATLAB functions with multidimensional arrays.

Good Programming Practice

Use multidimensional arrays to solve problems that are naturally multivariate in nature, such as aerodynamics and fluid flows.

8.3 Three-Dimensional Plots

MATLAB also includes a rich variety of three-dimensional plots that can be useful for displaying certain types of data. In general, three-dimensional plots are useful for displaying two types of data:

1. Two variables that are functions of the same independent variable, when you wish emphasize the importance of the independent variable.
2. A single variable that is a function of two independent variables.

8.3.1 Three-Dimensional Line Plots

A three-dimensional line plot can be created with the `plot3` function. This function is exactly like the two-dimensional `plot` function, except that each point is represented by x, y, and z values instead of just x and y values. The simplest form of this function is

```
plot(x,y,z);
```

where x, y, and z are equal-sized arrays containing the locations of data points to plot. Function `plot3` supports all the same line size, line style, and color options as `plot`, and you can use it immediately with the knowledge acquired in earlier chapters.

As an example of a three-dimensional line plot, consider the following functions

$$x(t) = e^{-0.2t}\cos 2t$$
$$y(t) = e^{-0.2t}\sin 2t \tag{8.16}$$

These functions might represent the decaying oscillations of a mechanical system in two dimensions, so x and y together represent the location of the system at any given time. Note that x and y are both functions of the *same* independent variable t.

We could create a series of (x,y) points and plot them using the two-dimensional function `plot` (see Figure 8.10a), but if we do so, the importance of time to the behavior of the system will not be obvious in the graph. The following statements create the two-dimensional plot of the location of the object shown in Figure 8.8a. It is not possible from this plot to tell how rapidly the oscillations are dying out.

```
t = 0:0.1:10;
x = exp(-0.2*t) .* cos(2*t);
y = exp(-0.2*t) .* sin(2*t);
plot(x,y);
title('\bfTwo-Dimensional Line Plot');
xlabel('\bfx');
ylabel('\bfy');
grid on;
```

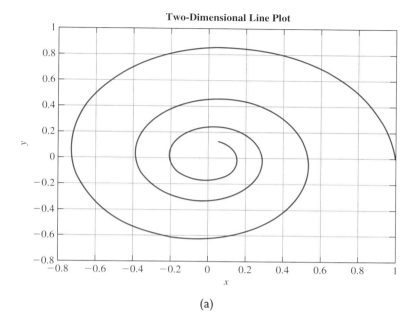

(a)

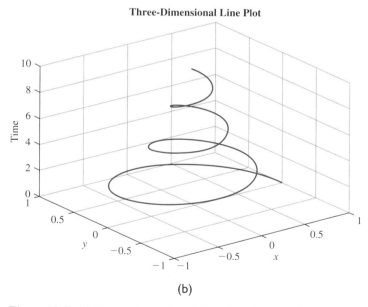

(b)

Figure 8.8 (a) A two-dimensional line plot showing the motion in (x,y) space of a mechanical system. This plot reveals nothing about the time behavior of the system. (b) A three-dimensional line plot showing the motion in (x,y) space versus time for the mechanical system. This plot clearly shows the time behavior of the system.

Instead, we could plot the variables with `plot3` to preserve the time information as well as the two-dimensional position of the object. The following statements will create a three-dimensional plot of Equations (8.16).

```
t = 0:0.1:10;
x = exp(-0.2*t) .* cos(2*t);
y = exp(-0.2*t) .* sin(2*t);
plot3(x,y,t);
title('\bfThree-Dimensional Line Plot');
xlabel('\bfx');
ylabel('\bfy');
zlabel('\bftime');
grid on;
```

The resulting plot is shown in Figure 8.8*b*. Note how this plot emphasizes time dependence of the two variables *x* and *y*.

8.3.2 Three-Dimensional Surface, Mesh, and Contour Plots

Surface, mesh, and contour plots are convenient ways to represent data that is a function of *two* independent variables. For example, the temperature at a point is a function of both the East-West location (*x*) and the North-South (*y*) location of the point. Any value that is a function of two independent variables can be displayed on a three-dimensional surface, mesh, or contour plot. The more common types of plots are summarized in Table 8.2, and examples of each plot are shown in Figure 8.9[1].

To plot data using one of these functions, a user must first create three equal-sized arrays. The three arrays must contain the *x*, *y*, and *z* values of every point to be plotted. The number of columns in each array will be equal to the number of *x* values

Table 8.2: Selected Mesh, Surface, and Contour Plot Functions

Function	Description
`mesh(x,y,z)`	This function creates a mesh or wireframe plot, where x is a two-dimensional array containing the *x* values of every point to display, y is a two-dimensional array containing the *y* values of every point to display, and z is a two-dimensional array containing the *z* values of every point to display.
`surf(x,y,z)`	This function creates a surface plot. Arrays x, y, and z have the same meaning as for a mesh plot.
`contour(x,y,z)`	This function creates a contour plot. Arrays x, y, and z have the same meaning as for a mesh plot.

---

[1]There are many variations on these basic plot types. Consult the MATLAB Help Browser documentation for a complete description of these variations.

Mesh Plot

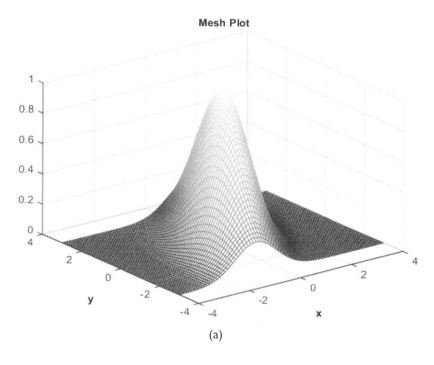

(a)

Surf Plot

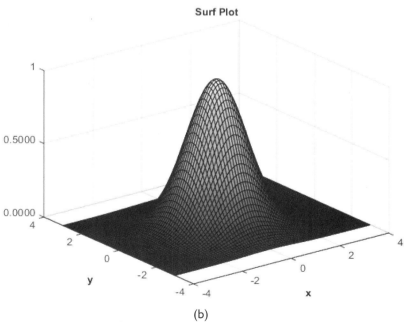

(b)

Figure 8.9 (a) A mesh plot of the function $z(x,y) = e^{-0.5[x^2 + 0.5(x-y)^2]}$. (b) A surface plot of the same function. [See color insert.]

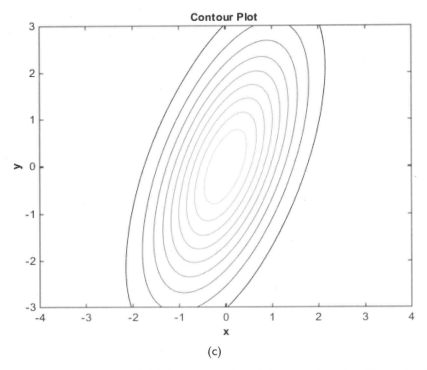

(c)

Figure 8.9 *(continued)* (c) A contour plot of the same function. [See color insert.]

to be plotted, and the number of rows in each array will be equal to the number of y values to be plotted. The first array will contain the x values of each (x,y,z) point to be plotted, the second array will contain the y values of each (x,y,z) point to be plotted, and third array will contain the z values of each (x,y,z) point to be plotted[2].

To understand this better, suppose that we wanted to plot the function

$$z(x,y) = \sqrt{x^2 + y^2} \tag{8.17}$$

for $x = 0$, 1, and 2, and for $y = 0$, 1, 2, and 3. Note that there are three values for x and four values for y, so we will need to calculate and plot a total of $3 \times 4 = 12$ values of z. These data points need to be organized as *three columns* (the number of x values) and *four rows* (the number of y values). Array 1 will contain the x values of each point to calculate, with the same value for all points in a given column, so array 1 will be:

$$\text{array1} = \begin{bmatrix} 1 & 2 & 3 \\ 1 & 2 & 3 \\ 1 & 2 & 3 \\ 1 & 2 & 3 \end{bmatrix}$$

---

[2]This is a very confusing aspect of MATLAB that usually causes trouble for new engineers. When we access arrays, we expect the first argument to specify the row number and the second argument to specify the column number. For some reason MATLAB has reversed this—the array of x arguments specify the number of columns and the array of y arguments specify the number of rows. This reversal has caused countless hours of frustration for beginning MATLAB users over the years.

Array 2 will contain the y values of each point to calculate, with the same value for all points in a given row, so array 2 will be:

$$\text{array2} = \begin{bmatrix} 1 & 1 & 1 \\ 2 & 2 & 2 \\ 3 & 3 & 3 \\ 4 & 4 & 4 \end{bmatrix}$$

Array 3 will contain the z values of each point based in the supplied x and y values. It can be calculated using Equation (8.17) for the supplied values.

$$\text{array3} = \begin{bmatrix} 1.4142 & 2.2361 & 3.1623 \\ 2.2361 & 2.8284 & 3.6056 \\ 3.1624 & 3.6056 & 4.2426 \\ 4.1231 & 4.4721 & 5.0000 \end{bmatrix}$$

The resulting function could then be plotted with the `surf` function as

```
surf(array1,array2,array3);
```

and the result will be as shown in Figure 8.10.

The arrays required for 3D plots can be created manually by using nested loops, or they can be created more easily using built-in MATLAB helper functions. To illustrate this, we will plot the same function twice, once using loops to create the arrays and once using the built-in MATLAB helper functions.

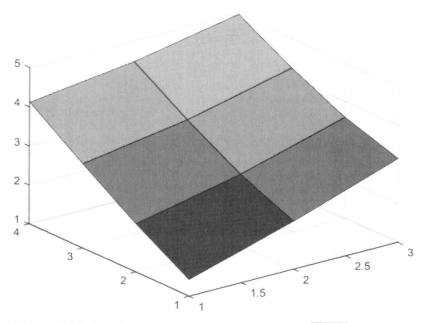

Figure 8.10 A surface plot of the function $z(x,y) = \sqrt{x^2 + y^2}$ for $x = 0, 1,$ and 2, and for $y = 0, 1, 2,$ and 3. [See color insert.]

Suppose that we wish to create a mesh plot of the function

$$z(x,y) = e^{-0.5[x^2+0.5(x-y)^2]} \tag{8.18}$$

over the interval $-4 \leq x \leq 4$ and $-3 \leq y \leq 3$ in steps of 0.1. To do this, we will need to calculate the value of z for all combinations of 61 different x values and 81 different y values. In 3D MATLAB plots, the number of x values corresponds to the number of columns in the z matrix of calculated data, and the number of y values corresponds to the number of *rows* in the z matrix, so the z matrix must contain 61 columns $\times$ 81 rows for a total 4941 values. The code to create the three arrays necessary for a mesh plot with nested loops is a follows:

```
% Get x and y values to calculate
x = -4:0.1:4;
y = -3:0.1:3;

% Preallocate the arrays for speed
array1 = zeros(length(y),length(x));
array2 = zeros(length(y),length(x));
array3 = zeros(length(y),length(x));

% Populate the arrays
for jj = 1:length(x)
    for ii = 1:length(y)
        array1(ii,jj) = x(jj);   % x value in columns
        array2(ii,jj) = y(ii);   % y value in rows
        array3(ii,jj) = ...
            exp(-0.5*(array1(ii,jj)^2+0.5*(array1(ii,jj)-array2(ii,jj))^2));
    end
end

% Plot the data
mesh(array1, array2, array3);
title('\bfMesh Plot');
xlabel('\bfx');
ylabel('\bfy');
zlabel('\bfz');
```

The resulting plot is shown in Figure 8.9a.

The MATLAB function meshgrid makes it much easier to create the arrays of x and y values required for these plots. The form of this function is

```
[arr1,arr2] = meshgrid(xstart:xinc:xend, ...
                       ystart:yinc:yend);
```

where xstart:xinc:xend specifies the x values to include in the grid, and ystart:yinc:yend specifies the y values to be included in the grid.

To create a plot, we can use meshgrid to create the arrays of x and y values and then evaluate the function to plot at each of those (x,y) locations. Finally, we call function mesh, surf, or contour to create the plot.

If we use `meshgrid`, it is much easier to create the 3D mesh plot shown in Figure 8.9*a*.

```
[array1,array2] = meshgrid(-4:0.1:4,-3:0.1:3);
array3 = exp(-0.5*(array1.^2+0.5*(array1-array2).^2));
mesh(array1, array2, array3);
title('\bfMesh Plot');
xlabel('\bfx');
ylabel('\bfy');
zlabel('\bfz');
```

Surface and contour plots may be created by substituting the appropriate function for the `mesh` function.

Good Programming Practice

Use the `meshgrid` function to simplify the creation of 3D `mesh`, `surf`, and `contour` plots.

The `mesh`, `surf`, and `contour` plots also have an alternative input syntax where the first argument is a vector of *x* values, the second argument is a vector of *y* values, and the third argument is a 2D array of data whose number of columns is equal to the number of elements in the *x* vector and whose number of rows is equal to the number of elements in the *y* vector. In this case, the plot function calls `meshgrid` internally to create the three 2D arrays instead of the engineer having to do so.

This is the way that the range-velocity space plot in Figure 7.10 was created. The range and velocity data were vectors, so the plot was created with the following commands:

```
load rd_space;
surf(range,velocity,amp);
xlabel('\bfRange (m)');
ylabel('\bfVelocity (m/s)');
zlabel('\bfAmplitude (dBm)');
title('\bfProcessed radar data containing targets and noise');
```

8.3.3 Creating Three-Dimensional Objects using Surface and Mesh Plots

Surface and mesh plots can be used to create plots of closed objects such as a sphere. To do this, we need to define a set of points representing the entire surface of the object and then plot those points using the `surf` or `mesh` function. For example, consider a simple object like a sphere. A sphere can be defined as the locus of all points that are a given distance r from the center, regardless of azimuth angle θ and elevation angle ϕ. The equation is

$$r = a \qquad (8.19)$$

where a is any positive number. In Cartesian space, the points on the surface of the sphere are defined by the following equations[3]

$$
\begin{aligned}
x &= r\cos\phi\cos\theta \\
y &= r\cos\phi\sin\theta \qquad (8.20) \\
z &= r\sin\phi
\end{aligned}
$$

where the radius r is a constant, the elevation angle ϕ varies from $-\pi/2$ to $\pi/2$, and the azimuth angle θ varies from $-\pi$ to π. A program to plot the sphere is shown below:

```
%  Script file: sphere.m
%
%  Purpose:
%    This program plots the sphere using the surf function.
%
%  Record of revisions:
%      Date          Engineer          Description of change
%      ====          ==========        =====================
%    06/02/14      S. J. Chapman        Original code
%
% Define variables:
%    n           -- Number of points in az and el to plot
%    r           -- Radius of sphere
%    phi         -- meshgrid list of elevation values
%    Phi         -- Array of elevation values to plot
%    theta       -- meshgrid list of azimuth values
%    Theta       -- Array of azimuth values to plot
%    x           -- Array of x point to plot
%    y           -- Array of y point to plot
%    z           -- Array of z point to plot

% Define the number of angles on the sphere to plot
% points at
n = 20;

% Calculate the points on the surface of the sphere
r = 1;
theta = linspace(-pi,pi,n);
phi = linspace(-pi/2,pi/2,n);
[theta,phi] = meshgrid(theta,phi);
```

---

[3]These are the equations that convert from spherical to rectangular coordinates, as we saw in Exercise 2.15.

```
% Convert to (x,y,z) values
x = r * cos(phi) .* cos(theta);
y = r * cos(phi) .* sin(theta);
z = r * sin(phi);

% Plot the sphere
figure(1)
surf (x,y,z);
title ('\bfSphere');
```

The resulting plot is shown in Figure 8.11.

The transparency of surface and patch objects on the current axes can be controlled with the `alpha` function. The `alpha` function takes the form

```
alpha(value);
```

where `value` is a number between 0 and 1. If the value is 0, all surfaces are transparent. If the value is 1, all surfaces are opaque. For any other value, the surfaces are partially transparent. For example, Figure 8.12 shows the sphere object after an alpha of 0.5 is selected. Note that we can now see through the outer surface of the sphere to the back side.

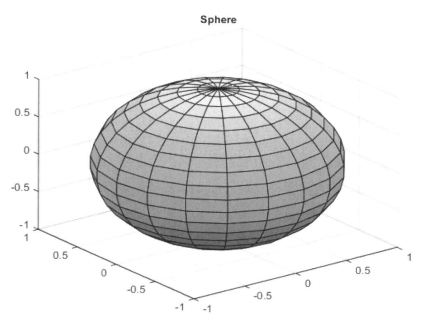

Figure 8.11 Three-dimensional plot of a sphere. [See color insert.]

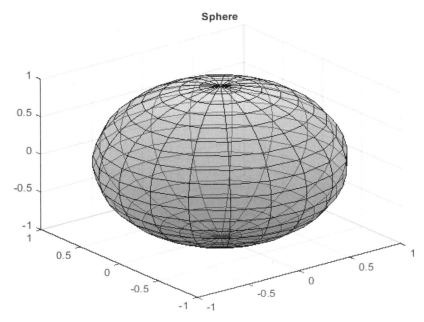

Figure 8.12 A partially transparent sphere, created with an alpha value of 0.5. [See color insert.]

8.4 Summary

MATLAB supports complex numbers as an extension of the double data type. They can be defined using the i or j, both of which are predefined to be $\sqrt{-1}$. Using complex numbers is straightforward, except that the relational operators >, >=, <, and <= only compare the *real parts* of complex numbers, not their magnitudes. They must be used with caution when working with complex values.

Multidimensional arrays are arrays with more than two dimensions. They may be created and used in a fashion similar to one-and two-dimensional arrays. Multidimensional arrays appear naturally in certain classes of physical problems.

MATLAB includes a rich variety of two- and three-dimensional plots. In this chapter, we introduced three-dimensional plots, including mesh, surface, and contour plots.

8.4.1 Summary of Good Programming Practice

The following guidelines should be adhered to:

1. Use multidimensional arrays to solve problems that are naturally multivariate in nature, such as aerodynamics and fluid flows.
2. Use the meshgrid function to simplify the creation of 3D mesh, surf, and contour plots.

8.4.2. MATLAB Summary

The following summary lists all of the MATLAB commands and functions described in this chapter, along with a brief description of each one.

abs	Returns absolute value (magnitude) of a number
alpha	Sets the transparency level of surface plots and patches.
angle	Returns the angle of a complex number in radians.
conj	Computes complex conjugate of a number.
contour	Creates a contour plot.
find	Finds indices and values of non-zero elements in a matrix.
imag	Returns the imaginary part of a complex number.
mesh	Creates a mesh plot.
meshgrid	Creates the (x, y) grid required for mesh, surface, and contour plots.
nonzeros	Returns a column vector containing the nonzero elements in a matrix.
plot(c)	Plots the real versus the imaginary part of a complex array.
real	Returns the real part of a complex number.
surf	Creates a surface plot.

8.5 Exercises

8.1 Write a function to_polar that accepts a complex number c and returns two output arguments containing the magnitude mag and angle theta of the complex number. The output angle should be in degrees.

8.2 Write a function to_complex that accepts two input arguments containing the magnitude mag and angle theta of a complex number c in degrees and returns the complex number c in rectangular form.

8.3 In a sinusoidal steady-state AC circuit, the voltage across a passive element (see Figure 8.13) is given by Ohm's law:

$$\mathbf{V} = \mathbf{I}Z \tag{8.21}$$

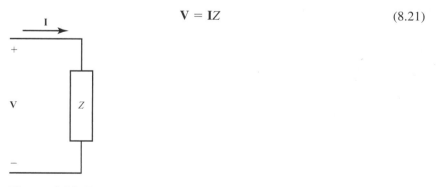

Figure 8.13 The voltage and current relationship on a passive AC circuit element.

where **V** is the voltage across the element, **I** is the current through the element, and Z is the impedance of the element. Note that all three of these values are complex and that these complex numbers are usually specified in the form of a magnitude at a specific phase angle expressed in degrees. For example, the voltage might be $\mathbf{V} = 120\angle30°$ V.

Write a program that reads the voltage across an element and the impedance of the element and calculates the resulting current flow. The input values should be given as magnitudes and angles expressed in degrees, and the resulting answer should be in the same form. Use the function `to_complex` from Exercise 8.2 to convert the numbers to rectangular form for the actual computation of the current, and the function `to_polar` from Exercise 8.1 to convert the answer into polar form for display.

8.4 Two complex numbers in polar form can be multiplied by calculating the product of their amplitudes and the sum of their phases. Thus, if $\mathbf{A}_1 = A_1\angle\theta_1$ and $\mathbf{A}_2 = A_2\angle\theta_2$, then $\mathbf{A}_1\mathbf{A}_2 = A_1A_2\angle\theta_1 + \theta_2$. Write a program that accepts two complex numbers in rectangular form and multiplies them using the above formula. Use the function `to_polar` from Exercise 8.1 to convert the numbers to polar form for the multiplication and the function `to_complex` from Exercise 8.2 to convert the answer into rectangular form for display. Compare the result with the answer calculated using MATLAB's built-in complex mathematics.

8.5 Series RLC Circuit Figure 8.14 shows a series *RLC* circuit driven by a sinusoidal AC voltage source whose value is $120\angle0°$ volts. The impedance of the inductor in this circuit is $Z_L = j2\pi fL$, where j is $\sqrt{-1}$, f is the frequency of the voltage source in hertz, and L is the inductance in henrys. The impedance of the capacitor in this circuit is $Z_C = -j\dfrac{1}{2\pi fC}$, where C is the capacitance in farads. Assume that $R = 100\ \Omega$, $L = 0.1$ mH, and $C = 0.25$ nF.

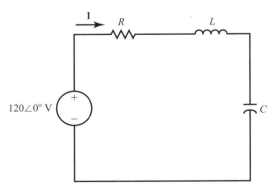

Figure 8.14 A series *RLC* circuit driven by a sinusoidal AC voltage source.

The current **I** flowing in this circuit is given by Kirchhoff's voltage law to be

$$I = \frac{120 \angle 0° \text{ V}}{R + j2\pi f L - j\dfrac{1}{2\pi f C}} \tag{8.22}$$

(a) Calculate and plot the magnitude of this current as a function of frequency as the frequency changes from 100 kHz to 10 MHz. Plot this information on both a linear and a log-linear scale. Be sure to include a title and axis labels.

(b) Calculate and plot the phase angle in degrees of this current as a function of frequency as the frequency changes from 100 kHz to 10 MHz. Plot this information on both a linear and a log-linear scale. Be sure to include a title and axis labels.

(c) Plot both the magnitude and phase angle of the current as a function of frequency on two sub-plots of a single figure. Use log-linear scales.

8.6 Write a function that will accept a complex number c, and plot that point on a Cartesian coordinate system with a circular marker. The plot should include both the x and y axes, plus a vector drawn from the origin to the location of c.

8.7 Plot the function $v(t) = 10e^{(-0.2+j\pi)t}$ for $0 \leq t \leq 10$ using the function plot(t,v). What is displayed on the plot?

8.8 Plot the function $v(t) = 10e^{(-0.2+j\pi)t}$ for $0 \leq t \leq 10$ using the function plot(v). What is displayed on the plot this time?

8.9 Create a polar plot of the function $v(t) = 10e^{(-0.2+j\pi)t}$ for $0 \leq t \leq 10$.

8.10 Plot of the function $v(t) = 10e^{(-0.2+j\pi)t}$ for $0 \leq t \leq 10$ using function plot3, where the three dimensions to plot are the real part of the function, the imaginary part of the function, and time.

8.11 **Euler's Equation** Euler's equation defines e raised to an imaginary power in terms of sinusoidal functions as follows:

$$e^{i\theta} = \cos\theta + j\sin\theta \tag{8.23}$$

Create a two-dimensional plot of this function as θ varies from 0 to 2π. Create a three-dimensional line plot using function plot3 as θ varies from 0 to 2π (the three dimensions are the real part of the expression, the imaginary part of the expression, and θ).

8.12 Create a mesh, surface plot, and contour plot of the function $z = e^{x+iy}$ for the interval $-1 \leq x \leq 1$ and $-2\pi \leq y \leq 2\pi$. In each case, plot the real part of z versus x and y.

8.13 **Electrostatic Potential** The electrostatic potential ("voltage") at a point a distance r from a point charge of value q is given by the equation

$$V = \frac{1}{4\pi\varepsilon_0}\frac{q}{r} \tag{8.24}$$

where V is in volts, ε_0 is the permeability of free space (8.85×10^{-12} farads/m), q is the charge in coulombs, and r is the distance from the point charge in meters. If q is positive, the resulting potential is positive; if q is negative, the resulting potential is negative. If more than one charge is present in the environment, the total potential at a point is the sum of the potentials from each individual charge. Suppose that four charges are located in a three-dimensional space as follows:

$$q_1 = 10^{-13} \text{ coul at point } (1,1,0)$$

$$q_2 = 10^{-13} \text{ coul at point } (1,-1,0)$$

$$q_3 = -10^{-13} \text{ coul at point } (-1,-1,0)$$

$$q_4 = 10^{-13} \text{ coul at point } (-1,1,0)$$

Calculate the total potential due to these charges at regular points on the plane $z = 1$ with the bounds $(10,10,1)$, $(10,-10,1)$, $(-10,-10,1)$, and $(-10,10,1)$. Plot the resulting potential three times using functions `surf`, `mesh`, and `contour`.

8.14 An ellipsoid of revolution is the solid analog of a two-dimensional ellipse. The equations for an ellipsoid of revolution rotated around the x axis are

$$x = a \cos \phi \cos \theta$$
$$y = b \cos \phi \sin \theta \qquad (8.25)$$
$$z = b \sin \phi$$

where a is radius along the x-axis, and b is the radius along the y- and z-axes. Plot an ellipsoid of revolution for $a = 2$ and $b = 1$.

8.15 Plot a sphere of radius 2 and an ellipsoid of revolution for $a = 1$ and $b = 0.5$ on the same axes. Make the sphere partially transparent so that the ellipsoid can be seen inside it.

Chapter 9

Additional Data Types

In earlier chapters, we were introduced to four fundamental MATLAB data types: `double`, `logical`, `char`, and function handles. In this chapter, we will learn additional details about some of these data types, and then we will study some additional MATLAB data types.

First, we will learn more about using the `char` data type and how to use strings in MATLAB programs.

Next, we will learn about some additional data types. The most common MATLAB data types are shown in Figure 9.1. We will learn about the `single` and integer types in this chapter and discuss the remaining ones on the figure later in this book.

MATLAB Data Types

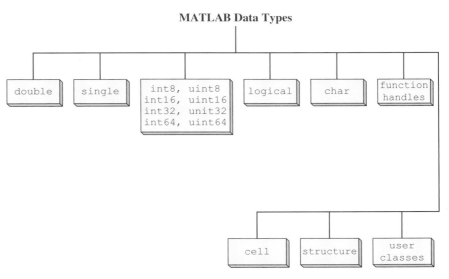

Figure 9.1 Common MATLAB data types.

9.1   Strings and String Functions

A MATLAB string is an array of type char. Each character is stored in two bytes of memory. By default, MATLAB uses the UTF-8 character set. The first 128 characters of this set are the same as the familiar ASCII character set, and the characters above that represent characters found in additional languages. Since MATLAB stores characters in two bytes of memory, it can represent the first 65,536 (= 2^16) characters of the UTF-8 character set, which covers most of the world's major languages.

A character variable is automatically created when a string is assigned to it. For example, the statement

```
str = 'This is a test';
```

creates a 14-element character array. The output of **whos** for this array is

```
» whos str
  Name      Size      Bytes     Class        Attributes
  str       1x14      28        char
```

A special function ischar can be used to check for character arrays. If a given variable is of type character, then ischar returns a true (1) value. If it is not, ischar returns a false (0) value.

The following subsections describe MATLAB functions useful for manipulating character strings.

9.1.1   String Conversion Functions

Variables may be converted from the char data type to the double data type using the double function. The output of the function is an array of double values, with each one containing the numerical value n. Thus if str is defined as

```
str = 'This is a test';
```

the statement double(str) yields the result:

```
» x = double(str)
x =
 Columns 1 through 12
   84 104 105 115  32 105 115  32  97  32 116 101
 Columns 13 through 14
 115 116
```

Variables can also be converted from the double data type to the char data type using the char function. If x is the 14-element array created above, then the statement char(x) yields the result:

```
» z = char(x)
z =
This is a test
```

This also works for non-English characters. For example, if x is defined as:

```
x = [945 946 947 1488];
```

then the statement char(x) yields the Greek characters α, β, and γ, followed by the Hebrew letter א (aleph):

```
» z = char(x)
z =
αβγא
```

9.1.2 Creating Two-Dimensional Character Arrays

It is possible to create two-dimensional character arrays, but *each row of such an array must have exactly the same length*. If one of the rows is shorter than the other rows, the character array is invalid and will produce an error. For example, the following statements are illegal because the two rows have different lengths.

```
name = ['Stephen J. Chapman';'Senior Engineer'];
```

The easiest way to produce two-dimensional character arrays is with the char function. This function will automatically pad all strings to the length of the largest input string.

```
» name = char('Stephen J. Chapman','Senior Engineer')
name =
Stephen J. Chapman
Senior Engineer
```

Two-dimensional character arrays can also be created with function strvcat, which is described below.

Good Programming Practice

Use the char function to create two-dimensional character arrays without worrying about padding each row to the same length.

9.1.3 Concatenating Strings

Function strcat concatenates two or more strings horizontally, ignoring any trailing blanks but preserving blanks within the strings. This function produces the result shown below

```
» result = strcat('String 1 ','String 2')
result =
String 1String 2
```

The result is `'String  1String  2'`. Note that the trailing blanks in the first string were ignored.

Function `strvcat` concatenates two or more strings vertically, automatically padding the strings to make a valid two-dimensional array. This function produces the result shown below

```
» result = strvcat('Long String 1 ','String 2')
result =
Long String 1
String 2
```

9.1.4 Comparing Strings

Strings and substrings can be compared in several ways:

- Two strings, or parts of two strings, can be compared for equality.
- Two individual characters can be compared for equality.
- Strings can be examined to determine whether each character is a letter or whitespace.

Comparing Strings for Equality

You can use four MATLAB functions to compare two strings as a whole for equality. They are:

- `strcmp`—determines if two strings are identical.
- `strcmpi`—determines if two strings are identical ignoring case.
- `strncmp`—determines if the first n characters of two strings are identical
- `strncmpi`—determines if the first n characters of two strings are identical ignoring case.

Function `strcmp` compares two strings, including any leading and trailing blanks, and returns a true (1) if the strings are identical[1]. Otherwise, it returns a false (0). Function `strcmpi` is the same as `strcmp`, except that it is not case sensitive (that is, it treats `'a'` as equal to `'A'`).

Function `strncmp` compares the first n characters of two strings, including any leading blanks, and returns a true (1) if the characters are identical. Otherwise, it returns a false (0). Function `strncmpi` is the same as `strncmp`, except that it is not case sensitive.

To understand these functions, consider the two strings:

```
str1 = 'hello';
str2 = 'Hello';
str3 = 'help';
```

---

[1]**Caution:** The behavior of this function is different from that of the `strcmp` in C. C programmers can be tripped up by this difference.

Strings `str1` and `str2` are not identical, but they differ only in the case of one letter. Therefore, `strcmp` returns false (0), while `strcmpi` returns true (1).

```
» c = strcmp(str1,str2)
c =
     0
» c = strcmpi(str1,str2)
c =
     1
```

Strings `str1` and `str3` are also not identical, and both `strcmp` and `strcmpi` will return a false (0). However, the first three characters of `str1` and `str3` *are* identical, so invoking `strncmp` with any value up to 3 returns a true (1):

```
» c = strncmp(str1,str3,2)
c =
1
```

Comparing Individual Characters for Equality and Inequality

You can use MATLAB relational operators on character arrays to test for equality *one character at a time*, as long as the arrays you are comparing have equal dimensions, or one is a scalar. For example, you can use the equality operator (`==`) to determine which characters in two strings match:

```
» a = 'fate';
» b = 'cake';
» result = a == b
result =
0 1 0 1
```

All of the relational operators (`>`, `>=`, `<`, `<=`, `==`, `~=`) compare the numerical position of the corresponding characters in the current character set.

Unlike C, MATLAB does not have an intrinsic function to define a "greater than" or "less than" relationship between two strings taken as a whole. We will create such a function in an example at the end of this section.

Categorizing Characters within a String

There are three functions for categorizing characters on a character-by-character basis inside a string:

- `isletter` determines if a character is a letter.
- `isspace` determines if a character is whitespace (blank, tab, or new line).
- `isstrprop('str', 'category')` is a more general function. It determines if a character falls into a user-specified category, such as alphabetic, alphanumeric, upper case, lower case, numeric, control, and so forth.

To understand these functions, let's create a string named `mystring`:

```
mystring = 'Room 23a';
```

We will use this string to test the categorizing functions.

Function `isletter` examines each character in the string, producing a `logical` output vector of the same length as `mystring` that contains a true (1) in each location corresponding to a letter of the alphabet, and a false (0) in the other locations. For example,

```
» a = isletter(mystring)
a =
1 1 1 1 0 0 0 1
```

The first four and the last elements in `a` are true (1) because the corresponding characters of `mystring` are letters of the alphabet.

Function `isspace` also examines each character in the string, producing a `logical` output vector of the same length as `mystring` that contains a true (1) in each location corresponding to whitespace, and a false (0) in the other locations. "Whitespace" is any character that separates tokens in MATLAB: tab, line feed, vertical tab, form feed, carriage return, and space, in addition to a number of other Unicode characters. For example,

```
» a = isspace(mystring)
a =
0 0 0 0 1 0 0 0
```

The fifth element in `a` is true (1) because the corresponding character of `mystring` is a space.

Function `isstrprop` is a more flexible replacement for `isletter`, `isspace`, and several other functions. This function has two arguments, `'str'` and `'category'`. The first argument is the string to characterize, and the second argument is the type of category to check for. Some possible categories are given in Table 9.1.

This function examines each character in the string, producing a `logical` output vector of the same length as the input string that contains a true (1) in each location that matches the category, and a false (0) in the other locations. For example, the following function checks to see which characters in `mystring` are numbers:

```
» a = isstrprop(mystring,'digit')
a =
0 0 0 0 0 1 1 0
```

Also, the following function checks to see which characters in `mystring` are lowercase letters:

```
» a = isstrprop(mystring,'lower')
a =
0 1 1 1 0 0 0 1
```

Table 9.1: Selected Categories for Function `isstrprop`

Category	Description
`'alpha'`	Returns true (1) for each character of the string that is alphabetic and false (0) otherwise.
`'alphanum'`	Returns true (1) for each character of the string that is alphanumeric and false (0) otherwise. [**Note:** This category is equivalent to function `isletter`.]
`'cntrl'`	Returns true (1) for each character of the string that is a control character and false (0) otherwise.
`'digit'`	Returns true (1) for each character of the string that is a number and false (0) otherwise.
`'graphic'`	Returns true (1) for each character of the string that is a graphic character and false (0) otherwise. Examples of non-graphic characters include space, line separator, paragraph separator, control characters, and certain other Unicode characters. All other characters return true for this category.
`'lower'`	Returns true (1) for each character of the string that is a lowercase letter and false (0) otherwise.
`'print'`	Returns true (1) for each character of the string that is either a graphic character or a space and false (0) otherwise.
`'punct'`	Returns true (1) for each character of the string that is a punctuation character and false (0) otherwise.
`'wspace'`	Returns true (1) for each character of the string that is whitespace and false (0) otherwise. [**Note:** This category replaces function `isspace`.]
`'upper'`	Returns true (1) for each character of the string that is an uppercase letter and false (0) otherwise.
`'xdigit'`	Returns true (1) for each character of the string that is a hexadecimal digit and false (0) otherwise.

9.1.5 Searching/Replacing Characters within a String

MATLAB provides several functions for searching and replacing characters in a string. Consider a string named `test`:

```
test = 'This is a test!';
```

Function `strfind(text,pattern)` returns the starting position of all occurrences of the characters in `pattern` within the string in `text`. For example, to find all occurrences of the string `'is'` inside `test`,

```
» position = strfind(test,'is')
position =
     3     6
```

The string `'is'` occurs twice within `test`, starting at positions 3 and 6.

Function `strmatch` is another matching function. This one looks at the beginning characters of the *rows* of a 2-D character array and returns a list of those rows that start with the specified character sequence. The form of this function is

```
result = strmatch(str,array);
```

For example, suppose that we create a 2-D character array with the function strvcat:

```
array = strvcat('maxarray','min value','max value');
```

Then the following statement will return the row numbers of all rows beginning with the letters 'max':

```
» result = strmatch('max',array)
result =
              1
              3
```

Function strrep performs the standard search-and-replace operation. It finds all occurrences of one string within another one and replaces them by a third string. The form of this function is

```
result = strrep(str,srch,repl)
```

where str is the string being checked, srch is the character string to search for, and repl is the replacement character string. For example,

```
» test = 'This is a test!'
» result = strrep(test,'test','pest')
result =
This is a pest!
```

The strtok function returns the characters before the first occurrence of a delimiting character in an input string. The default delimiting characters are the set of whitespace characters. The form of strtok is

```
[token,remainder] = strtok(string,delim)
```

where string is the input character string, delim is the (optional) set of delimiting characters, token is the first set of characters delimited by a character in delim, and remainder is the rest of the line. For example,

```
» [token,remainder] = strtok('This is a test!')
token =
This
remainder =
 is a test!
```

You can use the strtok function to parse a sentence into words; for example:

```
function all_words = words(input_string)
remainder = input_string;
all_words = ' ';
while (any(remainder))
    [chopped,remainder] = strtok(remainder);
    all_words = strvcat(all_words,chopped);
end
```

9.1.6 Uppercase and Lowercase Conversion

Functions upper and lower convert all of the alphabetic characters within a string to uppercase and lowercase respectively. For example,

```
» result = upper('This is test 1!')
result =
THIS IS TEST 1!
» result = lower('This is test 2!')
result =
this is test 2!
```

Note that the alphabetic characters were converted to the proper case, while the numbers and punctuation were unaffected.

9.1.7 Trimming Whitespace from Strings

There are two functions that trim leading and/or trailing whitespace from a string. Whitespace characters consists of the spaces, newlines, carriage returns, tabs, vertical tabs, and formfeeds.

Function deblank removes any extra *trailing* whitespace from a string, and function strtrim removes any extra *leading and trailing* whitespace from a string.

For example, the following statements create a 21-character string with leading and trailing whitespace. Function deblank trims the trailing whitespace characters in the string only, while function strtrim trims both the leading and the trailing whitespace characters.

```
» test_string = 'This is a test.'
test_string =
   This is a test.
» length(test_string)
ans =
    21
» test_string_trim1= deblank(test_string)
test_string_trim1 =
   This is a test.
» length(test_string_trim1)
ans =
    18
» test_string_trim2 = strtrim(test_string)
test_string_trim2 =
This is a test.
» length(test_string_trim2)
ans =
    15
```

9.1.8 Numeric-to-String Conversions

MATLAB contains several functions to convert numeric values into character strings. We have already seen two such functions, num2str and int2str. Consider a scalar x:

```
x = 5317;
```

By default, MATLAB stores the number x as a 1 × 1 double array containing the value 5317. The int2str (integer to string) function converts this scalar into a 1 × 4 char array containing the string '5317':

```
» x = 5317;
» y = int2str(x);
» whos
  Name        Size        Bytes       Class          Attributes

  x           1x1         8           double
  y           1x4         8           char
```

Function num2str converts a double value into a string, even if it does not contain an integer. It provides more control of the output string format than int2str. An optional second argument sets the number of digits in the output string or specifies an actual format to use. The format specifications in the second argument are similar to those used by fprintf. For example,

```
» p = num2str(pi)
p =
3.1416
» p = num2str(pi,7)
p =
3.141593
» p = num2str(pi,'%10.5e')
p =
3.14159e+000
```

Both int2str and num2str are handy for labeling plots. For example, the following lines use num2str to prepare automated labels for the *x*-axis of a plot:

```
function plotlabel(x,y)
plot(x,y)
str1 = num2str(min(x));
str2 = num2str(max(x));
out = ['Value of f from' str1 'to' str2];
xlabel(out);
```

There are also conversion functions designed to change numeric values into strings representing a decimal value in another base, such as a binary or hexadecimal representation. For example, the `dec2hex` function converts a decimal value into the corresponding hexadecimal string:

```
dec_num = 4035;
hex_num = dec2hex(dec_num)
hex_num =
FC3
```

Other functions of this type include `hex2num`, `hex2dec`, `bin2dec`, `dec2bin`, `base2dec`, and `dec2base`. MATLAB includes online help for all of these functions.

MATLAB function `mat2str` converts an array to a string that MATLAB can evaluate. This string is useful input for a function such as `eval`, which evaluates input strings just as if they were typed at the MATLAB command line. For example, if we define array a as

```
» a = [1 2 3; 4 5 6]
a =
     1     2     3
     4     5     6
```

then the function `mat2str` will return a string containing the result

```
» b = mat2str(a)
b =
[1 2 3; 4 5 6]
```

Finally, MATLAB includes a special function `sprintf` that is identical to function `fprintf`, except that the output goes into a character string instead of the Command Window. This function provides complete control over the formatting of the character string. For example,

```
» str = sprintf('The value of pi = %8.6f.',pi)
str =
The value of pi = 3.141593.
```

This function is extremely useful in creating complex titles and labels for plots.

9.1.9   String-to-Numeric Conversions

MATLAB also contains several functions to change character strings into numeric values. The most important of these functions are `eval`, `str2double`, and `sscanf`.

Function `eval` evaluates a string containing a MATLAB expression and returns the result. The expression can contain any combination of MATLAB functions, variables, constants, and operations. For example, the string a containing the characters `'2 * 3.141592'` can be converted to numeric form by the following statements:

```
» a = '2 * 3.141592';
» b = eval(a)
b =
    6.2832
» whos
  Name        Size        Bytes       Class        Attributes

  a           1x12        24          char
  b           1x1         8           double
```

Function `str2double` converts character strings into an equivalent `double` value[2]. For example, the string a containing the characters `'3.141592'` can be converted to numeric form by the following statements:

```
» a = '3.141592';
» b = str2double(a)
b =
    3.1416
```

Strings can also be converted to numeric form using the function `sscanf`. This function converts a string into a number according to a format conversion character. The simplest form of this function is

```
value = sscanf(string,format)
```

where `string` is the string to scan, and `format` specifies the type of conversion to occur. The two most common conversion specifiers for `sscanf` are `'%d'` for decimals and `'%g'` for floating-point numbers. This function will be covered in much greater detail in Chapter 11.

The following examples illustrate the use of `sscanf`.

```
» a = '3.141592';
» value1 = sscanf(a,'%g')
value1 =
    3.1416
» value2 = sscanf(a,'%d')
value2 =
    3
```

9.1.10 Summary

The common MATLAB string functions are summarized in Table 9.2.

---

[2]MATLAB also contains a function `str2num` that can convert a string into a number. For a variety of reasons mentioned in the MATLAB documentation, function `str2double` is better than function `str2num`. You should recognize function `str2num` when you see it, but always use function `str2double` in any new code that you write.

Table 9.2: Common MATLAB String Functions

Category	Function	Description
General	char	(1) Converts numbers to the corresponding character values. (2) Creates a 2D character array from a series of strings.
	double	Converts characters to the corresponding numeric codes.
	blanks	Creates a string of blanks.
	deblank	Removes trailing whitespace from a string.
	strtrim	Removes leading and trailing whitespace from a string.
String tests	ischar	Returns true (1) for a character array.
	isletter	Returns true (1) for letters of the alphabet.
	isspace	Returns true (1) for whitespace.
	isstrprop	Returns true (1) for characters matching the specified property.
String operations	strcat	Concatenates strings.
	strvcat	Concatenates strings vertically.
	strcmp	Returns true (1) if two strings are identical.
	strcmpi	Returns true (1) if two strings are identical, ignoring case.
	strncmp	Returns true (1) if first n characters of two strings are identical.
	strncmpi	Returns true (1) if first n characters of two strings are identical, ignoring case.
	findstr	Finds one string within another one.
	strjust	Justifies string.
	strmatch	Finds matches for string.
	strrep	Replaces one string with another.
	strtok	Finds token in string.
	upper	Converts string to uppercase.
	lower	Converts string to lowercase.
Number to string conversion	int2str	Converts integer to string.
	num2str	Converts number to string.
	mat2str	Converts matrix to string.
	sprintf	Writes formatted data to string.
String to number conversion	eval	Evaluates the result of a MATLAB expression.
	str2double	Converts string to a double value.
	str2num	Converts string to number.
	sscanf	Reads formatted data from string.
Base Number Conversion	hex2num	Converts IEEE hexadecimal string to double.
	hex2dec	Converts hexadecimal string to decimal integer.
	dec2hex	Converts decimal to hexadecimal string.
	bin2dec	Converts binary string to decimal integer.
	dec2bin	Converts decimal integer to binary string.
	base2dec	Converts base-B string to decimal integer.
	dec2base	Converts decimal integer to base-B string.

►Example 9.1—String Comparison Function

In C, function strmcp compares two strings according to the order of their characters in the UTF-8 character table (called the **lexicographic order** of the characters), and returns a −1 if the first string is lexicographically less than the second string, a 0 if the strings are equal, and a +1 if the first string is lexicographically greater than the second string. This function is extremely useful for such purposes as sorting strings in alphabetic order.

Create a new MATLAB function c_strcmp that compares two strings in a similar fashion to the C function and returns similar results. The function should ignore trailing blanks in doing its comparisons. Note that the function must be able to handle the situation where the two strings are of different lengths.

Solution

1. **State the problem**

   Write a function that will compare two strings str1 and str2 and return the following results:

   - −1   if str1 is lexicographically less than str2.
   - 0   if str1 is lexicographically equal to str2.
   - +1   if str1 is lexicographically greater than str2.

   The function must work properly if str1 and str2 do not have the same length, and the function should ignore trailing blanks.

2. **Define the inputs and outputs**

   The inputs required by this function are two strings, str1 and str2. The output from the function will be a −1, 0, or 1, as appropriate.

3. **Describe the algorithm**

   This task can be broken down into four major sections:

   ```
   Verify input strings
   Pad strings to be equal length
   Compare characters from beginning to end, looking
     for the first difference
   Return a value based on the first difference
   ```

   We will now break each of the above major sections into smaller, more detailed pieces. First, we must verify that the data passed to the function is correct. The function must have exactly two arguments, and the arguments must both be strings. The pseudocode for this step is:

   ```
   % Check for a legal number of input arguments.
   msg = nargchk(2,2,nargin)
   error(msg)
   % Check to see if the arguments are strings
   if either argument is not a string
      error('str1 and str2 must both be strings')
   else
      (add code here)
   end
   ```

Next, we must pad the strings to equal lengths. The easiest way to do this is to combine both strings into a 2D array using `strvcat`. Note that this step effectively results in the function ignoring trailing blanks, because both strings are padded out to the same length. The pseudocode for this step is:

```
% Pad strings
strings = strvcat(str1,str2)
```

Now we must compare each character until we find a difference and return a value based on that difference. One way to do this is to use relational operators to compare the two strings, creating an array of 0's and 1's. We can then look for the first 1 in the array, which will correspond to the first difference between the two strings. The pseudocode for this step is:

```
% Compare strings
diff = strings(1,:) ~= strings(2,:)
if sum(diff) == 0
    % Strings match
    result = 0
else
    % Find first difference
    ival = find(diff)
    if strings(1,ival) > strings(2,ival)
        result = 1
    else
        result = -1
    end
end
```

4. **Turn the algorithm into MATLAB statements.**

The final MATLAB code is shown below.

```
function result = c_strcmp(str1,str2)
%C_STRCMP Compare strings like C function "strcmp"
% Function C_STRCMP compares two strings and returns
% a -1 if str1 < str2, a 0 if str1 == str2, and a
% +1 if str1 > str2.

% Define variables:
%   diff       -- Logical array of string differences
%   msg        -- Error message
%   result     -- Result of function
%   str1       -- First string to compare
%   str2       -- Second string to compare
%   strings    -- Padded array of strings

%   Record of revisions:
%     Date         Programmer        Description of change
%     ====         ==========        =====================
%     02/25/14     S. J. Chapman     Original code
```

```
% Check for a legal number of input arguments.
msg = nargchk(2,2,nargin);
error(msg);

% Check to see if the arguments are strings
if ~(isstr(str1) & isstr(str2))
   error('Both str1 and str2 must be strings!')
else

   % Pad strings
   strings = strvcat(str1,str2);

   % Compare strings
   diff = strings(1,:) ~= strings(2,:);
   if sum(diff) == 0

      % Strings match, so return a zero!
      result = 0;
   else
      % Find first difference between strings
      ival = find(diff);
      if strings(1,ival(1)) > strings(2,ival(1))
         result = 1;
      else
         result = -1;
      end
   end
end
```

5. **Test the program.**

   Next, we must test the function using various strings.

```
» result = c_strcmp('String 1','String 1')
result =

          0
» result = c_strcmp('String 1','String 1')
result =

          0
» result = c_strcmp('String 1','String 2')
result =

        -1
» result = c_strcmp('String 1','String 0')
result =

         1
» result = c_strcmp('String','str')
result =

        -1
```

The first test returns a zero, because the two strings are identical. The second test also returns a zero, because the two strings are identical *except for trailing blanks*, and trailing blanks are ignored. The third test returns a −1, because the two strings first differ in position 8, and '1' < '2' at that position. The fourth test returns a 1, because the two strings first differ in position 8, and '1' > '0' at that position. The fifth test returns a −1, because the two strings first differ in position 1, and 'S' < 's' in the UTF-8 character sequence. This function appears to be working properly.

◀

Quiz 9.1

This quiz provides a quick check to see if you have understood the concepts introduced in Section 9.1. If you have trouble with the quiz, reread the section, ask your instructor, or discuss the material with a fellow student. The answers to this quiz are found in the back of the book.

For questions 1 through 9, determine whether these statements are correct. If they are, what is produced by each set of statements?

1.
```
str1 = 'This is a test!  ';
str2 = 'This line, too.';
res = strcat(str1,str2);
```

2.
```
str1 = 'Line 1';
str2 = 'line 2';
res = strcati(str1,str2);
```

3.
```
str1 = 'This is another test!';
str2 = 'This line, too.';
res = [str1; str2];
```

4.
```
str1 = 'This is another test!';
str2 = 'This line, too.';
res = strvcat(str1,str2);
```

5.
```
str1 = 'This is a test!  ';
str2 = 'This line, too.';
res = strncmp(str1,str2,5);
```

6.
```
str1 = 'This is a test!  ';
res = findstr(str1,'s');
```

7.
```
str1 = 'This is a test!  ';
str1(isspace(str1)) = 'x';
```

8.
```
str1 = 'aBcD 1234 !?';
res = isstrprop(str1,'alphanum');
```

9. ```
   str1 = 'This is a test!  ';
   str1(4:7) = upper(str1(4:7));
   ```

10. ```
    str1 = '  456  ';  % Note: Three blanks before & after
    str2 = '  abc  ';  % Note: Three blanks before & after
    str3 = [str1 str2];
    str4 = [strtrim(str1) strtrim(str2)];
    str5 = [deblank(str1) deblank(str2)];
    l1 = length(str1);
    l2 = length(str2);
    l3 = length(str3);
    l4 = length(str4);
    l5 = length(str4);
    ```

11. ```
    str1 = 'This way to the egress.';
    str2 = 'This way to the egret.'
    res = strncmp(str1,str2);
    ```

9.2 The `single` Data Type

Variables of type `single` are scalars or arrays of 32-bit *single-precision* floating-point numbers. They can hold real, imaginary, or complex values. Variables of type `single` occupy half the memory of variables of type `double`, but they have lower precision and a more limited range. The real and imaginary components of each `single` variable can be positive or negative numbers in the range 10^{-38} to 10^{38}, with 6 to 7 significant decimal digits of accuracy, plus the value 0.

The `single` function creates a variable of type `single`. For example, the following statement creates a variable of type `single` containing the value 3.1:

```
» var = single(3.1)
var =
    3.1000
» whos
  Name      Size      Bytes     Class       Attributes

  var       1x1       4         single
```

Once a `single` variable is created, it can be used in MATLAB operations just like a `double` variable. In MATLAB, an operation performed between a `single` value and a `double` value has a `single` result[3], so the result of the following statements will be of type `single`:

---

[3]CAUTION: This is unlike the behavior of any other computer language that the author has ever encountered. In every other language (Fortran, C, C++, Java, Basic, and so forth), the result of an operation between a `single` and a `double` would be of type `double`.
```

```
» b = 7;
» c = var * b
c =
   21.7000
» whos
  Name        Size        Bytes       Class       Attributes
   b          1x1         8           double
   c          1x1         4           single
   var        1x1         4           single
```

Values of type `single` can be used just like values of type `double` in most MATLAB operations. Built-in functions such as `sin`, `cos`, `exp`, and so forth all support the `single` data type, but some M-file functions may not support `single` values yet. As a practical matter, you will probably never use this data type. Its more limited range and precision make the results more sensitive to cumulative round-off errors or to exceeding the available range. You should only consider using this data type if you have enormous arrays of data that could not fit into your computer memory if they were saved in double precision.

Some MATLAB functions do no support the `single` data type. If you wish to, you can implement your own version of a function that supports `single` data. If you place this function in a directory named `@single` inside any directory on the MATLAB path, that function will be automatically used when the input arguments are of type `single`.

9.3 Integer Data Types

MATLAB also includes 8-, 16-, 32-, and 64-bit *signed* and *unsigned* integers. The data types are `int8`, `uint8`, `int16`, `uint16`, `int32`, `uint32`, `int64`, and `uint64`. The difference between a signed and an unsigned integer is the range of numbers represented by the data type. The number of values that can be represented by an integer depends on the number of bits in the integer:

$$\text{number of values} = 2^n \tag{9.1}$$

where n is the number of bits. An 8-bit integer can represent 256 values (2^8), a 16-bit integer can represent 65,536 values (2^{16}), and so forth. Signed integers use half of the available values to represent positive numbers and half for negative numbers, while unsigned integers use all of the available values to represent positive numbers. Therefore, the range of values that can be represented in the `int8` data type is -128 to 127 (a total of 256), while the range of values that can be represented in the `uint8` data type is 0 to 255 (a total of 256). Similarly, the range of values that can be represented in the `int16` data type is $-32,768$ to 32,767 (a total of 65,536), while the range of values that can be represented in the `uint16` data type is 0 to 65,535. The same idea applies to larger integer sizes.

Integer values are created by the `int8()`, `uint8()`, `int16()`, `uint16()`, `int32()`, `uint32()`, `int64()`, or `uint64()` functions. For example, the following statement creates a variable of type `int8` containing the value 3:

```
» var = int8(3)
var =
    3
» whos
  Name         Size         Bytes        Class         Attributes

  var          1x1          1            int8
```

Integers can also be created using the standard array creation functions, such as zeros, ones, and so forth, by adding a separate type option to the function. For example, we can create a 1000 × 1000 array of signed 8-bit integers as follows:

```
» array = zeros(1000,1000, 'int8');
» whos
  Name         Size          Bytes        Class        Attributes

  array        1000x1000     1000000      int8
```

Integers can be converted to other data types using the double, single, and char functions.

An operation performed between an integer value and a double value has an integer result[4], so the result of the following statements will be of type int8:

```
» b = 7;
» c = var * b
c =
    21
» whos
  Name         Size         Bytes        Class         Attributes

  b            1x1          8            double
  c            1x1          1            int8
  var          1x1          1            int8
```

MATLAB actually calculates this answer by converting the int8 to a double, doing the math in double precision, then rounding the answer to the nearest integer, and converting that value back to an int8. The same idea works for all types of integers.

MATLAB uses *saturating integer arithmetic*. If the result of an integer math operation would be larger than the largest possible value that can be represented in that data type, then the result will be the largest possible value. Similarly, if the result of an integer math operation would be smaller than the smallest possible value that can be represented in that data type, then the result will be the smallest possible

---

[4]CAUTION: This is unlike the behavior of any other computer language that the author has ever encountered. In every other language (Fortran, C, C++, Java, Basic, and so forth), the result of an operation between an integer and a double would be of type double.

value. For example, the largest possible value that can be represented in the `int8` data type is 127. The result of the operation `int8(100) + int8(50)` will be 127, because 150 is larger than 127, the maximum value that can be represented in the data type.

Some MATLAB functions do not support the various integer data types. If you wish to, you can implement your own version of a function that supports an integer data type. If you place this function in a directory named `@int8`, `@uint16`, and so forth inside any directory on the MATLAB path, that function will be automatically used when the input arguments are of the specified type.

It is unlikely that you will need to use the integer data type unless you are working with image data. If you do need more information, please consult the MATLAB documentation.

9.4 Limitations of the `single` and Integer Data Types

The `single` data type and integer data types have been around in MATLAB for a while, but they have been mainly used for purposes such as storing image data. MATLAB allows mathematical operations between values of the same type or between scalar `double` values and those types but not between different types of integers or between integers and `single` values. For example, you can add a `single` and a `double` or an integer and a `double`, but you cannot add a `single` and an integer.

```
» a = single(2.1)
a =
    2.1000
» b = int16(4)
b =
     4
» c = a + b
Error using  +
Integers can only be combined with integers of the
same class, or scalar doubles.
```

Unless you have some special need to manipulate images, you will probably never need to use either of these data types.

Good Programming Practice

Do not use the `single` or integer data types, unless you have a special need such as image processing.

Quiz 9.2

This quiz provides a quick check to see if you have understood the concepts introduced in Sections 9.2 through 9.4. If you have trouble with the quiz, reread the section, ask your instructor, or discuss the material with a fellow student. The answers to this quiz are found in the back of the book.

Determine whether the following statements are correct. If they are, what is produced by each set of statements?

```
1. a = uint8(12);
   b = int8(13);
   c = a + b;
2. a = single(1000);
   b = int8(10);
   c = a * b;
3. a = single([1 0;0 1]);
   b = [3 2; -2 3];
   c = a * b;
4. a = single([1 0;0 1]);
   b = [3 2; -2 3];
   c = a .* b;
```

9.5 Summary

String functions are functions designed to work with strings, which are arrays of type `char`. These functions allow a user to manipulate strings in a variety of useful ways, including concatenation, comparison, replacement, case conversion, and numeric-to-string and string-to-numeric type conversions.

The `single` data type consists of single-precision floating-point numbers. They are created using the `single` function. A mathematical operation between a `single` and a scalar `double` value produces a `single` result.

MATLAB includes signed and unsigned 8-, 16-, 32-, and 64-bit integers. The integer data types are the `int8`, `uint8`, `int16()`, `uint16`, `int32`, `uint32`, `int64`, and `uint64`. Each of these types is created using the corresponding function: `int8()`, `uint8()`, `int16()`, `uint16()`, `int32()`, `uint32()`, `int64()`, or `uint64()`. Mathematical operations (+, −, and so forth) can be performed on these data types; the result of an operation between an integer and a `double` has the same type as the integer. If the result of a mathematical operation is too large or too small to be expressed by an integer data type, the result is either the largest or smallest possible integer for that type.

9.5.1 Summary of Good Programming Practice

The following guidelines should be adhered to:

1. Use the `char` function to create two-dimensional character arrays without worrying about padding each row to the same length.

2. Use function `isstrprop` to determine the characteristics of each character in a string array.
3. Use multidimensional arrays to solve problems that are naturally multivariate in nature, such as aerodynamics and fluid flows.
4. Do not use the `single` or integer data types, unless you have a special need such as image processing.

9.5.2 MATLAB Summary

The following summary lists all of the MATLAB commands and functions described in this chapter, along with a brief description of each one.

base2dec	Converts base-B string to decimal integer.
bin2dec	Converts binary string to decimal integer.
blanks	Creates a string of blanks.
char	(1) Converts numbers to the corresponding character values. (2) Creates a 2D character array from a series of strings.
deblank	Removes trailing whitespace from a string.
dec2base	Converts decimal integer to base-B string.
dec2bin	Converts decimal integer to binary string.
double	Converts characters to the corresponding numeric codes.
findstr	Finds one string within another one.
hex2num	Converts hexadecimal string to `double`.
hex2dec	Converts hexadecimal string to decimal integer.
int2str	Converts integer to string.
ischar	Returns true (1) for a character array.
isletter	Returns true (1) for letters of the alphabet.
isreal	Returns true (1) if no element of array has an imaginary component.
isstrprop	Returns true (1) if a character has the specified property.
isspace	Returns true (1) for whitespace.
lower	Converts string to lowercase.
mat2str	Converts matrix to string.
num2str	Converts number to string.
sscanf	Reads formatted data from string.
str2double	Converts string to `double` value.
str2num	Converts string to number.
strcat	Concatenates strings.
strcmp	Returns true (1) if two strings are identical.
strcmpi	Returns true (1) if two strings are identical ignoring case.
strjust	Justifies string.
strncmp	Returns true (1) if first n characters of two strings are identical.

strncmpi	Returns true (1) if first n characters of two strings are identical ignoring case.
strmatch	Finds matches for string.
strtrim	Removes leading and trailing whitespace from a string.
strrep	Replaces one string with another.
strtok	Finds token in string.
strvcat	Concatenates strings vertically.
upper	Converts string to uppercase.

9.6   Exercises

9.1 Write a program that accepts an input string from the user and determines how many times a user-specified character appears within the string. (*Hint:* Look up the `'s'` option of the `input` function using the MATLAB Help Browser.)

9.2 Modify the previous program so that it determines how many times a user-specified character appears within the string without regard to the case of the character.

9.3 Write a program that accepts a string from a user with the `input` function, chops that string into a series of tokens, sorts the tokens into ascending order, and prints them out.

9.4 Write a program that accepts a series of strings from a user with the `input` function, sorts the strings into ascending order, and prints them out.

9.5 Write a program that accepts a series of strings from a user with the `input` function, sorts the strings into ascending order disregarding case, and prints them out.

9.6 MATLAB includes functions `upper` and `lower`, which shift a string to uppercase and lowercase respectively. Create a new function called `caps`, which capitalizes the first letter in each word, and forces all other letters to be lower case. (*Hint:* Take advantage of functions `upper`, `lower`, and `strtok`.)

9.7 Write a function that accepts a character string and returns a `logical` array with true values corresponding to each printable character that is *not* alphanumeric or whitespace (for example, $, %, #), and false values everywhere else.

9.8 Write a function that accepts a character string and returns a `logical` array with true values corresponding to each vowel and false values everywhere else. Be sure that the function works properly for both lower case and upper case characters.

9.9 By default, it is not possible to multiply a `single` value by an `int16` value. Write a function that accepts a `single` argument and an `int16` argument and multiplies them together, returning the resulting value as a `single`.

Chapter **10**

Sparse Arrays, Cell Arrays, and Structures

This chapter deals with a very useful feature of MATLAB: sparse arrays. Sparse arrays are a special type of array in which memory is only allocated for the non-zero elements in the array. They are an extremely useful and compact way to represent large arrays containing many zero values without wasting memory.

The chapter also includes an introduction to two additional data types: cell arrays and structures. A cell array is very a flexible type of array that can hold any sort of data. Each element of a cell array can hold any type of MATLAB data, and different elements within the same array can hold different types of data. They are used extensively in MATLAB Graphical User Interface (GUI) functions.

A structure is a special type of array with named subcomponents. Each structure can have any number of subcomponents, each with its own name and data type. Structures are the basis of MATLAB objects.

10.1 Sparse Arrays

We learned about ordinary MATLAB arrays in Chapter 2. When an ordinary array is declared, MATLAB creates a memory location for every element in the array. For example, the function a = eye(10) creates 100 elements arranged as a 10 × 10 structure. In this array, 90 of those elements are zero! This matrix requires 100 elements, but only 10 of them contain non-zero data. This is an example of a **sparse array** or **sparse matrix**. A sparse matrix is a large matrix in which the vast majority of the elements are zero.

```
» a = 2 * eye(10);
a =
     2    0    0    0    0    0    0    0    0    0
     0    2    0    0    0    0    0    0    0    0
     0    0    2    0    0    0    0    0    0    0
     0    0    0    2    0    0    0    0    0    0
     0    0    0    0    2    0    0    0    0    0
     0    0    0    0    0    2    0    0    0    0
     0    0    0    0    0    0    2    0    0    0
     0    0    0    0    0    0    0    2    0    0
     0    0    0    0    0    0    0    0    2    0
     0    0    0    0    0    0    0    0    0    2
```

Now suppose that we create another 10×10 matrix b defined as follows:

```
b =
     1    0    0    0    0    0    0    0    0    0
     0    2    0    0    0    0    0    0    0    0
     0    0    2    0    0    0    0    0    0    0
     0    0    0    1    0    0    0    0    0    0
     0    0    0    0    5    0    0    0    0    0
     0    0    0    0    0    1    0    0    0    0
     0    0    0    0    0    0    1    0    0    0
     0    0    0    0    0    0    0    1    0    0
     0    0    0    0    0    0    0    0    1    0
     0    0    0    0    0    0    0    0    0    1
```

If these two matrices are multiplied together, the result is

```
» c = a * b
c =
     2    0    0    0    0    0    0    0    0    0
     0    4    0    0    0    0    0    0    0    0
     0    0    4    0    0    0    0    0    0    0
     0    0    0    2    0    0    0    0    0    0
     0    0    0    0   10    0    0    0    0    0
     0    0    0    0    0    2    0    0    0    0
     0    0    0    0    0    0    2    0    0    0
     0    0    0    0    0    0    0    2    0    0
     0    0    0    0    0    0    0    0    2    0
     0    0    0    0    0    0    0    0    0    2
```

The process of multiplying these two sparse matrices together requires 1900 multiplications and additions, but most of the terms being added and multiplied are zeros, so it was largely wasted effort.

This problem gets worse rapidly as matrix size increases. For example, suppose that we were to generate two 200×200 sparse matrices a and b as follows:

```
a = 5 * eye(200);
b = 3 * eye(200);
```

Each matrix now contains 40,000 elements, of which 39,800 are zero! Furthermore, multiplying these two matrices together requires **15,960,000** additions and multiplications.

It should be apparent that storing and working with large sparse matrices, most of whose elements are zero, is a serious waste of both computer memory and CPU time. Unfortunately, many real-world problems naturally create sparse matrices, so we need some efficient way to solve problems involving them.

A large electric power system is an excellent example of a real-world problem involving sparse matrices. Large electric power systems can have a thousand or more electrical busses at generating plants and transmission and distribution substations. If we wish to know the voltages, currents, and power flows in the system, we must first solve for the voltage at every bus. For a 1000-bus system, this involves the simultaneous solution of 1000 equations in 1000 unknowns, which is equivalent to inverting a matrix with 1,000,000 elements. Solving this matrix requires millions of floating point operations.

However, each bus in the power system is probably connected to an average of only two or three other busses. Therefore, 996 of the 1000 terms in each row of the matrix will be zeros, and most of the operations involved in inverting the matrix will be additions and multiplications by zeros. The calculation of the voltages and currents in this power system would be much simpler and more efficient if the zeros could be ignored in the solution process.

10.1.1 The `sparse` Attribute

MATLAB has a special version of the `double` data type that is designed to work with sparse arrays. In this special version of the `double` data type, *only the non-zero elements of an array are allocated memory locations*, and the array is said to have the "sparse" attribute. An array with the sparse attribute actually saves three values for each non-zero element: the value of the element itself, and the row and column numbers where the element is located. Even though three values must be saved for each non-zero element, this approach is *much* more memory efficient than allocating full arrays if a matrix has only a few non-zero elements.

To illustrate the use of sparse matrices, we will create a 10 × 10 identity matrix:

```
» a = eye(10)
a =
     1    0    0    0    0    0    0    0    0    0
     0    1    0    0    0    0    0    0    0    0
     0    0    1    0    0    0    0    0    0    0
     0    0    0    1    0    0    0    0    0    0
     0    0    0    0    1    0    0    0    0    0
     0    0    0    0    0    1    0    0    0    0
     0    0    0    0    0    0    1    0    0    0
     0    0    0    0    0    0    0    1    0    0
     0    0    0    0    0    0    0    0    1    0
     0    0    0    0    0    0    0    0    0    1
```

If this matrix is converted to a sparse matrix using the `sparse` function, the results are:

```
» as = sparse(a)
as =
    (1,1)        1
    (2,2)        1
    (3,3)        1
    (4,4)        1
    (5,5)        1
    (6,6)        1
    (7,7)        1
    (8,8)        1
    (9,9)        1
   (10,10)       1
```

Note that the data in the sparse matrix is a list of row and column addresses, followed by the non-zero data value at that point. This is a very efficient way to store data as long as most of the matrix is zero. However, if there are many non-zero elements, it can take up even more space than the full matrix because of the need to store the addresses.

If we examine arrays `a` and `as` with the `whos` command, the results are:

```
» whos
  Name      Size        Bytes     Class       Attributes

  a         10x10       800       double
  as        10x10       248       double        sparse
```

The `a` array occupies 800 bytes because there are 100 elements with 8 bytes of storage each. The `as` array occupies 248 bytes because there are 10 non-zero elements with 8 bytes of storage each plus 20 array indices occupying 8 bytes each and 8 bytes of overhead. Note that the sparse array occupies much less memory than the full array.

The `issparse` function can be used to determine whether or not a given array is sparse. If an array is sparse, then `issparse(array)` returns true (1).

The power of the sparse data type can be seen by considering a 1000×1000 matrix z with an average of 4 non-zero elements per row. If this matrix is stored as a full matrix, it will require 8,000,000 bytes of space. On the other hand, if it is converted to a sparse matrix, the memory usage will drop dramatically.

```
» zs = sparse(z);
» whos
  Name      Size          Bytes       Class       Attributes

  z         1000x1000     8000000     double
  zs        1000x1000       72008     double        sparse
```

Generating Sparse Matrices

MATLAB can generate sparse matrices by converting a full matrix into a sparse matrix with the `sparse` function or by directly generating sparse matrices with the MATLAB functions `speye`, `sprand`, and `sprandn`, which are the sparse equivalents of `eye`, `rand`, and `randn`. For example, the expression a = `speye(4)` generates a 4 × 4 sparse matrix.

```
» a = speye(4)
a =
    (1,1)       1
    (2,2)       1
    (3,3)       1
    (4,4)       1
```

The expression b = `full(a)` converts the sparse matrix into a full matrix.

```
» b = full(a)
b =
    1    0    0    0
    0    1    0    0
    0    0    1    0
    0    0    0    1
```

Working with Sparse Matrices

Once a matrix is sparse, individual elements can be added to it or deleted from it, using simple assignment statements. For example, the following statement generates a 4 × 4 sparse matrix and then adds another non-zero element to it.

```
» a = speye(4)
a =
    (1,1)       1
    (2,2)       1
    (3,3)       1
    (4,4)       1
» a(2,1) = -2
a =
    (1,1)       1
    (2,1)      -2
    (2,2)       1
    (3,3)       1
    (4,4)       1
```

MATLAB allows full and sparse matrices to be freely mixed and used in any combination. The result of an operation between a full matrix and a sparse matrix may be either a full matrix or a sparse matrix, depending on which result is the most efficient. Essentially any matrix technique that is supported for full matrices is also available for sparse matrices.

A few of the common sparse matrix functions are listed in Table 10.1.

Table 10.1: Common MATLAB Sparse Matrix Functions

Function	Description
Create Sparse Matrices	
speye	Creates a sparse identity matrix.
sprand	Creates a sparse uniformly distributed random matrix.
sprandn	Creates a sparse normally distributed random matrix.
Full-to-Sparse Conversion Functions	
sparse	Converts a full matrix into a sparse matrix.
full	Converts a sparse matrix into a full matrix.
find	Finds indices and values of non-zero elements in a matrix.
Working with Sparse Matrices	
nnz	Number of nonzero matrix elements.
nonzeros	Returns a column vector containing the nonzero elements in a matrix.
nzmax	Returns the number of nonzero storage elements in an array.
spones	Replaces nonzero sparse matrix elements with ones.
spalloc	Allocates space for a sparse matrix.
issparse	Returns 1 (true) for sparse matrix.
spfun	Applies function to non-zero matrix elements.
spy	Visualizes sparsity pattern as a plot.

►Example 10.1—Solving Simultaneous Equations with Sparse Matrices

To illustrate the ease with which sparse matrices can be used in MATLAB, we will solve the following simultaneous system of equations with both full and sparse matrices.

$$1.0x_1 + 0.0x_2 + 1.0x_3 + 0.0x_4 + 0.0x_5 + 2.0x_6 + 0.0x_7 - 1.0x_8 = 3.0$$
$$0.0x_1 + 1.0x_2 + 0.0x_3 + 0.4x_4 + 0.0x_5 + 0.0x_6 + 0.0x_7 + 0.0x_8 = 2.0$$
$$0.5x_1 + 0.0x_2 + 2.0x_3 + 0.0x_4 + 0.0x_5 + 0.0x_6 - 1.0x_7 + 0.0x_8 = -1.5$$
$$0.0x_1 + 0.0x_2 + 0.0x_3 + 2.0x_4 + 0.0x_5 + 1.0x_6 + 0.0x_7 + 0.0x_8 = 1.0$$
$$0.0x_1 + 0.0x_2 + 1.0x_3 + 1.0x_4 + 1.0x_5 + 0.0x_6 + 0.0x_7 + 0.0x_8 = -2.0$$
$$0.0x_1 + 0.0x_2 + 0.0x_3 + 1.0x_4 + 0.0x_5 + 1.0x_6 + 0.0x_7 + 0.0x_8 = 1.0$$
$$0.5x_1 + 0.0x_2 + 0.0x_3 + 0.0x_4 + 0.0x_5 + 0.0x_6 + 1.0x_7 + 0.0x_8 = 1.0$$
$$0.0x_1 + 1.0x_2 + 0.0x_3 + 0.0x_4 + 0.0x_5 + 0.0x_6 + 0.0x_7 + 1.0x_8 = 1.0$$

Solution To solve this problem, we will create full matrices of the equation coefficients and then convert them to sparse form using the `sparse` function. Then we will solve the equation both ways, comparing the results and the memory required.

The script file to perform these calculations is shown below.

```
%   Script file: simul.m
%
%   Purpose:
%       This program solves a system of 8 linear equations in 8
%       unknowns (a*x = b), using both full and sparse matrices.
%
%   Record of revisions:
%       Date          Programmer          Description of change
%       ====          ==========          =====================
%       03/03/14      S. J. Chapman       Original code
%
% Define variables:
%   a               -- Coefficients of x (full matrix)
%   as              -- Coefficients of x (sparse matrix)
%   b               -- Constant coefficients (full matrix)
%   bs              -- Constant coefficients (sparse matrix)
%   x               -- Solution (full matrix)
%   xs              -- Solution (sparse matrix)

% Define coefficients of the equation a*x = b for
% the full matrix solution.
a =     [1.0  0.0  1.0  0.0  0.0  2.0  0.0 -1.0; ...
         0.0  1.0  0.0  0.4  0.0  0.0  0.0  0.0; ...
         0.5  0.0  2.0  0.0  0.0  0.0 -1.0  0.0; ...
         0.0  0.0  0.0  2.0  0.0  1.0  0.0  0.0; ...
         0.0  0.0  1.0  1.0  1.0  0.0  0.0  0.0; ...
         0.0  0.0  0.0  1.0  0.0  1.0  0.0  0.0; ...
         0.5  0.0  0.0  0.0  0.0  0.0  1.0  0.0; ...
         0.0  1.0  0.0  0.0  0.0  0.0  0.0  1.0];

b =     [3.0  2.0 -1.5  1.0 -2.0  1.0  1.0  1.0]';

% Define coefficients of the equation a*x = b for
% the sparse matrix solution.
as = sparse(a);
bs = sparse(b);

% Solve the system both ways
disp ('Full matrix solution:');
x = a\b
```

```
disp ('Sparse matrix solution:');
xs = as\bs

% Show workspace
disp('Workspace contents after the solutions:')
whos
```

When this program is executed, the results are:

```
» simul
Full matrix solution:
x =
    0.5000
    2.0000
   -0.5000
   -0.0000
   -1.5000
    1.0000
    0.7500
   -1.0000
Sparse matrix solution:
xs =
   (1,1)         0.5000
   (2,1)         2.0000
   (3,1)        -0.5000
   (5,1)        -1.5000
   (6,1)         1.0000
   (7,1)         0.7500
   (8,1)        -1.0000

Workspace contents after the solutions:
    Name       Size      Bytes     Class      Attributes
      a         8x8        512      double
      as        8x8        392      double     sparse
      b         8x1         64      double
      bs        8x1        144      double     sparse
      x         8x1         64      double
      xs        8x1        128      double     sparse
```

The answers are the same for both solutions. Note that the sparse solution does not contain a solution for x_4 because that value is zero, and zeros aren't carried in a sparse matrix! Also, note that the sparse form of matrix b actually takes up more space than the full form. This happens because the sparse representation must store the indices as well as the values in the arrays, so it is less efficient if most of the elements in an array are non-zero.

10.2 Cell Arrays

A **cell array** is a special MATLAB array whose elements are *cells*, containers that can hold other MATLAB arrays. For example, one cell of a cell array might contain an array of real numbers, another an array of strings, and yet another a vector of complex numbers (see Figure 10.1).

In programming terms, each element of a cell array is a *pointer* to another data structure, and those data structures can be of different types. Figure 10.2 illustrates this concept. Cell arrays are great ways to collect information about a problem since all of the information can be kept together and accessed by a single name.

Cell arrays use braces "{}" instead of parentheses "()" for selecting and displaying the contents of cells. This difference is due to the fact that *cell arrays contain data structures instead of data*. Suppose that the cell array a is defined as shown in Figure 10.2. Then the contents of element a(1,1) is a data structure containing a 3×3 array of numeric data, and a reference to a(1,1) displays the *contents* of the cell, which is the data structure.

```
» a(1,1)
ans =
      [3x3 double]
```

By contrast, a reference to a{1,1} displays *the contents of the data item contained in the cell.*

cell 1,1	cell 1,2
$\begin{bmatrix} 1 & 3 & -7 \\ 2 & 0 & 6 \\ 0 & 5 & 1 \end{bmatrix}$	'This is a text string.'
cell 2,1	**cell 2,2**
$\begin{bmatrix} 3+i4 & -5 \\ -i10 & 3-i4 \end{bmatrix}$	[ ]

Figure 10.1 The individual elements of a cell array may point to real arrays, complex arrays, string, other cell arrays, or even empty arrays.

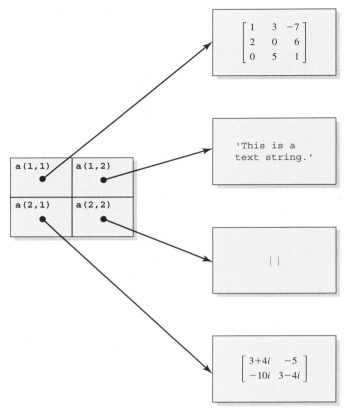

Figure 10.2 Each element of a cell array holds a *pointer* to another data structure, and different cells in the same cell array can point to different types of data structures.

```
» a{1,1}
ans =
      1     3    -7
      2     0     6
      0     5     1
```

In summary, the notation a(1,1) refers to the contents of cell a(1,1) (which is a data structure), while the notation a{1,1} refers to the contents of the data structure within the cell.

Programming Pitfalls

Be careful not to confuse "()" with "{}" when addressing cell arrays. They are very different operations!

10.2.1 Creating Cell Arrays

Cell arrays can be created in two ways:

- By using assignment statements.
- By preallocating a cell array using the `cell` function.

The simplest way to create a cell array is to directly assign data to individual cells, one cell at a time. However, preallocating cell arrays is more efficient, so you should preallocate really large cell arrays.

Allocating Cell Arrays Using Assignment Statements

You can assign values to cell arrays one cell at a time using assignment statements. There are two ways to assign data to cells, known as **content indexing** and **cell indexing**.

Content indexing involves placing braces "{ }" around the cell subscripts, together with cell contents in ordinary notation. For example, the following statements create the 2 × 2 cell array in Figure 10.2:

```
a{1,1} = [1 3 -7; 2 0 6; 0 5 1];
a{1,2} = 'This is a text string.';
a{2,1} = [3+4*i -5; -10*i 3 - 4*i];
a{2,2} = [];
```

This type of indexing defines the *contents of the data structure contained in a cell.*

Cell indexing involves placing braces "{ }" around the data to be stored in a cell, together with cell subscripts in ordinary subscript notation. For example, the following statements create the 2 × 2 cell array in Figure 10.2:

```
a(1,1) = {[1 3 -7; 2 0 6; 0 5 1]};
a(1,2) = {'This is a text string.'};
a(2,1) = {[3+4*i -5; -10*i 3 - 4*i]};
a(2,2) = {[]};
```

This type of indexing *creates a data structure containing the specified data and then assigns that data structure to a cell.*

These two forms of indexing are completely equivalent, and they may be freely mixed in any program.

⚡ Programming Pitfalls

Do not attempt to create a cell array with the same name as an existing numeric array. If you do this, MATLAB will assume that you are trying to assign cell contents to an ordinary array, and it will generate an error message. Be sure to clear the numeric array before trying to create a cell array with the same name.

Preallocating Cell Arrays with the `cell` Function

The `cell` function allows you to preallocate empty cell arrays of the specified size. For example, the following statement creates an empty 2×2 cell array.

```
a = cell(2,2);
```

Once a cell array is created, you can use assignment statements to fill values in the cells.

10.2.2 Using Braces { } as Cell Constructors

It is possible to define many cells at once by placing all of the cell contents between a single set of braces. Individual cells on a row are separated by commas, and rows are separated by semicolons. For example, the following statement creates a 2×3 cell array:

```
b = {[1 2], 17, [2;4]; 3-4*i, 'Hello', eye(3)}
```

10.2.3 Viewing the Contents of Cell Arrays

MATLAB displays the data structures in each element of a cell array in a condensed form that limits each data structure to a single line. If the entire data structure can be displayed on the single line, it is. Otherwise, a summary is displayed. For example, cell arrays a and b would be displayed as:

```
» a
a =
    [3x3 double]      [1x22 char]
    [2x2 double]               []
» b
b =
          [1x2 double]      [   17]      [2x1 double]
    [3.0000- 4.0000i]      'Hello'      [3x3 double]
```

Note that MATLAB *is displaying the data structures*, complete with brackets or apostrophes, not the entire contents of the data structures.

If you would like to see the full contents of a cell array, use the `celldisp` function. This function displays *the contents of the data structures in each cell*.

```
» celldisp(a)
a{1,1} =
     1      3     -7
     2      0      6
     0      5      1
a{2,1} =
   3.0000 + 4.0000i   -5.0000
        0 -10.0000i    3.0000 - 4.0000i
```

```
a{1,2} =
This is a text string.
a{2,2} =
      []
```

For a high-level graphical display of the structure of a cell array, use function `cellplot`. For example, the `cellplot(b)` function produces the plot shown in Figure 10.3.

10.2.4 Extending Cell Arrays

If a value is assigned to a cell array element that does not currently exist, the element will be automatically created, and any additional cells necessary to preserve the shape of the array will be automatically created. For example, suppose that array a has been defined to be a 2 × 2 cell array as shown in Figure 10.1. If the following statement is executed,

```
a{3,3} = 5
```

the cell array will be automatically extended to 3 × 3, as shown in Figure 10.4.

Preallocating cell arrays with the `cell` function is much more efficient than extending the arrays elements one at a time using assignment statements. When a new element is added to an existing array as we did above, MATLAB must create a

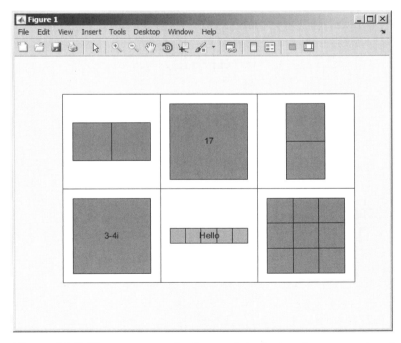

Figure 10.3 The structure of cell array b is displayed as a nested series of boxes by function `cellplot`.

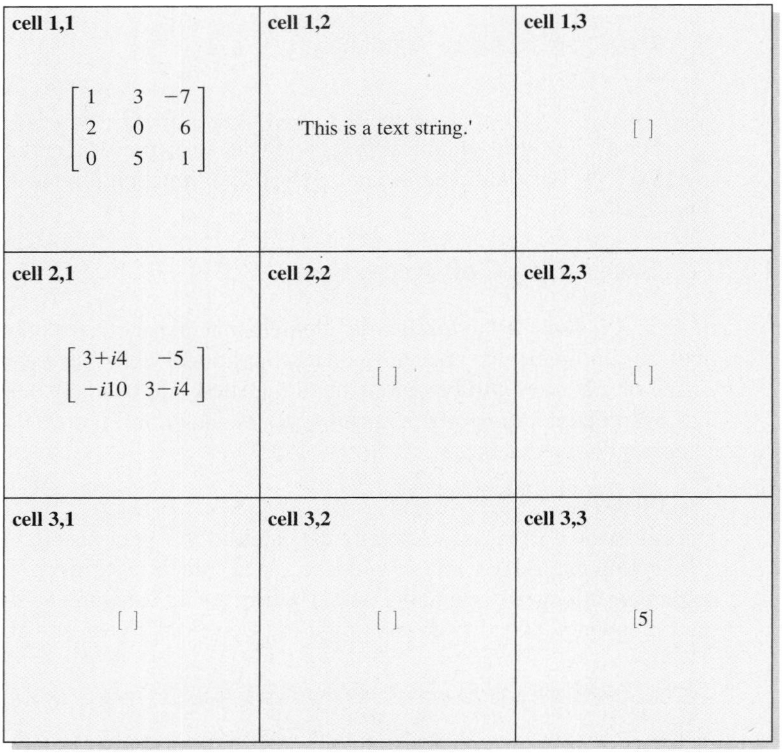

cell 1,1	cell 1,2	cell 1,3
$\begin{bmatrix} 1 & 3 & -7 \\ 2 & 0 & 6 \\ 0 & 5 & 1 \end{bmatrix}$	'This is a text string.'	[ ]
cell 2,1	cell 2,2	cell 2,3
$\begin{bmatrix} 3+i4 & -5 \\ -i10 & 3-i4 \end{bmatrix}$	[ ]	[ ]
cell 3,1	cell 3,2	cell 3,3
[ ]	[ ]	[5]

Figure 10.4 The result of assigning a value to $a\{3,3\}$. Note that four other empty cells were created to preserve the shape of the cell array.

new array large enough to include this new element, copy the old data into the new array, add the new value to the array, and then delete the old array. This can cost extra time. Instead, you should always allocate the largest size cell array that you will need and then add values to it, an element at a time. If you do that, only the new element needs to be added—the rest of the array can remain undisturbed.

The program shown below illustrates the advantages of preallocation. It creates a cell array containing 200,000 strings added one at a time, with and without preallocation.

```
%  Script file: test_preallocate.m
%
%  Purpose:
%    This program tests the creation of cell arrays with and
%    without preallocation.
%
%  Record of revisions:
%     Date            Engineer         Description of change
%     ====            ==========       =====================
%    03/04/14       S. J. Chapman      Original code
```

```
%
% Define variables:
%   a             -- Cell array
%   maxvals       -- Maximum values in cell array

% Create array without preallocation
clear all
maxvals = 200000;
tic
for ii = 1:maxvals
   a{ii} = ['Element ' int2str(ii)];
end
disp(['Elapsed time without preallocation = ' num2str(toc)]);

% Create array with preallocation
clear all
maxvals = 200000;
tic
a = cell(1,maxvals);
for ii = 1:maxvals
   a{ii} = ['Element ' int2str(ii)];
end
disp(['Elapsed time with preallocation    = ' num2str(toc)]);
```

When this program is executed on my computer, the results are as shown below. The advantages of preallocation are visible[1].

```
» test_preallocate
Elapsed time without preallocation = 8.0332
Elapsed time with preallocation    = 7.6763
```

Good Programming Practice

Always preallocate all cell arrays before assigning values to the elements of the array. This practice greatly increases the execution speed of a program.

10.2.5 Deleting Cells in Arrays

To delete an entire cell array, use the `clear` command. Subsets of cells may be deleted by assigning an empty array to them. For example, assume that a is the 3×3 cell array defined above.

---

[1]In earlier versions of MATLAB, the difference in performance was much more dramatic. This operation has been improved in recent versions of MATLAB by allocating extra variables in chunks instead of one at a time.

```
» a
a =
    [3x3 double]      [1x22 char]        []
    [2x2 double]             []          []
            []              []          [5]
```

It is possible to delete the entire third row with the statement

```
» a(3,:) = []
a =
    [3x3 double]      [1x22 char]        []
    [2x2 double]             []          []
```

10.2.6 Using Data in Cell Arrays

The data stored inside the data structures within a cell array may be used at any time, with either content indexing or cell indexing. For example, suppose that a cell array c is defined as

```
c = {[1 2;3 4], 'dogs'; 'cats', i}
```

The contents of the array stored in cell c(1,1) can be accessed as follows

```
» c{1,1}
ans =
        1       2
        3       4
```

and the contents of the array in cell c(2,1) can be accessed as follows

```
» c{2,1}
ans =
cats
```

Subsets of a cell's contents can be obtained by concatenating the two sets of subscripts. For example, suppose that we would like to get the element (1, 2) from the array stored in cell c(1,1) of cell array c. To do this, we would use the expression c{1,1}(1,2), which says: select element (1, 2) from the contents of the data structure contained in cell c(1,1).

```
» c{1,1}(1,2)
ans =
        2
```

10.2.7 Cell Arrays of Strings

It is often convenient to store groups of strings in a cell array instead of storing them in rows of a standard character array, because each string in a cell array can have a different length, while every row of a standard character array must have an identical length. This fact means that *strings in cell arrays do not have to be padded with blanks.*

Cell arrays of strings can be created in one of two ways. Either the individual strings can be inserted into the array with brackets, or else function `cellstr` can be used to convert a 2-D string array into a cell array of strings.

The following example creates a cell array of strings by inserting the strings into the cell array one at a time, and displays the resulting cell array. Note that the individual strings can be of different lengths.

```
» cellstring{1} = 'Stephen J. Chapman';
» cellstring{2} = 'Male';
» cellstring{3} = 'SSN 999-99-9999';
» cellstring
    'Stephen J. Chapman'    'Male'    'SSN 999-99-9999'
```

Function `cellstr` creates a cell array of strings from a 2-D string array. Consider the character array

```
» data = ['Line 1          ';'Additional Line']
data =
Line 1
Additional Line
```

This 2 × 15 character array can be converted into a cell array of strings with the `cellstr` function as follows:

```
» c = cellstr(data)
c =
    'Line 1'
    'Additional Line'
```

and it can be converted back to a standard character array using function `char`

```
» newdata = char(c)
newdata =
Line 1
Additional Line
```

The `iscellstr` function tests to see if a cell array is a cell array of strings. This function returns true (1) if every element of a cell array is either empty or contains a string, and returns false (0) otherwise.

10.2.8 The Significance of Cell Arrays

Cell arrays are extremely flexible since any amount of any type of data can be stored in each cell. As a result, cell arrays are used in many internal MATLAB data structures. We must understand them in order to use many features of Handle Graphics and the Graphical User Interfaces.

In addition, the flexibility of cell arrays makes them regular features of functions with variable numbers of input arguments and output arguments. A special input argument, `varargin`, is available within user-defined MATLAB functions to support variable numbers of input arguments. This argument appears as the last item in an input argument list, and it returns a cell array, so *a single dummy input argument can support any number of actual arguments*. Each actual argument becomes one element of the cell array returned by varargin. If it is used, varargin must be the *last* input argument in a function—after all of the required input arguments.

For example, suppose that we are writing a function that may have any number of input arguments. This function could be implemented as shown:

```
function test1(varargin)
disp(['There are' int2str(nargin) 'arguments.']);
disp('The input arguments are:');
disp(varargin);

end % function test1
```

When this function is executed with varying numbers of arguments, the results are:

```
» test1
There are 0 arguments.
The input arguments are:
» test1(6)
There are 1 arguments.
The input arguments are:
    [6]
» test1(1,'test 1',[1 2;3 4])
There are 3 arguments.
The input arguments are:
    [1]      'test 1'      [2x2 double]
```

As you can see, the arguments become a cell array within the function.

A sample function making use of variable numbers of arguments is shown below. Function plotline accepts an arbitrary number of 1×2 row vectors, with each vector containing the (x, y) position of one point to plot. The function plots a line connecting all of the (x, y) values together. Note that this function also accepts an optional line specification string and passes that specification on to the plot function.

```
function plotline(varargin)
%PLOTLINE Plot points specified by [x,y] pairs.
% Function PLOTLINE accepts an arbitrary number of
% [x,y] points and plots a line connecting them.
% In addition, it can accept a line specification
% string and pass that string on to function plot.

% Define variables:
%    ii          -- Index variable
%    jj          -- Index variable
%    linespec    -- String defining plot characteristics
%    msg         -- Error message
%    varargin    -- Cell array containing input arguments
%    x           -- x values to plot
%    y           -- y values to plot
```

```
%   Record of revisions:
%       Date              Engineer          Description of change
%       ====              ==========        =====================
%       03/18/14      S. J. Chapman         Original code

% Check for a legal number of input arguments.
% We need at least 2 points to plot a line...
msg = nargchk(2,Inf,nargin);
error(msg);

% Initialize values
jj = 0;
linespec = '';

% Get the x and y values, making sure to save the line
% specification string, if one exists.
for ii = 1:nargin

   % Is this argument an [x,y] pair or the line
   % specification?
   if ischar(varargin{ii})

      % Save line specification
      linespec = varargin{ii};

   else

      % This is an [x,y] pair.  Recover the values.
      jj = jj + 1;
      x(jj) = varargin{ii}(1);
      y(jj) = varargin{ii}(2);

   end
end

% Plot function.
if isempty(linespec)
   plot(x,y);
else
   plot(x,y,linespec);
end
```

When this function is called with the arguments shown below, the resulting plot is shown in Figure 10.5. Try the function with different numbers of arguments and see for yourself how it behaves.

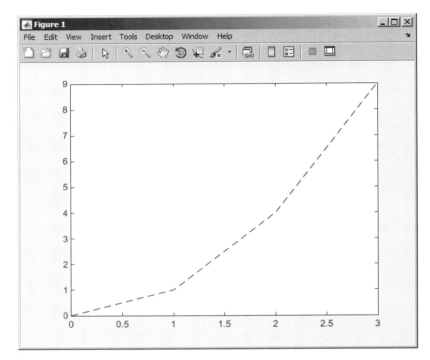

Figure 10.5 The plot produced by function `plotline`.

```
plotline([0 0],[1 1],[2 4],[3 9],'k--');
```

There is also a special output argument, `varargout`, to support variable numbers of output arguments. This argument appears as the last item in an output argument list, and it returns a cell array. Therefore, *a single dummy output argument can support any number of actual arguments*. Each actual argument becomes one element of the cell array stored in `varargout`.

If it is used, `varargout` must be the *last* output argument in a function, after all of the required input arguments. The number of values to be stored in `varargout` can be determined from function `nargout`, which specifies the number of actual output arguments for any given function call.

A sample function, `test2`, is shown below. This function detects the number of output arguments expected by the calling program, using the function `nargout`. It returns the number of random values in the first output argument and then fills the remaining output arguments with random numbers taken from a Gaussian distribution. Note that the function uses `varargout` to hold the random numbers, so that there can be an arbitrary number of output values.

```
function [nvals,varargout] = test2(mult)
% nvals is the number of random values returned
% varargout contains the random values returned
```

```
nvals = nargout - 1;
for ii = 1:nargout-1
   varargout{ii} = randn * mult;
end
```

When this function is executed, the results are as shown below.

```
» test2(4)
ans =
    -1
» [a b c d] = test2(4)
a =
     3
b =
   -1.7303
c =
   -6.6623
d =
    0.5013
```

Good Programming Practice

Use cell array arguments `varargin` and `varargout` to create functions that support varying numbers of input and output arguments.

10.2.9 Summary of `cell` Functions

The common MATLAB cell functions are summarized in Table 10.2.

Table 10.2: Common MATLAB Cell Functions

Function	Description
cell	Predefines a cell array structure.
celldisp	Displays contents of a cell array.
cellplot	Plots the structure of a cell array.
cellstr	Converts a 2D character array to a cell array of strings.
char	Converts a cell array of strings into a 2D character array.
iscellstr	Function that returns true of a cell array is a cell array of strings.
strjoin	Combines the elements of a cell array of strings into a single string, with a single space between each input string.

10.3 Structure Arrays

An *array* is a data type in which there is a name for the whole data structure, but individual elements within the array are only known by number. Thus the fifth element in the array named `arr` would be accessed as `arr(5)`. All of the individual elements in an array must be of the *same* type.

A *cell array* is a data type in which there is a name for the whole data structure, but individual elements within the array are only known by number. However, the individual elements in the cell array may be of *different* types.

In contrast, a **structure** is a data type in which each individual element has a name. The individual elements of a structure are known as **fields**, and each field in a structure may have a different type. The individual fields are addressed by combining the name of the structure with the name of the field, separated by a period.

Figure 10.6 shows a sample structure named `student`. This structure has five fields, called `name`, `addr1`, `city`, `state`, and `zip`. The field called "name" would be addressed as `student.name`.

A **structure array** is an array of structures. Each structure in the array will have identically the same fields, but the data stored in each field can differ. For example, a class could be described by an array of the structure `student`. The first student's name would be addressed as `student(1).name`, the second student's city would be addressed as `student(2).city`, and so forth.

10.3.1 Creating Structure Arrays

Structure arrays can be created in two ways.

- A field at a time using assignment statements
- All at once using the `struct` function

Building a Structure with Assignment Statements

You can build a structure, one field at a time, using assignment statements. Each time that data is assigned to a field, that field is automatically created. For example, the structure shown in Figure 10.6 can be created with the following statements.

```
» student.name = 'John Doe';
» student.addr1 = '123 Main Street';
» student.city = 'Anytown';
» student.state = 'LA';
» student.zip = '71211'
student =
     name: 'John Doe'
    addr1: '123 Main Street'
     city: 'Anytown'
    state: 'LA'
      zip: '71211'
```

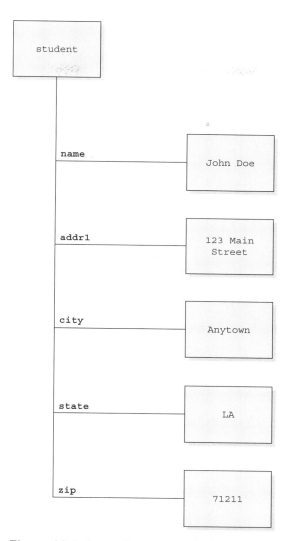

Figure 10.6 A sample structure. Each element within the structure is called a field, and each field is addressed by name.

A second student can be added to the structure by adding a subscript to the structure name (*before* the period).

```
» student(2).name = 'Jane Q. Public'
student =
1x2 struct array with fields:
    name
    addr1
    city
    state
    zip
```

student is now a 1×2 array. Note that when a structure array has more than one element, only the field names are listed, not their contents. The contents of each element can be listed by typing the element separately in the Command Window:

```
» student(1)
ans =
     name: 'John Doe'
    addr1: '123 Main Street'
     city: 'Anytown'
    state: 'LA'
      zip: '71211'
» student(2)
ans =
     name: 'Jane Q. Public'
    addr1: []
     city: []
    state: []
      zip: []
```

Note that *all of the fields of a structure are created for each array element whenever that element is defined*, even if they are not initialized. The uninitialized fields will contain empty arrays, which can be initialized with assignment statements at a later time.

The field names used in a structure can be recovered at any time, using the fieldnames function. This function returns a list of the field names in a cell array of strings and is very useful for working with structure arrays within a program.

Creating Structures with the struct Function

The struct function allows you to preallocate a structure or an array of structures. The basic form of this function is

```
str_array = struct('field1',val1,'field2',val2, ...)
```

where the arguments are field names and their initial values. With this syntax, function struct initializes every field to the specified value.

To preallocate an entire array with the struct function, simply assign the output of the struct function to the *last value* in the array. All of the values before that will be automatically created at the same time. For example, the statements shown below create an array containing 1000 structures of type student.

```
student(1000) = struct('name',[],'addr1',[], ...
                       'city',[],'state',[],'zip',[])
student =
1x1000 struct array with fields:
    name
    addr1
    city
    state
    zip
```

All of the elements of the structure are preallocated, which will speed up any program using the structure.

There is another version of the `struct` function that will preallocate an array and at the same time assign initial values to all of its fields. You will be asked to do this in an end-of-chapter exercise.

10.3.2 Adding Fields to Structures

If a new field name is defined for any element in a structure array, the field is automatically added to all of the elements in the array. For example, suppose that we add some exam scores to Jane Public's record:

```
» student(2).exams = [90 82 88]
student =
1x2 struct array with fields:
    name
    addr1
    city
    state
    zip
    exams
```

There is now a field called `exams` in every record of the array, as shown below. This field will be initialized for `student(2)` and will be an empty array for all other students until appropriate assignment statements are issued.

```
» student(1)
ans =
     name: 'John Doe'
    addr1: '123 Main Street'
     city: 'Anytown'
    state: 'LA'
      zip: '71211'
    exams: []
» student(2)
ans =
     name: 'Jane Q. Public'
    addr1: []
     city: []
    state: []
      zip: []
    exams: [90 82 88]
```

10.3.3 Removing Fields from Structures

A field may be removed from a structure array using the `rmfield` function. The form of this function is:

```
struct2 = rmfield(str_array,'field')
```

where `str_array` is a structure array, `'field'` is the field to remove, and `struct2` is the name of a new structure with that field removed. For example, we can remove the `'zip'` field from structure array `student` with the following statement:

```
» stu2 = rmfield(student,'zip')
stu2 =
1x2 struct array with fields:
    name
    addr1
    city
    state
    exams
```

10.3.4 Using Data in Structure Arrays

Now let's assume that structure array `student` has been extended to include three students, and all data has been filled in as shown in Figure 10.7. How do we use the data in this structure array?

To access the information in any field of any array element, just name the array element followed by a period and the field name:

```
» student(2).addr1
ans =
P. O. Box 17
» student(3).exams
ans =
    65    84    81
```

To access an individual item within a field, add a subscript after the field name. For example, the second exam of the third student is

```
» student(3).exams(2)
ans =
    84
```

The fields in a structure array can be used as arguments in any function that supports that type of data. For example, to calculate `student(2)`'s exam average, we could use the function

```
» mean(student(2).exams)
ans =
    86.6667
```

To extract the values from a given field across multiple array elements, simply place the structure and field name inside a set of brackets. For example, we can get access to an array of zip codes with the expression [`student.zip`]:

```
» [student.zip]
ans =
    71211       68888       10018
```

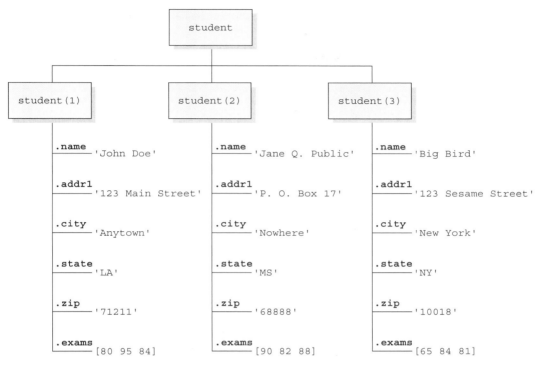

Figure 10.7 The student array with three elements and all fields filled in.

Similarly, we can get the average of *all* exams from *all* students with the function mean([student.exams]).

```
» mean([student.exams])
ans =
    83.2222
```

10.3.5 The getfield and setfield Functions

Two MATLAB functions are available to make structure arrays easier to use in programs. The getfield function gets the current value stored in a field, and the setfield function inserts a new value into a field. The structure of the getfield function is

```
f = getfield(array, {array_index}, 'field', {field_index})
```

where the field_index is optional, and array_index is optional for a 1 × 1 structure array. The function call corresponds to the statement

```
f = array(array_index).field(field_index);
```

but it can be used, even if the engineer doesn't know the names of the fields in the structure array at the time the program is written.

For example, suppose that we needed to write a function to read and manipulate the data in an unknown structure array. This function could determine the field names in the structure, using a call to `fieldnames`, and then it could read the data using the `getfield` function. To read the zip code of the second student, the function would be

```
» zip = getfield(student,{2},'zip')
zip =
        68888
```

Similarly, a program could modify values in the structure using function `setfield`. The structure of function `setfield` is

```
f = setfield(array,{array_index},'field',{field_index},value)
```

where `f` is the output structure array, the `field_index` is optional, and `array_index` is optional for a 1×1 structure array. The function call corresponds to the statement

```
array(array_index).field(field_index) = value;
```

10.3.6 Dynamic Field Names

There is an alternate way to access the elements of a structure: **dynamic field names**. A dynamic field name is a string enclosed in parentheses at a location where a field name is expected. For example, the name of student 1 can be retrieved with either static or dynamic field names as shown below:

```
» student(1).name              % Static field name
ans =
John Doe
» student(1).('name')          % Dynamic field name
ans =
John Doe
```

Dynamic field names perform the same function as static field names, but *dynamic field names can be changed during program execution*. This allows a user to access different information in the same function within a program.

For example, the following function accepts a structure array and a field name, calculating the average of the values in the specified field for all elements in the structure array. It returns that average (and optionally the number of values averaged) to the calling program.

```
function [ave, nvals] = calc_average(structure,field)
%CALC_AVERAGE Calculate the average of values in a field.
% Function CALC_AVERAGE calculates the average value
% of the elements in a particular field of a structure
% array. It returns the average value and (optionally)
% the number of items averaged.
```

```
% Define variables:
%    arr          -- Array of values to average
%    ave          -- Average of arr
%    ii           -- Index variable
%
%  Record of revisions:
%      Date            Engineer          Description of change
%      ====            ==========        =====================
%    03/04/14       S. J. Chapman       Original code
%
% Check for a legal number of input arguments.
msg = nargchk(2,2,nargin);
error(msg);

% Create an array of values from the field
arr = [];
for ii = 1:length(structure)
   arr = [arr structure(ii).(field)];
end

% Calculate average
ave = mean(arr);

% Return number of values averaged
if nargout == 2
   nvals = length(arr);
end
```

A program can average the values in different fields by simply calling this function multiple times with different structure names and different field names. For example, we can calculate the average values in fields exams and zip as follows:

```
» [ave,nvals] = calc_average(student,'exams')
ave =
   83.2222
nvals =
      9
» ave = calc_average(student,'zip')
ave =
      50039
```

10.3.7 Using the `size` Function with Structure Arrays

When the size function is used with a structure array, it returns the size of the structure array itself. When the size function is used with a *field* from a particular element in a structure array, it returns the size of that field instead of the size of the whole array. For example,

```
» size(student)
ans =
     1     3
» size(student(1).name)
ans =
     1     8
```

10.3.8 Nesting Structure Arrays

Each field of a structure array can be of any data type, including a cell array or a structure array. For example, the following statements define a new structure array as a field under array `student` to carry information about each class that the student in enrolled in.

```
student(1).class(1).name = 'COSC 2021'
student(1).class(2).name = 'PHYS 1001'
student(1).class(1).instructor = 'Mr. Jones'
student(1).class(2).instructor = 'Mrs. Smith'
```

After these statements are issued, `student(1)` contains the following data. Note the technique used to access the data in the nested structures.

```
» student(1)
ans =
      name: 'John Doe'
     addr1: '123 Main Street'
      city: 'Anytown'
     state: 'LA'
       zip: '71211'
     exams: [80 95 84]
     class: [1x2 struct]
» student(1).class
ans =
1x2 struct array with fields:
     name
     instructor
» student(1).class(1)
ans =
          name: 'COSC 2021'
     instructor: 'Mr. Jones'
» student(1).class(2)
ans =
          name: 'PHYS 1001'
     instructor: 'Mrs. Smith'
» student(1).class(2).name
ans =
PHYS 1001
```

10.3.9 Summary of `structure` Functions

The common MATLAB structure functions are summarized in Table 10.3.

Table 10.3: Common MATLAB Structure Functions

`fieldnames`	Returns a list of field names in a cell array of strings.
`getfield`	Gets current value from a field.
`rmfield`	Removes a field from a structure array.
`setfield`	Sets a new value into a field.
`struct`	Predefines a structure array.

Quiz 10.1

This quiz provides a quick check to see if you have understood the concepts introduced in Sections 10.1 through 10.3. If you have trouble with the quiz, reread the section, ask your instructor, or discuss the material with a fellow student. The answers to this quiz are found in the back of the book.

1. What is a sparse array? How does it differ from a full array? How can you convert from a sparse array to a full array and vice versa?
2. What is a cell array? How does it differ from an ordinary array?
3. What is the difference between content indexing and cell indexing?
4. What is a structure? How does it differ from ordinary arrays and cell arrays?
5. What is the purpose of `varargin`? How does it work?
6. Given the definition of array a shown below, what will be produced by each of the following sets of statements? (*Note:* some of these statements may be illegal. If a statement is illegal, explain why.)

    ```
    a{1,1} = [1 2 3; 4 5 6; 7 8 9];
    a(1,2) = {'Comment line'};
    a{2,1} = j;
    a{2,2} = a{1,1} - a{1,1}(2,2);
    ```
    (a) `a(1,1)`
    (b) `a{1,1}`
    (c) `2*a(1,1)`
    (d) `2*a{1,1}`
    (e) `a{2,2}`
    (f) `a(2,3) = {[-17; 17]}`
    (g) `a{2,2}(2,2)`

7. Given the definition of structure array b shown below, what will be produced by each of the following sets of statements? (*Note:* some of these statements may be illegal. If a statement is illegal, explain why.)

    ```
    b(1).a = -2*eye(3);
    b(1).b = 'Element 1';
    ```

```
b(1).c = [1 2 3];
b(2).a = [b(1).c' [-1; -2; -3] b(1).c'];
b(2).b = 'Element 2';
b(2).c = [1 0 -1];
```

(a) `b(1).a - b(2).a`
(b) `strncmp(b(1).b,b(2).b,6)`
(c) `mean(b(1).c)`
(d) `mean(b.c)`
(e) `b`
(f) `b(1).('b')`
(g) `b(1)`

►Example 10.2—Polar Vectors

A vector is a mathematical quantity that has both a magnitude and a direction. It can be represented as a displacement along the x and y axes in rectangular coordinates or by a distance r at an angle θ in polar coordinates (see Figure 10.8). The relationships among x, y, r, and θ are given by the following equations:

$$x = r \cos \theta \qquad (10.1)$$

$$y = r \sin \theta \qquad (10.2)$$

$$r = \sqrt{x^2 + y^2} \qquad (10.3)$$

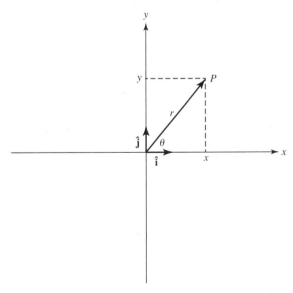

Figure 10.8 Relationship between the rectangular (x, y) description and the polar (r, θ) description of a vector.

$$\theta = \tan^{-1} \frac{y}{x} \tag{10.4}$$

where $\tan^{-1}(\ )$ is the two-argument inverse tangent function `atan2(y, x)`, whose output is defined over the range $-\pi \le \theta \le \pi$.

A vector in rectangular format can be represented as a structure having the fields x and y, for example

```
rect.x = 3;
rect.y = 4;
```

and a vector in polar format can be represented as a structure having the fields r and `theta` (where `theta` is in degrees), for example

```
polar.r = 5;
polar.theta = 36.8699;
```

Write a pair of functions that convert a vector in rectangular format to a vector in polar format, and vice versa.

Solution We will create two functions, `to_rect`, and `to_polar`.

Function `to_rect` must accept a vector in polar format and convert it into rectangular format, using Equations (10.1) and (10.2). This function will identify a vector in polar format because it will be stored in a structure having fields r and `theta`. If the input parameter is not a structure having fields r and `theta`, the function should generate an error and quit. The output from the function will be a structure having fields x and y.

The `to_polar` function must accept a vector in rectangular format and convert it into rectangular format, using Equations (10.3) and (10.4). This function will identify a vector in rectangular format because it will be stored in a structure having fields x and y. If the input parameter is not a structure having fields x and y, the function should generate an error and quit. The output from the function will be a structure having fields r and `theta`.

The calculation for r can use Equation (10.3) directly, but the calculation for `theta` needs to use the MATLAB function `atan2(y,x)`, because Equation (10.4) only produces output over the range $-\dfrac{\pi}{2} < \theta < \dfrac{\pi}{2}$, while the `atan2` function is valid in all four quadrants of the circle. Consult the MATLAB Help System for details of the operation of function `atan2`.

1. **State the problem**

   Assume that a polar vector is stored in a structure having fields r and `theta` (where `theta` is in degrees), and a rectangular vector is stored in a structure having fields x and y. Write a `to_rect` function to convert a polar vector to rectangular format, and a `to_polar` function to convert a rectangular vector into polar format.

2. **Define the inputs and outputs**

   The input to function `to_rect` is a vector in polar format stored in a structure with elements r and `theta`, and the output is a vector in rectangular format stored in a structure with elements x and y.

The input to function to_polar is a vector in rectangular format stored in a structure with elements x and y, and the output is a vector in rectangular format stored in a structure with elements r and theta.

3. **Design the algorithm**

The pseudocode for function to_rect is

```
Check to see that elements r and theta exist
out.x ← in.r * cos(in.theta * pi/180)
out.y ← in.r * sin(in.theta * pi/180)
```

Note that we have to convert the angle in degrees into an angle in radians before applying the sine and cosine functions.

The pseudocode for the to_polar function is

```
Check to see that elements r and theta exist
out.r ← sqrt(in.x.^2 + in.y.^2)
out.theta ← atan2(in.y,in.x) * 180/pi
```

Note that we have to convert the angle in radians into an angle in degrees before saving it in theta.

4. **Turn the algorithm into MATLAB statements.**

The final MATLAB functions are shown below.

```
function out = to_rect(in)
%TO_RECT Convert a vector from polar to rect
% Function TO_RECT converts a vector from polar
% coordinates to rectangular coordinates.
%
% Calling sequence:
%   out = to_rect(in)

% Define variables:
%   in    -- Structure containing fields r and theta (in degrees)
%   out   -- Structure containing fields x and y

%   Record of revisions:
%      Date          Programmer          Description of change
%      ====          ==========          =====================
%      09/01/14      S. J. Chapman       Original code

% Check for valid input
if ~isfield(in,'r') || ~isfield(in,'theta')
   error('Input argument does not contain fields ''r'' and ''theta''')
else
```

```
   % Calculate output.
   out.x = in.r * cos(in.theta * pi/180);
   out.y = in.r * sin(in.theta * pi/180);
end

function out = to_polar(in)
%TO_POLAR Convert a vector from rect to polar
% Function TO_POLAR converts a vector from rect
% coordinates to polar coordinates.
%
% Calling sequence:
%    out = to_rect(in)

% Define variables:
%   in    -- Structure containing fields x and y
%   out   -- Structure containing fields r and theta (in degrees)

% Record of revisions:
%     Date            Programmer          Description of change
%     ====            ==========          =====================
%     09/10/14        S. J. Chapman       Original code

% Check for valid input
if ~isfield(in,'x') || ~isfield(in,'y')
  error('Input argument does not contain fields ''x'' and ''y''')
else

   % Calculate output.
   out.r       = sqrt(in.x .^2 + in.y .^2);
   out.theta = atan2(in.y,in.x) * 180/pi;
end
```

5. **Test the program.**

   To test this program, we will use the example of a 3-4-5 right triangle. If the rectangular coordinates of the vector are $(x, y) = (3,4)$, then the polar form of the vector is

   $$r = \sqrt{3^2 + 4^2} = 5$$

   $$\theta = \tan^{-1}\frac{4}{3} = 53.13°$$

   When this program is executed, the results are.

   ```
   » v.x = 3;
   » v.y = 4;
   » out1 = to_polar(v)
   out1 =
            r: 5
        theta: 53.1301
   ```

```
» out2 = to_rect(out1)
out2 =
    x: 3
    y: 4
```

Going to polar coordinates and then back to rectangular coordinates produced the same results that we started with. ◀

10.4 Summary

Sparse arrays are special arrays in which memory is allocated only for non-zero elements. Three values are saved for each non-zero element—a row number, a column number, and the value itself. This form of storage is much more efficient than arrays for the situation where only a tiny fraction of the elements are non-zero. MATLAB includes functions and intrinsic calculations for sparse arrays, so they can be freely and transparently mixed with full arrays.

Cell arrays are arrays whose elements are *cells*, containers that can hold other MATLAB arrays. Any sort of data may be stored in a cell, including structure arrays and other cell arrays. They are a very flexible way to store data and are used in many internal MATLAB Graphical User Interface functions.

Structure arrays are a data type in which each individual element is given a name. The individual elements of a structure are known as fields, and each field in a structure may have a different type. The individual fields are addressed by combining the name of the structure with the name of the field, separated by a period. Structure arrays are useful for grouping together all of the data related to a particular person or thing into a single location.

10.4.1 Summary of Good Programming Practice

The following guidelines should be adhered to:

1. Always preallocate all cell arrays before assigning values to the elements of the array. This practice greatly increases the execution speed of a program.
2. Use cell array arguments `varargin` and `varargout` to create functions that support varying numbers of input and output arguments.

10.4.2 MATLAB Summary

The following summary lists all of the MATLAB commands and functions described in this chapter, along with a brief description of each one.

`cell`	Predefines a cell array structure.
`celldisp`	Displays contents of a cell array.
`cellplot`	Plots structure of a cell array.
`cellstr`	Converts a 2-D character array to a cell array of strings.

(continued)

char	Converts a cell array of strings into a 2-D character array.
fieldnames	Returns a list of field names in a cell array of strings.
figure	Creates a new figure/makes figure current.
iscellstr	Function that returns true of a cell array is a cell array of strings.
getfield	Gets current value from a field.
rmfield	Removes a field from a structure array.
setfield	Sets new value into a field.
strjoin	Combines the elements of a cell array of strings into a single string, with a single space between each input string.
uiimport	Imports data to MATLAB from a file created by an external program.

10.5 Exercises

10.1 Write a MATLAB function that will accept a cell array of strings and sort them into ascending order, according to the lexicographic order of the UTF-8 character set. (*Hint:* Look up function strcmp in the MATLAB Help System.)

10.2 Write a MATLAB function that will accept a cell array of strings and sort them into ascending order according to *alphabetical order*. (This implies that you must treat A and a as the same letter.) (*Hint:* Look up function strcmpi in the MATLAB Help System.)

10.3 Create a function that accepts any number of numeric input arguments and sums up all of individual elements in the arguments. Test your function by passing it the four arguments $a = 10$, $b = \begin{bmatrix} 4 \\ -2 \\ 2 \end{bmatrix}$, $c = \begin{bmatrix} 1 & 0 & 3 \\ -5 & 1 & 2 \\ 1 & 2 & 0 \end{bmatrix}$, and $d = \begin{bmatrix} 1 & 5 & -2 \end{bmatrix}$.

10.4 Modify the function of the previous exercise so that it can accept either ordinary numeric arrays or cell arrays containing numeric values. Test your function by passing it the two arguments a and b, where $a = \begin{bmatrix} 1 & 4 \\ -2 & 3 \end{bmatrix}$, $b\{1\} = \begin{bmatrix} 1 & 5 & 2 \end{bmatrix}$, and $b\{2\} = \begin{bmatrix} 1 & -2 \\ 2 & 1 \end{bmatrix}$.

10.5 Create a structure array containing all of the information needed to plot a data set. At a minimum, the structure array should have the following fields:

- x_data   *x*-data (one or more data sets in separate cells)
- y_data   *y*-data (one or more data sets in separate cells)
- type   linear, semilogx, and so forth
- plot_title   plot title
- x_label   *x*-axis label
- y_label   *y*-axis label
- x_range   *x*-axis range to plot
- y_range   *y*-axis range to plot

You may add additional fields that would enhance your control of the final plot.

After this structure array is created, create a MATLAB function that accepts an array of this structure and produces one plot for each structure in the array. The function should apply intelligent defaults if some data fields are missing. For example, if the `plot_title` field is an empty matrix, then the function should not place a title on the graph. Think carefully about the proper defaults before starting to write your function!

To test your function, create a structure array containing the data for three plots of three different types, and pass that structure array to your function. The function should correctly plot all three data sets in three different figure windows.

10.6 Define a structure `point` containing two fields x and y. The x field will contain the *x*-position of the point, and the y field will contain the *y*-position of the point. Then write a function `dist3` that accepts two points and returns the distance between the two points on the Cartesian plane. Be sure to check the number of input arguments in your function.

10.7 Write a function that will accept a structure as an argument, and return two cell arrays containing the names of the fields of that structure and the data types of each field. Be sure to check that the input argument is a structure, and generate an error message if it is not.

10.8 Write a function that will accept a structure array of `student` as defined in this chapter, and calculate the final average of each one assuming that all exams have equal weighting. Add a new field to each array to contain the final average for that student, and return the updated structure to the calling program. Also, calculate and return the final class average.

10.9 Write a function that will accept two arguments, the first a structure array and the second a field name stored in a string. Check to make sure that these input arguments are valid. If they are not valid, print out an error message. If they are valid and the designated field is a string, concatenate all of the strings in the specified field of each element in the array and return the resulting string to the calling program.

10.10 Calculating Directory Sizes Function `dir` returns the contents of a specified directory. The `dir` command returns a structure array with four fields, as shown below:

```
» d = dir('chap10')
d =
36x1 struct array with fields:
    name
    date
    bytes
    isdir
```

The `name` field contains the names of each file, `date` contains the last modification date for the file, `bytes` contains the size of the file in bytes, and `isdir` is 0 for conventional files and 1 for directories. Write a function that accepts a directory name and path and returns the total size of all files in the directory, in bytes.

10.11 **Recursion**   A function is said to be *recursive* if the function calls itself. Modify the function created in Problem 10.10 so that it calls itself when it finds a subdirectory and sums up the size of all files in the current directory plus all subdirectories.

10.12 Look up function `struct` in the MATLAB Help Browser, and learn how to pre-allocate a structure and simultaneously initialize all of the elements in the structure array to the same value. Then create a 2000 element array of type `student`, with the values in every array element initialized with the fields shown below:

```
 name: 'John Doe'
addr1: '123 Main Street'
 city: 'Anytown'
state: 'LA'
  zip: '71211'
```

10.13 **Vector Addition**   Write a function that will accept two vectors defined in either rectangular or polar coordinates (as defined in Example 10.2), add them, and save the result in rectangular coordinates.

10.14 **Vector Subtraction**   Write a function that will accept two vectors defined in either rectangular or polar coordinates (as defined in Example 10.2), subtract them, and save the result in rectangular coordinates.

10.15 **Vector Multiplication**   If two vectors are defined in polar coordinates so that $\mathbf{v}_1 = r_1 \angle \theta_1$ and $\mathbf{v}_2 = r_2 \angle \theta_2$, then the product of the two vectors $\mathbf{v}_1 \mathbf{v}_2 = r_1 r_2 \angle \theta_1 + \theta_2$. Write a function that will accept two vectors defined in either rectangular or polar coordinates (as defined in Example 10.2), perform the multiplication, and save the result in polar coordinates.

10.16 **Vector Division**   If two vectors are defined in polar coordinates so that $\mathbf{v}_1 = r_1 \angle \theta_1$ and $\mathbf{v}_2 = r_2 \angle \theta_2$, then $\dfrac{\mathbf{v}_1}{\mathbf{v}_2} = \dfrac{r_1}{r_2} \angle \theta_1 - \theta_2$. Write a function that will accept two vectors defined in either rectangular or polar coordinates (as defined in Example 10.2), perform the division, and save the result in polar coordinates.

10.17 **Distance Between Two Points**   If $\mathbf{v}_1$ is the distance from the origin to point P_1 and $\mathbf{v}_2$ is the distance from the origin to point P_2, then the distance between the two points will be $|\mathbf{v}_1 - \mathbf{v}_2|$. Write a function that will accept two vectors defined in either rectangular or polar coordinates (as defined in Example 10.2) and that returns the distance between the two.

10.18 **Function Generators**   Generalize the function generator of Exercise 7.22 to handle polynomials of arbitrary dimension. Test it by creating function handles and plots the same way that you did in Exercise 7.22. (*Hint:* Use `varagrin`.)

Chapter

11

Input/Output Functions

In Chapter 2, we learned how to load and save MATLAB data using the `load` and `save` commands, and how to write out formatted data using the `fprintf` function. In this chapter we will learn more about MATLAB's input/output capabilities. First, we will learn about `textread` and `textscan`, two very useful functions for reading text data from a file. Then, we will spend a bit more time examining the `load` and `save` commands. Finally, we will look at the other file I/O options available in MATLAB.

Those readers familiar with C will find much of this material very familiar. However, be careful—there are subtle differences between MATLAB and C functions that can trip you up.

11.1 The `textread` Function

The `textread` function reads text files that are formatted into columns of data, where each column can be of a different type, and stores the contents of each column in a separate output array. This function is *very* useful for importing tables of data printed out by other applications.

The form of the `textread` function is

```
[a,b,c,...] = textread(filename,format,n)
```

where `filename` is the name of the file to open, `format` is a string containing a description of the type of data in each column, and `n` is the number of lines to read. (If `n` is missing, the function reads to the end of the file.) The format string contains the same types of format descriptors as function `fprintf`. Note that the number of output arguments must match the number of columns that you are reading.

For example, suppose that file `test_input.dat` contains the following data:

```
James    Jones   O+    3.51    22    Yes
Sally    Smith   A+    3.28    23    No
```

This data could be read into a series of arrays with the following function:

```
[first,last,blood,gpa,age,answer] = ...
          textread('test_input.dat','%s %s %s %f %d %s')
```

When this command is executed, the results are:

```
» [first,last,blood,gpa,age,answer] = ...
            textread('test_input.dat','%s %s %s %f %d %s')
first =
    'James'
    'Sally'
last =
    'Jones'
    'Smith'
blood =
    'O+'
    'A+'
gpa =
    3.5100
    3.2800
age =
    42
    28
answer =
    'Yes'
    'No'
```

This function can also skip selected columns by adding an asterisk to the corresponding format descriptor (for example, `%*s`). The following statement reads only the `first`, `last`, and `gpa` from the file:

```
» [first,last,gpa] = ...
            textread('test_input.dat','%s %s %*s %f %*d %*s')
first =
    'James'
    'Sally'
last =
    'Jones'
    'Smith'
gpa =
    3.5100
    3.2800
```

The `textread` function is much more useful and flexible than the `load` command. The `load` command assumes that all of the data in the input file is of a single type—it cannot support different types of data in different columns. In addition, it stores all of the data into a single array. In contrast, the `textread` function allows each column to go into a separate variable, which is *much* more convenient when working with columns of mixed data.

Function `textread` has a number of additional options that increase its flexibility. Consult the MATLAB on-line documentation for details of these options.

11.2 More about the `load` and `save` Commands

The `save` command saves MATLAB workspace data to disk, and the `load` command loads data from disk into the workspace. The `save` command can save data either in a special binary format called a MAT file or in an ordinary text file. The form of the `save` command is

```
save filename [content] [options]
```

where *content* specifies the data to be saved and *options* specifies how to save it.

The `save` command all by itself saves all of the data in the current workspace to a file named `matlab.mat` in the current directory. If a file name is included, the data will be saved in file "`filename.mat`". If a list of variables is included at the *content* position, then only those particular variables will be saved.

For example, suppose that a workspace contains a 1000-element double array `x` and a character string `str`. We can save these two variables to a MAT file with the following command:

```
save test_matfile x str
```

This command creates a MAT file with the name `test_matfile.mat`. The contents of this file can be examined with `-file` option of the `whos` command:

```
» whos -file test_matfile.mat
  Name      Size       Bytes    Class      Attributes

  str       1x11          22    char
  x         1x1000      8000    double
```

The content to be saved can be specified in several ways, as described in Table 11.1.

The more important options supported by the `save` command are shown in Table 11.2; a complete list can be found in the MATLAB on-line documentation.

The `load` command can load data from MAT files or from ordinary text files. The form of the `load` command is

```
load filename [options] [content]
```

The command `load` all by itself loads all of the data in file `matlab.mat` into the current workspace. If a file name is included, the data will be loaded from that file

Table 11.1: Ways of Specifying `save` Command Content

Values for `content`	Description
`<nothing>`	Saves all data in current workspace.
`varlist`	Saves only the values in the variable list.
`-regexp exprlist`	Saves all variables that match any of the regular expressions in the expression list.
`-struct s`	Saves as individual variables all fields of the scalar structure s.
`-struct s fieldlist`	Saves as individual variables only the specified fields of structure s.

Table 11.2: Selected `save` Command Options

Option	Description
`'-mat'`	Saves data in MAT-file format (default).
`'-ascii'`	Saves data in space-separated text format with 8 digits of precision.
`'-ascii','-tabs'`	Saves data in tab-separated text format with 8 digits of precision.
`'-ascii','-double'`	Saves data in tab-separated text format with 16 digits of precision.
`-append`	Adds the specified variables to an existing MAT file.
`-v4`	Saves the MAT file in a format readable by MATLAB version 4 or later.
`-v6`	Saves the MAT file in a format readable by MATLAB versions 5 and 6 or later.
`-v7`	Saves the MAT file in a format readable by MATLAB versions 7 through 7.2 or later.
`-v7.3`	Saves the MAT file in a format readable by MATLAB versions 7.3 or later.

name. If specific variables are included in the content list, then only those variables will be loaded from the file. For example,

```
load                 % Loads entire content of matlab.mat
load mydat.mat       % Loads entire content of mydat.mat
load mydat.mat a b c % Loads only a, b, and c from mydat.mat
```

The options supported by the `load` command are shown in Table 11.3.

Although it is not immediately obvious, the `save` and `load` commands are the most powerful and useful I/O commands in MATLAB. Among their advantages are:

1. These commands are very easy to use.
2. MAT files are *platform independent*. A MAT file written on any type of computer that supports MATLAB can be read on any other computer

Table 11.3: `load` Command Options

Option	Description
`-mat`	Treats file as a MAT file (default if file extension is `mat`).
`-ascii`	Treats file as a space-separated text file (default if file extension is *not* `mat`).

supporting MATLAB. This format transfers freely among PCs, Macs, and Linux. Also, the Unicode character encoding ensures that character strings will be preserved properly across platforms.

3. MAT files are efficient users of disk space, using only the amount of memory required for each data type. They store the full precision of every variable—no precision is lost due to conversion to and from text format. MAT files can also be compressed to save even more disk space.

4. MAT files preserve all of the information about each variable in the workspace, including its class, name, and whether or not it is global. All of this information is lost in other types of I/O. For example, suppose that the workspace contains the following information:

```
» whos
   Name        Size        Bytes       Class       Attributes

   a           10x10       800         double
   b           10x10       800         double
   c           2x2         32          double
   string      1x14        28          char
   student     1x3         888         struct
```

If this workspace is saved with the command `save workspace.mat`, a file named `workspace.mat` will be created. When this file is loaded, all of the information will be restored, including the type of each item and whether or not it is global.

A disadvantage of these commands is that the MAT file format is unique to MATLAB and cannot be used to share data with other programs. The `-ascii` option can be used if you wish to share data with other programs, but it has serious limitations[1].

 Good Programming Practice

Unless you must exchange data with non-MATLAB programs, always use the `load` and `save` commands to save data sets in MAT file format. This format is efficient and transportable across MATLAB implementations, and it preserves all details of all MATLAB data types.

///

[1]This statement is only partially true. Modern MAT files are in HDF5 format, which is an industry standard, and there are free tools and packages in C++, Java, and so forth that can read data in this format.

The `save -ascii` command will not save cell or structure array data at all, and it converts string data to numbers before saving it. The `load -ascii` command will only load space- or tab-separated data with an equal number of elements on each row, and it will place all of the data into a single variable with the same name as the input file. If you need anything more elaborate (saving and loading strings, cells, structure arrays, and so on in formats suitable for exchanging with other programs), then it will be necessary to use the other file I/O commands described in this chapter.

If the file name and the names of the variables to be loaded or saved are in strings, then you should use the function forms of the `load` and `save` commands. For example, the following fragment of code asks the user for a file name and saves the workspace in that file.

```
filename = input('Enter save file name:','s');
save (filename,'-mat');
```

11.3 An Introduction to MATLAB File Processing

To use files within a MATLAB program, we need some way to select the desired file and to read from or write to it. MATLAB has a very flexible method to read and write files, whether they are on disk, memory stick, or some other device attached to the computer. This mechanism is known as the **file id** (sometimes known as **fid**). The file id is a number assigned to a file when it is opened, and is used for all reading, writing, and control operations on that file. The file id is a positive integer. Two file id's are always open—file id 1 is the standard output device (`stdout`) and file id 2 is the standard error (`stderr`) device for the computer on which MATLAB is executing. Additional file id's are assigned as files are opened, and released as files are closed.

Several MATLAB functions may be used to control disk file input and output. The file I/O functions are summarized in Table 11.4.

File id's are assigned to disk files or devices using the `fopen` statement, and are detached from them using the `fclose` statement. Once a file is attached to a file id using the `fopen` statement, we can read and write to that file using MATLAB file input and output statements. When we are through with the file, the `fclose` statement closes the file and makes the file id invalid. The `frewind` and `fseek` statements may be used to change the current reading or writing position in a file while it is open.

Data can be written to and read from files in two possible ways: as binary data or as formatted character data. Binary data consists of the actual bit patterns that are used to store the data in computer memory. Reading and writing binary data is very efficient, but a user cannot read the data stored in the file. Data in formatted files is translated into characters that can be read directly by a user. However, formatted I/O operations are slower and less efficient than binary I/O operations. We will discuss both types of I/O operations later in this chapter.

Table 11.4: MATLAB Input/Output Functions

Category	Function	Description
Load/Save Workspace	load	Loads workspace
	save	Saves workspace
File Opening and Closing	fopen	Opens file
	fclose	Closes file
Binary I/O	fread	Reads binary data from file
	fwrite	Writes binary data to file
Formatted I/O	fscanf	Reads formatted data from file
	fprintf	Writes formatted data to file
	fgetl	Reads line from file, discards newline character
	fgets	Reads line from file, keeps newline character
File Positioning, Status, and Miscellaneous	delete	Deletes file
	exist	Checks for the existence of a file
	ferror	Inquires file I/O error status
	feof	Tests for end-of-file
	fseek	Sets file position
	ftell	Checks file position
	frewind	Rewinds file
Temporary Files	tempdir	Gets temporary directory name
	tempname	Gets temporary file name

11.4 File Opening and Closing

The file opening and closing functions, `fopen` and `fclose`, are described below.

11.4.1 The `fopen` Function

The `fopen` function opens a file and returns a file id number for use with the file. The basic forms of this statement are

```
fid = fopen(filename, permission)
[fid, message] = fopen(filename, permission)
[fid, message] = fopen(filename, permission, format)
[fid, message] = fopen(filename, permission, format, encoding)
```

where *filename* is a string specifying the name of the file to open, *permission* is a character string specifying the mode in which the file is opened, *format* is an optional string specifying the numeric format of the data in the file, and *encoding* is the character encoding to use for subsequent read and write operations. If the open is successful, `fid` will contain a positive integer after this statement is executed, and `message` will be

Table 11.5: `fopen` File Permissions

File Permission	Meaning
`'r'`	Opens an existing file for reading only (default)
`'r+'`	Opens an existing file for reading and writing
`'w'`	Deletes the contents of an existing file (or creates a new file) and opens it for writing only
`'w+'`	Deletes the contents of an existing file (or creates a new file) and opens it for reading and writing
`'a'`	Opens an existing file (or creates a new file) and opens it for writing only, appending to the end of the file.
`'a+'`	Opens an existing file (or creates a new file) and opens it for reading and writing, appending to the end of the file.
`'W'`	Writes without automatic flushing (special command for tape drives)
`'A'`	Appends without automatic flushing (special command for tape drives)

an empty string. If the open fails, `fid` will contain a −1 after this statement is executed, and `message` will be a string explaining the error. If a file is opened for reading and it is not in the current directory, MATLAB will search for it along the MATLAB search path.

The possible permission strings are shown in Table 11.5.

On some platforms, such as PCs, it is important to distinguish between text files and binary files. If a file is to be opened in text mode, then a t should be added to the permissions string (for example, `'rt'` or `'rt+'`). If a file is to be opened in binary mode, a b may be added to the permissions string (for example, `'rb'`), but this is not actually required since files are opened in binary mode by default. This distinction between text and binary files does not exist on UNIX or Linux computers, so the t or b is never needed on those systems.

The *format* string in the `fopen` function specifies the numeric format of the data stored in the file. This string is only needed when transferring files between computers with incompatible numeric data formats, so it is rarely used. A few of the possible numeric formats are shown in Table 11.6; see the MATLAB Language Reference Manual for a complete list of possible numeric formats.

The *encoding* string in the `fopen` function specifies the type of character encoding to be used in the file. This string is only needed when not using the default character encoding, which is UTF-8. Examples of legal character encodings include `'UTF-8'`, `'ISO-8859-1'`, and `'windows-1252'`. See the MATLAB Language Reference Manual for a complete list of possible encodings.

There are also two forms of this function that provide information rather than open files. The function

```
fids = fopen('all')
```

returns a row vector containing a list of all file id's for currently open files (except for `stdout` and `stderr`). The number of elements in this vector is equal to the number of open files. The function

```
[filename, permission, format] = fopen(fid)
```

Table 11.6: `fopen` Numeric Format Strings

File Permission	Meaning
`'native'` or `'n'`	Numeric format for the machine MATLAB is executing on (default)
`'ieee-le'` or `'l'`	IEEE floating point with little-endian byte ordering
`'ieee-be'` or `'b'`	IEEE floating point with big-endian byte ordering
`'ieee-le.l64'` or `'a'`	IEEE floating point with little-endian byte ordering and 64-bit-long data type
`'ieee-le.b64'` or `'s'`	IEEE floating point with big-endian byte ordering and 64-bit-long data type

returns the file name, permission string, and numeric format for an open file specified by file id.

Some examples of correct `fopen` functions are shown below.

Case 1: Opening a Binary File for Input

The function below opens a file named `example.dat` for binary input only.

```
fid = fopen('example.dat','r')
```

The permission string is `'r'`, indicating that the file is to be opened for reading only. The string could have been `'rb'`, but this is not required because binary access is the default case.

Case 2: Opening a File for Text Output

The functions below open a file named `outdat` for text output only.

```
fid = fopen('outdat','wt')
```

or

```
fid = fopen('outdat','at')
```

The `'wt'` permissions string specifies that the file is a new text file; if it already exists, then the old file will be deleted and a new empty file will be opened for writing. This is the proper form of the `fopen` function for an *output file* if we want to replace preexisting data.

The `'at'` permissions string specifies that we want to append to an existing text file. If it already exists, then it will be opened and new data will be appended to the currently existing information. This is the proper form of the `fopen` function for an *output file* if we don't want to replace preexisting data.

Case 3: Opening a Binary File for Read/Write Access

The function below opens a file named `junk` for binary input and output.

```
fid = fopen('junk','r+')
```

The function below also opens the file for binary input and output.

```
fid = fopen('junk','w+')
```

The difference between the first and the second statements is that the first statement requires the file to exist before it is opened, while the second statement will delete any preexisting file.

Good Programming Practice

Always be careful to specify the proper permissions in `fopen` statements, depending on whether you are reading from or writing to a file. This practice will help prevent errors, such as accidentally overwriting data files that you want to keep.

It is important to check for errors after you attempt to open a file. If the `fid` is −1, then the file failed to open. You should report this problem to the user, and allow him or her to either select another file or else quit the program.

Good Programming Practice

Always check the status after a file open operation to make sure that it is successful. If the file open fails, tell the user and provide a way to recover from the problem.

11.4.2 The `fclose` Function

The `fclose` function closes a file. Its form is

```
status = fclose(fid)
status = fclose('all')
```

where `fid` is a file id and `status` is the result of the operation. If the operation is successful, `status` will be 0, and if it is unsuccessful, `status` will be −1.

The form `status = fclose('all')` closes all open files except for `stdout` (`fid = 1`) and `stderr` (`fid = 2`). It returns a status of 0 if all files close successfully, and −1 otherwise.

11.5 Binary I/O Functions

The binary I/O functions, `fwrite` and `fread`, are described below.

11.5.1 The `fwrite` Function

The `fwrite` function writes binary data in a user-specified format to a file. Its form is

```
count = fwrite(fid,array,precision)
count = fwrite(fid,array,precision,skip)
count = fwrite(fid,array,precision,skip,format)
```

where `fid` is the file id of a file opened with the `fopen` function, `array` is the array of values to write out, and `count` is the number of values written to the file.

MATLAB writes out data in *column order*, which means that the entire first column is written out, followed by the entire second column, and so forth. For example, if $\text{array} = \begin{bmatrix} 1 & 2 \\ 3 & 4 \\ 5 & 6 \end{bmatrix}$, then the data will be written out in the order 1, 3, 5, 2, 4, 6.

The optional *precision* string specifies the format in which the data will be output. MATLAB supports both platform-independent precision strings, which are the same for all computers that MATLAB runs on, and platform-dependent precision strings that vary among different types of computers. *You should only use the platform-independent strings*, and those are the only forms presented in this book.

For convenience, MATLAB accepts some C and Fortran data type equivalents for the MATLAB precision strings. If you are a C or Fortran programmer, you may find it more convenient to use the names of the data types in the language that you are most familiar with.

The possible platform-independent precisions are presented in Table 11.7. All of these precisions work in units of bytes, except for `'bitN'` or `'ubitN'`, which work in units of bits.

Table 11.7: Selected **MATLAB** Precision Strings

MATLAB Precision String	C/Fortran Equivalent	Meaning
`'char'`	`'char*1'`	8-bit characters
`'schar'`	`'signed char'`	8-bit signed character
`'uchar'`	`'unsigned char'`	8-bit unsigned character
`'int8'`	`'integer*1'`	8-bit integer
`'int16'`	`'integer*2'`	16-bit integer
`'int32'`	`'integer*4'`	32-bit integer
`'int64'`	`'integer*8'`	64-bit integer
`'uint8'`	`'integer*1'`	8-bit unsigned integer
`'uint16'`	`'integer*2'`	16-bit unsigned integer
`'uint32'`	`'integer*4'`	32-bit unsigned integer
`'uint64'`	`'integer*8'`	64-bit unsigned integer
`'float32'`	`'real*4'`	32-bit floating point
`'float64'`	`'real*8'`	64-bit floating point
`'bitN'`		N-bit signed integer, $1 \leq N \leq 64$
`'ubitN'`		N-bit unsigned integer, $1 \leq N \leq 64$

The optional argument *skip* specifies the number of bytes to skip in the output file before each write. This option is useful for placing values at certain points in fixed-length records. Note that if *precision* is a bit format like `'bitN'` or `'ubitN'`, skip is specified in bits instead of bytes.

The optional argument *format* is an optional string specifying the numeric format of the data in the file, as shown in Table 11.6.

11.5.2 The `fread` Function

The `fread` function reads binary data in a user-specified format from a file, and returns the data in a (possibly different) user-specified format. Its form is

```
[array,count] = fread(fid,size,precision)
[array,count] = fread(fid,size,precision,skip)
[array,count] = fread(fid,size,precision,skip,format)
```

where `fid` is the file id of a file opened with the `fopen` function, `size` is the number of values to read, `array` is the array to contain the data, and `count` is the number of values read from the file.

The optional argument `size` specifies the amount of data to be read from the file. There are three versions of this argument:

- n—Read exactly n values. After this statement, `array` will be a column vector containing n values read from the file.
- Inf—Read until the end of the file. After this statement, `array` will be a column vector containing all of the data until the end of the file.
- [n m] —Read exactly n × m values, and format the data as an n × m array.

If `fread` reaches the end of the file and the input stream does not contain enough bits to write out a complete array element of the specified precision, `fread` pads the last byte or element with zero bits until the full value is obtained. If an error occurs, reading is done up to the last full value.

The *precision* argument specifies both the format of the data on the disk and the format of the data array to be returned to the calling program. The general form of the precision string is

'disk_precision => array_precision'

where `disk_precision` and `array_precision` are both one of the precision strings found in Table 11.7. The `array_precision` value can be defaulted. If it is missing, then the data is returned in a `double` array. There is also a shortcut form of this expression if the disk precision and the array precision are the same: `'*disk_precision'`.

A few examples of `precision` strings are shown below:

`'single'`	Reads data in single precision format from disk and returns it in a `double` array.
`'single=>single'`	Reads data in single precision format from disk and returns it in a `single` array.

`'*single'`	Reads data in single precision format from disk and returns it in a `single` array (a shorthand version of the previous string).
`'double=>real*4'`	Reads data in double precision format from disk and returns it in a `single` array.

The optional argument *skip* specifies the number of bytes to skip in the output file before each write. This option is useful for placing values at certain points in fixed-length records. Note that if *precision* is a bit format like v`'bitN'` or `'ubitN'`, skip is specified in bits instead of bytes.

The optional argument *format* is an optional string specifying the numeric format of the data in the file, as shown in Table 11.6.

▶ Example 11.1—Writing and Reading Binary Data

The example script file shown below creates an array containing 10,000 random values, opens a user-specified file for writing only, writes the array to disk in 64-bit floating-point format, and closes the file. It then opens the file for reading and reads the data back into a 100×100 array. It illustrates the use of binary I/O operations.

```
% Script file: binary_io.m
%
% Purpose:
%   To illustrate the use of binary i/o functions.
%
% Record of revisions:
%    Date          Programmer         Description of change
%    ====          ==========         =====================
%   03/21/14     S. J. Chapman        Original code
%
% Define variables:
%   count      -- Number of values read / written
%   fid        -- File id
%   filename   -- File name
%   in_array   -- Input array
%   msg        -- Open error message
%   out_array  -- Output array
%   status     -- Operation status
% Prompt for file name
filename = input('Enter file name:','s');

% Generate the data array
out_array = randn(1,10000);

% Open the output file for writing.
[fid,msg] = fopen(filename,'w');
```

```
% Was the open successful?
if fid > 0

    % Write the output data.
    count = fwrite(fid,out_array,'float64');

    % Tell user
    disp([int2str(count)'values written...']);

    % Close the file
    status = fclose(fid);

else

    % Output file open failed.  Display message.
    disp(msg);

end

% Now try to recover the data.  Open the
% file for reading.
[fid,msg] = fopen(filename,'r');

% Was the open successful?
if fid > 0

    % Write the output data.
    [in_array, count] = fread(fid,[100 100],'float64');

    % Tell user
    disp([int2str(count) 'values read...']);

    % Close the file
    status = fclose(fid);

else

    % Input file open failed.  Display message.
    disp(msg);

end
```

When this program is executed, the results are

```
» binary_io
Enter file name: testfile
10000 values written...
10000 values read...
```

An 80,000-byte file named testfile was created in the current directory. This file is 80,000 bytes long because it contains 10,000 64-bit values, and each value occupies 8 bytes.

QUIZ 11.1

This quiz provides a quick check to see if you have understood the concepts introduced in Sections 11.1 through 11.5. If you have trouble with the quiz, reread the section, ask your instructor, or discuss the material with a fellow student. The answers to this quiz are found in the back of the book.

1. Why is the `textread` function especially useful for reading data created by programs written in other languages?
2. What are the advantages and disadvantages of saving data in a MAT file?
3. What MATLAB functions are used to open and close files? What is the difference between opening a binary file and opening a text file?
4. Write the MATLAB statement to open a preexisting file named `myinput.dat` for appending new text data.
5. Write the MATLAB statements required to open an unformatted input file for reading only. Check to see if the file exists and generate an appropriate error message if it doesn't.

For questions 6 and 7, determine whether the MATLAB statements are correct or not. If they are in error, specify what is wrong with them.

```
6. fid = fopen('file1','rt');
   array = fread(fid,Inf)
   fclose(fid);

7. fid = fopen('file1','w');
   x = 1:10;
   count = fwrite(fid,x);
   fclose(fid);
   fid = fopen('file1','r');
   array = fread(fid,[2 Inf])
   fclose(fid);
```

11.6 Formatted I/O Functions

The formatted I/O functions are described below.

11.6.1 The `fprintf` Function

The `fprintf` function writes formatted data in a user-specified format to a file. Its form is

```
count = fprintf(fid,format,val1,val2,...)
fprint(format,val1,val2,...)
```

where `fid` is the file id of a file to which the data will be written, and `format` is the format string controlling the appearance of the data. If `fid` is missing, the data

The Components of a Format Specifier

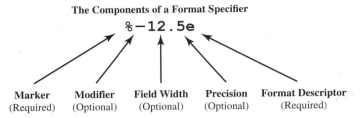

Figure 11.1 The structure of a typical format specifier.

is written to the standard output device (the Command Window). This is the form of `fprintf` that we have been using since Chapter 2.

The format string specifies the alignment, significant digits, field width, and other aspects of output format. It can contain ordinary alphanumeric characters along with special sequences of characters that specify the exact format in which the output data will be displayed. The structure of a typical format string is shown in Figure 11.1. A single % character always marks the beginning of a format—if an ordinary % sign is to be printed out, then it must appear in the format string as %%. After the % character, the format can have a flag, a field width and precision specifier, and a conversion specifier. The % character and the conversion specifier are always required in any format, while the field and field width and precision specifier are optional.

The possible conversion specifiers are listed in Table 11.8, and the possible flags are listed in Table 11.9. If a field width and precision are specified in a format, then the number before the decimal point is the field width, which is the number of characters used to display the number. The number after the decimal point is the precision, which is the minimum number of significant digits to display after the decimal point.

Table 11.8: Format Conversion Specifiers for `fprintf`

Specifier	Description
`%c`	Single character
`%d`	Decimal notation (signed)
`%e`	Exponential notation (using a lowercase e as in `3.1416e+00`)
`%E`	Exponential notation (using an uppercase E as in `3.1416E+00`)
`%f`	Fixed-point notation
`%g`	The more compact of `%e` or `%f`. Insignificant zeros do not print.
`%G`	Same as `%g`, but using an uppercase E
`%o`	Octal notation (unsigned)
`%s`	String of characters
`%u`	Decimal notation (unsigned)
`%x`	Hexadecimal notation (using lowercase letters a–f)
`%X`	Hexadecimal notation (using uppercase letters A–F)

Table 11.9: Format Flags

Flag	Description
Minus sign (-)	Left-justifies the converted argument in its field (Example: %-5.2d). If this flag is not present, the argument is right-justified.
+	Always print a + or - sign (Example: %+5.2d).
0	Pad argument with leading zeros instead of blanks (Example: %05.2d).

Table 11.10: Escape Characters in Format Strings

Escape Sequences	Description
\n	New line
\t	Horizontal tab
\b	Backspace
\r	Carriage return
\f	Form feed
\\	Print an ordinary backslash (\) symbol
\'' or ''	Print an apostrophe or single quote
%%	Print an ordinary percent (%) symbol

In addition to ordinary characters and formats, certain special escape characters can be used in a format string. These special characters are listed in Table 11.10.

11.6.2 Understanding Format Conversion Specifiers

The best way to understand the wide variety of format conversion specifiers is by example, so we will now present several examples along with their results.

Case 1: Displaying Decimal Data

Decimal (integer) data is displayed with the %d format conversion specifier. The d may be preceded by a flag and a field width and precision specifier, if desired. If used, the precision specifier sets a minimum number of digits to display. If there are not enough digits, leading zeros will be added to the number.

If a non-decimal number is displayed with the %d conversion specifier, the specifier will be ignored and the number will be displayed in exponential format. For example,

```
fprintf('%6d\n',123.4)
```

produces the result 1.234000e+002.

Function	Result	Comment
`fprintf('%d\n',123)`	`----|----|` `123`	Displays the number using as many characters as required. For the number 123, three characters are required.
`fprintf('%6d\n',123)`	`----|----|` `   123`	Displays the number in a 6-character-wide field. By default the number is *right justified* in the field.
`fprintf('%6.4d\n',123)`	`----|----|` `  0123`	Displays the number in a 6-character-wide field using a minimum of 4 characters. By default the number is *right justified* in the field.
`fprintf('%-6.4d\n',123)`	`----|----|` `0123`	Displays the number in a 6-character-wide field using a minimum of 4 characters. The number is *left justified* in the field.
`fprintf('%+6.4d\n',123)`	`----|----|` ` +0123`	Displays the number in a 6-character-wide field using a minimum of 4 characters plus a sign character. By default the number is *right justified* in the field.

Case 2: Displaying Floating-Point Data

Floating-point data can be displayed with the `%e`, `%f`, or `%g` format conversion specifiers. They may be preceded by a flag and a field width and precision specifier, if desired. If the specified field width is too small to display the number, it is ignored. Otherwise, the specified field width is used.

Function	Result	Comment
`fprintf('%f\n',123.4)`	`----|----|` `123.400000`	Displays the number using as many characters as required. The default case for `%f` is to display 6 digits after the decimal place.
`fprintf('%8.2f\n',123.4)`	`----|----|` `  123.40`	Displays the number in an 8-character-wide field, with two places after the decimal point. The number is *right justified* in the field.
`fprintf('%4.2f\n',123.4)`	`----|----|` `123.40`	Displays the number in a 6-character-wide field. The width specification was ignored because it was too small to display the number.
`fprintf('%10.2e\n',123.4)`	`----|----|` ` 1.23e+002`	Displays the number in exponential format in a 10-character-wide field using 2 decimal places. By default the number is *right justified* in the field.
`fprintf('%10.2E\n',123.4)`	`----|----|` ` 1.23E+002`	The same but with a capital E for the exponent.

Case 3: Displaying Character Data

Character data may be displayed with the %c or %s format conversion specifiers. They may be preceded by field width specifier, if desired. If the specified field width is too small to display the number, it is ignored. Otherwise, the specified field width is used.

Function	Result	Comment
fprintf('%c\n','s')	----|----| s	Displays a single character.
fprintf('%s\n','string')	----|----| string	Displays the character string.
fprintf('%8s\n','string')	----|----| string	Displays the character string in an 8-character-wide field. By default the string is *right justified* in the field.
fprintf('%-8s\n','string')	----|----| string	Displays the character string in an 8-character-wide field. The string is *left justified* in the field.

11.6.3 How Format Strings Are Used

The fprintf function contains a format string followed by zero or more values to print out. When the fprintf function is executed, the list of output values associated with the fprintf function is processed together with the format string. The function begins at the left end of the variable list and the left end of the format string, and scans from left to right, associating the first value in the output list with the first format descriptor in the format string, and so on. The variables in the output list must be of the same type and in the same order as the format descriptors in the format, or unexpected results may be produced. For example, if we attempt to display a floating-point number such as 123.4 with a %c or %d descriptor, the descriptor is ignored totally and the number is printed in exponential notation.

Programming Pitfalls

Make sure that there is a one-to-one correspondence between the types of the data in an fprintf function and the types of the format conversion specifiers in the associated format string, or your program will produce unexpected results.

As the program moves from left to right through the variable list of an fprintf function, it also scans from left to right through the associated format string. Format strings are scanned according to the following rules:

1. *Format strings are scanned in order from left to right.* The first format conversion specifier in the format string is associated with the first value in the output list of the fprintf function, and so forth. The type of each format conversion specifier must match the type of the data being output. In the example shown below, specifier %d is associated with variable a, %f with variable b, and %s with variable c. Note that the specifier types match the data types.

   ```
   a = 10; b = pi; c = 'Hello';
   fprintf('Output: %d %f %s\n',a,b,c);
   ```

2. If the scan reaches the end of the format string before the fprintf function runs out of values, the program starts over *at the beginning of the format string*. For example, the statements

   ```
   a = [10 20 30 40];
   fprintf('Output = %4d %4d\n',a);
   ```

   will produce the output

   ```
   ----|----|----|----|
   Output =   10   20
   Output =   30   40
   ```

   When the function reaches the end of the format string after printing a(2), it starts over at the beginning of the string to print a(3) and a(4).

3. If the fprintf function runs out of variables before the end of the format string, *the use of the format string stops at the first format conversion specifier without a corresponding variable or at the end of the format string, whichever comes first.* For example, the statements

   ```
   a = 10; b = 15; c = 20;
   fprintf('Output = %4d\nOutput = %4.1f\n',a,b,c);
   ```

   will produce the output

   ```
   Output = 10
   Output = 15.0
   Output = 20
   Output = »
   ```

   The use of the format string stops at %4.1f, which is the first unmatched format conversion specifier. On the other hand, the statements

   ```
   voltage = 20;
   fprintf('Voltage = %6.2f kV.\n',voltage);
   ```

   will produce the output

   ```
   Voltage = 20.00 kV,
   ```

   since there are no unmatched format conversion specifiers, and the use of the format stops at the end of the format string.

11.6.4 The `sprintf` Function

The `sprintf` function is exactly like `fprintf`, except that it writes formatted data to a character string instead of a file. Its form is

```
string = sprint(format,val1,val2,...)
```

where `fid` is the file id of a file to which the data will be written and `format` is the format string controlling the appearance of the data. This function is very useful for creating formatted data that can be displayed within a program.

▶ **Example 11.2—Generating a Table of Information**

A good way to illustrate the use of `fprintf` functions is to generate and print out a table of data. The example script file shown below generates the square roots, squares, and cubes of all integers between 1 and 10 and presents the data in a table with appropriate headings.

```
% Script file: create_table.m
%
% Purpose:
%    To create a table of square roots, squares, and
%    cubes.
%
% Record of revisions:
%      Date             Programmer            Description of change
%      ====             ==========            =====================
%    03/22/14         S. J. Chapman          Original code
%
% Define variables:
%    cube           -- Cubes
%    ii             -- Index variable
%    square         -- Squares
%    square_roots   -- Square roots
%    out            -- Output array

% Print the title of the table.
fprintf('Table of Square Roots, Squares, and Cubes\n\n');

% Print column headings
fprintf('Number    Square Root    Square    Cube\n');
fprintf('======    ===========    ======    ====\n');

% Generate the required data
ii = 1:10;
square_root = sqrt(ii);
square = ii.^2;
cube = ii.^3;
```

```
% Create the output array
out = [ii' square_root' square' cube'];

% Print the data
for ii = 1:10
    fprintf ('%2d    %11.4f    %6d    %8d\n',out(ii,:));
end
```

When this program is executed, the result is

```
» table
Table of Square Roots,  Squares,  and Cubes
  Number     Square Root     Square     Cube
  ======     ===========     ======     ====
     1         1.0000           1          1
     2         1.4142           4          8
     3         1.7321           9         27
     4         2.0000          16         64
     5         2.2361          25        125
     6         2.4495          36        216
     7         2.6458          49        343
     8         2.8284          64        512
     9         3.0000          81        729
    10         3.1623         100       1000
```

11.6.5 The `fscanf` Function

The `fscanf` function reads formatted data in a user-specified format from a file. Its form is

```
array = fscanf(fid,format)
[array, count] = fscanf(fid,format,size)
```

where `fid` is the file id of a file from which the data will be read, `format` is the format string controlling how the data is read, and `array` is the array that receives the data. The output argument `count` returns the number of values read from the file.

The optional argument *size* specifies the amount of data to be read from the file. There are three versions of this argument:

- n—Read exactly n values. After this statement, `array` will be a column vector containing n values read from the file.
- `Inf`—Read until the end of the file. After this statement, `array` will be a column vector containing all of the data until the end of the file.
- [n m]—Read exactly n × m values, and format the data as an n × m array.

Table 11.11: Format Conversion Specifiers for `fscanf`

Specifier	Description
`%c`	Reads a single character. This specifier reads any character including blanks, new lines, and so on.
`%Nc`	Reads *N* characters.
`%d`	Reads a decimal number (ignores blanks).
`%e` `%f` `%g`	Reads a floating-point number (ignores blanks).
`%i`	Reads a signed integer (ignores blanks).
`%s`	Reads a string of characters. The string is terminated by blanks or other special characters such as new lines.

The format string specifies the format of the data to be read. It can contain ordinary characters along with format conversion specifiers. The `fscanf` function compares the data in the file with the format conversion specifiers in the format string. As long as the two match, `fscanf` converts the value and stores it in the output array. This process continues until the end of the file or until the amount of data in *size* has been read, whichever comes first.

If the data in the file does not match the format conversion specifiers, the operation of `fscanf` stops immediately.

The format conversion specifiers for `fscanf` are basically the same as those for `fprintf`. The most common specifiers are shown in Table 11.11.

To illustrate the use of `fscanf`, we will attempt to read a file called `x.dat` containing the following values on two lines:

```
10.00    20.00
30.00    40.00
```

1. If the file is read with the statement
   ```
   [z, count] = fscanf(fid,'%f');
   ```
   variable z will be the column vector $\begin{bmatrix} 10 \\ 20 \\ 30 \\ 40 \end{bmatrix}$ and `count` will be 4.

2. If the file is read with the statement

   ```
   [z, count] = fscanf(fid,'%f',[2 2]);
   ```
   variable z will be the array $\begin{bmatrix} 10 & 30 \\ 20 & 40 \end{bmatrix}$ and `count` will be 4.

3. Next, let's try to read this file as decimal values. If the file is read with the statement

   ```
   [z, count] = fscanf(fid,'%d',Inf);
   ```

variable z will be the single value 10 and count will be 1. This happens because the decimal point in the 10.00 does not match the format conversion specifier, and fscanf stops at the first mismatch.

4. If the file is read with the statement

```
[z, count] = fscanf(fid,'%d.%d',[1 Inf]);
```

variable z will be the row vector [10 0 20 0 30 0 40 0] and count will be 8. This happens because the decimal point is now matched in the format conversion specifier and the numbers on either side of the decimal point are interpreted as separate integers.

5. Now let's try to read the file as individual characters. If the file is read with the statement

```
[z, count] = fscanf(fid,'%c');
```

variable z will be a row vector containing every character in the file, including all spaces and newline characters! Variable count will be equal to the number of characters in the file.

6. Finally, let's try to read the file as a character string. If the file is read with the statement

```
[z, count] = fscanf(fid,'%s');
```

variable z will be a row vector containing the 20 characters 10.0020.0030.0040.00, and count will be 4. This happens because the string specifier ignores white space, and the function found four separate strings in the file.

11.6.6 The fgetl Function

The fgetl function reads the next line *excluding the end-of-line characters* from a file as a character string. Its form is

```
line = fgetl(fid)
```

where fid is the file id of a file from which the data will be read, and line is the character array that receives the data. If fgetl encounters the end of a file, the value of line is set to −1.

11.6.7 The fgets Function

The fgets function reads the next line *including the end-of-line characters* from a file as a character string. Its form is

```
line = fgets(fid)
```

where fid is the file id of a file from which the data will be read, and line is the character array that receives the data. If fgets encounters the end of a file, the value of line is set to −1.

11.7 Comparing Formatted and Binary I/O Functions

Formatted I/O operations produce formatted files. A **formatted file** contains recognizable characters, numbers, and so forth that are stored as ordinary text. These files are easy to distinguish, because we can see the characters and numbers in the file when we display them on the screen or print them on a printer. However, to use data in a formatted file, a MATLAB program must translate the characters in the file into the internal data format used by the computer. Format conversion specifiers provide the instructions for this translation.

Formatted files have the advantages that we can readily see what sort of data they contain, and it is easy to exchange data between different types of programs using them. However, they also have disadvantages. A program must do a good deal of work to convert a number between the computer's internal representation and the characters contained in the file. All of this work is just wasted effort if we are going to be reading the data back into another MATLAB program. Also, the internal representation of a number usually requires much less space than the corresponding representation of the number found in a formatted file. For example, the internal representation of a 64-bit floating-point value requires 8 bytes of space. The character representation of the same value would be $\pm$d.ddddddddddddddE$\pm$ee, which requires 21 bytes of space (one byte per character). Thus, storing data in character format is inefficient and wasteful of disk space.

Unformatted files (or **binary files**) overcome these disadvantages by copying the information from the computer's memory directly to the disk file with no conversions at all. Since no conversions occur, no computer time is wasted formatting the data. In MATLAB, binary I/O operations are *much* faster than formatted I/O operations because there is no conversion. Furthermore, the data occupies a much smaller amount of disk space. On the other hand, unformatted data cannot be examined and interpreted directly by humans. In addition, it usually cannot be moved between different types of computers, because those types of computers have different internal ways to represent integers and floating-point values.

Formatted and unformatted files are compared in Table 11.12. In general, formatted files are best for data that people must examine, or data that may have to be

Table 11.12: Comparison of Formatted and Unformatted Files

Formatted Files	Unformatted Files
Can display data on output devices.	Cannot display data on output devices.
Can easily transport data between different computers.	Cannot easily transport data between computers with different internal data representations.
Requires a relatively large amount of disk space.	Requires relatively little disk space.
Slow: requires a lot of computer time.	Fast: requires little computer time.
Truncation or rounding errors possible in formatting.	No truncation or rounding errors.

moved between different programs on different computers. Unformatted files are best for storing information that will not need to be examined by human beings, and that will be created and used on the same type of computer. Under those circumstances, unformatted files are both faster and occupy less disk space.

〈/〉 Good Programming Practice

Use formatted files to create data that must be readable by humans or that must be transferable between programs on computers of different types. Use unformatted files to efficiently store large quantities of data that do not have to be directly examined and that will remain on only one type of computer. Also, use unformatted files when I/O speed is critical.

///

►Example II.3—Comparing Formatted and Binary I/O

The program shown below compares the time required to read and write a 10,000 element array, using both formatted and binary I/O operations. Note that each operation is repeated 10 times and the average time is reported.

```
% Script file: compare.m
%
% Purpose:
%    To compare binary and formatted I/O operations.
%    This program generates an array of 10,000 random
%    values and writes it to disk both as a binary and
%    as a formatted file.
%
% Record of revisions:
%      Date          Programmer          Description of change
%      ====          ==========          =====================
%    03/22/14      S. J. Chapman         Original code
%
% Define variables:
%    count      -- Number of values read / written
%    fid        -- File id
%    in_array   -- Input array
%    msg        -- Open error message
%    out_array  -- Output array
%    status     -- Operation status
%    time       -- Elapsed time in seconds
```

```matlab
%%%%%%%%%%%%%%%%%%%%%%%%%%%%%%%%%%%%%%%%%%%%%%%%
% Generate the data array.
%%%%%%%%%%%%%%%%%%%%%%%%%%%%%%%%%%%%%%%%%%%%%%%%
out_array = randn(1,100000);

%%%%%%%%%%%%%%%%%%%%%%%%%%%%%%%%%%%%%%%%%%%%%%%%
% First, time the binary output operation.
%%%%%%%%%%%%%%%%%%%%%%%%%%%%%%%%%%%%%%%%%%%%%%%%
% Reset timer
tic;

% Loop for 10 times
for ii = 1:10

    % Open the binary output file for writing.
    [fid,msg] = fopen('unformatted.dat','w');

    % Write the data
    count = fwrite(fid,out_array,'float64');

    % Close the file
    status = fclose(fid);

end

% Get the average time
time = toc / 10;
fprintf ('Write time for unformatted file = %6.3f\n',time);

%%%%%%%%%%%%%%%%%%%%%%%%%%%%%%%%%%%%%%%%%%%%%%%%
% Next, time the formatted output operation.
%%%%%%%%%%%%%%%%%%%%%%%%%%%%%%%%%%%%%%%%%%%%%%%%
% Reset timer
tic;

% Loop for 10 times
for ii = 1:10

    % Open the formatted output file for writing.
    [fid,msg] = fopen('formatted.dat','wt');

    % Write the data
    count = fprintf(fid,'%23.15e\n',out_array);

    % Close the file
    status = fclose(fid);
end
```

```matlab
% Get the average time
time = toc / 10;
fprintf ('Write time for formatted file = %6.3f\n',time);

%%%%%%%%%%%%%%%%%%%%%%%%%%%%%%%%%%%%%%%%%%%%%%%%%%%
% Time the binary input operation.
%%%%%%%%%%%%%%%%%%%%%%%%%%%%%%%%%%%%%%%%%%%%%%%%%%%
% Reset timer
tic;

% Loop for 10 times
for ii = 1:10

    % Open the binary file for reading.
    [fid,msg] = fopen('unformatted.dat','r');

    % Read the data
    [in_array, count] = fread(fid,Inf,'float64');

    % Close the file
    status = fclose(fid);

end

% Get the average time
time = toc / 10;
fprintf ('Read time for unformatted file =  %6.3f\n',time);

%%%%%%%%%%%%%%%%%%%%%%%%%%%%%%%%%%%%%%%%%%%%%%%%%%%%
% Time the formatted input operation.
%%%%%%%%%%%%%%%%%%%%%%%%%%%%%%%%%%%%%%%%%%%%%%%%%%%%

% Reset timer
tic;

% Loop for 10 times
for ii = 1:10

    % Open the formatted file for reading.
    [fid,msg] = fopen('formatted.dat','rt');

    % Read the data
    [in_array, count] = fscanf(fid,'%f',Inf);

    % Close the file
    status = fclose(fid);
end
```

```
% Get the average time
time = toc / 10;
fprintf ('Read time for formatted file =    %6.3f\n',time);
```

When this program is executed in MATLAB R2014b, the results are:

```
» compare
Write time for unformatted file =    0.001
Write time for formatted file =    0.095
Read time for unformatted file =    0.002
Read time for formatted file =    0.139
```

The files written to disk are shown below.

```
D:\book\matlab\chap8>dir *.dat
Volume in drive C is SYSTEM
Volume Serial Number is 0866-1AC5

Directory of c:\book\matlab\5e\rev1\chap11

09/09/2014  07:01 PM    <DIR>          .
09/09/2014  07:01 PM    <DIR>          ..
09/09/2014  07:01 PM           250,000 formatted.dat
09/09/2014  07:01 PM            80,000 unformatted.dat
            4 File(s)          330,000 bytes
            2 Dir(s)   181,243,170,816 bytes free
```

Note that the write time for the formatted file was almost 100 times slower than the write time for the unformatted file, and the read time for the formatted file was about 70 times slower than the read time for the unformatted file. Furthermore, the formatted file was 3 times larger than the unformatted file.

It is clear from these results that unless you *really* need formatted data, binary I/O operations are the preferred way to save data in MATLAB.

◄

QUIZ 11.2

This quiz provides a quick check to see if you have understood the concepts introduced in Sections 11.6 and 11.7. If you have trouble with the quiz, reread the section, ask your instructor, or discuss the material with a fellow student. The answers to this quiz are found in the back of the book.

1. What is the difference between unformatted (binary) and formatted I/O operations?
2. When should formatted I/O be used? When should unformatted I/O be used?
3. Write the MATLAB statements required to create a table that contains the sine and cosine of x for $x = 0, 0.1\pi, ..., \pi$. Be sure to include a title and label on the table.

For questions 4 and 5, determine whether the MATLAB statements are correct or not. If they are in error, specify what is wrong with them.

4. ```
a = 2*pi;
b = 6;
c = 'hello';
fprintf(fid,'%s %d %g\n',a,b,c);
```

5. ```
data1 = 1:20;
data2 = 1:20;
fid = fopen('xxx','w+');
fwrite(fid,data1);
fprintf(fid,'%g\n',data2);
```

11.8 File Positioning and Status Functions

As we stated previously, MATLAB files are sequential—they are read in order from the first record in the file to the last record in the file. However, we sometimes need to read a piece of data more than once or to process a whole file more than once during a program. How can we skip around within a sequential file?

The MATLAB function `exist` can determine whether or not a file exists before it is opened. There are two functions to tell us where we are within a file once it is opened: `feof` and `ftell`. In addition, there are two functions to help us move around within the file: `frewind` and `fseek`.

Finally, MATLAB includes a function, `ferror`, that provides a detailed description of cause of I/O errors when they occur. We will now explore these five functions, looking at `ferror` first because it can be used with all of the other functions.

11.8.1 The `exist` Function

The MATLAB function `exist` checks for the existence of a variable in a workspace, a built-in function, or a file in the MATLAB search path. The forms of the `ferror` function are:

```
ident = exist('item');
ident = exist('item','kind');
```

If `'item'` exists, this function returns a value based on its type. The possible results are shown in Table 11.13.

The second form of the `exist` function restricts the search for an item to a specified kind. The legal types are `'var'`, `'file'`, `'builtin'`, and `'dir'`.

The `exist` function is very important because we can use it to check for the existence of a file before it is overwritten by `fopen`. The permissions `'w'` and `'w+'` delete the contents in an existing file when they open it. Before a programmer allows `fopen` to delete an existing file, he or she should check with the user to confirm that the file really should be deleted.

Table 11.13: Values Returned by the `exist` Function

Value	Meaning
0	Item not found
1	Item is a variable in the current workspace
2	Item is an M-file or a file of unknown type
3	Item is a MEX file
4	Item is a MDL file
5	Item is a built-in function
6	Item is a P file
7	Item is a directory
8	Item is a Java class

►Example 11.4—Opening an Output File

The program shown gets an output file name from the user and checks to see if it exists. If it exists, the program checks to see if the user wants to delete the existing file or to append the new data to it. If the file does not exist, then the program simply opens the output file.

```
% Script file: output.m
%
% Purpose:
%   To demonstrate opening an output file properly.
%   This program checks for the existence of an output
%   file. If it exists,the program checks to see if
%   the old file should be deleted, or if the new data
%   should be appended to the old file.
%
% Record of revisions:
%     Date          Programmer        Description of change
%     ====          ==========        =====================
%   03/24/14      S. J. Chapman       Original code
%
% Define variables:
%   fid           -- File id
%   out_filename  -- Output file name
%   yn            -- Yes/No response

% Get the output file name.
out_filename = input('Enter output filename: ','s');
```

```
% Check to see if the file exists.
if exist(out_filename,'file')

   % The file exists
   disp('Output file already exists.');
   yn = input('Keep existing file? (y/n) ','s');

   if yn == 'n'
      fid = fopen(out_filename,'wt');
   else
      fid = fopen(out_filename,'at');
   end

else

   % File doesn't exist
   fid = fopen(out_filename,'wt');

end
% Output data
fprintf(fid,'%s\n',date);

% Close file
fclose(fid);
```

When this program is executed, the results are:

```
» output
Enter output filename: xxx          (Non-existent file)
» type xxx

23-Mar-2014

» output
Enter output filename: xxx
Output file already exists.
Keep existing file? (y/n) y          (Keep current file)
» type xxx

23-Mar-2014
23-Mar-2014                          (Note new data added)

» output
Enter output filename: xxx
Output file already exists.
Keep existing file? (y/n) n          (Replace current file)
» type xxx

23-Mar-2014
```

The program appears to be functioning correctly in all three cases.

Good Programming Practice

Do not overwrite an output file without confirming that the user would like to delete the preexisting information.

11.8.2 The `ferror` Function

The MATLAB I/O system has several internal variables, including a special error indicator that is associated with each open file. This error indicator is updated by every I/O operation. The `ferror` function gets the error indicator and translates it into an easy-to-understand character message. The forms of the `ferror` function are:

```
message = ferror(fid)
message = ferror(fid,'clear')
[message,errnum] = ferror(fid)
```

This function returns the most recent error message (and optionally error number) associated with the file attached to `fid`. It may be called at any time after any I/O operation to get a more detailed description of what went wrong. If this function is called after a successful operation, the message will be '. . ' and the error number will be 0.

The `'clear'` argument clears the error indicator for a particular file id.

11.8.3 The `feof` Function

The `feof` function tests to see if the current file position is at the end of the file. The form of the `feof` function is:

```
eofstat = feof(fid)
```

This function returns a logical true (1) if the current file position is at the end of the file, and logical false (0) otherwise.

11.8.4 The `ftell` Function

The `ftell` function returns the current location of the file position indicator for the file specified by `fid`. The position is a nonnegative integer specified in bytes from the beginning of the file. A returned value of −1 for position indicates that the query was unsuccessful. If this happens, use `ferror` to determine why the request failed. The form of the `ftell` function is:

```
position = ftell(fid)
```

11.8.5 The `frewind` Function

The `frewind` function allows a programmer to reset a file's position indicator to the beginning of the file. The form of the `frewind` function is:

```
frewind(fid)
```

This function does not return status information.

11.8.6 The `fseek` Function

The `fseek` function allows a programmer to set a file's position indicator to an arbitrary location within a file. The form of the `fseek` function is:

```
status = fseek(fid,offset,origin)
```

This function repositions the file position indicator in the file with the given `fid` to the byte with the specified `offset` relative to `origin`. The `offset` is measured in bytes, with a positive number indicating motion towards the end of the file and a negative number indicating motion towards the head of the file. The `origin` is a string that can have one of three possible values.

- `'bof'`—This is the beginning of the file.
- `'cof'`—This is the current position within the file.
- `'eof'`—This is the end of the file.

The returned `status` is zero if the operation is successful and -1 if the operation fails. If the returned status is -1, use `ferror` to determine why the request failed.

As an example of using `fseek` and `ferror` together, consider the following statements.

```
[fid,msg] = fopen('x','r');
status = fseek(fid,-10,'bof');
if status ~= 0
   msg = ferror(fid);
   disp(msg);
end
```

These commands open a file and attempt to set the file pointer 10 bytes before the beginning of the file. Since this is impossible, `fseek` returns a status of -1, and `ferror` gets an appropriate error message. When these statements are executed, the result is an informative error message:

```
Offset is bad - before beginning-of-file.
```

▶Example 11.5—Fitting a Line to a Set of Noisy Measurements

In Example 5.6, we learned how to perform a fit of a noisy set of measurements (x,y) to a line of the form

$$y = mx + b \qquad (11.1)$$

The standard method for finding the regression coefficients m and b is the method of least squares. This method is named "least squares" because it produces the line $y = mx + b$ for which the sum of the squares of the differences between the observed *y* values and the predicted *y* values is as small as possible. The slope of the least squares line is given by

$$m = \frac{\left(\sum xy\right) - \left(\sum x\right)\bar{y}}{\left(\sum x^2\right) - \left(\sum x\right)\bar{x}} \tag{11.2}$$

and the intercept of the least squares line is by

$$b = \bar{y} - m\bar{x} \tag{11.3}$$

where

$\sum x$ is the sum of the *x* values
$\sum x^2$ is the sum of the squares of the *x* values
$\sum xy$ is the sum of the products of the corresponding *x* and *y* values
$\bar{x}$ is the mean (average) of the *x* values
$\bar{y}$ is the mean (average) of the *y* values

Write a program that will calculate the least-squares slope *m* and *y*-axis intercept *b* for a given set of noisy measured data points (*x*,*y*) *that are to be found in an input data file*.

Solution

1. **State the problem**
   Calculate the slope *m* and intercept *b* of a least-squares line that best fits an input data set consisting of an arbitrary number of (*x*,*y*) pairs. The input (*x*,*y*) data resides in a user-specified input file.

2. **Define the inputs and outputs**
   The inputs required by this program are pairs of points (*x*,*y*), where *x* and *y* are real quantities. Each pair of points will be located on a separate line in the input disk file. The number of points in the disk file is not known in advance.

   The outputs from this program are the slope and intercept of the least-squares fitted line, plus the number of points going into the fit.

3. **Describe the algorithm**
   This program can be broken down into four major steps:

   ```
   Get the name of the input file and open it
   Accumulate the input statistics
   Calculate the slope and intercept
   Write out the slope and intercept
   ```

   The first major step of the program is to get the name of the input file and to open the file. To do this, we will have to prompt the user to enter the name of the input file. After the file is opened, we must check to see that the open was successful. Next, we must read the file and keep track of the number of values entered, plus the sums $\sum x$, $\sum y$, $\sum x^2$, and $\sum xy$. The pseudocode for these steps is:

```
        Initialize n, sum_x, sum_x2, sum_y, and sum_xy to 0
        Prompt user for input file name
        Open file 'filename'
        Check for error on open
        if no error
           Read x, y from file 'filename'
           while not at end-of-file
              n ← n + 1
              sum_x ← sum_x + x
              sum_y ← sum_y + y
              sum_x2 ← sum_x2 + x^2
              sum_xy ← sum_xy + x*y
              Read x, y from file 'filename'
           end
           (further processing)
        end
```

Next, we must calculate the slope and intercept of the least-squares line. The pseudocode for this step is just the MATLAB versions of Equations (11.2) and (11.3).

```
x_bar ← sum_x / n
y_bar ← sum_y / n
slope ← (sum_xy - sum_x*y_bar) / (sum_x2 - sum_x*x_bar)
y_int ← y_bar - slope * x_bar
```

Finally, we must write out the results.
```
Write out slope 'slope' and intercept 'y_int'.
```

4. **Turn the algorithm into MATLAB statements.**
The final MATLAB program is shown below.

```
% Script file: lsqfit.m
%
% Purpose:
%   To perform a least-squares fit of an input data set
%   to a straight line, and print out the resulting slope
%   and intercept values.  The input data for this fit
%   comes from a user-specified input data file.
%
% Record of revisions:
%     Date            Programmer          Description of change
%     ====            ==========          =====================
%   03/24/14        S. J. Chapman        Original code
%
% Define variables:
%   count      -- number of values read
%   filename   -- Input file name
%   fid        -- File id
```

```
%    msg          -- Open error message
%    n            -- Number of input data pairs (x,y)
%    slope        -- Slope of the line
%    sum_x        -- Sum of all input X values
%    sum_x2       -- Sum of all input X values squared
%    sum_xy       -- Sum of all input X*Y values
%    sum_y        -- Sum of all input Y values
%    x            -- An input X value
%    x_bar        -- Average X value
%    y            -- An input Y value
%    y_bar        -- Average Y value
%    y_int        -- Y-axis intercept of the line

% Initialize sums
n = 0; sum_x = 0; sum_y = 0; sum_x2 = 0; sum_xy = 0;

% Prompt user and get the name of the input file.
disp('This program performs a least-squares fit of an');
disp('input data set to a straight line. Enter the name');
disp('of the file containing the input (x,y) pairs: ');
filename = input(' ','s');

% Open the input file
[fid,msg] = fopen(filename,'rt');

% Check to see if the open failed.
if fid < 0

   % There was an error--tell user.
   disp(msg);

else

   % File opened successfully. Read the (x,y) pairs from
   % the input file. Get first (x,y) pair before the
   % loop starts.
   [in,count] = fscanf(fid,'%g %g',2);

   while ~feof(fid)
      x = in(1);
      y = in(2);
      n        = n + 1;                  %
      sum_x    = sum_x + x;              % Calculate
      sum_y    = sum_y + y;              % statistics
      sum_x2   = sum_x2 + x.^2;          %
      sum_xy   = sum_xy + x * y;         %
```

```
    % Get next (x,y) pair
    [in,count] = fscanf(fid,'%f',[1 2]);

end

% Close the file
fclose(fid);

% Now calculate the slope and intercept.
x_bar = sum_x / n;
y_bar = sum_y / n;
slope = (sum_xy - sum_x*y_bar) / (sum_x2 - sum_x*x_bar);
y_int = y_bar - slope * x_bar;

% Tell user.
fprintf('Regression coefficients for the least-squares line:\n');
fprintf('   Slope (m)       = %12.3f\n',slope);
fprintf('   Intercept (b)   = %12.3f\n',y_int);
fprintf('   No of points    = %12d\n',n);

end
```

5. **Test the program.**

To test this program, we will try a simple data set. For example, if every point in the input data set actually falls along a line, then the resulting slope and intercept should be exactly the slope and intercept of that line. Thus the data set

```
1.1   1.1
2.2   2.2
3.3   3.3
4.4   4.4
5.5   5.5
6.6   6.6
7.7   7.7
```

should produce a slope of 1.0 and an intercept of 0.0. If we place these values in a file called input1, and run the program, the results are:

```
» lsqfit
This program performs a least-squares fit of an
input data set to a straight line. Enter the name
of the file containing the input (x,y) pairs:
input1
Regression coefficients for the least-squares line:
   Slope (m)       = 1.000
   Intercept (b)   = 0.000
   No of points    =      7
```

Now let's add some noise to the measurements. The data set becomes

```
1.1    1.01
2.2    2.30
3.3    3.05
4.4    4.28
5.5    5.75
6.6    6.48
7.7    7.84
```

If these values are placed in a file called `input2` and the program is run on that file, the results are:

```
» lsqfit
This program performs a least-squares fit of an
input data set to a straight line. Enter the name
of the file containing the input (x,y) pairs:
input2
Regression coefficients for the least-squares line:
   Slope (m)     = 1.024
   Intercept (b) = -0.120
   No of points  =      7
```

If we calculate the answer by hand, it is easy to show that the program gives the correct answers for our two test data sets. The noisy input data set and the resulting least-squares fitted line are shown in Figure 11.2.

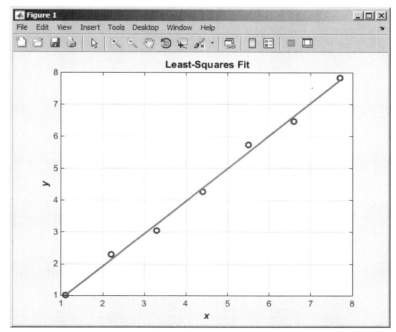

Figure 11.2 A noisy input data set and the resulting least-squares fitted line.

11.9 The `textscan` Function

The `textscan` function reads text files that are formatted into columns of data, where each column can be of a different type, and stores the contents into the columns of a cell array. This function is *very* useful for importing tables of data printed out by other applications. It is basically similar to `textread`, except that it is faster and more flexible.

The form of the `textscan` function is

```
a = textscan(fid, 'format')
a = textscan(fid, 'format', N)
a = textscan(fid, 'format', param, value, ...)
a = textscan(fid, 'format', N, param, value, ...)
```

where `fid` is the file id of a file that has already been opened with `fopen`, `format` is a string containing a description of the type of data in each column, and `n` is the number of times to use the format specifier. (If `n` is -1 or is missing, the function reads to the end of the file.) The format string contains the same types of format descriptors as the `fprintf` function. Note that there is only one output argument, with all of the values returned in a cell array. The cell array will contain a number of elements equal to the number of format descriptors to read.

For example, suppose that file `test_input1.dat` contains the following data:

```
James    Jones    O+    3.51    22    Yes
Sally    Smith    A+    3.28    23    No
Hans     Carter   B-    2.84    19    Yes
Sam      Spade    A+    3.12    21    Yes
```

This data could be read into a cell array with the following function:

```
fid = fopen('test_input1.dat','rt');
a = textscan(fid,'%s %s %s %f %d %s',-1);
fclose(fid);
```

When this command is executed, the results are:

```
» fid = fopen('test_input1.dat','rt');
» a = textscan(fid,'%s %s %s %f %d %s',-1)
a =
   {4x1 cell}    {4x1 cell}    {4x1 cell}    [4x1 double]
   [4x1 int32]    {4x1 cell}
» a{1}
ans =
    'James'
    'Sally'
    'Hans'
    'Sam'
» a{2}
ans =
    'Jones'
    'Smith'
    'Carter'
    'Spade'
```

```
» a{3}
ans =
     'O+'
     'A+'
     'B-'
     'A+'
>> a{4}
ans =
     3.5100
     3.2800
     2.8400
     3.1200
» fclose(fid);
```

This function can also skip selected columns by adding an asterisk to the corresponding format descriptor (for example, %*s). For example, the following statements read only the first name, last name, and gpa from the file:

```
fid = fopen('test_input1.dat','rt');
a = textscan(fid,'%s %s %*s %f %*d %*s',-1);
fclose(fid);
```

Function textscan is similar to function textread, but it is more flexible and faster. The advantages of textscan include:

1. The textscan function offers better performance than textread, making it a better choice when reading large files.
2. With textscan, you can start reading at any point in the file. When the file is opened with fopen, you can move to any position in the file with fseek and begin the textscan at that point. The textread function requires that you start reading from the beginning of the file.
3. Subsequent textscan operations start reading the file at a point where the last textscan left off. The textread function always begins at the start of the file, regardless of any prior textread operations.
4. Function textscan returns a single cell array regardless of how many fields you read. With textscan, you don't need to match the number of output arguments with the number of fields being read, as you would with textread.
5. Function textscan offers more choices in how the data being read is converted.

Function textscan has a number of additional options that increase its flexibility. Consult the MATLAB online documentation for details of these options.

Good Programming Practice

Use function textscan in preference to textread to import text data in column format from programs written in other languages or exported from applications, such as spreadsheets.

11.10 Function `uiimport`

Function `uiimport` is a GUI-based way to import data from a file or from the clipboard. This command takes the forms

```
uiimport
structure = uiimport;
```

In the first case, the imported data is inserted directly into the current MATLAB workspace. In the second case, the data is converted into a structure and saved in variable `structure`.

When the command `uiimport` is typed, the Import Wizard is displayed in a window (see Figure 11.3). The user can then select the file that he or she would like to import from and the specific data within that file. Many different formats are supported—a partial list is given in Table 11.14. In addition, data can be imported from almost *any* application by saving the data on the clipboard. This flexibility can be very useful when you are trying to get data into MATLAB for analysis.

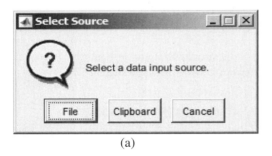

(a)

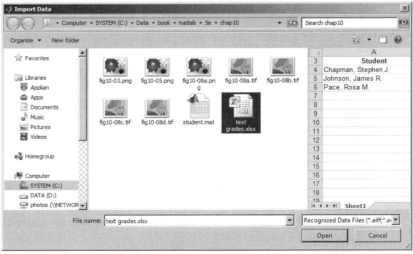

(b)

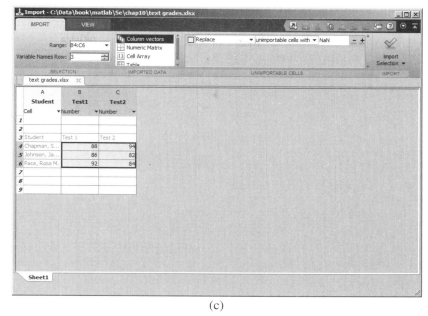

(c)

Figure 11.3 Using `uiimport`: (a) The Import Wizard first prompts the user to select a data source. (b) The Import Wizard after a file is selected but not yet loaded. (c) After a data file has been selected, one or more data arrays are created, and their contents can be examined. The user selects the ones to import, and clicks Import Selection.

Table 11.14: Selected File Formats Supported by `uiimport`

File Extension	Meaning
`*.gif`	Image files
`*.jpg`	
`*.jpeg`	
`*.jp2`	
`*.jpf`	
`*.jpx`	
`*.j2c`	
`*.j2k`	
`*.ico`	
`*.png`	
`*.pcx`	
`*.tif`	
`*.tiff`	
`*.bmp`	
`*.mat`	MATLAB data files
`*.cur`	Cursor format
`*.hdf`	Hierarchical Data Format file
`*.h5`	

(continued)

Table 11.14: Selected File Formats Supported by `uiimport` (Continued)

File Extension	Meaning
`*.au`	Sound files
`*.flac`	
`*.ogg`	
`*.snd`	
`*.wav`	
`*.avi`	Video files
`*.mov`	
`*.mpg`	
`*.mp4`	
`*.wmv`	
`*.xml`	XML files
`*.csv`	Spreadsheet files
`*.xls`	
`*.xlsx`	
`*.xlsm`	
`*.wk1`	
`*.txt`	Text files
`*.dat`	
`*.dlm`	
`*.tab`	

11.11 Summary

In Chapter 11, we presented an introduction file I/O operations. Many MATLAB I/O functions are quite similar to C functions, but there are differences in some details.

The `textread` and `textscan` functions can be used to import text data in column format from programs written in other languages or exported from applications such as spreadsheets. Of these two functions, `textscan` is preferred, because it is more flexible and faster than `textread`.

The `load` and `save` commands using MAT files are very efficient, are transportable across MATLAB implementations, and preserve full precision, data types, and global status for all variables. MAT files should be used as the first-choice method of I/O, unless data must be shared with other applications or must be readable by humans.

There are two types of I/O statements in MATLAB: binary and formatted. Binary I/O statements store or read data in unformatted files, and formatted I/O statements store or read data in formatted files.

MATLAB files are opened with the `fopen` function and closed with the `fclose` function. Binary reads and writes are performed with the `fread` and `fwrite` functions, while formatted reads and writes are performed with the `fscanf` and `fprintf` functions. Functions `fgets` and `fgetl` simply transfer a line of text from a formatted file into a character string.

The `exist` function can be used to determine if a file exists before it is opened. This is useful to ensure that existing data is not accidentally overwritten.

It is possible to move around within a disk file using the `frewind` and `fseek` functions. The `frewind` function moves the current file position to the beginning of the file, while the `fseek` function moves the current file position to a point a specified number of bytes ahead or behind a reference point. The reference point may be the current file position, the beginning of the file, or the end of the file.

MATLAB includes a GUI-based tool called `uiimport`, which allows users to import data into MATLAB from files created by many other programs in a wide variety of formats.

11.11.1 Summary of Good Programming Practice

The following guidelines should be adhered to when working with MATLAB I/O functions.

1. Unless you must exchange data with non-MATLAB programs, always use the `load` and `save` commands to save data sets in MAT file format. This format is efficient and transportable across MATLAB implementations, and it preserves all details of all MATLAB data types.
2. Always be careful to specify the proper permissions in `fopen` statements, depending on whether you are reading from or writing to a file. This practice will help prevent errors such as accidentally overwriting data files that you want to keep.
3. Always check the status after a file open operation to make sure that it is successful. If the file open fails, tell the user and provide a way to recover from the problem.
4. Use formatted files to create data that must be readable by humans or that must be transferable between programs on computers of different types. Use unformatted files to efficiently store large quantities of data that do not have to be directly examined and that will remain on only one type of computer. Also, use unformatted files when I/O speed is critical.
5. Do not overwrite an output file without confirming that the user would like to delete the preexisting information.
6. Use function `textscan` in preference to `textread` to import text data in column format from programs written in other languages or exported from applications, such as spreadsheets.

11.11.2 MATLAB Summary

The following summary lists all of the MATLAB commands and functions described in this chapter, along with a brief description of each one.

`exist`	Checks for the existence of a file
`fclose`	Closes file
`feof`	Tests for end-of-file

ferror	Inquires file I/O error status
fgetl	Reads a line from file and discards newline character
fgets	Reads a line from file and keeps newline character
fopen	Opens file
fprintf	Writes formatted data to file
fread	Reads binary data from file
frewind	Rewinds file
fscanf	Reads formatted data from file
fseek	Sets file position
ftell	Checks file position
fwrite	Writes binary data to a file
sprintf	Writes formatted data to a character string
textread	Reads data of various types organized in column format from a text file, and store the data in each column in separate variables
textscan	Reads data of various types organized in column format from a text file and stores the data in a cell array
uiimport	Starts a GUI tool for importing data

11.12 Exercises

11.1 What is the difference between binary and formatted I/O? Which MATLAB functions perform each type of I/O?

11.2 **Table of Logarithms** Write a MATLAB program to generate a table of the base-10 logarithms between 1 and 10 in steps of 0.1.

Table of Logarithms

	X.0	X.1	X.2	X.3	X.4	X.5	X.6	X.7	X.8	X.9
1.0	0.000	0.041	0.079	0.114	...					
2.0	0.301	0.322	0.342	0.362	...					
3.0	...									
4.0	...									
5.0	...									
6.0	...									
7.0	...									
8.0	...									
9.0	...									
10.0	...									

11.3 Write a MATLAB program that reads the time in seconds since the start of the day (this value will be somewhere between 0 and 86400.) and prints a character string containing time in the form HH:MM:SS using the 24-hour clock convention. Use the proper format converter to ensure that leading zeros are preserved in the MM and SS fields. Also, be sure to check the input number of seconds for validity, and write an appropriate error message if an invalid number is entered.

11.4 Gravitational Acceleration The acceleration due to the Earth's gravity at any height h above the surface of the Earth is given by the equation

$$g = -G \frac{M}{(R + h)^2} \tag{11.4}$$

where G is the gravitational constant (6.672×10^{-11} N m^2 / kg^2), M is the mass of the Earth (5.98×10^{24} kg), R is the mean radius of the Earth (6371 km), and h is the height above the Earth's surface. If M is measured in kg and R and h in meters, then the resulting acceleration will be in units of meters per second squared. Write a program to calculate the acceleration due to the Earth's gravity in 500 km increments at heights from 0 km to 40,000 km above the surface of the Earth. Print out the results in a table of height versus acceleration with appropriate labels, including the units of the output values. Plot the data as well.

11.5 The program in Example 11.5 illustrated the use of formatted I/O commands to read (x,y) pairs of data from disk. However, this could also be done with the `load -ascii` function. Rewrite this program to use `load` instead of the formatted I/O functions. Test your rewritten program to confirm that it gives the same answers as Example 11.5.

11.6 Rewrite the program in Example 11.5 to use the `textread` function instead of the formatted I/O functions.

11.7 Rewrite the program in Example 11.5 to use the `textscan` function instead of the formatted I/O functions. How difficult was it to use `textscan`, compared to using `textread`, `load -ascii`, or the formatted I/O functions?

11.8 Write a program that reads an arbitrary number of real values from a user-specified input data file, rounds the values to the nearest integer, and writes the integers out to a user-specified output file. Make sure that the input file exists, and if not, tell the user and ask for another input file. If the output file exists, ask the user whether or not to delete it. If not, prompt for a different output file name.

11.9 Table of Sines and Cosines Write a program to generate a table containing the sine and cosine of θ for θ between 0° and 90°, in 1° increments. The program should properly label each of the columns in the table.

11.10 File `int.dat` (available at the book's website) contains 25 integer values in `'int8'` format. Write a program that reads these values into a `single` array using the function `fread`.

11.11 Interest Calculations Suppose that you have a sum of money P in an interest-bearing account at a local bank (P stands for *present value*). If the bank pays you interest on the money at a rate of i percent per year and compounds the interest monthly, the amount of money that you will have in the bank after n months is given by the equation

$$F = P \left(1 + \frac{i}{1200} \right)^n \tag{11.5}$$

where F is the future value of the account and $\frac{i}{12}$ is the monthly percentage interest rate (the extra factor of 100 in the denominator converts the interest rate from percentages to fractional amounts). Write a MATLAB program that will read an initial amount of money P and an annual interest rate i, and will calculate and write out a table showing the future value of the account every month for the next 5 years. The table should be written to an output file called `'interest'`. Be sure to properly label the columns of your table.

11.12 Write a program to read a set of integers from an input data file, and locate the largest and smallest values within the data file. Print out the largest and smallest values, together with the lines on which they were found. Assume that you do not know the number of values in the file before the file is read.

11.13 Create a 400 × 400 element `double` array x, and fill it with random data using the function `rand`. Save this array to a MAT file `x1.dat`, and then save it again to a second MAT file `x2.dat` using the `-compress` option. How do the sizes of the two files compare?

11.14 **Means** In Exercise 5.34, we wrote a MATLAB program that calculated the arithmetic mean (average), rms average, geometric mean, and harmonic mean for a set of numbers. Modify that program to read an arbitrary number of values from an input data file, and calculate the means of those numbers. To test the program, place the following values into an input data file and run the program on that file: 1.0, 2.0, 5.0, 4.0, 3.0, 2.1, 4.7, 3.0.

11.15 **Converting Radians to Degrees/Minutes/Seconds** Angles are often measured in degrees (°), minutes ('), and seconds ("), with 360 degrees in a circle, 60 minutes in a degree, and 60 seconds in a minute. Write a program that reads angles in radians from an input disk file and converts them into degrees, minutes, and seconds. Test your program by placing the following four angles expressed in radians into an input file, and reading that file into the program: 0.0, 1.0, 3.141593, 6.0.

11.16 Create a data set in some other program on your computer, such as Microsoft Word, Microsoft Excel, a text editor, and so on. Copy the data set to the clipboard using the Windows or UNIX copy function, and then use the function `uiimport` to load the data set into MATLAB.

User-Defined Classes and Object-Oriented Programming

Since the beginning of this book, we have been using MATLAB to write **procedural programs**. In procedural programs, the programmer breaks a problem down into a set of functions (or procedures), where each function is a recipe (an algorithm) to perform some part of the total problem. These functions work together to solve the total problem. The key idea in procedural programming is the procedure, the description of how a task is accomplished. Data is passed to the procedure as input arguments, and the results of the calculation are returned as output arguments. For example, we might write a procedure to solve a set of simultaneous linear equations, and then use that procedure over and over again with different input data sets.

The other major programming paradigm is called **object-oriented programming**. In object-oriented programming, the problem is broken down into a series of objects that interact with other objects to solve the total problem. Each object contains a series of **properties**, which are the characteristics of the object, and a set of **methods**, which define the behaviors of the object.

This chapter introduces MATLAB user-defined classes and object-oriented programming. It teaches the basic concepts behind object-oriented programming, and then shows how MATLAB implements those features[1].

This chapter is an appropriate lead-in to the following chapters on handle graphics and graphical user interfaces, since all graphics in MATLAB are implemented as objects.

---

[1]MATLAB does not provide a full implementation of object-oriented programming in the traditional computer science sense. Such key object-oriented concepts as polymorphism are not implemented in MATLAB, so people with prior object-oriented programming experience will find that some familiar features are missing.

12.1 An Introduction to Object-Oriented Programming

Object-oriented programming (OOP) is the process of programming by modeling objects in software. The principal features of OOP are described in the subsections of Section 12.1, and then the MATLAB implementation of these features is described in subsequent sections of the chapter.

12.1.1 Objects

The physical world is full of objects: cars, pencils, trees, and so on. Any real object can be characterized by two different aspects: its *properties* and its *behavior*. For example, a car can be modeled as an object. A car has certain properties (color, speed, direction of motion, available fuel, and so forth) and certain behaviors (starting, stopping, turning, and so forth).

In the software world, an **object** is a software component whose structure is like that of objects in the real world. Each object consists of a combination of data (called **properties** or **instance variables**) and behaviors (called **methods**). The properties are variables describing the essential characteristics of the object, while the methods describe how the object behaves and how the properties of the object can be modified. Thus, a software object is a bundle of variables and related methods.

A software object is often represented as shown in Figure 12.1. The object can be thought of as a cell, with a central nucleus of variables (properties) and an outer layer of methods that form an interface between the object's variables and the outside world. The nucleus of data is hidden from the outside world by the outer layer of methods. The object's variables are said to be *encapsulated* within the object, meaning that no code outside of the object can see or directly manipulate them. Any access to the object's data must be through calls to the object's methods.

The ordinary methods in a MATLAB object are formally known as **instance methods** to distinguish them from static methods (described later in Section 12.1.4).

Typically, encapsulation is used to hide the implementation details of an object from other objects in the program. If the other objects in the program cannot see or

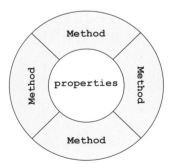

Figure 12.1 An object may be represented as a nucleus of data (properties) surrounded and protected by methods, which implement the object's behavior and form an interface between the properties and the outside world.

change the internal state of an object, they cannot introduce bugs by accidentally modifying the object's state. If other objects want to change the value of a property, they have to call one of the object's methods to make that change. The method can verify that the new data is valid before it is used to update the property.

In addition, changes to the internal operation of the object will not affect the operation of the other objects in a program. As long as the interface to the outer world is unchanged, the implementation details of an object can change at any time without affecting other parts of the program.

Encapsulation provides two primary benefits to software developers:

Modularity—An object can be written and maintained independently of the source code for other objects. Therefore, the object can be easily re-used and passed around in the system.

Information Hiding—An object has a public interface (the calling sequence of its methods) that other objects can use to communicate with it. However, the object's instance variables are not directly accessible to other objects. Therefore, if the public interface is not changed, an object's variables and methods can be changed at any time without introducing side-effects in the other objects that depend on it.

12.1.2 Messages

Objects communicate by passing messages back and forth among themselves. These messages are the method calls. For example, if object A wants object B to perform some action for it, it calls one of object B's methods. This method can then perform some act to either modify or use the properties stored in object B (see Figure 12.2).

Each message has three components, which provide all the information necessary for the receiving object to perform the desired action:

1. A reference pointing to the object to which the message is addressed. In MATLAB, this reference is known as a **handle**.
2. The name of the method to perform on that object.
3. Any parameters needed by the method.

An object's behavior is expressed through its methods, so message passing supports all possible interactions between objects.

12.1.3 Classes

Classes are the software blueprints from which objects are made. A class is a software construct that specifies the number and type of properties to be included in an object, and the methods that will be defined for the object. Each component of a class is known as a **member**. The two types of members are **properties**, which specify the data values defined by the class, and **methods**, which specify the operations on those properties. For example, suppose that we wish to create a class to represent students in a university. Such a class could have three properties describing a student, one for the student's name, one for the student's social security number, and one for the

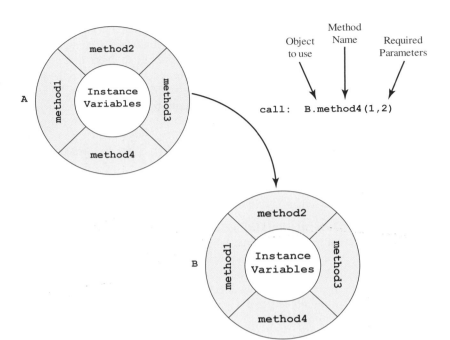

Figure 12.2 If object A wants object B to perform some action for it, it calls one of object B's methods. The call contains three parts: a reference to the object to use, the name of the method within the object which will do the work, and any required parameters. Note that the names of the object and method are separated by a period.

student's address. In addition, it could have methods allowing a program to use or modify the student's information, or to use the student's information in other ways. If there were 1000 students in the university, we could create 1000 objects from class `Student`, with each object having its own unique copy of the properties (name, ssn, address) but with all objects sharing a common set of methods describing how to use the properties.

Note that a class is a *blueprint* for an object, not an object itself. The class describes what an object will look and behave like once it is created. Each object is created or *instantiated* in memory from the blueprint provided by a class, and many different objects can be instantiated from the same class. For example, Figure 12.3 shows a class `Student`, together with three objects a, b, and c created from that class. Each of the three objects has its own copies of the properties `name`, `ssn`, and `address`, while sharing a single set of methods to modify them.

12.1.4 Static Methods

As we described above, each object created from a class receives its own copies of all the instance variables defined in the class, but they all share the same methods.

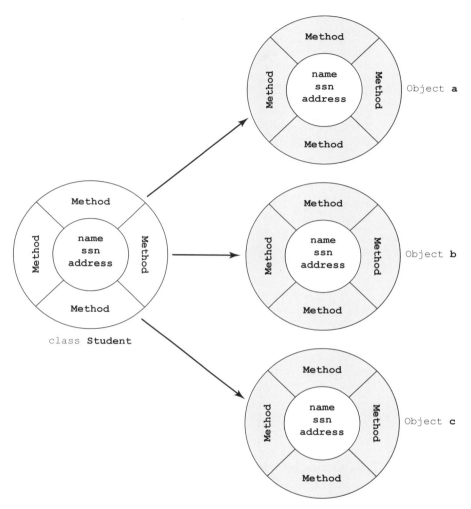

Figure 12.3 Many objects can be instantiated from a single class. In this example, three objects a, b, and c have been instantiated from class Student.

When a method is used with object a, it modifies the data in object a. When the same method is used with object b, it modifies the data in object b, and so forth.

It is also possible to define **static methods**. Static methods are methods that exist independently of any objects defined from the class. These methods do not access instance variables or invoke instance methods.

Static methods are declared using the Static attribute in the method definition. Static methods can be used without ever instantiating (creating) an object from the class in which they are defined. They are used by typing the class name followed by a period and followed by the method name. Static methods are often used for utility calculations that are independent of the data in any particular object.

12.1.5 Class Hierarchy and Inheritance

All classes in an object-oriented language are organized in a **class hierarchy**, with the highest level classes being very general in behavior and lower-level ones becoming more specific. Each lower-level class is based on and derived from a higher-level class, and the lower-level classes *inherit both the properties and the methods* of the class from which it is derived. A new class starts with all of the non-private properties and methods of the class on which it is based, and the programmer then adds the additional variables and methods necessary for the new class to perform its function.

The class on which a new class is based is referred to as a **superclass**, and the new class is referred to as a **subclass**. The new subclass can itself become the superclass for another new subclass. A subclass normally adds instance variables and instance methods of its own, so a subclass is generally larger than its superclass. In addition, it can **override** some methods of its superclass, replacing the method in the superclass with a different one having the same name. This changes the subclass's behavior from that of its superclass. Because a subclass is more specific than its superclass, it represents a smaller group of objects.

For example, suppose that we define a class called Vector2D to contain two-dimensional vectors. Such a class would have two instance variables x and y to contain the *x* and *y* components of the 2D vectors, and it would need methods to manipulate the vectors such as adding two vectors, subtracting two vectors, calculating the length of a vector, and so forth. Now suppose that we need to create a class called Vector3D to contain three-dimensional vectors. If this class is based on Vector2D, then it will automatically inherit instance variables x and y from its superclass, so the new class will only need to define a variable z (see Figure 12.4). The new class will also override the methods used to manipulate 2D vectors to allow them to work properly with 3D vectors.

12.1.6 Object-Oriented Programming

Object-oriented programming (OOP) is the process of programming by modeling objects in software. In OOP, a programmer examines the problem to be solved and tries to break it down into identifiable objects, each of which contains certain data and has certain behaviors. Sometimes these objects will correspond to physical objects in nature, and sometimes that will be purely abstract software constructs. The data identified by the programmer will become the properties of corresponding classes, and the behaviors of the objects will become the methods of the classes.

Once the objects making up the problem have been identified, the programmer identifies the type of data to be stored as properties in each object and the exact calling sequence of each method needed to manipulate the data.

The programmer can then develop and test the classes in the model one at a time. As long as the *interfaces* between the classes (the calling sequence of the methods) are unchanged, each class can be developed and tested without needing to change any other part of the program.

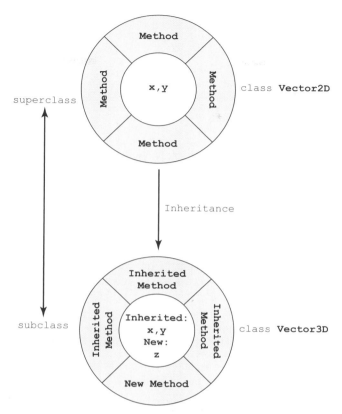

Figure 12.4 An example of inheritance. Class `Vector2D` has been defined to handle two-dimensional vectors. When class `Vector3D` is defined as a subclass of `Vector2D`, it inherits the instance variables `x` and `y`, as well as many methods. The programmer then adds a new instance variable `z` and new methods to the ones inherited from the superclass.

12.2 The Structure of a MATLAB Class

The major components (class members) of a MATLAB class are (see Figure 12.5):

1. **Properties**. Properties define the instance variables that will be created when an object is instantiated from a class. Instance variables are the data encapsulated inside an object. A new set of instance variables is created each time that an object is instantiated from the class.
2. **Methods**. Methods implement the behaviors of a class. Some methods may be explicitly defined in a class, while other methods may be inherited from superclasses of the class.
3. **Constructor**. Constructors are special methods that specify how to initialize an object when it instantiated. The arguments of the constructor include values to use in initializing the properties. Constructors are easy to identify

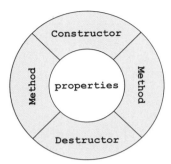

Figure 12.5 A MATLAB class contains properties to store information, methods to modify and perform calculations with the properties, a constructor to initialize the object when it is created, and (optionally) a destructor to release resources when it is deleted.

because they have the same name as the class that they are initializing, and the only output argument is the object constructed.

4. **Destructor**. Destructors are special methods that clean up the resources (open files, etc.) used by an object just before it is destroyed. Just before an object is destroyed, it makes a call to a special method named **delete** if it exists. The only input argument is the object to be destroyed, and there must be no output argument. Many classes do not need a delete method at all.

The members of a class, whether variables or methods, are accessed by referring to an object created from the class with the **access operator**, also known as the **dot operator**. For example, suppose that a class MyClass contains an instance variable a and a method processA. If a reference to object of this class is named obj, then the instance variable in obj would be accessed as obj.a, and the method would be accessed as obj.processA().

12.2.1 Creating a Class

In MATLAB, a class is declared using a classdef keyword. The class definition starts with a classdef keyword and ends with an end statement. Inside the class definition are one or more blocks defining the properties and methods associated with the class. Properties are defined in one or more blocks that begin with a properties keyword and end with an end statement. Methods are defined in one or more blocks that begin with a methods keyword and end with an end statement.

The simplest form of a class definition is

```
classdef (Attributes) ClassName < SuperClass

   properties (Attributes)
      PropertyName1
      PropertyName2
      ...
   end
```

```
methods (Attributes)
   function [obj = ] methodName(obj,arg1,arg2, ...)
      ...
   end

end
```

Here, ClassName is the name of the new class, and the optional value SuperClass is the name of the superclass it is derived from (if the class has a superclass). The properties blocks declare properties, which will be instance variables when an object is created from the class. The methods blocks declare the names and calling arguments for the methods associated with the class. (Note that for some types of methods, the body of the methods can be further down the file, or even in another file.)

For example, the following code declares a very simple class called vector containing two properties x and y, and no methods:

```
classdef vector

   properties
      x;               % X value of vector
      y;               % Y value of vector
   end

end
```

This class is saved in a file named vector.m.

An object of class vector is instantiated by the following assignment statement:

```
» a = vector
a =
  vector with properties:

    x: []
    y: []
```

This assignment created an object of class vector, containing two instance variables corresponding to the properties x and y, which are initially empty. Values can be assigned to the properties using the dot operator:

```
» a.x = 2;
» a.y = 3;
» a
a =
  vector with properties:

    x: 2
    y: 3
```

Values can also be accessed through the dot operator:

```
» a.x
ans =
     2
```

If another object of class vector is instantiated, the instance variables x and y in the new object are completely different from the ones in the first object, and they can hold different values.

```
» b = point;
» b.x = -2;
» b.y = 9;
» a
a =
   point with properties:

     x: 2
     y: 3
» b
b =
   point with properties:

     x: -2
     y: 9
```

12.2.2 Adding Methods to a Class

Methods can be added to a class by defining them in a methods block within the class definition. We will now add three methods to class vector: a constructor and two ordinary instance methods.

A constructor is a method that initializes objects of the class when they are instantiated. Note that when the objects of class vector were created earlier, their instance variables (properties) were empty. A constructor allows objects to be created with initial data stored in the instance variables.

A constructor is a method that has the *same name as the class*. There can be any number of input arguments to a constructor, but the single output of the constructor is an object of the type being created. An example constructor for the vector class is:

```
% Declare the constructor
function v = vector(a,b)
    v.x = a;
    v.y = b;
end
```

This constructor accepts two input values a and b, and uses them to initialize the instance variables x and y when the object is instantiated.

It is important to design the constructor for a class so that it can work as a **default constructor** as well as a constructor with input arguments. Some MATLAB functions can call class constructors with no arguments under certain circumstances, and this will cause a crash unless the constructor is designed to deal with that case. We normally do this by using the `nargin` function to check for the presence of input arguments, and using default values if the input arguments are missing. A version of the vector class constructor that also supports the default case is shown below:

```
% Declare the constructor
function v = vector(a,b)
   if nargin < 2
      v.x = 0;
      v.y = 0;
   else
      v.x = a;
      v.y = b;
   end
end
```

Good Programming Practice

Define a constructor for a class to initialize the data in objects of that class when they are instantiated. Be sure to support a default case (one without arguments) in the constructor design.

Instance methods are methods that use or modify the instance variables stored in objects created from the class. They are functions with a special syntax. The first argument of each function must be the object that the instance methods are defined in. In object-oriented programming, the current object passed in as the first argument is often called `this`, meaning "this object." If the methods modify the data in the object, they also must return the modified object as an output[2].

We will now add two sample instance methods to this class. Method `length` returns the length of the vector, calculated from the equation

$$length = \sqrt{x^2 + y^2} \tag{12.1}$$

where x and y are the instance variables in the class. This is an example of a method that works with the instance variables but does not modify them. Since the instance variables were not modified, the object is not returned as an output from the function.

---

[2]The requirement to return modified objects is true if the objects are created from *value classes*, and not true if they are created from *handle classes*. Both of these class types will be defined later in the chapter, and this distinction will become clear then.

```
% Declare a method to calculate the length
% of the vector.
function result = length(this)
    result = sqrt(this.x.^2 + this.y.^2);
end
```

Method add sums the contents of the current vector object this and another vector object obj2, with the result stored in output object obj. This is an example of a method that creates a new vector object, which is returned as an output from the function. Note that this method uses the default constructor to create the output vector object before performing the addition.

```
% Declare a method to add two vectors together
function this = add(this,obj2)
    obj = vector();
    obj.x = this.x + obj2.x;
    obj.y = this.y + obj2.y;
end
```

The vector class with these methods added is:

```
classdef vector

    properties
        x;              % X value of vector
        y;              % Y value of vector
    end

    methods

        % Declare the constructor
        function v = vector(a,b)
            if nargin < 2
                v.x = 0;
                v.y = 0;
            else
                v.x = a;
                v.y = b;
            end
        end

        % Declare a method to calculate the length
        % of the vector.
        function result = length(this)
            result = sqrt(this.x.^2 + this.y.^2);
        end
```

```
% Declare a method to add two vectors together
function obj = add(this,obj2)
    obj = vector();
    obj.x = this.x + obj2.x;
    obj.y = this.y + obj2.y;
end

end

end
```

When an instance method in a MATLAB object is called, *the hidden object* `this` *is not included in the calling statement*. It is understood that the object named before the dot is the one to be passed to the method. For example, the method `length` above is defined to take an object of class `vector` as an input argument. However, that object is not explicitly included when the method is called. If `ob` is an object of type `vector`, then the length would be calculated as `ob.length` or `ob.length()`. The object itself is not included as an explicit input argument in the method call.

⟨/⟩ Good Programming Practice

When an instance method is invoked, do not include the object in the method's list of calling arguments.

///

The following examples show how to create three objects of type `vector` using the new constructor. Note that the objects are now instantiated with the initial data in the instance variables instead of being empty.

```
» a = vector(3,4)
a =
  vector with properties:

    x: 3
    y: 4
» b = vector(-12,5)
b =
  vector with properties:

    x: -12
    y: 5
» c = vector
c =
  vector with properties:

    x: 0
    y: 0
```

The `length` method calculates the length of each vector from the data in the instance variables:

```
» a.length
ans =
     5
» b.length()
ans =
     13
» c.length()
ans =
     0
```

Note that the method can be invoked either with or without the empty parentheses.

Finally, the `add` method adds two objects of the vector type according to the definition defined in the method:

```
» c = a.add(b)
c =
  vector with properties:

     x: -9
     y: 9
```

12.2.3 Listing Class Types, Properties, and Methods

The `class`, `properties`, and `methods` functions can be used to get the type of a class and a list of all the public properties and methods declared in the class. For example, if `a` is the vector object declared in the previous section, then the `class` function will return the class of the object, the `properties` function will return the list of public properties in the class, and the `methods` function will return the list of public methods in the class. Note that the constructor also appears in the method list.

```
» class(a)
ans =
vector
» properties(a)
Properties for class vector:
     x
     y
» methods(a)

Methods for class vector:

add      length  vector
```

12.2.4 Attributes

Attributes are modifiers that change the behavior of classes, properties, or methods. They are defined in parentheses after the `classdef`, `properties`, and `methods` statements in the class definition. We will discuss property and method attributes in this section; class attributes are discussed in later sections.

Property attributes modify the behavior of properties defined in a class. The general form of a properties declaration with attributes is

```
properties (Attribute1 = value1, Attribute2 = value2, ...)
   ...
end
```

The attributes will affect the behavior of all properties defined within the code block. Note that sometimes some properties need different attributes than others in the same class. In that case, just define two or more `properties` blocks with different attributes, and declare each property in the block containing its required attributes.

```
    properties (Attribute1 = value1)
       ...
    end

    properties (Attribute2 = value2)
       ...
    end
```

A list of some selected property attributes is given in Table 12.1. These attributes will all be discussed later on in the chapter.

The following example class contains three properties: a, b, and c. Properties a and b are declared to have public access, and property c is declared to have public read access and private write access. This means that when an object is instantiated from this class, it will be possible to both examine and modify instance variables a and b from outside the object. However, instance variable c can be examined but *not* modified from outside the object.

```
classdef test1

   % Sample class illustrating access control using attributes

   properties (Access=public)
      a;           % Public access
      b;           % Public access
   end

   properties (GetAccess=public, SetAccess=private)
      c;           % Read only
   end
```

(This listing continues on page 477)

Table 12.1: Selected `property` Attributes

Property	Type	Description
Access	Enumeration: Possible values are `public`, `protected`, or `private`	This property controls access to this property, as follows: ■ `public`—This property can be read and written from any part of the program. ■ `private`—This property can be read and written only by methods within the current class. ■ `protected`—This property can be read and written only by methods within the current class or one of its subclasses. Setting this attribute is equivalent to setting both `GetAccess` and `SetAccess` for a property.
Constant	Logical: default = false	If true, then the corresponding properties are constants, defined once. Every object instantiated from this class inherits the same constants.
GetAccess	Enumeration: Possible values are `public`, `protected`, or `private`	This property controls read access to this property, as follows: ■ `public`—This property can be read and written from any part of the program. ■ `private`—This property can be read and written only by methods within the current class. ■ `protected`—This property can be read and written only by methods within the current class or one of its subclasses.
Hidden	Logical: default = false	If true, this property will not be displayed in a property list.
SetAccess	Enumeration: Possible values are `public`, `protected`, or `private`	This property controls write access to this property, as follows: ■ `public`—This property can be read and written from any part of the program. ■ `private`—This property can be read and written only by methods within the current class. ■ `protected`—This property can be read and written only by methods within the current class or one of its subclasses.

```
methods

    % Declare the constructor
    function obj = test1(a,b,c)
        obj.a = a;
        obj.b = b;
        obj.c = c;
    end
end

end
```

When we create an object of this class, the constructor initializes all of its instance variables.

```
» obj1 = test1(1,2,3)
obj1 =
    test1 with properties:

        a: 1
        b: 2
        c: 3
```

It is possible to examine and modify the value of a from outside the object.

```
» obj1.a
ans =
    1
» obj1.a = 10
obj1 =
    test1 with properties:

        a: 10
        b: 2
        c: 3
```

It is possible to examine but *not* to modify the value of c from outside the object.

```
» obj1.c
ans =
    3
» obj1.c = -2
You cannot set the read-only property 'c' of test1.
```

This is a very important feature of objects. If the properties of a class are set to have private access, then those properties can be modified only by instance methods inside the class. These methods can be used to check input values for validity before allowing them to be used, making sure that no illegal values are assigned to the properties.

⟨/⟩ Good Programming Practice

Use the access control attributes to protect properties from being set to invalid values.

///

Method attributes modify the behavior of methods defined in a class. The general form a methods declaration with attributes is

```
methods (Attribute1 = value1, Attribute2 = value2, ...)
   ...
end
```

The attributes will affect the behavior of all methods defined within the code block. Note that sometimes some methods need different attributes than others in the same class. In that case, just define two or more `methods` blocks with different attributes, and declare each property in the block containing its required attributes.

```
methods (Attribute1 = value1)
   ...
end

methods (Attribute2 = value2)
   ...
end
```

A list of some selected property attributes is given in Table 12.2. These attributes will all be discussed later on in the chapter.

Table 12.2: Selected `method` Attributes

Property	Type	Description
Access	Enumeration: Possible values are public, protected, or private	This property controls access to this property, as follows: ■ public—This property can be read and written from any part of the program. ■ private—This property can be read and written only by methods within the current class. ■ protected—This property can be read and written only by methods within the current class or one of its subclasses.
Hidden	Logical: default = false	If true, this property will not be displayed in a property list.
Sealed	Logical: default = false	If true, this method cannot be redefined in a subclass.
Static	Logical: default = false	If true, these methods do not depend on objects of this class and do not require the object as an input argument.

12.3 Value Classes versus Handle Classes

MATLAB supports two kinds of classes: **value classes** and **handle classes**. If one object of a value class type is assigned to another variable, MATLAB *copies* the original object, and there are now two objects in memory. Each of the two objects can be changed separately without affecting each other (see Figure 12.6*a*). By contrast, if an object of a handle class is assigned to another variable, MATLAB *copies a reference* (a **handle**) to the class, and the two variables contain handles that point to the *same* object in memory (see Figure 12.6*b*). With a handle class, a change made using one handle will also be seen when using the other one, because they both point to the same object in memory.

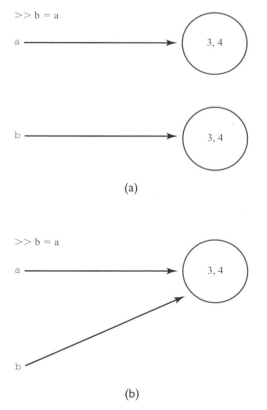

Figure 12.6 (a) When an object of a value class is assigned to a new variable, MATLAB makes an independent copy of the object and assigns it to the new variable. Variables a and b point to independent objects. (b) When an object of a handle class is assigned to a new variable, MATLAB copies the reference (or handle) of the object, and assigns it to the new variable. Both variables a and b point to the *same* object.

12.3.1 Value Classes

The `vector` class that we developed in Section 12.2.2 is an example of a value class. If we create a vector and then assign it to a new variable, MATLAB makes a copy of the object and assigns it to the new variable.

```
» a = vector(3,4)
a =
   vector with properties:

     x: 3
     y: 4
» b = a
b =
   vector with properties:

     x: 3
     y: 4
```

We can show that these two variables are different by assigning different values to one of them.

```
» b.x = -1;
» b.y = 0;
» a
a =
   vector with properties:

     x: 3
     y: 4
» b
b =
   vector with properties:

     x: -1
     y: 0
```

Note that changing variables in one of the objects did not affect the other one at all.

If one object of a value class type is assigned to another variable, MATLAB *copies* the original object, and there are now two objects in memory. Each of the two objects can be changed separately without affecting each other (see Figure 12.6a). Furthermore, if one of the objects is deleted, the other one is unaffected because it is an independent object.

In MATLAB, we create value classes by defining a class that *is not* a subclass of the `handle` object. The `vector` class is a value class because the class definition does not inherit from `handle`.

Value classes are typically used to store data values for use in calculations. For example, the `double`, `single`, `int32`, and other standard MATLAB data types are all really value classes.

Objects made from value classes can be deleted when they are no longer needed using the `clear` command. For example, when the statements described earlier in this section have been executed, objects a and b are in memory:

```
» whos
  Name          Size                Bytes  Class       Attributes

    a           1x1                  120   vector
    b           1x1                  120   vector
```

If we now issue the command `clear a`, object a will be removed from memory:

```
» clear a
» whos
  Name          Size                Bytes  Class       Attributes

    b           1x1                  120   vector
```

The command `clear all` would have removed all of the objects from memory.

12.3.2 Handle Classes

A **handle class** is a class that inherits directly or indirectly from the `handle` super-class. These classes use a reference (a handle) to point to the object in memory. When a variable of a handle class is assigned to another variable, the handle is copied, *not the object itself.* Thus, after copying we have two handles both pointing to the same object in memory (see Figure 12.6*b*).

A handle to a handle class object can be used to access or modify that object. Since the handle can be copied and passed to various functions, multiple parts of a program can have access to the object at the same time.

A handle class version of the vector class is shown below. This class is a handle class because the new class is a subclass of the superclass `handle`. Note that the superclass that a class is based on is specified in the `classdef` statement by a < symbol followed by the superclass name. This syntax means that the new class being defined is a subclass of the specified superclass, and it inherits the superclass's properties and methods. Here, `vector_handle` is a subclass of class `handle`.

```
% The vector as a handle class
classdef vector_handle < handle

    properties
        x;          % X value of vector
        y;          % Y value of vector
    end

    methods
```

```
        % Declare the constructor
        function this = handle_vector(a,b)
            this.x = a;
            this.y = b;
        end

        % Declare a method to calculate the length
        % of the vector.
        function result = length(this)
            result = sqrt(this.x.^2 + this.y.^2);
        end

        % Declare a method to add two vectors together
        function add(this,obj2)
            this.x = this.x + obj2.x;
            this.y = this.y + obj2.y;
        end

    end

end
```

There are two key differences in the handle version of this class. First, the class is declared to be a subclass of handle in the class definition. Second, methods modifying an object of this class do not return the modified object as a calling argument.

The *value class* version of the add method was:

```
% Declare a method to add two vectors together
function obj = add(this,obj2)
    obj = vector();
    obj.x = this.x + obj2.x;
    obj.y = this.y + obj2.y;
end
```

This method receives *copies of two objects* as input arguments, the current object and another object of the same class. The method creates a new output object and uses the two input objects to calculate the output values. When the method ends, only the new output argument obj is returned from the function. Note that the values of the input vectors this and obj2 are not modified by this operation.

In contrast, the *handle class* version of the add method is:

```
% Declare a method to add two vectors together
function add(this,obj2)
    this.x = this.x + obj2.x;
    this.y = this.y + obj2.y;
end
```

This method receives *handles to the two objects* as input arguments, the current object and another object of the same class. The method performs calculations using the handles, which point to the *original objects*, not copies of the objects. The two vectors are added together, with the result stored in the original vector object (`this`). The results of those calculations are automatically saved in the original object, so no output argument needs to be returned from the function. Unlike the value class case, the value of the original vector is modified here.

If we create a vector using the `vector_handle` class, and then assign it to a new variable, MATLAB makes a copy of the *object handle* and assigns it to the new variable.

```
» a = vector_handle(3,4)
a =
   vector_handle with properties:

      x: 3
      y: 4
» b = a
b =
   vector with properties:

      x: 3
      y: 4
```

We can show that these two variables are the same by assigning different values to one of them and seeing that the new values also show up in the other one.

```
» b.x = -1;
» b.y = 0;
» a
a =
   vector_handle with properties:

      x: -1
      y: 0
» b
b =
   vector_handle with properties:

      x: -1
      y: 0
```

Changing the instance variables using one of the handles has affected the results seen using all handles because they all point to the same physical object in memory.

Objects made from handle classes are automatically deleted by MATLAB when there are no handles left that point to them. For example, following two statements create two `vector_handle` objects:

```
» a = vector_handle(3,4);
» b = vector_handle(-4,3);
» whos
```

```
Name      Size      Bytes     Class           Attributes

a         1x1       112       vector_handle
b         1x1       112       vector_handle
```

If we now execute the statement

```
» a = b;
```

both handles a and b now point to the original object allocated using handle b. The object that was originally allocated using handle a is no longer accessible, because no handle to it exists anymore, and MATLAB will automatically delete that object.

A user can delete a handle object at any time using the **delete** function with *any* handle pointing to that object. After the delete function is called, all the handles that pointed to that object are still in memory, but they no longer point to any object. The object that they had pointed to has been removed.

```
» delete(a)
» whos
  Name      Size      Bytes     Class           Attributes

  a         1x1       104       vector_handle
  b         1x1       104       vector_handle

» a
a =
  handle to deleted vector_handle
» b
b =
  handle to deleted vector_handle
```

The handles themselves can be removed using the clear command.

Handle classes are traditionally used for objects that perform some function in a program, such as writing to a file. There can be only one object that opens and writes to the file, because the file can only be opened once. However, *many* parts of the program can have handles to that object, and so they can pass data to the object to write to the file.

Handle classes are the type of classes traditionally meant by the term "object-oriented programming," and the special features of object-oriented programming such as exceptions, listeners, and so forth are only applicable to handle classes. Most of the discussions in the remainder of this chapter refer to handle classes only.

12.4 Destructors: The **delete** Method

If a class includes a method called delete with a single input argument of the object's type, MATLAB will call this method to clean up resources used by the object just before it is deleted. For example, if the object has a file open, the delete method would be designed to close the file before the object is deleted.

We can use this fact to observe when objects are deleted from memory. If we create a class with a `delete` method and write a message to the command window from that method, we can tell exactly when an object is destroyed. For example, suppose that we add a `delete` method to the `vector_handle` class.

```
% Modified vector_handle class
classdef vector_handle < handle

    properties
        x;              % X value of vector
        y;              % Y value of vector
    end

    methods

        . . .

        % Declare a destructor
        function delete(this)
            disp('Object destroyed.');
        end

    end

end
```

Note that the `delete` method is optional and is not present in most classes. It is normally included if a class has some resources (such as files) open that need to be released before the object is destroyed.

The `clear` command deletes the *handle* to an object, not the object itself. However, the object is sometimes destroyed automatically in this case. If we create an object of the `vector_handle` class, and then `clear` the handle to it, the object will be automatically destroyed because there is no longer a reference to it.

```
» a = vector_handle(1,2);
» clear a
Object destroyed.
```

On the other hand, if we create an object of this class and assign its handle to another variable, there will be *two* handles to the object. In this case, clearing one will *not* cause the object to be destroyed because there is still a valid handle to it.

```
» a = vector_handle(1,2);
» b = a;
» clear a
```

We can now see the difference between `clear` and `delete`. The `clear` command deletes a handle, whereas the `delete` command deletes an object. The `clear` command *may* cause the object to be deleted, too, but only if there is no other handle to the object.

Good Programming Practice

Define a `delete` method to close files or delete other resources before an object is destroyed.

12.5 Access Methods and Access Controls

In object-oriented programming, it is normally a good idea to prevent the parts of a program outside of an object from seeing or modifying the object's instance variables. If the outside parts of the program could directly modify an instance variable, they might assign improper or illegal values to the variable, and that could break the program. For example, we could define a vector as follows:

```
» a = vector_handle(3,4)
a =
   vector_handle with properties:

      x: 3
      y: 4
```

It would be perfectly possible for some part of the program to assign a string to the numeric instance variable x:

```
» a.x = 'junk'
a =
   vector_handle with properties:

      x: 'junk'
      y: 4
```

The `vector` class is depending on the x and y properties containing `double` values. If a string is assigned instead, the methods associated with the class will fail. Thus, something done in another part of the program could break this vector object!

To prevent this from happening, we want to ensure that other parts of the program cannot modify the instance variables in a method. MATLAB supports two ways to accomplish this:

1. Access Methods
2. Access Controls

Both techniques are described in the following sections.

12.5.1 Access Methods

It is possible to protect properties from being modified inappropriately by using special **access methods** to save and retrieve data from the properties. If they are defined, MATLAB will always call access methods whenever attempts are made to use or change the properties in an object[3], and the access methods can verify the data before allowing it to be used. It appears to the user as if the properties can be freely read and written, but in fact a "hidden" method is run in each case that can check to make sure that the data is valid.

Access methods can be written to ensure that only valid data is set or retrieved, thus preventing other parts of the program from breaking the object. For example, they can ensure that the data is of the right type, that it is in the right range, and that any specified subscripts are within the valid range of the data.

Access methods have special names that allow MATLAB to identify them. The name is always get or set followed by a period and the property name to access. To save a value in property PropertyName, we would create a special method called set.PropertyName. To get a value from property PropertyName, we would create a special method called get.PropertyName. If methods with these names are defined in a methods block without attributes, then the corresponding method will be called automatically whenever a property is accessed. The access method will perform checks on the data before it is used.

For example, let's create a set method for the x property of the vector_handle class. This set method will check to see if the input value is of type double using the isa function. (The isa function checks to see if the first argument is of the type specified in the second argument and returns true if it is.) In this case, if the input value is of type double, the function will return true, and the value will be assigned to x. Otherwise, a warning will be printed out, and x will be unchanged. The method is:

```
methods % no attributes

    function set.x(this,value)
        if isa(value,'double')
            this.x = value;
        else
            disp('Invalid value assigned to x ignored');
        end
    end

end
```

If this set method is included in the class vector_handle, the attempt to assign a string to variable x will cause an error, and the assignment will not occur:

```
» a = vector_handle(3,4)
a =
  vector_handle with properties:

    x: 3
    y: 4
```

---

[3]There are a few exceptions. The access methods are not called for changes within the access methods (to prevent recursion), are not called for assignments in the constructor, and are not called when setting a property to its default value.

```
» a.x = 'junk'
Invalid value assigned to x ignored
a =
   vector_handle with properties:

     x: 3
     y: 4
```

Good Programming Practice

Use access methods to protect class properties from being modified in inappropriate ways. The access methods should check for valid data types, subscript ranges, and so forth before allowing a property to be modified.

12.5.2 Access Controls

In object-oriented programming, it is often customary to declare some important class properties to have `private` or `protected` access, so that they cannot be modified directly by any parts of the program outside of the class. This will force other parts of the program to use the class's methods to interact with it, instead of trying to directly modify the class properties. The methods thus serve as an interface between the object and the rest of the program, hiding the internals of the class.

This idea of information hiding is one key to object-oriented programming. If the internals of a class are hidden from the rest of the program and are accessible only through interface methods, the internals of the class can be modified without breaking the rest of the program, as long as the calling sequences of the interface methods remain unchanged.

A good example of a `private` property would be the file id in a file writer class. If a file writer object has opened a file, the file id used to write to that file should be hidden so that no other part of the program can see it and use it to write to the file independently.

Good Programming Practice

Set the access controls to restrict access to properties that should be `private` in a class.

Note that defining an access method is almost equivalent to setting a class property access to `private` or `protected` and can serve the same purpose. If an access method is defined for a property, then the method will filter access to the property, which is the key goal of declaring `private` or `protected` access.

12.5.3 Example: Creating a Timer Class

To consolidate the lessons we have learned so far, we will now create a class that serves as a stopwatch or elapsed timer.

▶Example 12.1—Timer Class

When developing software, it is often useful to be able to determine how long a particular part of a program takes to execute. This measurement can help us locate the "hot spots" in the code, the places where the program is spending most of its time, so that we can try to optimize them. This is usually done with an *elapsed time calculator*[4]. This object measures the time difference between now and when the object was created or last reset. Create a sample class called `MyTimer` to implement an elapsed time calculator.

Solution An elapsed time calculator makes a great sample class, because it is so simple. It is analogous to a physical stopwatch. A stopwatch is an object that measures the elapsed time between a push on a start button and a push on a stop button (often they are the same physical button). The basic actions (methods) performed on a physical stopwatch are:

1. A button push to reset and start the timer.
2. A button push to stop the timer and display the elapsed time.

Internally, the stopwatch must remember the time of the first button push in order to calculate the elapsed time.

Similarly, an elapsed time class needs to contain the following components:

1. A method to store the start time of the timer (`startTime`). This method will not require any input parameters from the calling program, and will not return any results to the calling program.
2. A method to return the elapsed time since the last start (`elapsedTime`). This method will not require any input parameters from the calling program, but it will return the elapsed time in seconds to the calling program.
3. A property to store the time that the timer started running, for use by the elapsed time method.

In addition, the class must have a constructor to initialize the instance variable when an object is instantiated. The constructor will initialize the `startTime` to be the time when the object was created.

The timer class must be able to determine the current time whenever one of its methods is called. In MATLAB, the function `clock` returns the date and time as an array of 6 integers, corresponding to the current year, month, day, hour, minute, and second, respectively. We will convert the last three of these values into the number of seconds since the start of the day and use that value in the timer calculations. The basic equation is

$$second\_in\_day = 3600h + 60m + s \qquad (12.2)$$

---

[4]MATLAB includes the built-in functions `tic` and `toc` for this purpose.

where *h* is the number of hours, *m* is the number of minutes, and *s* is the number of seconds at the current time.

We will implement the timer class in a series of steps, defining the properties, constructor, and methods in succession.

1. **Define the properties**

   The timer class must contain a single property called savedTime, which contains the time at which the object was created or the last time at which startTimer method was called. This property will have private access, so that no code outside the class can modify it.

   The property is declared in a property block with private access, as follows:

```
classdef MyTimer < handle

    properties (Access = private)
        savedTime;      % Time of creation or last reset
    end

    (methods)

end
```

2. **Create the constructor**

   The constructor for a class is automatically called by MATLAB when an object is instantiated from the class. The constructor must initialize the instance variables of the class, and may perform other functions as well (such as opening files, etc.). In this class, the constructor will initialize the savedTime value to the time at which the MyTimer object is created.

   A constructor is created within a methods block. The constructor looks just like any other method, except that it has *exactly* the same name (including capitalization) as the class that it is defined in and has only one output argument—the object created. The constructor for the Timer class is shown below:

```
% Constructor
function this = MyTimer()
    % Initialize object to current time
    timvec = clock;
    this.savedTime = 3600*timvec(4) + 60*timvec(5) + timvec(6);
end
```

3. **Create the methods**

   The class must also include two methods to reset the timer and to read the elapsed time. Method resetTimer() simply resets the start time in the instance variable savedTime.

```
% Reset timer
function resetTimer(this)
   % Reset object to current time
   timvec = clock;
   this.savedTime = 3600*timvec(4) + 60*timvec(5) + timvec(6);
end
```

Method `elapsedTime()` returns the elapsed time since the start of the timer in seconds.

```
% Calculate elapsed time
function dt = elapsedTime (this)
   % Get the current time
   timvec = clock;
   timeNow = 3600*timvec(4) + 60*timvec(5) + timvec(6);

   % Now calculate elapsed time
   dt = timeNow - this.savedTime;
end
```

The resulting `MyTimer` class is shown in Figure 12.7, and the final code for this class is shown below:

```
classdef MyTimer < handle
   % Timer to measure elapsed time since object creation or last reset
   properties (Access = private)
      savedTime;      % Time of creation or last reset
   end

   methods (Access = public)

      % Constructor
      function this = MyTimer()
         % Initialize object to current time
         timvec = clock;
         this.savedTime = 3600*timvec(4) + 60*timvec(5) + timvec(6);
      end
```

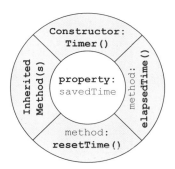

Figure 12.7 The `Timer` class.

```
      % Reset timer
      function resetTimer(this)
          % Reset object to current time
          timvec = clock;
          this.savedTime = 3600*timvec(4) + 60*timvec(5) + timvec(6);
      end

      % Calculate elapsed time
      function dt = elapsedTime(this)
          % Get the current time
          timvec = clock;
          timeNow = 3600*timvec(4) + 60*timvec(5) + timvec(6);

          % Now calculate elapsed time
          dt = timeNow - this.savedTime;
      end
   end
end
```

4. **Test the class.**

To test this class, we will write a script file that creates a `MyTimer` object, performs some calculations, and measures the resulting elapsed time. In this case we will create and solve a 1000×1000 system of simultaneous equations and a $10,000 \times 10,000$ system of simultaneous equations, timing each solution with a `MyTimer` object. The `MyTimer` object will be created just before the first set of equations is solved. After the first solution, the script will call method `elapsedTime()` to determine the time taken to solve the system of equations.

Next, the timer will be reset using method `resetTimer()`, the second set of equations will be solved, and the script will again use `elapsedTime()` to determine the time taken.

```
% Program to test the MyTimer class

% Create the timer object
t = MyTimer();

% Solve a 1000 x 1000 set of simultaneous equations
A = rand(1000,1000);
b = rand(1000,1);
x = A\b;

% Get the elapsed time
disp(['The time to solve a 1000 x 1000 set of equations is ' ...
      num2str(t.elapsedTime())]);

% Reset the timer
t.resetTimer();
```

```
% Solve a 10000 x 10000 set of simultaneous equations
A = rand(10000,10000);
b = rand(10000,1);
x = A\b;

% Get the elapsed time
disp(['The time to solve a 10000 x 10000 set of equations is ' ...
      num2str(t.elapsedTime())]);
```

When this script is executed, the results are

```
» test_timer
The time to solve a 1000 x 1000 set of equations is 0.063
The time to solve a 10000 x 10000 set of equations is 13.026
```

The timer class appears to be working as desired.

12.5.4 Notes on the `MyTimer` Class

This section contains a few notes about the operation of the `MyTimer` class, and of classes in general.

First, note that the `MyTimer` class saves its start time in the property `savedTime`. Each time that an object is instantiated from a class, it receives its *own copy* of all instance variables defined in the class. Therefore, many `MyTimer` objects could be instantiated and used simultaneously in a program, and *they will not interfere with each other* because each timer has its own private copy of the instance variable `savedTime`.

Also, notice that the blocks defining the properties and methods in the class are all declared with either a `public` or `private` attribute. Any property or method declared with the `public` attribute can be accessed from other classes in the program. Any property or method declared with the `private` attribute is only accessible to methods of the object in which it is defined.

In this case, the property `savedTime` is declared `private`, so it cannot be seen or modified by any method outside of the object in which it is defined. Since no method outside of `MyTimer` can see `savedTime`, it is not possible for some other part of the program to accidentally modify the value stored there and so mess up the elapsed time measurement. The only way that a program can utilize the elapsed time measurement is through the `public` methods `resetTimer()` and `elapsedTime()`. You should normally declare all properties within your classes to be `private` (or otherwise protect them with access methods).

Also, note that the formula that calculates elapsed time in seconds in this class [Equation (12.2)] resets at midnight each day, so this timer would fail if it ran over midnight. You will be asked in an end-of-chapter problem to modify this equation so that the timer works properly over longer periods of time.

12.6 Static Methods

Static methods are methods within a class that do not perform calculations on individual objects instantiated from the class. They typically perform "utility" calculations that might be needed by the instance methods within the class, or that might be needed in other parts of the program. Because these methods do not modify the properties of the class, they do *not* include an object of the class as the first input argument the way that instance methods do.

The static methods defined within a class can be used without instantiating an object from the class first, so they can be called from the class constructor while an object is being created. If the static methods have `public` access, they can also be called from other parts of the program without creating an object first.

Static methods are declared by adding a `Static` attribute to the `methods` block in which they are declared. They can be accessed without creating an instance to the class first by naming the class name followed by a period and the method name. Alternately, if an object created from the class exists, then the static methods can be accessed by the object reference followed by a period and the method name.

As an example, suppose that we are creating a class that works with angles and, as a part of this class, we would like to have methods that convert from degrees to radians and from radians to degrees. These methods do not involve the properties defined in the class, and they could be declared as static methods as follows:

```
classdef Angle
    ...
    methods(Static, Access = public)
        function out = deg2Rad(in)
            out = in * pi / 180;
        end

        function out = rad2Deg(in)
            out = in * 180 / pi;
        end
    end
    ...
end
```

These static methods could be accessed from inside and outside the class because their access is `public`. They would be invoked using the class name followed by a dot and the method name: `Angle.deg2Rad()` and `Angle.rad2Deg()`.

If an object of the `Angle` class is created as follows:

```
a = Angle();
```

then the static methods could also be called using the instance object name: `a.deg2Rad()` and `a.rad2Deg()`.

Good Programming Practice

Use static methods to implement utility calculations within a class.

12.7 Defining Class Methods in Separate Files

So far, we have defined all of the methods in a class within a `methods` block in the class definition. This certainly works, but if the methods are very large and there are many of them, the class definition could be thousands of lines long! MATLAB supports an alternate way to declare the methods in a class without having to force all the methods to be in a single file.

Suppose that we wanted to create a class called `MyClass`. If we create a directory called `@MyClass` and place it in a directory on the MATLAB path, MATLAB will assume that all of the contents of that directory are components of class `MyClass`.

The directory *must* contain a file called `MyClass.m` that contains the class definition. The class definition must contain the definition of the properties and methods in the class but does not have to contain all of the method implementations. The signature of each method (the calling sequence and return values) must be declared in a `methods` block, but the actual functions can be declared in separate files.

The following example defines a class `MyClass` with three properties a, b, and c, and two instance methods `calc1` and `calc2`. The `methods` block contains the *signature* of the two methods (the number of input arguments and output arguments), but not the methods themselves.

```
classdef MyClass
    ...
    Properties (Access = private)
        a;
        b;
        c;
    End

    methods(Access = public)
        function output = calc1(this);
        function output = calc2(this, arg1, arg2);
    end
end
```

There must then be two separate files `calc1.m` and `calc2.m` in the same directory that would contain the function definitions to implement the methods. File `calc1.m` would contain the definition of function `calc1`:

```
function output = calc1(this);
    ...
    ...
end
```

and file `calc2.m` would contain the definition of function `calc2`:

```
function output = calc2(this, arg1, arg2);
   ...
   ...
end
```

The directory `@MyClass` would contain the following files:

```
@MyClass\MyClass.m
@MyClass\calc1.m
@MyClass\calc2.m
```

Note that certain methods *must* be in the file with the class definition. These methods include

1. The constructor method
2. The destructor method (`delete`)
3. Any method that has a dot in the method name, such as `get` and `set` access methods.

All other methods can be declared in a class definition `methods` block, but actually defined in separate files in the same subdirectory.

12.8 Overriding Operators

MATLAB implements the standard mathematical operators such as addition, subtraction, multiplication, and division as methods with special names defined in the class that defines a data type. For example, `double` is a built-in MATLAB class that contains a single double-precision floating point property. This class includes a set of methods to implement addition, subtraction, and so forth for two objects of this class. When a user defines two double variables a and b and then adds them together, MATLAB really calls the method `plus(a,b)` defined in the `double` class.

MATLAB allows programmers to define operators for their own user-defined classes as well. As long as a method with the right name and number of calling arguments is defined in the class, MATLAB will call that method when it encounters the appropriate operation between two objects of the class. If the class includes a `plus(a,b)` method, then that method will be automatically called when the expression a + b is evaluated with a and b being objects of that class. This is sometimes called **operator overloading**, because we are giving the standard operators a new definition.

User-defined classes have a higher precedence than built-in MATLAB classes, so mixed operations between user-defined classes and built-in classes are evaluated by the method defined in the user-defined class. For example, if a is a `double` variable and b is an object of a user-defined class, then the expression a + b will be evaluated by the `plus(a,b)` method in the user-defined class. Be aware of this—you must be sure that your methods can handle both objects of the defined class and built-in classes like `double`.

Table 12.3 lists the names and signatures of each method associated with a MATLAB operator. Each method defined in the table accepts objects of the class it is defined in (plus possibly double objects as well), and returns an object of the same class. Note that a user-defined class does *not* need to implement all of these operators. It can implement none, all, or any subset that makes sense for the problem being solved.

Table 12.3: Selected MATLAB Operators and Associated Functions

Operation	Method to Define	Description
a + b	plus(a,b)	Binary addition
a – b	minus(a,b)	Binary subtraction
-a	uminus(a)	Unary minus
+1	uplus(a)	Unary plus
a .* b	times(a,b)	Element-wise multiplication
a * b	mtimes(a,b)	Matrix multiplication
a ./ b	rdivide(a,b)	Right element-wise division
a .\ b	ldivide(a,b)	Left element-wise division
a / b	mrdivide(a,b)	Matrix right division
a \ b	mldivide(a,b)	Matrix left division
a .^ b	power(a,b)	Element-wise power
a ^ b	mpower(a,b)	Matrix power
a < b	lt(a,b)	Less than
a > b	gt(a,b)	Greater than
a <= b	le(a,b)	Less than or equal to
a >= b	ge(a,b)	Greater than or equal to
a ~= b	ne(a,b)	Not equal
a == b	eq(a,b)	Equal
a & b	and(a,b)	Logical AND
a | b	or(a,b)	Logical OR
~a	not(a)	Logical NOT
a:d:b	colon(a,d,b)	Colon operator
a:b	colon(a,b)	
a'	ctranspose(a)	Complex conjugate transpose
a.'	transpose(a)	Matric transpose
command window output	display(a)	Display method
[a b]	horzcat(a,b,…)	Horizontal concatenation
[a; b]	vertcat(a,b,…)	Vertical concatenation
a(s1,s2,…,sn)	subsref(a,s)	Subscripted reference
a(a1,…sn) = b	subsasgn(a,s,b)	Subscripted assignment
b(a)	subsindx(a)	Subscript index

►Example 12.2—Vector Class

Create a handle class called `Vector3D` that holds a three-dimensional vector. The class will define properties x, y, and z, and should implement a constructor and the plus, minus, equal, and not equal operators for objects of this class.

Solution This class will have the three properties x, y, and z, with `public` access. (Note that this is *not* a good idea for serious classes—we should implement access methods for the class. We will keep this as simple as possible in this example and then make the class better in end-of-chapter exercises.) It will implement a constructor and the operators plus, minus, equal, and not equal operators for objects of this class.

The constructor for this class will implement both a default constructor and one that provides initial values.

1. **Define the properties**

   The `Vector3D` class must contain three properties called x, y, and z. The properties are declared in a `property` block with `public` access, as follows:

   ```
   % Declare the Vector 3D class
   classdef Vector3D < handle

       properties (Access = public)
           x;              % X value of vector
           y;              % Y value of vector
           z;              % Z value of vector
       end

   end
   ```

2. **Create the constructor**

   The constructor for this class must initialize a `Vector3D` object with the supplied input data, and it must also be able to function as a default constructor if no arguments are supplied. We will use the `nargin` function to distinguish these two cases. The constructor for the `Vector3D` class is shown below:

   ```
   % Declare the constructor
   function this = Vector3D(a,b,c)
       if nargin < 3

           % Default constructor
           this.x = 0;
           this.y = 0;
           this.z = 0;
       else

           % Constructor with input variables
           this.x = a;
   ```

```
            this.y = b;
            this.z = c;
        end
    end
```

3. **Create the methods**

   The class must also include four methods to implement the operators +, -, ==, and ~=. The plus and minus methods will return an object of the Vector3D type, and the equal and not equal methods will return a logical result. The plus and minus methods are implemented by defining an output vector and then adding or subtracting the two vectors term-by-term, saving the result in the output vector. The equal and not equal methods consist of comparing the two vectors term-by-term. The resulting Vector3D class is shown below:

```
% Declare the Vector 3D class
classdef Vector3D < handle

    properties (Access = public)
        x;              % X value of vector
        y;              % Y value of vector
        z;              % Z value of vector
    end

    methods (Access = public)

        % Declare the constructor
        function this = Vector3D(a,b,c)
            if nargin < 3

                % Default constructor
                this.x = 0;
                this.y = 0;
                this.z = 0;

            else

                % Constructor with input variables
                this.x = a;
                this.y = b;
                this.z = c;
            end
        end

        % Declare a method to add two vectors
        function obj = plus(objA,objB)
            obj = Vector3D;
```

```
                obj.x = objA.x + objB.x;
                obj.y = objA.y + objB.y;
                obj.z = objA.z + objB.z;
            end

            % Declare a method to subtract two vectors
            function obj = minus(objA,objB)
                obj = Vector3D;
                obj.x = objA.x - objB.x;
                obj.y = objA.y - objB.y;
                obj.z = objA.z - objB.z;
            end

            % Declare a method to check for equivalence
            function result = eq(objA,objB)
                result = (objA.x == objB.x) && ...
                         (objA.y == objB.y) && ...
                         (objA.z == objB.z);
            end

            % Declare a method to check for non-equivalence
            function result = ne(objA,objB)
                result = (objA.x ~= objB.x) || ...
                         (objA.y ~= objB.y) || ...
                         (objA.z ~= objB.z);
            end
        end

    end
```

4. **Test the class.**
   To test this class, we will create two Vector3D objects, and then add them, subtract them, and compare them for equality and inequality.

```
» a = Vector3D(1,2,3)
a =
  Vector3D with properties:

    x: 1
    y: 2
    z: 3
» b = Vector3D(-3,2,-1)
b =
  Vector3D with properties:

    x: -3
    y: 2
```

```
        z: -1
» c = a + b
c =
  Vector3D with properties:

    x: -2
    y: 4
    z: 2
» d = a - b
d =
  Vector3D with properties:

    x: 4
    y: 0
    z: 4
» eq = a == b
eq =
    0
» ne = a ~= b
ne =
    1
» whos
  Name      Size       Bytes     Class          Attributes

  a         1x1        112       Vector3D
  b         1x1        112       Vector3D
  c         1x1        112       Vector3D
  d         1x1        112       Vector3D
  eq        1x1          1       logical
  ne        1x1          1       logical
```

Note from the output of the whos statement that the sum and difference of vectors a and b are also vectors of the same type, and the equality/inequality tests yield logical results.

◀

12.9 Events and Listeners

Events are notices that an object broadcasts when something happens, such as a property value changing or a user entering data on the keyboard or clicking a button with a mouse. **Listeners** are objects that execute a callback method when notified that an event of interest has occurred. Programs use events to communicate things that happen to objects, and respond to these events by executing the listener's callback function. They are used extensively to create callbacks in Graphical User Interfaces (GUIs), as we shall see in Chapter 14.

Only handle classes can define events and listeners—they do not work for value classes.

The events produced by a class are defined in an `events` block as a part of the class definition, similar to the `properties` and `methods` blocks. Events are triggered by calling the `notify` function in a method. The calling syntax for this function is

```
notify(obj,'EventName');
notify(obj,'EventName',data);
```

This function notifies listeners that the event `'EventName'` is occurring in the specified object. The optional argument `'data'` is an object of class `event.EventData` containing additional information about the event. By default, it contains the source of the event and the name of the event, but this information can be extended as described in the MATLAB documentation.

Listeners are MATLAB functions that listen for specific events and then trigger a specified callback function when the event occurs. Listeners can be created and associated with an event using the `addlistener` method.

```
lh = addlistener(obj,'EventName',@CallbackFunction)
```

where `obj` is a handle to the object creating the event, `'EventName'` is the name of the event, and `@CallbackFunction` is a handle to the function to call when the event occurs. The return argument `lh` is a handle to the listener object.

A simple example of declaring events in a class is shown below. This is a version of the `Vector3D` class that defines a `CreateEvent` and a `DestroyEvent`. The `CreateEvent` is published in the constructor when an object is created, and the `DestroyEvent` is published in the `delete` method when the object is destroyed.

Note that listener objects are created for each event in the constructor when the object is created.

```
% Declare the Vector 3D class that generates events
classdef Vector3D < handle

   properties (Access = public)
      x;          % X value of vector
      y;          % Y value of vector
      z;          % Z value of vector
   end

   events
      CreateEvent;    % Create object event
      DestroyEvent;   % Destroy object event
   end

   methods (Access = public)
```

```
            % Declare the constructor
            function this = Vector3D(a,b,c)

                % Add event listeners when the object is created
                addlistener(this,'CreateEvent',@createHandler);
                addlistener(this,'DestroyEvent',@destroyHandler);

                % Notify about the create event
                notify(this,'CreateEvent');

                if nargin < 3

                    % Default constructor
                    this.x = 0;
                    this.y = 0;
                    this.z = 0;

                else

                    % Constructor with input variables
                    this.x = a;
                    this.y = b;
                    this.z = c;
                end
            end

            ...
            ...
            ...

            % Declare the destructor
            function delete(this);
                notify(this,'DestroyEvent');
            end

        end

    end
```

The callback functions specified in the listeners are shown below:

```
function createHandler(eventSrc,eventData)
    disp('In callback createHandler:');
    disp(['Object of type ' class(eventData.Source) ' created.']);
    disp(['eventData.EventName = ' eventData.EventName]);
    disp(' ');
end
```

```
function destroyHandler(eventSrc,eventData)
   disp('In callback destroyHandler:');
   disp(['Object of type ' class(eventData.Source) ' destroyed.']);
   disp(['eventData.EventName = ' eventData.EventName]);
   disp(' ');
end
```

When objects of this type are created and destroyed, we will see corresponding callbacks occurring:

```
» a = Vector3D(1,2,3);
In callback createHandler:
Object of type Vector3D created.
eventData.EventName = CreateEvent

» b = Vector3D(3,2,1);
In callback createHandler:
Object of type Vector3D created.
eventData.EventName = CreateEvent

» a = b;
In callback destroyHandler:
Object of type Vector3D destroyed.
eventData.EventName = DestroyEvent
```

If it is saved, the handle to the listener object can be used to temporarily disable or permanently remove the callback. If `lh` is the handle to the listener object, the callback can be temporarily disabled by setting the `enable` property to false.

```
lh.enable = false;
```

The callback can be permanently removed by deleting the listener object entirely

```
delete(lh);
```

12.9.1 Property Events and Listeners

All handle classes have four special events associated with each property: `PreSet`, `PostSet`, `PreGet`, and `PostGet`. The `PreSet` property is set just before a property is updated, and the `PostSet` property is set just after the property is updated. The `PreGet` property is set just before a property is read, and the `PostGet` property is set just after the property is read.

These events are enabled if the `SetObservable` attribute is enabled and disabled if it is not present. For example, if a property is declared as

```
properties (SetObservable)
   myProp;                    % My property
end
```

then the four events described above will be declared before and after that property is read or written. If listeners are attached to these properties, the callbacks will occur before and after the property is accessed.

12.10 Exceptions

`Exceptions` are interruptions to the normal flow of program execution due to errors in the code. When an error occurs that a method cannot recover from by itself, it collects information about the error (what the error was, what line it occurred on, and the calling stack describing how program execution got to that point). It bundles this information into a `MException` object, and then **throws the exception**.

A `MException` object contains the following properties:

- **identifier**—The identifier is a string describing the error in a hierarchical way, with the component causing the error followed by a mnemonic string describing the error, separated by colons. Combining the component name with the mnemonic guarantees that the identifier for each error will be unique.
- **message**—This is a string containing a text description of the error.
- **stack**—This property contains an array of structures specifying the calling path to the location of the error, the name of the function, and the line number where the error occurred.
- **cause**—If there are secondary exceptions related to the main one, the additional information about the other exceptions is stored in the cause property.

As an example of an exception, let's create a set of functions, with the first one calling the second one, and the second one calling the plot command `surf` without calling arguments. This is illegal, so `surf` will throw an exception.

```
function fun1()
   try
      fun2;
   catch ME

      id = ME.identifier
      msg = ME.message
      stack = ME.stack
      cause = ME.cause

      % Display the stack
      for ii = 1:length(stack)
         stack(ii)
      end
   end
end

function fun2;
   surf;
end
```

When this function is executed, the results are:

```
» fun1
id =
MATLAB:narginchk:notEnoughInputs
msg =
Not enough input arguments.
stack =
3x1 struct array with fields:
    file
    name
    line
cause =
    {}
ans =
    file: 'C:\Program Files\MATLAB\R2014b\toolbox\matlab\graph3...'
    name: 'surf'
    line: 49
ans =
    file: 'C:\Data\book\matlab\5e\chap12\fun1.m'
    name: 'fun2'
    line: 22
ans =
    file: 'C:\Data\book\matlab\5e\chap12\fun1.m'
    name: 'fun1'
    line: 3
```

Note that the id string combines the component and the specific error mnemonic. The message contains a plain English description of the error, and the stack contains a structure array of the files, names, and line numbers that lead to the error. The cause is not used because there were no other errors.

This error message can be displayed in a convenient form using the getReport() method of the MException class. This will return a brief text summary of the error.

```
» ME.getReport()

Error using surf (line 49)
Not enough input arguments.
Error in fun1>fun2 (line 22)
    surf;
Error in fun1 (line 3)
        fun2;
```

12.10.1 Creating Exceptions in Your Own Programs

If you write a MATLAB function that cannot function properly (perhaps it doesn't have all the data required) and you can detect the error, you should create a MException object describing the error and throwing an exception. The MException object would be created using the constructor

```
ME = MException(identifier,string);
```

where the identifier is a string of the form component:mnemonic and the string is a text string describing the error. When the data is stored in ME, then the function should throw the error using the command

```
throw(ME);
```

This command will terminate the currently running function and return control to the calling function. The throw function sets the stack field of the exception object before returning to the caller, so the exception contains the complete stack trace to the location where the error occurred.

12.10.2 Catching and Fixing Exceptions

If an exception is thrown in a function, execution will stop and return to the caller. If the caller does not handle the exception, execution will stop and return to that function's caller, and so forth, all the way back to the command window. If the exception is still not handled, the error will be printed out in the command window using the MException.last method, and the program will stop executing. The output of MException.last looks like the output of the MException.getReport() method that we saw earlier.

An exception can be handled at any level in the calling stack by a try / catch structure. If the error occurs in a try clause of a function and an exception occurs, then control will transfer to the catch clause with the exception argument. If the function can fix the error, it should do so. If it cannot fix the error, it should pass the exception on to the next higher caller in the calling tree using a rethrow(ME) function. This function is similar to the original throw function, except that it does not modify the stack trace. This leaves the stack still pointing at the lower level where the error really occurred.

The following example shows the same two functions calling surf, but with try / catch structures in fun1 and fun2.

```
function fun1()
   try
      fun2;
   catch ME
      disp('Catch in fun1:');
      ME.getReport()
      rethrow(ME);
   end
end

function fun2
   try
```

```
            surf;
        catch ME
            disp('Catch in fun2:');
            ME.getReport()
            rethrow(ME);
        end
end
```

When fun1 is executed, we can see that fun2 catches and displays the error and then rethrows it. Then fun1 catches and displays the error and then rethrows it. After that the error reaches the command window, and the program stops.

```
» fun1
Catch in fun2:
ans =
Error using surf (line 49)
Not enough input arguments.
Error in fun1>fun2 (line 13)
        surf;
Error in fun1 (line 3)
        fun2;
Catch in fun1:
ans =
Error using surf (line 49)
Not enough input arguments.
Error in fun1>fun2 (line 13)
        surf;
Error in fun1 (line 3)
        fun2;
Error using surf (line 49)
Not enough input arguments.
Error in fun1>fun2 (line 13)
        surf;
Error in fun1 (line 3)
        fun2;
```

12.11   Superclasses and Subclasses

All handle classes form a part of a class hierarchy. Every handle class except handle is a subclass of some other class, and the class inherits both properties and methods from its parent class. The class can add additional properties and methods and can also override the behavior of methods inherited from its parent class.

Any class above a specific class in the class hierarchy is known as a **superclass** of that class. The class just above a specific class in the hierarchy is known as the *immediate superclass* of the class. Any class below a specific class in the class hierarchy is known as a **subclass** of that class.

Any subclass inherits the public properties and methods of the parent class. The methods defined in a parent class can be **overridden** in a subclass, and the behavior of the modified method will be used for objects of that subclass. If a method is defined in a superclass and is not overridden in the subclass, then the method defined in the superclass will be used by objects of the subclass whenever the method is called.

12.11.1 Defining Superclasses and Subclasses

For example, suppose that we were to create a class Shape, describing the characteristics of a two-dimensional shape. This class would include properties containing the area and perimeter of the shape. However, there are many different types of shapes, with different ways to calculate the area and perimeter for each shape. For example, we could create two subclasses of Shape called EquilateralTriangle and Square, with different methods for calculating the shape properties (see Figure 12.8). Both of these subclasses would inherit all of the common information and methods from Shape (area, perimeter, etc.), but would override the methods used to calculate the properties.

Objects of either the EquilateralTriangle *or* Square *classes may be treated as objects of the* Shape *class*, and so forth for any additional classes up the inheritance hierarchy. An object of the EquilateralTriangle class is also an object of the Shape class.

The MATLAB code for the Shape class is shown in Figure 12.9. This class includes two instance variables, area and perimeter. The class also defines a constructor, methods for calculating the area and perimeter of the shape, and a string method for providing a text description of the object.

Note that this class and the following subclasses also include debugging disp statements in each method, so that we can see exactly what code is executed when an object of a given class is created and used. These statements are labeled "For debugging only" in the following three figures.

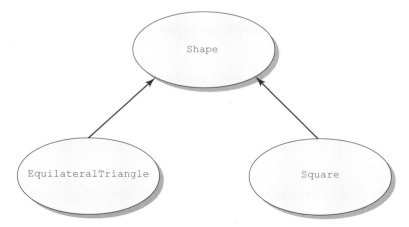

Figure 12.8 A simple inheritance hierarchy. Both EquilateralTriangle and Square inherit from Shape, and an object of either of their classes is also an object of the Shape class.

```matlab
classdef Shape < handle

    properties
        area;           % Area of shape
        perimeter;      % Perimeter of shape
    end

    methods

        % Declare the constructor
        function this = Shape()

            % For debugging only
            disp('In Shape constructor...');

            this.area = 0;
            this.perimeter = 0;

        end

        % Declare a method to calculate the area
        % of the shape.
        function calc_area(this)

            % For debugging only
            disp('In Shape method calc_area...');

            this.area = 0;
        end

        % Declare a method to calculate the perimeter
        % of the shape.
        function calc_perimeter(this)

            % For debugging only
            disp('In Shape method calc_perimeter...');

            this.perimeter = 0;
        end

        % Declare a method that returns info about
        % the shape.
        function string(this)

            % For debugging only
            disp('In Shape method string...');
```

Figure 12.9 The Shape class.

```
        str = ['Shape of class "' class(this) ...
                '", area ' num2str(this.area) ...
                ', and perimeter ' num2str(this.perimeter)];
        disp(str);
    end

end

end
```

Figure 12.9 (*continued*)

Notice that the methods calc_area and calc_perimeter produce zero values instead of valid results because the method of calculating the values will depend on the type of shape, and we don't know that information yet in this class.

The MATLAB code for the EquilateralTriangle subclass is shown in Figure 12.10. This class *inherits* the two instance variables, area and perimeter, and adds an additional instance variable len. It also overrides methods calc_area and calc_perimeter from the superclass so that they perform the proper calculations for an equilateral triangle.

$$area = \frac{\sqrt{3}}{4} \times len^2 \tag{12.3}$$

$$perimeter = 3 \times len \tag{12.4}$$

```
classdef EquilateralTriangle < Shape

    properties
        len;          % Length of side
    end

    methods

        % Declare the constructor
        function this = EquilateralTriangle(len)

            % For debugging only
            disp('In EquilateralTriangle constructor...');

            if nargin > 0
                this.len = len;
            end
            this.calc_area();
            this.calc_perimeter();

        end
```

Figure 12.10 The EquilateralTriangle class.

```
% Declare a method to calculate the area
% of the shape.
function calc_area(this)

    % For debugging only
    disp('In EquilateralTriangle method calc_area...');

    this.area = sqrt(3) / 4 * this.len.^2;
end

% Declare a method to calculate the perimeter
% of the shape.
function calc_perimeter(this)

    % For debugging only
    disp('In EquilateralTriangle method calc_perimeter...');

    this.perimeter = 3 * this.len;
end

    end

end
```

Figure 12.10 (*continued*)

A class is declared as a subclass of another class by including an < symbol followed by the superclass name. In this case, class EquilateralTriangle is a subclass of class Shape because of the < Shape clause on the classdef line. Therefore, this class inherits all of the non-private instance variables and methods from class Shape.

Class EquilateralTriangle defines a constructor to build objects of this class. When an object of a subclass is instantiated, *a constructor for its superclass is called either implicitly or explicitly before any other initialization is performed.* In the constructor of class EquilateralTriangle, the superclass constructor is called implicitly in the first line to initialize area and perimeter to their default values. (Any implicit call to a superclass constructor is always performed with no input parameters. If you need to pass parameters to the superclass constructor, then an explicit call must be used.) The superclass *must* be initialized either implicitly or explicitly before any subclass initialization can occur.

🔲 Good Programming Practice

When writing a subclass, call the superclass's constructor either implicitly or explicitly *as the first action in the subclass constructor.*

///

Note that both the constructor in `Shape` and the constructor in `EquilateralTriangle` contain `disp` statements that are printed out when the code is executed, so it will be possible to see that the superclass constructor is executed before the subclass constructor is executed.

This class also defines new methods `calc_area` and `calc_perimeter` that override the definitions given in the superclass. Since the method `string` is not re-defined, the one in the superclass `Shape` will also apply to any objects of the subclass `EquilateralTriangle`.

The MATLAB code for the `Square` subclass is shown in Figure 12.11. This class *inherits* the two instance variables, `area` and `perimeter`, and adds an additional instance variable `len`. It also overrides methods `calc_area` and `calc_perimeter` from the superclass so that they perform the proper calculations for a square.

$$area = len^2 \tag{12.5}$$

$$perimeter = 4 \times len \tag{12.6}$$

```
classdef Square < Shape

    properties
        len;            % Length of side
    end

    methods

        % Declare the constructor
        function this = Square(len)

            % For debugging only
            disp('In Square constructor...');

            this = this@Shape();
            if nargin > 0
                this.len = len;
            end
            this.calc_area();
            this.calc_perimeter();

        end

        % Declare a method to calculate the area
        % of the shape.
        function calc_area(this)

            % For debugging only
            disp('In Square method calc_area...');
```

Figure 12.11 The `Square` class.

```
        this.area = this.len.^2;
    end

    % Declare a method to calculate the perimeter
    % of the shape.
    function calc_perimeter(this)

        % For debugging only
        disp('In Square method calc_perimeter...');

        this.perimeter = 4 * this.len;
    end

    end

end
```

Figure 12.11 (*continued*)

Class Square defines a constructor to build objects of this class. In the constructor of class Square, the superclass constructor is called explicitly in the first line to initialize area and perimeter to their default values. If additional arguments were needed to initialize the Shape class, they could be added to the explicit call: this = this@Shape(arg1, arg2, ...).

12.11.2 Example Using Superclasses and Subclasses

To illustrate the use of these classes, we will create an object of the EquilateralTriangle class with sides of length 2:

```
» a = EquilateralTriangle(2)
In Shape constructor...
In EquilateralTriangle constructor...
In EquilateralTriangle method calc_area...
In EquilateralTriangle method calc_perimeter...
a =
   EquilateralTriangle with properties:

         len: 2
        area: 1.7321
   perimeter: 6
```

Notice that the superclass Shape constructor was called first to perform its initialization, followed by the EquilateralTriangle constructor. That constructor called methods calc_area and calc_perimeter from class Equilateral-Triangle to initialize the object.

The methods defined in this class can be found using the `methods` function:

```
» methods(a)

Methods for class EquilateralTriangle:

EquilateralTriangle    calc_perimeter
calc_area              string
```

Note that the methods defined in this class include the unique constructor `EquilateralTriangle`, the overridden methods `calc_area` and `calc_perimeter`, and the inherited method `string`.

The properties defined in this class can be found using the `properties` function:

```
» properties(a)
Properties for class EquilateralTriangle:
    len
    area
    perimeter
```

Note that the properties defined in this class include the inherited properties `area` and `perimeter`, plus the unique property `len`.

The class of this object is `EquilateralTriangle`:

```
» class(a)
ans =
EquilateralTriangle
```

However, a is also an object of any class that is a superclass of the object, as we can see using the `isa` function:

```
» isa(a,'EquilateralTriangle')
ans =
    1
» isa(a,'Shape')
ans =
    1
» isa(a,'handle')
ans =
    1
```

If the `calc_area` or `calc_perimeter` methods are called on the new object, the methods defined in class `EquilateralTriangle` will be used instead of the methods defined in class `Shape` because the ones defined in class `EquilateralTriangle` have overridden the superclass method.

```
» a.calc_area
In EquilateralTriangle method calc_area...
```

In contrast, if method `string` is called on the new object, the method defined in class `Shape` will be used because it is inherited by the subclass:

```
» a.string
In Shape method string...
Shape of class "EquilateralTriangle", area 1.7321, and perimeter 6
```

Similarly, we can create an object of the `Square` class with sides of length 2:

```
» b = Square(2)
In Square constructor...
In Shape constructor...
In Square method calc_area...
In Square method calc_perimeter...
b =
  Square with properties:

          len: 2
         area: 4
    perimeter: 8
```

This object is of class `Square`, which is a subclass of `Shape`, so the `string` method will also work with it.

```
» b.string
In Shape method string...
Shape of class "Square", area 4, and perimeter 8
```

►Example 12.3—File Writer Class

Create a `FileWriter` class that opens a file when an object is created, includes a method to write string data to the file, and automatically closes and saves the file when the object is destroyed. Include a feature that counts the number of times data has been written to the file, and a method to report that count. Use good programming practices in your design, including hiding the properties to make them inaccessible from outside the object. The class should throw exceptions in the event of errors, so that a program using the class could trap and respond to the errors.

Solution This class will require two properties, a file ID to access the file and a count to keep track of the number of writes to the file. The class will require four methods, as follows:

1. A constructor to create the object and open the file. The constructor must accept two arguments: a file name and an access mode (write or append).
2. A method to write an input string to a line in the file.
3. A method to return the number of writes so far.
4. `delete` (destructor) method to close and save the file when the object is destroyed. This method must have a single argument that is the type of the object to be destroyed.

The class should throw exceptions if it encounters errors during execution. These exceptions should cover the following errors:

1. There should be an exception in the constructor if no file name is supplied when the object is created.
2. There should be an exception in the constructor if the file name data is not a character string.
3. There should be an exception in the write method if the data to be written is not a character string.

Note that there can be no exceptions in the delete method—it is not allowed to throw anything.

1. **Define the properties**

   The FileWriter class must contain two properties called fid and numberOfWrites, which contain the file ID of the open file and the number of writes to that file so far. These properties will have private access, so that no code outside the class can modify them.

   The properties are declared in a property block with private access, as follows:

```
classdef FileWriter < handle


    % Property data is private to the class
    properties (Access = private)
        fid                    % File ID
        numberOfWrites         % Number of writes to file
    end % properties


    (methods)


end
```

2. **Create the constructor**

   The constructor for this class will check to see that a file name has been provided and that the value of the file name is a character string. It will also check to see if the file access type (w for write or a for append) is provided, and assume append mode if no value is supplied. If not, it should throw appropriate exceptions. Then it will open the file, checking that the open was valid. If not, it should throw an exception.

   The constructor for the FileWriter class is shown below:

```
% Constructor
function this = FileWriter(filename,access)

    % Check arguments
    if nargin == 0

        % No file name
```

```
        ME = MException('FileWriter:noFileName', ...
                    'No file name supplied');
        throw(ME);

    elseif nargin < 2

        % Assume append access by default
        access = 'a';

    end

    % Validate that filename contains a string
    if ~isa(filename,'char')

        % The input data is of an invalid type
        ME = MException('FileWriter:invalidFileNameString', ...
                    'Input filename is not a valid string');
        throw(ME);

    else

        % Open file and save File ID
        this.fid = fopen(filename,access);

        % Did the file open successfully?
        if this.fid <= 0

            % The input data is of an invalid type
            ME = MException('FileWriter:openFailed', ...
                        'Input file cannot be opened');
            throw(ME);

        end

        % Zero the number of writes
        this.numberOfWrites = 0;

    end

end
```

3. **Create the methods**

The class must also include two methods to write a string to the disk and to return the number of writes so far. Method `writeToFile` tests to see if a valid string has been supplied and writes it to the file. Otherwise, it throws an exception.

```
% Write string to file
function writeToFile(this,text_str)

    % Validate that the input parameter is a string
    if ~isa(text_str,'char')

        % The input data is of an invalid type
        ME = MException('FileWriter:writeToFile:invalidString', ...
                        'Input parameter is not a valid string');
        throw(ME);

    else

        % Open file and save File ID
        fprintf(this.fid,'%s\n',text_str);
        this.numberOfWrites = this.numberOfWrites + 1;

    end

end
```

Method getNumberOfWrites returns the number of writes to the file so far.

```
    % Get method for numberOfWrites
    function count = getNumberOfWrites(this)
        count = this.numberOfWrites;
    end
```

Finally, we need a destructor method delete to close the file when the object is destroyed.

```
% Destructor method to close file when object is destroyed
function delete(this)
    fclose(this.fid);
end
```

The resulting class is shown in Figure 12.12.

```
classdef FileWriter < handle

    % Property data is private to the class
    properties (Access = private)
        fid                     % File ID
        numberOfWrites          % Number of writes to file
    end % properties
```

Figure 12.12 The FileWriter class.

```matlab
% Declare methods in class
methods (Access = public)

   % Constructor
   function this = FileWriter(filename,access)

      % Check arguments
      if nargin == 0

         % No file name
         ME = MException('FileWriter:noFileName', ...
                         'No file name supplied');
         throw(ME);

      elseif nargin < 2

         % Assume append access by default
         access = 'a';

      end

      % Validate that filename contains a string
      if ~isa(filename,'char')

         % The input data is of an invalid type
         ME = MException('FileWriter:invalidFileNameString', ...
                         'Input filename is not a valid string');
         throw(ME);

      else

         % Open file and save File ID
         this.fid = fopen(filename,access);

         % Did the file open successfully?
         if this.fid <= 0

            % The input data is of an invalid type
            ME = MException('FileWriter:openFailed', ...
                            'Input file cannot be opened');
            throw(ME);

         end
```

Figure 12.12 (*continued*)

```
                % Zero the number of writes
                this.numberOfWrites = 0;

            end

        end

        % Write string to file
        function writeToFile(this,text_str)

            % Validate that the input parameter is a string
            if ~isa(text_str,'char')

                % The input data is of an invalid type
                ME = MException('FileWriter:writeToFile:invalidString', ...
                            'Input parameter is not a valid string');
                throw(ME);

            else

                % Open file and save File ID
                fprintf(this.fid,'%s\n',text_str);
                this.numberOfWrites = this.numberOfWrites + 1;

            end

        end

        % Get method for numberOfWrites
        function count = getNumberOfWrites(this)
            count = this.numberOfWrites;
        end

        % Finalizer method to close file when object is destroyed
        function delete(this)
            fclose(this.fid);
        end

    end   % methods

end % class
```

Figure 12.12 (*continued*)

4. **Test the class.**

To test this class, we will write a series of scripts that use the class to write to a file correctly and that illustrate various failure modes. The first test is of the class writing to a file and deleting any pre-existing file.

```
% This script tests the FileWriter in 'w', which
% deletes any pre-existing file.

% Create object
a = FileWriter('newfile.txt','w');

% Write three lines of text
a.writeToFile('Line 1');
a.writeToFile('Line 2');
a.writeToFile('Line 3');

% How many lines have been written?
disp([int2str(a.getNumberOfWrites()) ' lines have been written.']);

% Destroy the object
a.delete();

% Display data
type 'newfile.txt'
```

When this script is executed, the results are

```
» testFileWriter1
3 lines have been written.

Line 1
Line 2
Line 3
```

These results are correct.

The second test is of the class writing to a file appending to any pre-existing data.

```
% This script tests the FileWriter in 'a', which
% preserves any pre-existing file.

% Create object
a = FileWriter('newfile.txt','a');

% Write three lines of text
a.writeToFile('Line 1');
a.writeToFile('Line 2');
a.writeToFile('Line 3');
```

```
% How many lines have been written?
disp([int2str(a.getNumberOfWrites()) ' lines have been written.']);

% Destroy the object
a.delete();

% Display data
type 'newfile.txt'
```

When this script is executed, the results are

```
» testFileWriter2
3 lines have been written.

Line 1
Line 2
Line 3
Line 1
Line 2
Line 3
```

The three new lines were appending to the existing ones.
Now let's try a few error cases:

```
» a = FileWriter()
Error using FileWriter (line 21)
No file name supplied

» a = FileWriter(123)
Error using FileWriter (line 36)
Input filename is not a valid string

» a = FileWriter('newfile.txt');
» a.writeToFile(123);
Error using FileWriter/writeToFile (line 69)
Input parameter is not a valid string
```

This class appears to be working as desired.

Quiz 12.1

This quiz provides a quick check to see if you have understood the concepts introduced in Chapter 12. If you have trouble with the quiz, reread the section, ask your instructor, or discuss the material with a fellow student. The answers to this quiz are found in the back of the book.

1. What is a class? What is an object? Explain the difference between the two.
2. How do you create a user-defined class in MATLAB?

3. What are the principal components of a class?
4. What is a constructor? How can you distinguish a constructor from other methods in a class?
5. What is a destructor method? If it exists, when is a destructor executed?
6. What are events? What triggers an event? How can a program listen to and respond to events?
7. What are exceptions? When are exceptions thrown? How are they created? How are they handled by a program?
8. What is a subclass? Explain how a subclass is created from another class.

12.12 Summary

In Chapter 12, we have introduced the basics of object-oriented programming in MATLAB. An object is a software component whose structure is like that of objects in the real world. Each object consists of a combination of data (called properties) and behaviors (called methods). The properties are variables describing the essential characteristics of the object, while the methods describe how the object behaves and how the properties of the object can be modified.

Classes are the software blueprints from which objects are made. A class is a software construct that specifies the number and type of properties to be included in an object and the methods that will be defined for the object. Methods come in two varieties: instance methods and static methods. Instance methods perform calculations involving the properties of an object. In contrast, static methods perform calculations that do not involve the properties of an object. They can be used without creating objects from the class first, if desired.

Each class contains four types of components:

1. **Properties**. Properties define the instance variables that will be created when an object is instantiated from a class.
2. **Methods**. Methods implement the behaviors of a class.
3. **Constructor**. Constructors are special methods that specify how to initialize an object when it instantiated. They always have the same name as the class in which they are defined.
4. **Destructor**. Destructors are special methods that clean up the resources (open files, etc.) used by an object just before it is destroyed. They always have the name `delete`.

Classes are created using a `classdef` structure, and properties and methods are defined within `properties` and `methods` blocks within the `classdef` structure. There can be more than one `properties` and `methods` block within a class definition, with each one specifying properties or methods that have different attributes.

The behavior of classes, properties, and methods can be modified by specifying attributes associated with the block in which they are defined. Some of the more important possible attributes are given in Tables 12.1 and 12.2.

MATLAB supports two kinds of classes: **value classes** and **handle classes**. If one object of a value class type is assigned to another variable, MATLAB *copies* the original object, and there are now two objects in memory. In contrast, if an object of a handle class is assigned to another variable, MATLAB *copies a reference* (a **handle**) to the class, and the two variables contain handles that point to the *same* object in memory. Value classes are used to store and manipulate numeric and string data in MATLAB. Handle classes behave more like objects in other programming languages such as C++ and Java.

The data stored in the properties of a class can be protected from improper modification by using access methods and/or access controls. Access methods intercept assignment statements, using the properties, and check that the data is valid before allowing the assignment to occur. Access controls hide access to properties so that methods outside of an object cannot modify the properties directly.

It is possible to create custom definitions of operators such as +, −, *, and / so that they work with user-defined classes. This is done by defining methods in the class with standard names. If a method of the appropriate name is found within a class, it will be called when the corresponding operator is encountered in a program. For example, if the method `plus(a,b)` is defined in a class, then it will be called whenever two objects of that class are added together using the operator `a + b`.

Events are notices that an object broadcasts when something happens, such as a property value changing or a user entering data on the keyboard or clicking a button with a mouse. Listeners are objects that execute a callback method when notified that an event of interest has occurred. Programs use events to communicate things that happen to objects and respond to these events by executing the listener's callback function.

Exceptions are interruptions to the normal flow of program execution due to errors in the code. When an error occurs that a method cannot recover from by itself, it collects information about the error (what the error was, what line it occurred on, and the calling stack describing how program execution got to that point). It bundles this information into a `MException` object and then throws the exception. `try / catch` structures are used to capture and handle exceptions when they occur.

12.12.1 Summary of Good Programming Practice

The following guidelines should be adhered to when working with MATLAB classes.

1. Define a constructor for a class to initialize the data in objects of that class when they are instantiated. Be sure to support a default case (one without arguments) in the constructor design.
2. When an instance method is invoked, do not include the object in the method's list of calling arguments.
3. Use access methods to protect class properties from being modified in inappropriate ways. The access methods should check for valid data types, subscript ranges, and so forth before allowing a property to be modified.
4. Define a `delete` method to close files or delete other resources before an object is destroyed.
5. Use static methods to implement utility calculations within a class.

12.12.2 MATLAB Summary

The following summary lists all of the MATLAB commands and functions described in this chapter, along with a brief description of each one.

class	Returns the class of the input argument.
classdef	Keyword to mark the start of a class definition.
clear	Function to remove a reference to a handle object from memory. If there is no other reference to the object, the object will also be deleted.
delete	Function to remove an object of a handle class from memory.
delete	Method in a class that is called when the object is about to be destroyed.
events	Keyword to mark the start of an events block, which defined the events produced by a class.
isa	Function that tests to see if an object belongs to a particular class.
methods	Function that lists the non-hidden methods defined in a class.
methods	Keyword to mark the start of a methods block, which declares methods in a class.
MException	MATLAB exception class, which is created when an error occurs during MATLAB execution.
properties	Function that lists the non-hidden properties defined in a class.
properties	Keyword to mark the start of a properties block, which declares variables in a class.
try / catch block	Code structure used to track exceptions in MATLAB code.

12.13 Exercises

12.1 Demonstrate that multiple copies of the Timer class of Example 12.1 can function independently without interfering with each other. Write a program that creates a random 50×50 set of simultaneous equations and then solves the equations. Create three Timer objects as follows: one to time the equation creation process, one to time the equation solution process, and one to time the entire process (creation plus solution). Show that the three objects are functioning independently without interfering with each other.

12.2 Improve the Timer class of Example 12.1 so that it does not fail if it is timing objects over midnight. To do this, you will need to use function datenum, which converts a date and time into a serial date number that represents the years since year zero, including fractional parts. To calculate the elapsed time, represent the start time and elapsed time as serial date numbers and subtract the two values. The result will be elapsed time in years, which then needs to be converted to seconds for use in the Timer class. Create a static method to convert a date number in years into a date number in seconds, and use that method to convert both the start time and elapsed time in your calculations.

12.3 Create a handle class called `PolarComplex` containing a complex number represented in polar coordinates. The class should contain two properties called `magnitude` and `angle`, where angle is specified in radians. The class should include access methods to allow controlled access to the property values, as well as methods to add, subtract, multiply, and divide two `PolarComplex` objects. `PolarComplex` objects can be converted to rectangular form using the following equations:

$$c = a + bi = z \angle \theta \tag{12.7}$$

$$a = z \cos \theta \tag{12.8}$$

$$b = z \sin \theta \tag{12.9}$$

$$z = \sqrt{a^2 + b^2} \tag{12.10}$$

$$\theta = \tan^{-1} \frac{b}{a} \tag{12.11}$$

Complex numbers are best added and subtracted in rectangular form.

$$c_1 + c_2 = (a_1 + a_2) + (b_1 + b_2)i \tag{12.12}$$

$$c_1 - c_2 = (a_1 - a_2) + (b_1 - b_2)i \tag{12.13}$$

Complex numbers are best multiplied and divided in polar form.

$$c_1 \times c_2 = z_1 z_2 \angle \theta_1 + \theta_2 \tag{12.14}$$

$$\frac{c_1}{c_2} = \frac{z_1}{z_2} \angle \theta_1 - \theta_2 \tag{12.15}$$

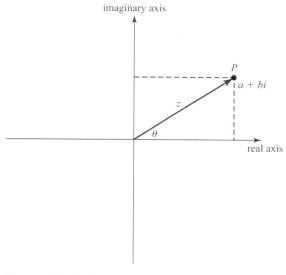

Figure 12.13 Representing a complex number in polar coordinates.

Create methods that add, subtract, multiply, and divide `PolarComplex` numbers based on Equations (12.12) through (12.15), designing them so that two objects can be manipulated with ordinary math symbols. Include static methods to convert back and forth from rectangular to polar form for use with these calculations.

12.4 Three-Dimensional Vectors The study of the dynamics of objects in motion in three dimensions is an important area of engineering. In the study of dynamics, the position and velocity of objects, forces, torques, and so forth are usually represented by three-component vectors $\mathbf{v} = x\,\hat{\mathbf{i}} + y\,\hat{\mathbf{j}} + z\,\hat{\mathbf{k}}$, where the three components (x, y, z) represent the projection of the vector $\mathbf{v}$ along the x, y, and z axes respectively, and $\hat{\mathbf{i}}$, $\hat{\mathbf{j}}$, and $\hat{\mathbf{k}}$ are the unit vectors along the x, y, and z axes (see Figure 12.14). The solutions of many mechanical problems involve manipulating these vectors in specific ways.

The most common operations performed on these vectors are:

1. **Addition.** Two vectors are added together by separately adding their x, y, and z components. If $\mathbf{v}_1 = x_1\,\hat{\mathbf{i}} + y_1\,\hat{\mathbf{j}} + z_1\,\hat{\mathbf{k}}$ and $\mathbf{v}_2 = x_2\,\hat{\mathbf{i}} + y_2\,\hat{\mathbf{j}} + z_2\,\hat{\mathbf{k}}$, then $\mathbf{v}_1 + \mathbf{v}_2 = (x_1 + x_2)\,\hat{\mathbf{i}} + (y_1 + y_2)\,\hat{\mathbf{j}} + (z_1 + z_2)\,\hat{\mathbf{k}}$.

2. **Subtraction.** Two vectors are subtracted by separately subtracting their x, y, and z components. If $\mathbf{v}_1 = x_1\,\hat{\mathbf{i}} + y_1\,\hat{\mathbf{j}} + z_1\,\hat{\mathbf{k}}$ and $\mathbf{v}_2 = x_2\,\hat{\mathbf{i}} + y_2\,\hat{\mathbf{j}} + z_2\,\hat{\mathbf{k}}$, then $\mathbf{v}_1 - \mathbf{v}_2 = (x_1 - x_2)\,\hat{\mathbf{i}} + (y_1 - y_2)\,\hat{\mathbf{j}} + (z_1 - z_2)\,\hat{\mathbf{k}}$.

3. **Multiplication by a Scalar.** A vector is multiplied by a scalar by separately multiplying each component by the scalar. If $\mathbf{v} = x\,\hat{\mathbf{i}} + y\,\hat{\mathbf{j}} + z\,\hat{\mathbf{k}}$, then $a\mathbf{v} = ax\,\hat{\mathbf{i}} + ay\,\hat{\mathbf{j}} + az\,\hat{\mathbf{k}}$.

4. **Division by a Scalar.** A vector is divided by a scalar by separately dividing each component by the scalar. If $\mathbf{v} = x\,\hat{\mathbf{i}} + y\,\hat{\mathbf{j}} + z\,\hat{\mathbf{k}}$, then
$$\frac{\mathbf{v}}{a} = \frac{x}{a}\,\hat{\mathbf{i}} + \frac{y}{a}\,\hat{\mathbf{j}} + \frac{z}{a}\,\hat{\mathbf{k}}.$$

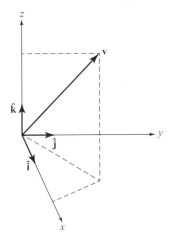

Figure 12.14 A three-dimensional vector.

5. **The Dot Product.** The dot product of two vectors is one form of multiplication operation performed on vectors. It produces a scalar that is the sum of the products of the vector's components. If $\mathbf{v}_1 = x_1\,\hat{\mathbf{i}} + y_1\,\hat{\mathbf{j}} + z_1\,\hat{\mathbf{k}}$ and $\mathbf{v}_2 = x_2\,\hat{\mathbf{i}} + y_2\,\hat{\mathbf{j}} + z_2\,\hat{\mathbf{k}}$, then the dot product of the vectors is $\mathbf{v}_1 \cdot \mathbf{v}_2 = x_1 x_2 + y_1 y_2 + z_1 z_2$.

6. **The Cross Product.** The cross product is another multiplication operation that appears frequently between vectors. The cross product of two vectors is another vector whose direction is perpendicular to the plane formed by the two input vectors. If $\mathbf{v}_1 = x_1\,\hat{\mathbf{i}} + y_1\,\hat{\mathbf{j}} + z_1\,\hat{\mathbf{k}}$ and $\mathbf{v}_2 = x_2\,\hat{\mathbf{i}} + y_2\,\hat{\mathbf{j}} + z_2\,\hat{\mathbf{k}}$, then the cross product of the two vectors is defined as $\mathbf{v}_1 \times \mathbf{v}_2 = (y_1 z_2 - y_2 z_1)\,\hat{\mathbf{i}} + (z_1 x_2 - z_2 x_1)\,\hat{\mathbf{j}} + (x_1 y_2 - x_2 y_1)\,\hat{\mathbf{k}}$.

7. **Magnitude.** The magnitude of a vector is defined as $\mathbf{v} = \sqrt{x^2 + y^2 + z^2}$.

Create a *value* class called `Vector3D`, having three properties x, y, and z. Define a constructor to create vector objects from three input values. Define get and put access methods for each property, and define methods to perform the seven vector operations defined above. Be sure to design the methods so that they work with operator overloading when possible. Then, create a program to test all of the functions of your new class.

12.5 If no exceptions are thrown within a `try` block, where does execution continue after the `try` block is finished? If an exception is thrown within a `try` block and caught in a `catch` block, where does execution continue after the `catch` block is finished?

12.6 Modify the `FileWriter` class by adding new methods to write numeric data to the file as text strings, with one numeric value per line.

Chapter 13

Handle Graphics and Animation

In this chapter we will learn about a low-level way to manipulate MATLAB plots (called handle graphics), and about how to create animations and movies in MATLAB.

13.1 Handle Graphics

Handle graphics is the name of a set of low-level graphics functions that control the characteristics of graphical objects generated by MATLAB. The "handles" referred to are handles to objects from MATLAB graphical classes. These graphical classes are handle classes because they are subclasses of `handle`, and most of what we learned in Chapter 12 about handle classes applies to them.

The MATLAB graphics system has been replaced in Release 2014b. The new graphics system is sometimes referred to as "H2 Graphics"; it generally produces better quality plots than the older system. The discussion in this chapter will be about the new H2 Graphics system, but it will also describe those features of the new system that are backward compatible with older versions of MATLAB.

Handle graphics objects correspond to graphical features, such as figures, axes, lines, text boxes, and so forth. Each object has its own set of properties, which control when and how the object will be displayed on a plot. The various properties can be modified using the handles, as we will discuss in this chapter.

We have actually been using handle graphics indirectly since almost the beginning of the book. For example, we learned in Chapter 3 how to set extra properties when plotting lines, such as setting the line width:

```
plot(x,y,'LineWidth',2);
```

The `'LineWidth'` here was actually a property of the handle graphics object representing the line we are plotting, and the 2 is the value to be stored in that property.

Handle graphics properties and functions are very important to programmers since they allow them to have fine control of the appearance of the plots and graphs that they create. For example, it is possible to use handle graphics to turn on a grid on the *x*-axis only, or to choose a line color like orange, which is not supported by the standard `LineSpec` option of the `plot` command. In addition, handle graphics enable a programmer to create graphical user interfaces (GUIs) for programs, as we will see in the next chapter.

This chapter introduces the structure of the MATLAB graphics system, and explains how to control the properties of graphical objects to create a desired display.

13.2 The MATLAB Graphics System

The MATLAB graphics system is based on a hierarchical system of core **graphics objects**, each of which can be accessed by a **handle** that refers to the object[1]. Each graphics object is derived from a handle class, and each class represents some feature of a graphical plot, such as a figure, a set of axes, a line, a text string, and so forth. Each class includes special **properties** that describe the object, and changing those properties changes how the particular object will be displayed. For example, a `line` is one type of graphical class. The properties defined in a `line` class include: *x*-data, *y*-data, color, line style, line width, marker type, and so forth. Modifying any of these properties will change the way that the line is displayed in a Figure Window.

Every component of a MATLAB graph is a graphical object. For example, each line, axes, and text string is a separate object with its own unique handle and characteristics. All graphical objects are arranged in a hierarchy with **parent objects** and **child objects**, as shown in Figure 13.1. In general, a child object is one that appears embedded within the parent object. For example, an axes object is embedded within a figure, and one or more line objects could be embedded within the axes object. When a child object is created, it inherits many of its properties from its parent.

The highest-level graphics object in MATLAB is the **root**, which can be thought of as the entire computer screen. A handle to the `root` object can be obtained from function `groot`, which stands for "Graphics Root Object." The graphics root object is created automatically when MATLAB starts up, and it is always present until the program is shut down. The properties associated with the root object are the defaults that apply to all MATLAB windows.

Under the root object, there can be one or more Figure Windows, or just **figures**. Each `figure` is a separate window on the computer screen that can display graphical data, and each figure has its own properties. The properties associated with a `figure` include color, color map, paper size, paper orientation, pointer type, and so forth.

---

[1] Before MATLAB R2014b, graphical object handles were `double` values returned from functions that created the objects. The root was object 0, figures were objects 1, 2, 3, and so forth, and other graphical objects had handles with non-integer values. In MATLAB R2014b and later, the new "H2 graphics" system has been enabled. In this system, graphical object handles are actual handles to MATLAB classes, with access to the public properties of the class. This chapter describes the new graphics system, but much of it will work in the older versions of MATLAB as well for backward compatibility.

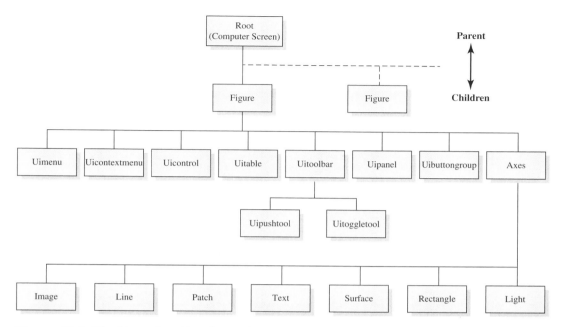

Figure 13.1 The hierarchy of handle graphics objects.

Each figure can contain eight types of objects: uimenus, uicontextmenus, uicontrols, uitoolbars, uipanels, uitables, uibuttongroups, and axes. Uimenus, uicontextmenus, uicontrols, uitoolbars, uipanels, uitables, and uibuttongroups are special graphics objects used to create graphical user interfaces—they will be described in the next chapter. Axes are regions within a figure where data is actually plotted. There can be more than one set of axes in a single figure.

Each set of axes can contain as many lines, text strings, patches, and so forth as necessary to create the plot of interest.

13.3 Object Handles

When a graphics object is created, the creating function returns a handle to the object. For example, the function call

```
» hndl = figure;
```

creates a new figure and returns the handle of that figure in variable hndl. The key public properties of the object can be displayed by typing its name in the Command Window.

```
» hndl
hndl =
  Figure (1) with properties:
```

```
  Number: 1
    Name: ''
   Color: [0.940000000000000 0.940000000000000 0.940000000000000]
Position: [680 678 560 420]
   Units: 'pixels'
```

Show <u>all properties</u>

If the user then clicks on the Show all properties line, the full list of 64 public properties for this figure object will be displayed.

Notice that one of the properties of the figure object is `Number`. This property contains the figure number, which is the value that was called a "handle" on the older pre-Release 2014b graphics system. The number of the root object is always 0, and the number of each figure object is normally a small positive integer, such as 1, 2, 3, ... The numbers associated with all other graphics objects are arbitrary floating-point values.

The handle graphics system includes many functions to get and set properties in objects. These functions are all designed to accept either the actual handle to an object or the `number` property from that handle. This makes the H2 Graphics system backward compatible with older MATLAB programs.

There are MATLAB functions available to get the handles of figures, axes, and other objects. For example, the function `gcf` returns the handle of the currently selected figure, `gca` returns the handle of the currently selected axes within the currently selected figure, and `gco` returns the handle of the currently selected object. These functions will be discussed in more detail later.

By convention, handles are usually stored in variables that begin with the letter h. This practice helps us to recognize handles in MATLAB programs.

13.4 Examining and Changing Object Properties

Object properties describe the data stored in a graphics object when it is instantiated. These properties control aspects of how that object behaves. Each property has a **property name** and an associated value. The property names are strings that are typically displayed in mixed case with the first letter of each word capitalized.

13.4.1 Changing Object Properties at Creation Time

When an object is created, all of its properties are automatically initialized to default values. These default values can be overridden at creation time by including `'PropertyName'`, `value` pairs in the object creation function[2]. For example, we saw in Chapter 3 that the width of a line could be modified in the `plot` command as follows.

```
plot(x,y,'LineWidth',2);
```

---

[2]Examples of object creation functions include `figure`, which creates a new figure, `axes`, which creates a new set of axes within a figure, and `line`, which creates a line within a set of axes. High-level functions such as `plot` are also object creation functions.

This function overrides the default `LineWidth` property with the value 2 at the time that the line object is created.

13.4.2 Changing Object Properties after Creation Time

The `public` properties of any object can be examined or modified at any time using one of three techniques:

1. Directly accessing the properties using standard object syntax, which is the object handle followed by a dot and the property name: `hndl.property`. (This technique only works for the new H2 Graphics system.)
2. Accessing the properties through `get` and `set` functions. (This technique works for both the old and the new graphics systems.)
3. Using the Property Editor.

The first two approaches are almost identical in operation.

13.4.3 Examining and Changing Properties Using Object Notation

Object properties can be examined using the object reference `handle.property`. If the command `"handle.property"` is typed at the command line, the corresponding property will be displayed. If only the object handle is typed in the Command Window, then MATLAB will display *all* the public properties of the object.

Object properties can also be changed using the object reference `handle.property`. The command

```
handle.property = value;
```

will set the property to the specified value if it is a legal selection for that property.

For example, suppose that we plotted the function $y(x) = x^2$ from 0 to 2 with the following statements:

```
x = 0:0.1:2;
y = x.^2;
hndl = plot(x,y);
```

The resulting plot is shown in Figure 13.2a. The handle of the plotted line is stored in `hndl`, and we can use it to examine or modify the properties of the line. Typing `hndl` on the command line will return a list of the object's properties.

```
» hndl
hndl =
  Line with properties:

        Color: [0 0.447000000000000 0.741000000000000]
    LineStyle: '-'
```

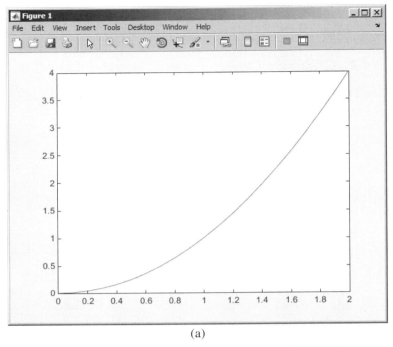

(a)

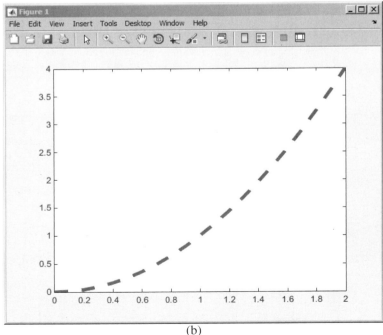

(b)

Figure 13.2 (a) Plot of the function $y = x^2$ using the default linewidth. (b) Plot of the function after modifying the `LineWidth` and `LineStyle` properties.

```
        LineWidth: 0.500000000000000
           Marker: 'none'
       MarkerSize: 6
  MarkerFaceColor: 'none'
            XData: [1x21 double]
            YData: [1x21 double]
            ZData: []
```

Note that the current line width is 0.5 pixels, and the current line style is a solid line. We can change the line width and the line style with the following sets of commands:

```
» hndl.LineWidth = 4;
» hndl.LineStyle = '--';
```

The plot after either command is issued is shown in Figure 13.2*b*.

Note that the property to be examined or set must be capitalized exactly as defined in the class, or it will not be recognized.

13.4.4 Examining and Changing Properties Using get/set Functions

Object properties can also be examined using the get function. The get function will also display the property. This function takes the form

```
value = get(handle,'PropertyName')
value = get(handle)
```

where value is the value contained in the specified property of the object whose handle is supplied. If only the handle is included in the get function call, the function returns a structure array in which the property names and values of *all* of the public properties are shown.

Object properties can be changed using the set function. The set function takes the form

```
set(handle,'PropertyName1',value1,...);
```

where there can be any number of 'PropertyName',value pairs in a single function.

For example, suppose that we plotted the function $y(x) = x^2$ from 0 to 2 with the following statements:

```
x = 0:0.1:2;
y = x.^2;
hndl = plot(x,y);
```

The resulting plot is shown in Figure 13.2*a*. The handle of the plotted line is stored in hndl, and we can use it to examine or modify the properties of the line. Calling the function get(hndl) will return all of the properties of this line in a structure, with each property name being an element of the structure.

```
» result = get(hndl)
result =
        AlignVertexCenters: 'off'
                Annotation: [1x1 matlab.graphics.eventdata.Annotation]
              BeingDeleted: 'off'
                BusyAction: 'queue'
             ButtonDownFcn: ''
                  Children: []
                  Clipping: 'on'
                     Color: [0 0.447000000000000 0.741000000000000]
                 CreateFcn: ''
                 DeleteFcn: ''
               DisplayName: ''
          HandleVisibility: 'on'
                   HitTest: 'on'
             Interruptible: 'on'
                 LineStyle: '-'
                 LineWidth: 0.500000000000000
                    Marker: 'none'
           MarkerEdgeColor: 'auto'
           MarkerFaceColor: 'none'
                MarkerSize: 6
                    Parent: [1x1 Axes]
                  Selected: 'off'
        SelectionHighlight: 'on'
                       Tag: ''
                      Type: 'line'
             UIContextMenu: []
                  UserData: []
                   Visible: 'on'
                     XData: [1x21 double]
                 XDataMode: 'manual'
               XDataSource: ''
                     YData: [1x21 double]
               YDataSource: ''
                     ZData: []
               ZDataSource: ''
```

Note that the current line width is 0.5 pixels, and the current line style is a solid line. We can change the line width and the line style with the following set function:

```
» set(hndl,'LineWidth',4,'LineStyle','--')
```

The plot after either command is issued is shown in Figure 13.2*b*; it is identical regardless of the method used to modify the line's properties.

The get/set functions have three significant advantages over object notation for examining and modifying graphics properties:

1. The get/set functions work with both the old and new graphics systems, so programs written using them will work in older versions of MATLAB.
2. The get/set functions will locate the proper properties and display or modify them even if the capitalization of a property is incorrect. This is not true for object notation. For example, the property 'LineWidth' must be capitalized exactly that way in object notation, but 'lineWidth' or 'linewidth' would also work in a get or set function.
3. When a property has an enumerated list of legal values, the function set(hndl,'property') will return a list of all possible legal values. The object notation will not do this. For example, the legal line styles of a line object are:

```
» set(hndl,'LineStyle')
    '-'
    '--'
    ':'
    '-.'
    'none'
```

13.4.5 Examining and Changing Properties Using the Property Editor

Either the direct access to object properties or the get and set functions can be very useful for programmers, because they can be directly inserted into MATLAB programs to modify a figure, based on a user's input. As we shall see in the next chapter, these functions are used extensively in GUI programming.

For the end user, however, it is often easier to change the properties of a MATLAB object interactively. The Property Editor is a GUI-based tool designed for this purpose. The Property Editor is started by first selecting the Edit button ( ) on the figure toolbar and then clicking on the object that you want to modify with the mouse. Alternately, the property editor can be started from the command line with or without a list of objects to edit:

```
propedit(HandleList);
propedit;
```

For example, the following statements will create a plot containing the line $y = x^2$ over the range 0 to 2, and open the Property Editor to allow the user to interactively change the properties of the line.

```
figure(2);
x = 0:0.1:2;
y = x.^2;
hndl = plot(x,y);
propedit(hndl);
```

The Property Editor invoked by these statements is shown in Figure 13.3. The Property Editor contains a series of panes that vary depending on the type of object being modified.

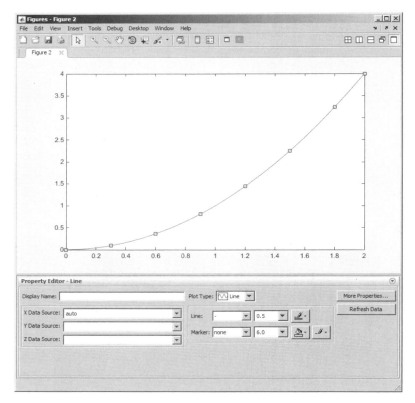

Figure 13.3 The Property Editor when editing a line object. Changes in style are immediately displayed on the figure as the object is edited.

►Example 13.1— Using Low-Level Graphics Commands

The function sinc(x) is defined by the equation

$$\text{sinc } x = \begin{cases} \dfrac{\sin x}{x} & x \neq 0 \\ 1 & x = 0 \end{cases} \tag{13.1}$$

Plot this function from $x = -3\pi$ to $x = 3\pi$. Use handle graphics functions to customize the plot as follows:

1. Make the figure background pink.
2. Use y-axis grid lines only (no x-axis grid lines).
3. Plot the function as a 2-point-wide solid orange line.

Solution To create this graph, we need to plot the function sinc x from $x = -3\pi$ to $x = 3\pi$ using the `plot` function. The plot function will return a handle for the line that we can save and use later.

After plotting the line, we need to modify the color of the *figure* object, the grid status of the *axes* object, and the color and width of the *line* object. These modifications require us to have access to the handles of the `figure`, `axes`, and `line` objects. The handle of the `figure` object is returned by the `gcf` function, the handle of the `axes` object is returned by the `gca` function, and the handle of the `line` object is returned by the `plot` function that created it.

The low-level graphics properties that need to be modified can be found by referring to the on-line MATLAB Help Browser documentation, under the topic Handle Graphics. They are the `'Color'` property of the current figure, the `'YGrid'` property of the current axes, and the `'LineWidth'` and `'Color'` properties of the line.

1. **State the problem**

   Plot the function sinc x from $x = -3\pi$ to $x = 3\pi$ using a figure with a pink background, y-axis grid lines only, and a 2-point-wide solid orange line.

2. **Define the inputs and outputs**

   There are no inputs to this program, and the only output is the specified figure.

3. **Describe the algorithm**

   This program can be broken down into three major steps.

   ```
   Calculate sinc(x)
   Plot sinc(x)
   Modify the required graphics object properties
   ```

   The first major step is to calculate sinc x from $x = -3\pi$ to $x = 3\pi$. This can be done with vectorized statements, but the vectorized statements will produce a NaN at $x = 0$, since the division of 0/0 is undefined. We must replace the NaN with a 1.0 before plotting the function. The detailed pseudocode for this step is:

   ```
   % Calculate sinc(x)
   x = -3*pi:pi/10:3*pi
   y = sin(x) ./ x

   % Find the zero value and fix it up.  The zero is
   % located in the middle of the x array.
   index = fix(length(y)/2) + 1
   y(index) = 1
   ```

   Next, we must plot the function, saving the handle of the resulting line for further modifications. The detailed pseudocode for this step is:

   ```
   hndl = plot(x,y);
   ```

   Now we must use handle graphics commands to modify the figure background, y-axis grid, and line width and color. Remember that the figure handle can be recovered with the function `gcf`, and the axis handle can be recovered with the function `gca`. The color pink can be created with the RGB vector [1 0.8 0.8], and the color orange can be created with the RGB vector [1 0.5 0]. The detailed pseudocode for this step is:

```
set(gcf,'Color',[1 0.8 0.8])
set(gca,'YGrid','on')
set(hndl,'Color',[1 0.5 0],'LineWidth',2)
```

4. **Turn the algorithm into MATLAB statements.**
   The final MATLAB program is shown below.

```
%   Script file: plotsinc.m
%
%   Purpose:
%     This program illustrates the use of handle graphics
%     commands by creating a plot of sinc(x) from -3*pi to
%     3*pi, and modifying the characteristics of the figure,
%     axes, and line using the "set" function.
%
%   Record of revisions:
%      Date          Programmer          Description of change
%      ====          ==========          =====================
%     04/02/14      S. J. Chapman        Original code
%
% Define variables:
%    hndl            -- Handle of line
%    x               -- Independent variable
%    y               -- sinc(x)

% Calculate sinc(x)
x = -3*pi:pi/10:3*pi;
y = sin(x) ./ x;

% Find the zero value and fix it up. The zero is
% located in the middle of the x array.
index = fix(length(y)/2) + 1;
y(index) = 1;


% Plot the function.
hndl = plot(x,y);

% Now modify the figure to create a pink background,
% modify the axis to turn on y-axis grid lines, and
% modify the line to be a 2-point-wide orange line.
set(gcf,'Color',[1 0.8 0.8]);
set(gca,'YGrid','on');
set(hndl,'Color',[1 0.5 0],'LineWidth',2);
```

5. **Test the program.**
   Testing this program is very simple—we just execute it and examine the
   resulting plot. The plot created is shown in Figure 13.4, and it does have the
   characteristics that we wanted.

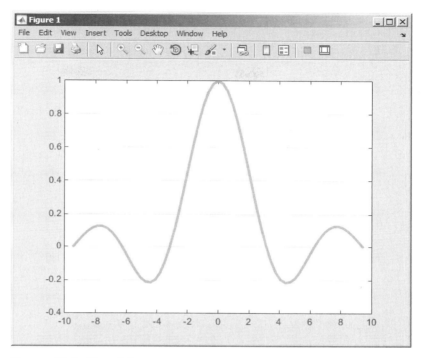

Figure 13.4 Plot of sinc *x* versus *x*.

You will be asked to modify this program to use object property notation in an end-of-chapter exercise.

13.5 Using `set` to List Possible Property Values

The `set` function can be used to provide lists of possible property values. If a `set` function call contains a property name but not a corresponding value, `set` returns a list of all of the legal choices for that property. For example, the command `set(hndl,'LineStyle')` will return a list of all legal line styles:

```
» set(hndl,'LineStyle')
ans =
    '-'
    '--'
    ':'
    '-.'
    'none'
```

This function shows that the legal line styles are `'-'`, `'--'`, `':'`, `'-.'`, and `'none'`, with the first choice as the default.

If the property does not have a fixed set of values, MATLAB returns an empty cell array:

```
» set(hndl,'LineWidth')
ans =
     {}
```

The function set(hndl) will return all of the possible choices for all of the properties of an object.

```
» xxx = set(hndl)
xxx =
                Color: {}
             EraseMode: {4x1 cell}
             LineStyle: {5x1 cell}
             LineWidth: {}
                Marker: {14x1 cell}
            MarkerSize: {}
       MarkerEdgeColor: {2x1 cell}
       MarkerFaceColor: {2x1 cell}
                 XData: {}
                 YData: {}
                 ZData: {}
         ButtonDownFcn: {}
              Children: {}
              Clipping: {2x1 cell}
             CreateFcn: {}
             DeleteFcn: {}
            BusyAction: {2x1 cell}
      HandleVisibility: {3x1 cell}
               HitTest: {2x1 cell}
         Interruptible: {2x1 cell}
              Selected: {2x1 cell}
    SelectionHighlight: {2x1 cell}
                   Tag: {}
         UIContextMenu: {}
              UserData: {}
               Visible: {2x1 cell}
                Parent: {}
           DisplayName: {}
             XDataMode: {2x1 cell}
           XDataSource: {}
           YDataSource: {}
           ZDataSource: {}
```

Any of the items in this list can be expanded to see the available list of options.

```
» xxx.EraseMode
ans =
    'normal'
    'background'
    'xor'
    'none'
```

13.6 User-Defined Data

In addition to the standard properties defined for a GUI object, a programmer can define special properties to hold program-specific data. These extra properties are a convenient way to store any kind of data that the programmer might wish to associate with the GUI object. Any amount of any type of data can be stored and used for any purpose.

User-defined data is stored in a manner similar to standard properties. Each data item has a name and a value. Data values are stored in an object with the `setappdata` function and retrieved from the object using the `getappdata` function.

The general form of `setappdata` is

```
setappdata(hndl,'DataName',DataValue);
```

where `hndl` is the handle of the object to store the data into, `'DataName'` is the name given to the data, and `DataValue` is the value assigned to that name. Note that the data value can be either numeric or a character string.

For example, suppose that we wanted to define two special data values, one containing the number of errors that have occurred on a particular figure, and the other containing a string describing the last error detected. Such data values could be given the names `'ErrorCount'` and `'LastError'`. If we assume that `h1` is the handle of the figure, then command to create these data items and initialize them would be:

```
setappdata(h1,'ErrorCount',0);
setappdata(h1,'LastError','No error');
```

Application data can be retrieved at any time using the function `getappdata`. The two forms of `getappdata` are

```
value = getappdata(hndl,'DataName');
struct = getappdata(hndl);
```

where `hndl` is the handle of the object containing the data, and `'DataName'` is the name of the data to be retrieved. If a `'DataName'` is specified, then the value associated with that data name will be returned. If it is not specified, then *all* user-defined data associated with that object will be returned in a structure. The names of the data items will be structure element names in the returned structure.

For the example given above, `getappdata` will produce the following results:

```
» value = getappdata(h1,'ErrorCount')
value =
    0
```

Table 13.1: Functions for Manipulating User-Defined Data

Function	Description
`setappdata(hndl,'DataName',DataValue)`	Stores `DataValue` in an item named `'DataName'` within the object specified by the handle `hndl`.
`value = getappdata(hndl,'DataName')` `struct = getappdata(hndl)`	Retrieves user-defined data from the object specified by the handle `hndl`. The first form retrieves the value associated with `'DataName'` only, and the second form retrieves all user-defined data.
`isappdata(hndl,'DataName')`	A logical function that returns a 1 if `'DataName'` is defined within the object specified by the handle `hndl`, and 0 otherwise.
`rmappdata(hndl,'DataName')`	Removes the user-defined data item named `'DataName'` from the object specified by the handle `hndl`.

```
» struct = getappdata(h1)
struct =
    ErrorCount: 0
     LastError: 'No error'
```

The functions associated with user-defined data are summarized in Table 13.1.

13.7 Finding Objects

Each new graphics object that is created has its own handle, and that handle is returned by the creating function. If you intend to modify the properties of an object that you create, then it is a good idea to save the handle for later use with get and set.

Good Programming Practice

If you intend to modify the properties of an object that you create, save the handle of that object so that its properties can be examined and modified later.

However, sometimes we might not have access to the handle. Suppose that we lost a handle for some reason. How can we examine and modify the graphics objects? MATLAB provides four special functions to help find the handles of objects.

- gcf—Returns the handle of the current *figure*.
- gca—Returns the handle of the current *axes* in the current *figure*.
- gco—Returns the handle of the current *object*.
- findobj—Finds a graphics object with a specified property value.

The function gcf returns the handle of the current figure. If no figure exists, gcf *will create one* and return its handle. The function gca returns the handle of the current axes within the current figure. If no figure exists or if the current figure exists but contains no axes, gca *will create a set of axes* and return its handle. The function gco has the form

```
h_obj = gco;
h_obj = gco(h_fig);
```

where h_obj is the handle of the object and h_fig is the handle of a figure. The first form of this function returns the handle of the *current object in the current figure*, while the second form of the function returns the handle of the *current object in a specified figure*.

The current object is defined as the last object clicked on with the mouse. This object can be any graphics object except the root. There will not be a current object in a figure until a mouse click has occurred within that figure. Before the first mouse click, function gco will return an empty array []. Unlike gcf and gca, gco does not create an object if it does not exist.

Once the handle of an object is known, we can determine the type of the object by examining its 'Type' property. The 'Type' property will be a character string, such as 'figure', 'line', 'text', and so forth.

```
h_obj = gco;
type = get(h_obj,'Type')
```

The easiest way to find an arbitrary MATLAB object is with the findobj function. The basic form of this function is

```
hndls = findobj('PropertyName1',value1,...)
```

This command starts at the root object, and searches the entire tree for all objects that have the specified values for the specified properties. Note that multiple property/value pairs may be specified, and findobj will only return the handles of objects that match *all* of them.

For example, suppose that we have created Figures 1 and 3. Then the function findobj('Type','figure') will return the results:

```
» h_fig = findobj('Type','figure')
h_fig =
  2x1 Figure array:

  Figure    (1)
  Figure    (3)
```

This form of the findobj function is very useful, but it can be slow since it must search through the entire object tree to locate any matches. If you must use an object multiple times, make only one call to findobj and save the handle for re-use.

Restricting the number of objects that must be searched can increase the execution speed of this function. This can be done with the following form of the function:

```
hndls = findobj(Srchhndls,'PropertyName1',value1,...)
```

Here, only the handles listed in array Srchhndls and their children will be searched to find the object. For example, suppose that you wanted to find all of the dashed lines in Figure 1. The command to do this would be:

```
hndls = findobj(1,'Type','line','LineStyle','--');
```

13.8 Selecting Objects with the Mouse

Function gco returns the handle of the current object, which is the last object clicked on by the mouse. Each object has a **selection region** associated with it, and any mouse click within that selection region is assumed to be a click on that object. This is very important for thin objects like lines or points—the selection region allows the user to be slightly sloppy in mouse position and still select the line. The width of and shape of the selection region varies for different types of objects. For instance, the selection region for a line is 5 pixels on either side of the line, while the selection region for a surface, patch, or text object is the smallest rectangle that can contain the object.

The selection region for an axes object is the area of the axes plus the area of the titles and labels. However, lines or other objects inside the axes have a higher priority, so to select the axes you must click on a point within the axes that is not near lines or text. Clicking on a figure outside of the axes region will select the figure itself.

What happens if a user clicks on a point that has two or more objects, such as the intersection of two lines? The answer depends on the **stacking order** of the objects. The stacking order is the order in which MATLAB selects objects. This order is specified by the order of the handles listed in the 'Children' property of a figure. If a click is in the selection region of two or more objects, the one with the highest position in the 'Children' list will be selected.

MATLAB includes a function called waitforbuttonpress that is sometimes used when selecting graphics objects. The form of this function is:

```
k = waitforbuttonpress
```

When this function is executed, it halts the program until either a key is pressed or a mouse button is clicked. The function returns 0 if it detects a mouse button click or 1 if it detects a key press.

The function can be used to pause a program until a mouse click occurs. After the mouse click occurs, the program can recover the handle of the selected object using the gco function.

►Example 13.2—Selecting Graphics Objects

The program shown below explores the properties of graphics objects and, incidentally, shows how to select objects using `waitforbuttonpress` and `gco`. The program allows objects to be selected repeatedly until a key press occurs.

```
%  Script file: select_object.m
%
%  Purpose:
%    This program illustrates the use of waitforbuttonpress
%    and gco to select graphics objects. It creates a plot
%    of sin(x) and cos(x) and then allows a user to select
%    any object and examine its properties. The program
%    terminates when a key press occurs.
%
%  Record of revisions:
%      Date          Programmer          Description of change
%      ====          ==========          =====================
%    04/02/14      S. J. Chapman        Original code
%
% Define variables:
%    details      -- Object details
%    h1           -- handle of sine line
%    h2           -- handle of cosine line
%    handle       -- handle of current object
%    k            -- Result of waitforbuttonpress
%    type         -- Object type
%    x            -- Independent variable
%    y1           -- sin(x)
%    y2           -- cos(x)
%    yn           -- Yes/No

% Calculate sin(x) and cos(x)
x = -3*pi:pi/10:3*pi;
y1 = sin(x);
y2 = cos(x);

% Plot the functions.
h1 = plot(x,y1);
set(h1,'LineWidth',2);
hold on;
h2 = plot(x,y2);

set(h2,'LineWidth',2,'LineStyle',':','Color','r');
title('\bfPlot of sin \itx \rm\bf and cos \itx');
xlabel('\bf\itx');
ylabel('\bfsin \itx \rm\bf and cos \itx');
legend('sine','cosine');
hold off;
```

```
% Now set up a loop and wait for a mouse click.
k = waitforbuttonpress;

while k == 0
   % Get the handle of the object
   handle = gco;

   % Get the type of this object.
   type = get(handle,'Type');

   % Display object type
   disp (['Object type = ' type '.']);

   % Do we display the details?
   yn = input('Do you want to display details? (y/n) ','s');

   if yn == 'y'
      details = get(handle);
      disp(details);
   end

   % Check for another mouse click
   k = waitforbuttonpress;
end
```

When this program is executed, it produces the plot shown in Figure 13.5. Experiment by clicking on various objects on the plot and seeing the resulting characteristics.

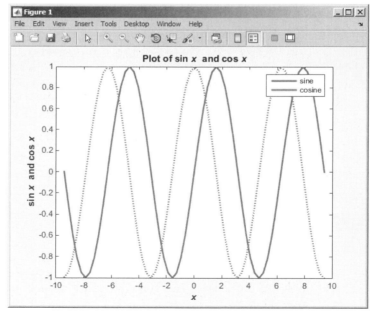

Figure 13.5 Plot of sin *x* and cos *x*.

13.9 Position and Units

Many MATLAB objects have a `'position'` property, which specifies the size and position of the object on the computer screen. This property differs slightly for different kinds of objects, as described below.

13.9.1 Positions of `figure` Objects

The `'position'` property for a figure specifies the location of that figure on the computer screen using a four-element row vector. The values in this vector are [left bottom width height], where left is the leftmost edge of the figure, bottom is the bottom edge of the figure, width is the width of the figure, and height is the height of the figure. These position values are in the units specified in the `'Units'` property for the object. For example, the position and units associated with a the current figure can be found as follows:

```
» get(gcf,'Position')
ans =
   176    204    672    504
» get(gcf,'Units')
ans =
pixels
```

This information specifies that the lower left corner of the current figure window is 176 pixels to the right and 204 pixels above the lower left corner of the screen, and the figure is 672 pixels wide by 504 pixels high. This is the drawable region of the figure, excluding borders, scrollbars, menus, and the figure title area.

The `'units'` property of a figure defaults to pixels, but it can be inches, centimeters, points, characters, or normalized coordinates. Pixels are screen pixels, which are the smallest rectangular shape that can be drawn on a computer screen. Typical computer screens will be at least 640 pixels wide $\times$ 480 pixels high, and screens can have more than 1000 pixels in each direction. Since the number of pixels varies from computer screen to computer screen, the size of an object specified in pixels will also vary.

Normalized coordinates are coordinates in the range 0 to 1, where the lower left corner of the screen is at (0,0) and the upper right corner of the screen is at (1,1). If an object position is specified in normalized coordinates, it will appear in the same relative position on the screen regardless of screen resolution. For example, the following statements create a figure and place it into the upper left quadrant of the screen on any computer, regardless of screen size[3].

```
h1 = figure(1)
set(h1,'units','normalized','position',[0 .5 .5 .45])
```

---

[3]The normalized height of this figure is reduced to 0.45 to allow room for the figure title and menu bar, both of which are above the drawing area.

Good Programming Practice

If you would like to place a window in a specific location, it is easier to place the window at the desired location using normalized coordinates, and the results will be the same regardless of the computer's screen resolution.

13.9.2 Positions of `axes` and `uicontrol` Objects

The position of `axes` and `uicontrol` objects is also specified by a four-element vector, but the object position is specified relative to the lower left-hand corner of the *figure* instead of the position of the screen. In general, the `'Position'` property of a child object is relative to the position of its parent.

By default, the positions of axes objects are specified in *normalized* units within a figure, with (0,0) representing the lower left-hand corner of the figure, and (1,1) representing the upper right-hand corner of the figure.

13.9.3 Positions of `text` Objects

Unlike other objects, `text` objects have a position property containing only two or three elements. These elements correspond to the *x*, *y*, and *z* values of the text object *within* an `axes` object. Note that these values are in the units being displayed on the axes themselves.

The position of the text object with respect to the specified point is controlled by the object's `HorizontalAlignment` and `VerticalAlignment` properties. The `HorizontalAlignment` can be {Left}, Center, or Right; and the `VerticalAlignment` can be Top, Cap, {Middle}, Baseline, or Bottom.

The size of `text` objects is determined by the font size and the number of characters being displayed, so there are no height and width values associated with them.

►Example 13.3—Positioning Objects within a Figure

As we mentioned earlier, axes positions are defined relative to the lower left-hand corner of the frame that they are contained in, while text object positions are defined within axes in the data units being displayed on the axes.

To illustrate the positioning of graphics objects within a figure, we will write a program that creates two overlapping sets of axes within a single figure. The first set of axes will display sin *x* versus *x* and will have a text comment attached to the display line. The second set of axes will display cos x versus x and will have a text comment in the lower left-hand corner.

A program to create the figure is shown below. Note that we are using the `figure` function to create an empty figure, and then we will use two `axes` functions to create the two sets of axes within the figure. The position of the `axes` functions is specified in normalized units within the figure, so the first set of axes, which starts at (0.05,0.05), is in the lower left-hand corner of the figure, and the second set of axes, which starts at (0.45,0.45), is in the upper right-hand corner of the figure. Each set of axes has the appropriate function plotted on it.

The first `text` object is attached to the first set of axes at position $(-\pi, 0)$, which is a point on the curve. The `'HorizontalAlignment'`, `'right'` property is selected, so the *attachment point* $(-\pi, 0)$ is on the *right hand side* of the text string. As a result, the text appears to the *left* of the attachment point in the final figure. (This can be confusing for new programmers!)

The second `text` object is attached to the second set of axes at position $(-7.5, -0.9)$, which is near the lower left-hand corner of the axes. This string uses the default horizontal alignment, which is `'left'`, so the attachment point $(-7.5, -0.9)$ is on the *left-hand side* of the text string. As a result, the text appears to the right of the attachment point in the final figure.

```
%   Script file: position_object.m
%
%   Purpose:
%     This program illustrates the positioning of graphics
%     graphics objects.  It creates a figure, and then places
%     two overlapping sets of axes on the figure.  The first
%     set of axes is placed in the lower left-hand corner of
%     the figure, and contains a plot of sin(x).  The second
%     set of axes is placed in the upper right hand corner of
%     the figure, and contains a plot of cos(x).  Then two
%     text strings are added to the axes, illustrating the
%     positioning of text within axes.
%
%   Record of revisions:
%       Date          Programmer         Description of change
%       ====          ==========         =====================
%     04/03/14        S. J. Chapman      Original code
%
% Define variables:
%   h1            -- Handle of sine line
%   h2            -- Handle of cosine line
%   ha1           -- Handle of first axes
%   ha2           -- Handle of second axes
```

```
%     x                -- Independent variable
%     y1               -- sin(x)
%     y2               -- cos(x)

% Calculate sin(x) and cos(x)
x = -2*pi:pi/10:2*pi;
y1 = sin(x);
y2 = cos(x);

% Create a new figure
figure;

% Create the first set of axes and plot sin(x).
% Note that the position of the axes is expressed
% in normalized units.
ha1 = axes('Position',[.05 .05 .5 .5]);
h1 = plot(x,y1);
set(h1,'LineWidth',2);
title('\bfPlot of sin \itx');
xlabel('\bf\itx');
ylabel('\bfsin \itx');
axis([-8 8 -1 1]);

% Create the second set of axes and plot cos(x).
% Note that the position of the axes is expressed
% in normalized units.
ha2 = axes('Position',[.45 .45 .5 .5]);
h2 = plot(x,y2);
set(h2,'LineWidth',2,'Color','r','LineStyle','--');
title('\bfPlot of cos \itx');
xlabel('\bf\itx');
ylabel('\bfsin \itx');
axis([-8 8 -1 1]);

% Create a text string attached to the line on the first
% set of axes.
axes(ha1);
text(-pi,0.0,'sin(x)\rightarrow','HorizontalAlignment','right');

% Create a text string in the lower left-hand corner
% of the second set of axes.
axes(ha2);
text(-7.5,-0.9,'Test string 2');
```

When this program is executed, it produces the plot shown in Figure 13.6. You should execute this program again on your computer, changing the size and/or location of the objects being plotted, and observing the results.

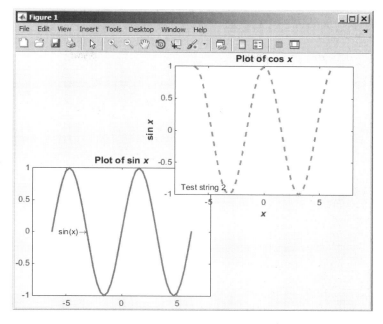

Figure 13.6 The output of program `position_object`.

13.10 Printer Positions

The `'Position'` and `'Units'` properties specify the location of a figure on the *computer screen*. There are also five other properties that specify the location of a figure on a sheet of paper *when it is printed*. These properties are summarized in Table 13.2 below.

Table 13.2: **Printing-related Figure Properties**

Option	Description
PaperUnits	Units for paper measurements: [{inches} | centimeters | normalized | points]
PaperOrientation	[{portrait} | landscape]
PaperPosition	A position vector of the form [left, bottom, width, height] where all units are as specified in PaperUnits.
PaperSize	A two-element vector containing the power size, for example [8.5 11]
PaperType	Sets paper type. Note that setting this property automatically updates the PaperSize property. [{usletter} | uslegal | A0 | A1 | A2 | A3 | A4 | A5 | B0 | B1 | B2 | B3 | B4 | B5 | arch-A | arch-B | arch-C | arch-D | arch-E | A | B | C | D | E | tabloid | <custom>]

For example, to set a plot to print out in landscape mode, on A4 paper, in normalized units, we could set the following properties:

```
set(hndl,'PaperType','A4')
set(hndl,'PaperOrientation','landscape')
set(hndl,'PaperUnits','normalized');
```

13.11 Default and Factory Properties

MATLAB assigns default properties to each object when it is created. If those properties are not what you want, then you must use `set` to select the desired values. If you needed to change a property in every object that you create, this process could become very tedious. For those cases, MATLAB allows you to modify the default property itself, so that all objects will inherit the correct value of the property when they are created.

When a graphics object is created, MATLAB looks for a default value for each property by examining the object's parent. If the parent sets a default value, that value is used. If not, MATLAB examines the parent's parent to see if that object sets a default value, and so on, back to the root object. MATLAB uses the *first* default value that it encounters when working back up the tree.

Default properties may be set at any point in the graphics object hierarchy that is *higher* than the level at which the object is created. For example, a default `figure` color would be set in the `root` object, and then all figures created after that time would have the new default color. On the other hand, a default `axes` color could be set in either the `root` object or the `figure` object. If the default `axes` color is set in the `root` object, it will apply to all new axes in all figures. If the default `axes` color is set in the `figure` object, it will apply to all new axes in the current figure only.

Default values are set using a string consisting of `'default'` followed by the object type and the property name. Thus the default figure color would be set with the property `'defaultFigureColor'` and the default axes color would be set with the property `'defaultAxesColor'`. Some examples of setting default values are shown below:

`set(groot,'defaultFigureColor','y')`	Yellow figure background—all new figures
`set(groot,'defaultAxesColor','r')`	Red axes background—all new axes in all figures
`set(gcf,'defaultAxesColor','r')`	Red axes background—all new axes in current figure only
`set(gca,'defaultLineLineStyle',':')`	Set default line style to dashed, in current axes only.

If you are working with existing objects, it is always a good idea to restore them to their existing condition after they are used. *If you change the default properties of an object in a function, save the original values and restore them before exiting the function.* For example, suppose that we wish to create a series of figures in normalized units. We could save and restore the original units as follows:

```
saveunits = get(groot,'defaultFigureUnits');
set(groot,'defaultFigureUnits','normalized');
...
<MATLAB statements>
...
set(groot,'defaultFigureUnits',saveunits);
```

If you want to customize MATLAB to use different default values at all times, then you should set the defaults in the root object every time that MATLAB starts up. The easiest way to do this is to place the default values into the startup.m file, which is automatically executed every time MATLAB starts. For example, suppose you always use A4 paper and you always want a grid displayed on your plots. Then you could set the following lines into startup.m:

```
set(groot,'defaultFigurePaperType','A4');
set(groot,'defaultFigurePaperUnits','centimeters');
set(groot,'defaultAxesXGrid','on');
set(groot,'defaultAxesYGrid','on');
set(groot,'defaultAxesZGrid','on');
```

There are three special value strings that are used with handle graphics: 'remove', 'factory', and 'default'. If you have set a default value for a property, the 'remove' value will remove the default that you set. For example, suppose that you set the default figure color to yellow:

```
set(groot,'defaultFigureColor','y');
```

The following function call will cancel this default setting and restore the previous default setting.

```
set(groot,'defaultFigureColor','remove');
```

The string 'factory' allows a user to temporarily override a default value and use the original MATLAB default value instead. For example, the following figure is created with the factory default color despite a default color of yellow being previously defined.

```
set(groot,'defaultFigureColor','y');
figure('Color','factory')
```

The string 'default' forces MATLAB to search up the object hierarchy until it finds a default value for the desired property. It uses the first default value that it finds. If it fails to find a default value, it uses the factory default value for that property. This use is illustrated below:

```
% Set default values
set(groot,'defaultLineColor','k');   % root default = black
set(gcf,'defaultLineColor','g'); % figure default = green

% Create a line on the current axes. This line is green.
hndl = plot(randn(1,10));
set(hndl,'Color','default');
pause(2);
```

```
% Now clear the figure's default and set the line color to the new
% default. The line is now black.
set(gcf,'defaultLineColor','remove');
set(hndl,'Color','default');
```

13.12 Graphics Object Properties

There are hundreds of different graphics object properties—far too many to discuss in detail here. The best place to find a complete list of graphics object properties is in the Help Browser distributed with MATLAB.

We have mentioned a few of the most important properties for each type of graphics object as we have needed them (`'LineStyle'`, `'Color'`, and so forth). A complete set of properties is given in the MATLAB Help Browser documentation under the descriptions of each type of object.

13.13 Animations and Movies

Handle graphics can be used to create animations in MATLAB. There are two possible approaches to this task:

1. Erasing and redrawing
2. Creating a movie

In the first case, the user draws a figure, and then updates the data in the figure regularly using handle graphics. Each time that the data is updated, the program will redraw the object with the new data, producing an animation. In the second case, the user draws a figure, captures a copy of the figure as a frame in a movie, redraws the figure, captures the new figure as the next frame in the movie, and so forth until the entire movie has been created.

13.13.1 Erasing and Redrawing

To create an animation by erasing and redrawing, the user first creates a plot, and then changes the data displayed in the plot by updating the line objects and so forth, using handle graphics. To see how this works, consider the function

$$f(x,t) = A(t) \sin x \tag{13.2}$$

where

$$A(t) = \cos t \tag{13.3}$$

For any given time t, this function will be the plot of a sine wave. However the amplitude of the sine wave will vary with time, so the plot will look different at different times.

The key to creating an animation is to save the handle associated with the line plotting the sine wave and then to update the `'YData'` property of that handle at each time step with the new y-axis data. Note that we won't have to change the x data because the x limits of the plot will be the same at any time.

An example program creating the sine wave that varies with time is shown below. In this program, we create the sine wave plot at time $t = 0$ and capture a handle hndl to the line object when it is created. Then the plot data is recalculated in a loop at each time step, and the line is updated using handle graphics.

Note the drawnow command in the update loop. This command causes the graphics to be rendered at the moment it is executed, which ensures that the display is updated each time new data is loaded into the line object.

Also, note that we have set the y-axis limits to be -1 to 1 using the handle graphics command set(gca,'YLim',[-1 1]). If the y-axis limits are not set, the scale of the plot will change with each update and the user will not be able to tell that the sine wave is getting larger and smaller.

Finally, note that there is a pause(0.1) command commented out in the program. If executed, this command would pause for 0.1 second after each update of the drawing. The pause command can be used in a program if the updates are occurring too fast when it executes (because a particular computer is very fast), and adjusting the delay time will allow the user to adjust the update rate.

```
%   Script file: animate_sine.m
%
%   Purpose:
%     This program illustrates the animation of a plot
%     by updating the data in the plot with time.
%
%   Record of revisions:
%       Date            Programmer          Description of change
%       ====            ==========          =====================
%     05/02/14        S. J. Chapman         Original code
%
% Define variables:
%   h1            -- Handle of line
%   a             -- Amplitude of sine function at an instant
%   x             -- Independent variable
%   y             -- a * cos(t) * sin(x)

% Calculate the times at which to plot the sine function
t = 0:0.1:10;

% Calculate sin(x) for the first time
a = cos(t(1));
x = -3*pi:pi/10:3*pi;
y = a * sin(x);

% Plot the function.
figure(1);
hndl = plot(x,y);
xlabel('\bfx');
ylabel('\bfAmp');
title(['\bfSine Wave Animation at t = ' num2str(t(1),'%5.2f')]);
```

```
% Set the size of the y axes
set(gca,'YLim',[-1 1]);

% Now do the animation
for ii = 2:length(t)

   % Pause for a moment
   drawnow;
   %pause(0.1);

   % Calculate sin(x) for the new time
   a = cos(t(ii));
   y = a * sin(x);

   % Update the line
   set(hndl, 'YData', y);

   % Update the title
title(['\bfSine Wave Animation at t = ' num2str(t(ii),'%5.2f')]);

end
```

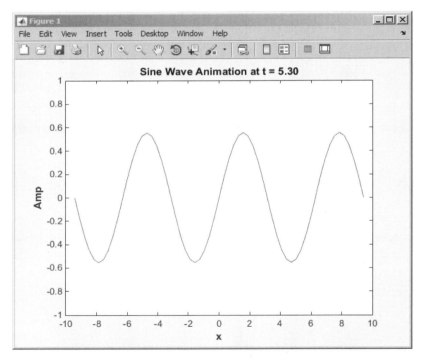

Figure 13.7 One snapshot from the sine wave animation.

When this program executes, the amplitude of the sine wave rises and falls. One snapshot from the animation is shown in Figure 13.7.

It is also possible to do animations of three-dimensional plots, as shown in the next example.

▶▶Example 13.4—Animating a Three-Dimensional Plot

Create a three-dimensional animation of the function

$$f(x,y,t) = A(t) \sin x \sin y \tag{13.4}$$

where

$$A(t) = \cos t \tag{13.5}$$

for time $t = 0$ s to $t = 10$ s in steps of 0.1 s.

Solution For any given time t, this function will be the plot of a two-dimensional sine wave varying in both x and y. However the amplitude of the sine wave will vary with time, so the plot will look different at different times.

This program will be similar to the variable sine wave example above, except that the plot itself will be a 3D surface plot, and the z data needs to be updated at each time step instead of the y data. The original 3D `surf` plot is created by using `meshgrid` to create the arrays of x and y values, evaluating Equation (13.4) at all the points on the grid and plotting the `surf` function. After that, Equation (13.4) is reevaluated at each time step, and the `'ZData'` property of the surf object is updated using handle graphics.

```
%   Script file: animate_sine_xy.m
%
%   Purpose:
%     This program illustrates the animation of a 3D plot
%     by updating the data in the plot with time.
%
%   Record of revisions:
%       Date            Programmer          Description of change
%       ====            ==========          =====================
%     06/02/14        S. J. Chapman        Original code
%
% Define variables:
%   h1              -- Handle of line
%   a               -- Amplitude of sine function at an instant
%   array1          -- Meshgrid output for x values
```

```
%    array2         -- Meshgrid output for y values
%    x              -- Independent variable
%    y              -- Independent variable
%    z              -- cos(t) * sin(x) * sin(y)
% Calculate the times at which to plot the sine function
t = 0:0.1:10;

% Calculate sin(x)*sin(y) for the first time
a = cos(t(1));
[array1,array2] = meshgrid(-3*pi:pi/10:3*pi,-3*pi:pi/10:3*pi);
z = a .* sin(array1) .* sin(array2);

% Plot the function.
figure(1);
hndl = surf(array1,array2,z);
xlabel('\bfx');
ylabel('\bfy');
zlabel('\bfAmp');
title(['\bfSine Wave Animation at t = ' num2str(t(1),'%5.2f')]);

% Set the size of the z axes
set(gca,'ZLim',[-1 1]);

% Now do the animation
for ii = 2:length(t)

   % Pause for a moment
   drawnow;
   %pause(0.1);

   % Calculate sine(x) for the new time
   a = cos(t(ii));
   z = a .* sin(array1) .* sin(array2);

   % Update the line
   set(hndl, 'ZData', z);

   % Update the title
   title(['\bfSine Wave Animation at t = ' num2str(t(ii),'%5.2f')]);
end
```

When this program executes, the amplitude of the 2D sine waves on the surface rises and falls with time. One snapshot from the animation is shown in Figure 13.8.

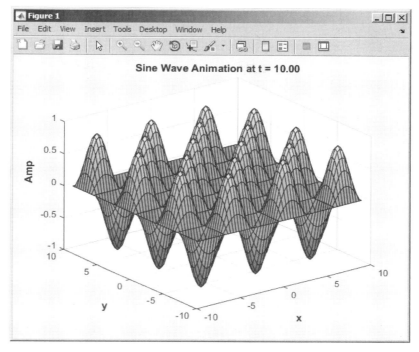

Figure 13.8 One snapshot from the 3D sine wave animation. [See color insert.]

13.13.2 Creating a Movie

The second approach to animations is to create a MATLAB movie. A MATLAB movie is a set of images of a figure that have been captured in a movie object, which can be saved to disk and played back at some future time without actually having to redo all of the calculations that created the plots in the first place. Because the calculations do not have to be performed again, the movie can sometimes run faster and with less jerkiness than the original program that did the calculations and plots[4].

A movie is stored in a MATLAB structure array, with each frame of the movie being one element of the structure array. Each frame of a movie is captured using a special function called getframe after the data in the plot has been updated, and it is played back using the movie command.

---

[4]Sometimes the erase and redraw method is faster than the movie—it depends on how much calculation is required to create the data to be displayed.

A version of the 2D sine plotting program that creates a MATLAB movie is shown below. The statements that create and play back the movie are highlighted in bold face.

```
%   Script file: animate_sine_xy_movie.m
%
%   Purpose:
%     This program illustrates the animation of a 3D plot
%     by creating and playing back a movie.
%
%   Record of revisions:
%       Date          Programmer        Description of change
%       ====          ==========        =====================
%     06/02/14      S. J. Chapman       Original code
%
% Define variables:
%   h1              -- Handle of line
%   a               -- Amplitude of sine function at an instant
%   array1          -- Meshgrid output for x values
%   array2          -- Meshgrid output for y values
%   m               -- Index of movie frames
%   movie           -- The movie
%   x               -- Independent variable
%   y               -- Independent variable
%   z               -- cos(t) * sin(x) * sin(y)

% Clear out any old data
clear all;

% Calculate the times at which to plot the sine function
t = 0:0.1:10;

% Calculate sin(x)*sin(y) for the first time
a = cos(t(1));
[array1,array2] = meshgrid(-3*pi:pi/10:3*pi,-3*pi:pi/10:3*pi);
z = a .* sin(array1) .* sin(array2);

% Plot the function.
figure(1);
hndl = surf(array1,array2,z);
xlabel('\bfx');
ylabel('\bfy');
zlabel('\bfAmp');
title(['\bfSine Wave Animation at t = ' num2str(t(1),'%5.2f')]);

% Set the size of the z axes
set(gca,'ZLim',[-1 1]);

% Capture the first frame of the movie
m = 1
M(m) = getframe;
```

```
% Now do the animation
for ii = 2:length(t)

    % Pause for a moment
    drawnow;
    %pause(0.1);

    % Calculate sine(x) for the new time
    a = cos(t(ii));
    z = a .* sin(array1) .* sin(array2);

    % Update the line
    set(hndl, 'ZData', z);

    % Update the title
    title(['\bfSine Wave Animation at t = ' num2str(t(ii),'%5.2f')]);
    % Capture the next frame of the movie
    m = m + 1;
    M(m) = getframe;
end

% Now we have the movie, so play it back twice
movie(M,2);
```

When this program is executed, you will see the scene played three times. The first time is while the movie is being created, and the next two times are while it is being played back.

13.14 Summary

Every element of a MATLAB plot is a graphics object. Each object is identified by a unique handle, and each object has many properties associated with it, which affect the way the object is displayed.

MATLAB objects are arranged in a hierarchy with **parent objects** and **child objects**. When a child object is created, it inherits many of its properties from its parent.

The highest-level graphics object in MATLAB is the root, which can be thought of as the entire computer screen. This object is accessed using function groot. Under the root there can be one or more Figure Windows. Each figure is a separate window on the computer screen that can display graphical data, and each figure has its own properties.

Each figure can contain seven types of objects: uimenus, uicontextmenus, uicontrols, uitoolbars, uipanels, uitables, uibuttongroups, and axes. Uimenus, uicontextmenus, uicontrols, uitoolbars, uipanels, and uibuttongroups are special graphics objects used to create graphical user interfaces—they will be described in the next chapter. Axes are regions within a figure where data is actually plotted. There can be more than one set of axes in a single figure.

Each set of axes can contain as many lines, text strings, patches, and so forth as necessary to create the plot of interest.

Public graphics object properties can be accessed and changed, using either the object syntax (object.property) or get and set methods. The object syntax only works for MATLAB Release R2014b and later. The `get` and `set` methods work for earlier versions of MATLAB, as well.

The handles of the current figure, current axes, and current object may be recovered with the `gcf`, `gca`, and `gco` functions, respectively. The properties of any object may be examined and modified using the `get` and `set` functions.

There are literally hundreds of properties associated with MATLAB graphics functions, and the best place to find the details of these functions is the MATLAB online documentation.

MATLAB animations can be created by erasing and redrawing objects using handle graphics to update the contents of the objects, or else by creating movies.

13.14.1 Summary of Good Programming Practice

The following guidelines should be adhered to when working with MATLAB handle graphics.

1. If you intend to modify the properties of an object that you create, save the handle of that object so that its properties can be examined and modified later.
2. If possible, restrict the scope of your searches with `findobj` to make them faster.
3. If you would like to place a window in a specific location, it is easier to place the window at the desired location using normalized coordinates, and the results will be the same regardless of the computer's screen resolution.

13.14.2 MATLAB Summary

The following summary lists all of the MATLAB commands and functions described in this chapter, along with a brief description of each one.

`axes`	Creates a new axes/makes axes current.
`figure`	Creates a new figure/makes figure current.
`findobj`	Finds an object based on one or more property values.
`gca`	Gets handle of current axes.
`gcf`	Gets handle of current figure.
`gco`	Gets handle of current object.
`get`	Gets object properties.
`getappdata`	Gets user-defined data in an object.
`groot`	Returns a handle to the `root` object.
`isappdata`	Tests to see if an object contains user-defined data with the specified name.
`rmappdata`	Removes user-defined data from an object.
`set`	Sets object properties.
`setappdata`	Stores user-defined data in an object.
`waitforbuttonpress`	Pauses program, waiting for a mouse click or keyboard input.

13.15 Exercises

13.1 What is meant by the term "handle graphics"?

13.2 Use the MATLAB Help System to learn about the `Name` and `NumberTitle` properties of a `figure` object. Create a figure containing a plot of the function $y(x) = e^x$ for $-2 \leq x \leq 2$. Change the properties mentioned above to suppress the figure number and to add the title "Plot Window" to the figure.

13.3 Write a program that modifies the default figure color to orange and the default line width to 3.0 points. Then create a figure plotting the ellipse defined by the equations

$$x(t) = 10 \cos t$$
$$y(t) = 6 \ \sin t \tag{13.6}$$

from $t = 0$ to $t = 2\pi$. What color and width was the resulting line?

13.4 Use the MATLAB Help System to learn about the `CurrentPoint` property of an `axes` object. Use this property to create a program that creates an `axes` object and that plots a line connecting the locations of successive mouse clicks within the axes. Use the function `waitforbuttonpress` to wait for mouse clicks and update the plot after each click. Terminate the plot when a keyboard press occurs.

13.5 Modify the program created in Example 13.1 to specify properties using MATLAB object syntax instead of `get/set` functions.

13.6 Use the MATLAB Help System to learn about the `CurrentCharacter` property of a `figure` object. Modify the program created in Exercise 13.4 by testing the `CurrentCharacter` property when a keyboard press occurs. If the character typed on the keyboard is a "c" or "C," change the color of the line being displayed. If the character typed on the keyboard is an "s" or "S," change the line style of the line being displayed. If the character typed on the keyboard is a "w" or "W," change the width of the line being displayed. If the character typed on the keyboard is an "x" or "X," terminate the plot. (Ignore all other input characters.)

13.7 Create a MATLAB program that plots the functions

$$x(t) = \cos \frac{t}{\pi}$$
$$x(t) = 2 \ \sin \frac{t}{2\pi} \tag{13.7}$$

for the range $-2 \leq t \leq 2$. The program should then wait for mouse clicks, and if the mouse has clicked on one of the two lines, the program should change the line's color randomly from a choice of red, green, blue, yellow, cyan, magenta, or black. Use the function `waitforbuttonpress` to wait for mouse clicks, and update the plot after each click. Use the function `gco` to determine the object clicked on, and use the `Type` property of the object to determine if the click was on a line.

13.8 The `plot` function plots a line and returns a handle to that line. This handle can be used to get or set the line's properties after it has been created. Two of a line's properties are `XData` and `YData`, which contain the x- and y-values currently plotted. Write a program that plots the function

$$x(t) = \cos(2\pi t - \theta) \tag{13.8}$$

between the limits $-1.0 \leq t \leq 1.0$, and saves the handle of the resulting line. The angle θ is initially 0 radians. Then, re-plot line over and over with $\theta = \pi/10$ rad, $\theta = 2\pi/10$ rad, $\theta = 3\pi/10$ rad, and so forth up to $\theta = 2\pi$ rad. To re-plot the line, use a `for` loop to calculate the new values of x and t, and update the line's `XData` and `YData` properties using MATLAB object syntax. Pause 0.5 seconds between each update, using MATLAB's `pause` command.

13.9 Create a data set in some other program on your computer, such as Microsoft Word, Microsoft Excel, a text editor, and so on. Copy the data set to the clipboard using the Windows or Unix copy function, and then use function `uiimport` to load the data set into MATLAB.

13.10 **Wave Patterns** In the open ocean under circumstances where the wind is blowing steadily in the direction of wave motion, successive wavefronts tend to be parallel. The height of the water at any point might be represented by the equation

$$h(x,y,t) = A \cos\left(\frac{2\pi}{T}t - \frac{2\pi}{L}x\right) \tag{13.9}$$

where T is the period of the waves in seconds, L is the spacing between wave peaks, and t is current time. Assume that the wave period is 4 s and the spacing between wave peaks is 12 m. Create an animation of this wave pattern for a region of -300 m $\leq x \leq 300$ m and -300 m $\leq y \leq 300$ m over a time of $0 \leq t \leq 20$ s using erase and redraw.

13.11 **Wave Patterns** Create a movie of the wave patterns from Exercise 13.10, and replay the movie.

13.12 **Generating a Rotating Magnetic Field** The fundamental principle of AC electric machine operation is that *if a three-phase set of currents, each of equal magnitude and differing in phase by 120°, flows in a three-phase winding, then it will produce a rotating magnetic field of constant magnitude.*" The three-phase winding consists of three separate windings spaced 120° degrees apart around the surface of the machine. Figure 13.9 shows three windings $a\text{-}a'$, $b\text{-}b'$, and $c\text{-}c'$ in a stator, with a magnetic field **B** coming out of each set of windings. The magnitude and direction of the magnetic flux density out of each set of windings is

$$\mathbf{B}_{aa'}(t) = B_M \sin \omega t \angle 0° \text{ T}$$

$$\mathbf{B}_{bb'}(t) = B_M \sin(\omega t - 120°) \angle 120° \text{ T} \tag{13.10}$$

$$\mathbf{B}_{cc'}(t) = B_M \sin(\omega t - 240°) \angle 240° \text{ T}$$

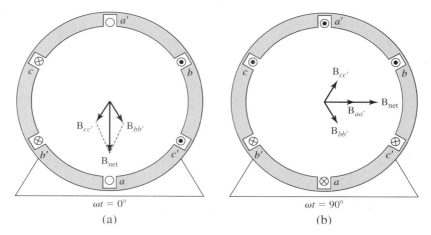

Figure 13.9 Snapshot of the total magnetic field inside a three-phase ac motor at (a) time $\omega t = 0°$; (b) $\omega t = 90°$.

The magnetic field from winding a-a' is oriented to the right (at $0°$). The magnetic field from winding b-b' is oriented at an angle of $120°$, and the magnetic field from winding b-b' is oriented at an angle of $240°$.

The total magnetic field at any time is

$$\mathbf{B}_{net}(t) = \mathbf{B}_{aa'}(t) + \mathbf{B}_{bb'}(t) + \mathbf{B}_{cc'}(t) \tag{13.11}$$

At time $\omega t = 0°$, the magnetic fields add together as shown in Figure 13.9a so that the net field is down. At time $\omega t = 90°$, the magnetic fields add together as shown in Figure 13.9b so that the net field is to the right. Note that the net field has the same amplitude but is rotated at a different angle.

Write a program that creates an animation of this rotating magnetic field, showing that the net magnetic field is constant in amplitude but rotating in angle with time.

13.13 Saddle Surface A saddle surface is a surface that curves upward in one dimension and downward in the orthogonal dimension so that it looks like a saddle. The following equation defines a saddle surface

$$z = x^2 - y^2 \tag{13.12}$$

Plot this function and demonstrate that it has a saddle shape.

Chapter **14**

Graphical User Interfaces

A Graphical User Interface (GUI) is a pictorial interface to a program. A good GUI can make programs easier to use by providing them with a consistent appearance and with intuitive controls like pushbuttons, edit boxes, list boxes, sliders, menus, and so forth. The GUI should behave in an understandable and predictable manner, so that a user knows what to expect when he or she performs an action. For example, when a mouse click occurs on a pushbutton, the GUI should initiate the action described on the label of the button.

This chapter contains an introduction to the basic elements of the MATLAB GUIs. It does not contain a complete description of components or GUI features, but it does provide us with the basics required to create functional GUIs for your programs.

14.1 How a Graphical User Interface Works

A graphical user interface provides the user with a familiar environment in which to work. It contains pushbuttons, toggle buttons, lists, menus, text boxes, and so forth, all of which are already familiar to the user, so that he or she can concentrate on the purpose of the application instead of the mechanics involved in doing things. However, GUIs are harder for the programmer, because a GUI-based program must be prepared for mouse clicks (or possibly keyboard input) for any GUI element at any time. Such inputs are known as **events**, and a program that responds to events is said to be *event driven*.

The three principal elements required to create a MATLAB Graphical User Interface are:

1. **Components**. Each item on a MATLAB GUI (pushbuttons, labels, edit boxes, and so forth) is a graphical component. The types of components include graphical **controls** (pushbuttons, toggle buttons, edit boxes, lists, sliders, and so forth) static elements (text boxes), **menus, toolbars**, and **axes**. Graphical controls and text boxes are created by the function `uicontrol`, and menus are created by the functions `uimenu`, and `uicontextmenu`. Toolbars are created by function `uitoolbar`. Tables are created by function `uitable`. Axes, which are used to display graphical data, are created by the function `axes`.

2. **Containers**. The components of a GUI must be arranged within a **container**, which is a window on the computer screen. The most common container is a **figure**. A figure is an window on the computer screen that has a title bar along the top, and that can optionally have menus attached. In the past, figures have been created automatically whenever we plotted data. However, empty figures can be created with the function `figure`, and they can be used to hold any combination of components and other containers.

   The other types of containers are **panels** (created by the function `uipanel`) and **button groups** (created by the function `uibuttongroup`). Panels can contain components or other containers, but they do not have a title bar and cannot have menus attached. Button groups are special panels that can manage groups of radio buttons or toggle buttons to ensure that no more than one button in the group is on at any time.

3. **Callbacks**. Finally, there must be some way to perform an action if a user clicks a mouse on a button or types information on a keyboard. A mouse click or a key press is an **event**, and the MATLAB program must respond to each event if the program is to perform its function. (We discussed events in Chapter 12.) For example, if a user clicks on a button, then that event must cause the MATLAB code that implements the function of the button to be executed. The code executed in response to an event is known as a **callback**. There must be a callback to implement the function of each graphical component on the GUI.

The basic GUI elements are summarized in Table 14.1, and some sample elements are shown in Figure 14.1. We will be studying examples of these elements, and then build working GUIs from them.

Table 14.1: Some Basic GUI Components

Component	Created By	Description
		Containers
Figure	`figure`	Creates a figure, which is container that can hold components and other containers. Figures are separate windows that have title bars and can have menus.
Panel	`uipanel`	Creates a panel, which is container that can hold components and other containers. Unlike figures, panels do not have title bars or menus. Panels can be placed inside figures or other panels.

(*continued*)

Table 14.1: Some Basic GUI Components (*Continued*)

Component	Created By	Description
Button Group	`uibuttongroup`	Creates a button group, which is a special kind of panel. Button groups automatically manage groups of radio buttons or toggle buttons to ensure that only one item of the group is on at any given time.
		Graphical Controls
Pushbutton	`uicontrol`	A graphical component that implements a pushbutton. It triggers a callback when clicked with a mouse.
Toggle Button	`uicontrol`	A graphical component that implements a toggle button. A toggle button is either "on" or "off", and it changes state each time that it is clicked. Each mouse button click also triggers a callback.
Radio Button	`uicontrol`	A radio button is a type of toggle button that appears as a small circle with a dot in the middle when it is "on". Groups of radio buttons are used to implement mutually exclusive choices. Each mouse click on a radio button triggers a callback.
Check Box	`uicontrol`	A checkbox is a type of toggle button that appears as a small square with a check mark in it when it is "on". Each mouse click on a check box triggers a callback.
Edit Box	`uicontrol`	An edit box displays a text string, and allows the user to modify the information displayed. A callback is triggered when the user presses the `Enter` key, or when the user clicks on a different object with the mouse.
List Box	`uicontrol`	A list box is a graphical control that displays a series of text strings. A user may select one of the text strings by single- or double-clicking on them. A callback is triggered when the user selects a string.
Popup Menus	`uicontrol`	A popup menu is a graphical control that displays a series of text strings in response to a mouse click. When the popup menu is not clicked on, only the currently selected string is visible.
Slider	`uicontrol`	A slider is a graphical control to adjust a value in a smooth, continuous fashion by dragging the control with a mouse. Each slider change triggers a callback.
Table	`uitable`	Creates a table of data.
		Static Elements
Text Field	`uicontrol`	Creates a label, which is a text string located at a point on the figure. Text fields never trigger callbacks.
		Menus, Toolbars, Axes
Menu Items	`uimenu`	Creates a menu item. Menu items trigger a callback when a mouse button is released over them.
Context Menus	`uicontextmenu`	Creates a context menu, which is a menu that appears over a graphical object when a user right-clicks the mouse on that object.
Toolbar	`uitoolbar`	Creates a toolbar, which is a bar across the top of the figure containing quick-access buttons.
Toolbar Pushbutton	`uipushtool`	Creates a pushbutton to go in a toolbar.
Toolbar Toggle Button	`uitoggletool`	Creates a toggle button to go in a toolbar.
Axes	`axes`	Creates a new set of axes to display data on. Axes never trigger callbacks.

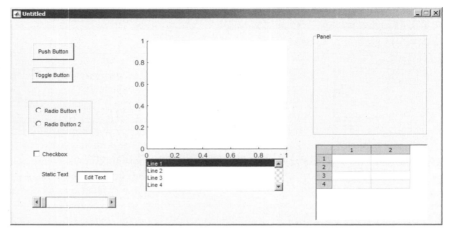

Figure 14.1 A Figure Window showing examples of MATLAB GUI elements. From top to bottom and left to right, the elements are: (1) a pushbutton; (2) a toggle button in the "on" state; (3) two radio buttons within a button group; (4) a check box; (5) a label and an edit box; (6) a slider; (7) a set of axes; (8) a list box; (9) a panel; and (10) a table.

14.2 Creating and Displaying a Graphical User Interface

MATLAB Graphical User Interfaces are created using a tool called guide, the GUI Development Environment. This tool allows a programmer to lay out the GUI, selecting and aligning the GUI components to be placed in it. Once the components are in place, the programmer can edit their properties: name, color, size, font, text to display, and so forth. When guide saves the GUI, it creates a working program including skeleton functions that the programmer can modify to implement the behavior of the GUI.

When guide is executed, it creates the Layout Editor, shown in Figure 14.2. The large grey area with grid lines is the *layout area*, where a programmer can lay out the GUI. The Layout Editor window has a palate of GUI components along the left-hand side of the layout area. A user can create any number of GUI components by first clicking on the desired component, and then dragging its outline in the layout area. The top of the window has a toolbar with a series of useful tools that allow the user to distribute and align GUI components, modify the properties of GUI components, add menus to GUIs, and so forth.

The basic steps required to create a MATLAB GUI are:

1. Decide what elements are required for the GUI, and what the function of each element will be. Make a rough layout of the components by hand on a piece of paper.
2. Use the MATLAB tool called guide (GUI Development Environment) to lay out the components on a figure. The size of the figure, and the alignment and spacing of components on the figure, can be adjusted using the tools built into guide.

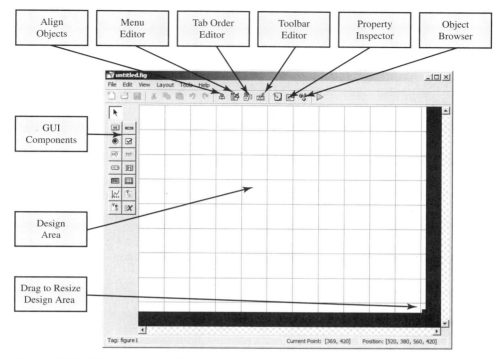

Figure 14.2 The guide tool window.

3. Use a MATLAB tool called the Property Inspector (built into guide) to give each component a name (a "tag"), and to set the characteristics of each component, such as its color, the text it displays, and so forth.
4. Save the figure to a file. When the figure is saved, two files will be created on disk with the same name but different extents. The fig file contains the GUI layout and the components of the GUI, while the M-file contains the code to load the figure, and also skeleton callback functions for each GUI element.
5. Write code to implement the behavior associated with each callback function.

As an example of these steps, let's consider a simple GUI that contains a single pushbutton and a single text string. Each time that the pushbutton is clicked, the text string will be updated to show the total number of clicks since the GUI started.

Step 1: The design of this GUI is very simple. It contains a single pushbutton and a single text field. The callback from the pushbutton will cause the number displayed in the text field to increase by one each time that the button is pressed. A rough sketch of the GUI is shown in Figure 14.3.

Step 2: To lay out the components on the GUI, run the MATLAB function guide. When guide is executed, it creates the window shown in Figure 14.2.

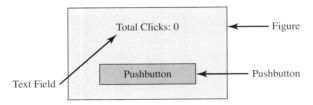

Figure 14.3 Rough layout for a GUI containing a single pushbutton and a single text field.

First, we must set the size of the layout area, which will become the size of the final GUI. We do this by dragging the small square on the lower right hand corner of the layout area until it has the desired size and shape. Then, click on the "pushbutton" button in the list of GUI components, and create the shape of the pushbutton in the layout area. Finally, click on the "text" button in the list of GUI components, and create the shape of the text field in the layout area. The resulting figure after these steps is shown in Figure 14.4. We could now adjust the alignment of these two elements using the Alignment Tool, if desired.

Step 3: To set the properties of the pushbutton, click on the button in the layout area and then select "Property Inspector" (▣) from the toolbar. Alternately, right-click on the button and select "Property Inspector" from the popup menu. The Property Inspector window, shown in Figure 14.5, will appear. Note that this window lists every property available for the pushbutton and allows us to set each value using a GUI interface. The Property Inspector performs the same function as the `get` and `set` functions introduced in Chapter 13, but in a much more convenient form.

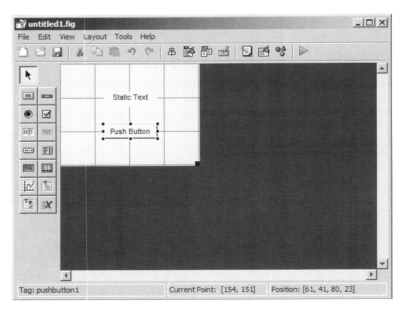

Figure 14.4 The completed GUI layout within the `guide` window.

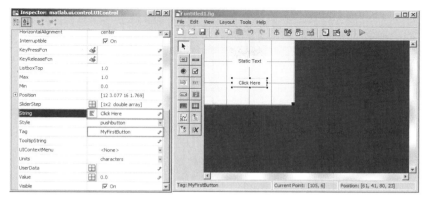

Figure 14.5 The Property Inspector showing the properties of the pushbutton. Note that the String is set to `Click Here`, and the Tag is set to `MyFirstButton`.

For the pushbutton, we may set many properties such as color, size, font, text alignment, and so forth. However, we *must* set two properties: the `String` property, which contains the text to be displayed, and the `Tag` property, which is the name of the pushbutton. In this case, the `String` property will be set to `Click Here`, and the `Tag` property will be set to `MyFirstButton`.

For the text field, we *must* set two properties: the `String` property, which contains the text to be displayed, and the `Tag` property, which is the name of the text field. This name will be needed by the callback function to locate and update the text field. In this case, the `String` property will be set to `Total Clicks: 0`, and the `Tag` property defaulted to `MyFirstText`. The layout area after these steps is shown in Figure 14.6*a*.

It is possible to set the properties of the figure itself by clicking on a clear spot in the Layout Editor, and using the Property Inspector to examine and set the figure's properties. Although not required, it is a good idea to set the figure's `Name` property. The string in the `Name` property will be displayed in the title bar of the resulting GUI when it is executed. In this program, we will set the `Name` to `MyFirstGUI`.

It is also a good idea to check and set the GUI options at this time. Select the "Tools > GUI Options" menu item, and the GUI shown in Figure 14.6*b* will appear. The key options settable on this GUI are:

1. **Resize behavior**—This popup menu allows the designer to specify whether the GUI is fixed size or variable size. If it is variable size, all GUI elements can scale proportionally, or else the GUI can execute a callback function to re-layout the components when it changes size.
2. **Command-line accessibility**—This specifies whether this GUI becomes the current figure when a callback is executing. The default is that it does, so function `gcf` will point to the GUI during callback execution. You will probably never need to change this option.
3. **Generate FIG file and MATLAB file**—This radio button specifies that `Guide` should generate both a figure file with the layout of the GUI and an M-file that would create the GUI and handle the callbacks. This option should always be set.

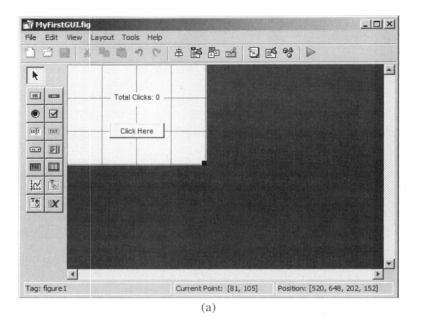

(a)

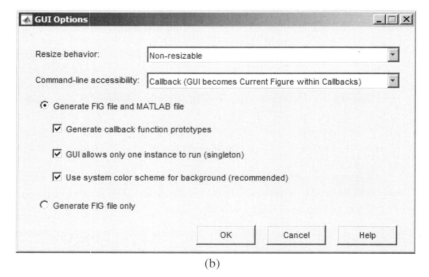

(b)

Figure 14.6 (a) The design area after the properties of the pushbutton and the text field have been modified. (b) Setting the GUI options.

4. **Generate callback function prototypes**—This checkbox specifies that Guide should generate the skeletons of all callback functions, so that the programmer only has to fill in the functions. This option should always be set.

5. **GUI allows only one instance to run**—This checkbox specifies that only a single copy of the GUI should be allowed to run at a time. If this box is ticked, MATLAB will re-use the same GUI each time it is needed instead of creating a new copy.
6. **Use system color scheme for background**—This checkbox specifies that Guide should make the background color of the GUI match the color of the operating system it is running on. If this switch is on, the background color of the GUI will automatically adjust if the same program is run on different types of computers (say, a Windows PC and a Mac). This option should always be set.

Step 4: We will now save the layout area under the name MyFirstGUI. Select the "File/Save As" menu item, type the name MyFirstGUI as the file name, and click "Save". This action will automatically create two files, MyFirstGUI.fig and MyFirstGUI.m. The figure file contains the actual GUI that we have created. The M-file contains code that loads the figure file and creates the GUI, plus a skeleton callback function for each active GUI component.

At this point, we have a complete GUI, but one that does not yet do the job it was designed to do. You can start this GUI by typing MyFirstGUI in the Command Window, as shown in Figure 14.7. If the button is clicked on this GUI, nothing happens.

The M-file automatically created by guide is shown in Figure 14.8. This file contains the main function MyFirstGUI, plus local functions to specify the behavior of the active GUI components. The file contains a *dummy callback*

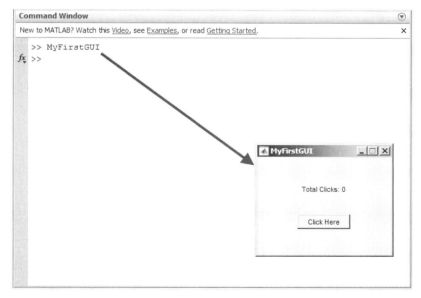

Figure 14.7 Typing MyFirstGUI in the Command Window starts the GUI.

```
function varargout = MyFirstGUI(varargin)  ◄─────────── Main Function
% MYFIRSTGUI MATLAB code for MyFirstGUI.fig
%      MYFIRSTGUI, by itself, creates a new MYFIRSTGUI or raises the existing singleton*.
%
%      H = MYFIRSTGUI returns the handle to a new MYFIRSTGUI or the handle to
%      the existing singleton*.
%
%      MYFIRSTGUI('CALLBACK',hObject,eventData,handles,...) calls the local
%      function named CALLBACK in MYFIRSTGUI.M with the given input arguments.
%
%      MYFIRSTGUI('Property','Value',...) creates a new MYFIRSTGUI or raises the
%      existing singleton*. Starting from the left, property value pairs are
%      applied to the GUI before MyFirstGUI_OpeningFcn gets called. An
%      unrecognized property name or invalid value makes property application
%      stop.  All inputs are passed to MyFirstGUI_OpeningFcn via varargin.
%
%      *See GUI Options on GUIDE's Tools menu. Choose "GUI allows only one
%      instance to run (singleton)".
%
% See also: GUIDE, GUIDATA, GUIHANDLES

% Edit the above text to modify the response to help MyFirstGUI

% Last Modified by GUIDE v2.5 27-Aug-2014 16:04:03

% Begin initialization code - DO NOT EDIT
gui_Singleton = 1;
gui_State = struct('gui_Name',        mfilename, ...
                   'gui_Singleton',   gui_Singleton, ...
                   'gui_OpeningFcn',  @MyFirstGUI_OpeningFcn, ...
                   'gui_OutputFcn',   @MyFirstGUI_OutputFcn, ...
                   'gui_LayoutFcn',   [] , ...
                   'gui_Callback',    []);
if nargin && ischar(varargin{1})
    gui_State.gui_Callback = str2func(varargin{1});
end
```

```
if nargout
    [varargout{1:nargout}] = gui_mainfcn(gui_State, varargin{:});
else
   gui_mainfcn(gui_State, varargin{:});
end
% End initialization code - DO NOT EDIT


% --- Executes just before MyFirstGUI is made visible.
function MyFirstGUI_OpeningFcn(hObject, eventdata, handles, varargin)
% This function has no output args, see OutputFcn.
% hObject       handle to figure
% eventdata   reserved - to be defined in a future version of MATLAB
% handles       structure with handles and user data (see GUIDATA)
% varargin    command line arguments to MyFirstGUI (see VARARGIN)

% Choose default command line output for MyFirstGUI
handles.output = hObject;

% Update handles structure
guidata(hObject, handles);

% UIWAIT makes MyFirstGUI wait for user response (see UIRESUME)
% uiwait(handles.figure1);

% --- Outputs from this function are returned to the command line.
function varargout = MyFirstGUI_OutputFcn(hObject, eventdata, handles)
% varargout     cell array for returning output args (see VARARGOUT);
% hObject       handle to figure
% eventdata   reserved - to be defined in a future version of MATLAB
% handles       structure with handles and user data (see GUIDATA)

% Get default command line output from handles structure
varargout{1} = handles.output;

% --- Executes on button press in MyFirstButton.
function MyFirstButton_Callback(hObject, eventdata, handles)
% hObject       handle to MyFirstButton (see GCBO)
% eventdata   reserved - to be defined in a future version of MATLAB
% handles       structure with handles and user data (see GUIDATA)
```

Figure Opening Function

Data Output Function

Button Callback Function

Figure 14.8 The M-file for MyFirstGUI, automatically created by guide.

function for every active GUI component that you defined. In this case, the only active GUI component was the pushbutton, so there is a callback function called `MyFirstButton_Callback`, which is executed when the user clicks on the button.

If function `MyFirstGUI` is called *without* arguments, then the function displays the GUI contained in file `MyFirstGUI.fig`. If function `MyFirstGUI` is called *with* arguments, then the function assumes that the first argument is the name of a local function (subfunction), and it calls that local function using `feval`, passing the other arguments on to that subfunction.

Each callback function handles events from a single GUI component. If a mouse click (or keyboard input for edit fields) occurs on the GUI component, then the component's callback function will be automatically called by MATLAB. The name of the callback function will be the value in the `Tag` property of the GUI component plus the characters "`_Callback`". Thus, the callback function for `MyFirstButton` will be named `MyFirstButton_Callback`.

M-files created by `guide` contain callbacks for each active GUI component, but these callbacks don't do anything yet.

Step 5: Now, we need to write the callback subfunction code for the pushbutton. This function will include a `persistent` variable that can be used to count the number of clicks that have occurred. When a click occurs on the pushbutton, MATLAB will call the function `MyFirstGUI` with the string `'MyFirstButton_Callback'` as the first argument. Then function `MyFirstGUI` will call subfunction `MyFirstButton_Callback`, as shown in Figure 14.9. This function should increase the count of clicks by one, create a new text string containing the count, and store the new string in the `String` property of the text field `MyFirstText`. A function to perform this step is shown below:

```
function MyFirstButton_Callback(hObject, eventdata, handles)
% hObject    handle to MyFirstButton (see GCBO)
% eventdata  reserved - to be defined in a future version of MATLAB
% handles    structure with handles and user data (see GUIDATA)

% Declare and initialize variable to store the count
persistent count
if isempty(count)
   count = 0;
end
% Update count
count = count + 1;

% Create new string
str = sprintf('Total Clicks: %d',count);

% Update the text field
set (handles.MyFirstText,'String',str);
```

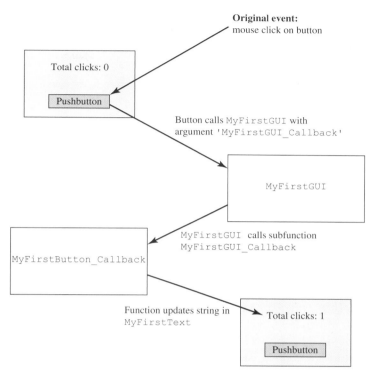

Figure 14.9 Event handling in program `MyFirstGUI`. When a user clicks on the button with the mouse, the function `MyFirstGUI` is called automatically with the argument `'MyFirstButton_Callback'`. Function `MyFirstGUI` in turn calls subfunction `MyFirstButton_Callback`. This function increments `count`, and then saves the new count in the text field on the GUI.

Note that this function declares a persistent variable `count`, and initializes it to zero. Each time that the function is called, it increments `count` by 1 and creates a new string containing the count. Then, it updates the string displayed in the text field `MyFirstText`.

The resulting program is executed by typing `MyFirstGUI` in the Command Window. When the user clicks on the button, MATLAB automatically calls function `MyFirstGUI` with `MyFirstButton_Callback` as the first argument, and function `MyFirstGUI` calls subfunction `MyFirstButton_Callback`. This function increments variable `count` by one, and updates the value displayed in the text field. The resulting GUI after three button pushes is shown in Figure 14.10.

Good Programming Practice

Use `guide` to lay out a new GUI, and use the Property Inspector to set the initial properties of each component such as the text displayed on the component, the color of the component, and the name of the callback function, if required.

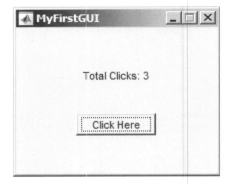

Figure 14.10 The resulting program after three button pushes.

Good Programming Practice

After creating a GUI with guide, manually edit the resulting M-file to add comments describing its purpose and components, and to implement the behavior of callbacks.

14.2.1 A Look Under the Hood

Figure 14.8 shows the M-file that was automatically generated by guide for MyFirstGUI. We will now examine this M-file more closely to understand how it works.

First, let's look at the main function declaration itself. Note that this function uses varargin to represent its input arguments, and varargout to represent its output results. As we learned in Chapter 10, function varargin can represent an arbitrary number of input arguments, and function varargout can represent a varying number of output arguments. Therefore, *a user can call function* MyFirstGUI *with any number of arguments*.

The main function begins with a series of comments that serve as the help message displayed when the user types "help MyFirstGUI". You should edit these comments to reflect the actual function of your program.

Next, the main function creates a structure called gui_State. The code to create this structure is shown below:

```
gui_Singleton = 1;
gui_State = struct('gui_Name',        mfilename, ...
                   'gui_Singleton',  gui_Singleton, ...
                   'gui_OpeningFcn', @MyFirstGUI_OpeningFcn, ...
                   'gui_OutputFcn',  @MyFirstGUI_OutputFcn, ...
```

```
                             'gui_LayoutFcn',  [] , ...
                             'gui_Callback',   []);
if nargin && ischar(varargin{1})
    gui_State.gui_Callback = str2func(varargin{1});
end
```

The structure contains some control information, plus function handles for some of the local functions in the file. Other MATLAB GUI functions use these function handles to call the local functions from outside of the M-file. Note that the first argument is converted into a callback function handle using str2func, if it exists.

The value gui_Singleton specifies whether there can be one or more simultaneous copies of the GUI. If gui_Singleton is 1, then there can be only one copy of the GUI. If gui_Singleton is 0, then there can be many simultaneous copies of the GUI. Guide sets this based on the selection made in the GUI options page, as described in Step 3 of the GUI creation process.

The main function calls the MATLAB function gui_mainfcn, and passes the gui_State structure and all of the input arguments to it. Function gui_mainfcn is a built-in MATLAB function. It actually does the work of creating the GUI, or calling the appropriate local function in response to a callback.

If the user calls MyFirstGUI *without* arguments, function gui_mainfcn loads the GUI from the figure file MyFirstGUI.fig. Then, function gui_mainfcn creates a structure containing the handles of all the objects in the current figure, and calls the file opening function MyFirstGUI_OpeningFcn, which stores that structure as application data in the figure.

```
% Update handles structure
guidata(hObject, handles);
```

Function guihandles saves a structure containing handles to all of the objects within the specified figure in the figure object. The element names in the structure correspond to the Tag properties of each GUI component, and the values are the handles of each component. For example, the handle structure returned in MyFirstGUI.m is

```
» handles = guihandles(fig)
handles =
             figure1: [1x1 Figure]
         MyFirstText: [1x1 UIControl]
       MyFirstButton: [1x1 UIControl]
```

There are three GUI components in this figure—the figure itself, plus a text field, and a pushbutton. Function guidata saves the handles structure as application data in the figure, using the setappdata function that we studied in Chapter 13.

Function gui_OpeningFcn provides a way for the programmer to customize the GUI before showing it to the user. Note the commented uiwait statement. If the programmer uncomments this line, MATLAB will lock and wait for input from the

GUI before continuing. It is also possible to make other customizations here, such as changing background colors, and so forth.

Finally, function `gui_mainfcn` calls the output function `MyFirstGUI_OutputFcn` to return the handles structure to the M-file that created the GUI. This structure gives the calling M-files the information required to work with the GUI programmatically.

After the GUI is created, the user can interact with it using the mouse and (possibly) the keyboard. When the user clicks on an active GUI element, MATLAB calls `MyFirstGUI` with the name of the GUI element's callback function in the first argument. If `MyFirstGUI` is called *with* arguments, function `gui_mainfcn` calls the callback function using this function handle. The callback executes and responds to the mouse click or keyboard input, as appropriate.

Figure 14.11 summarizes the operation of `MyFirstGUI` on first and subsequent calls.

14.2.2 The Structure of a Callback Subfunction

Every callback subfunction has the standard form

```
function ComponentTag_Callback(hObject, eventdata, handles)
```

where `ComponentTag` is the name of the component generating the callback (the string in its `Tag` property). The arguments of this subfunction are:

- **hObject**—The handle of the parent figure.
- **eventdata**—A currently unused (in MATLAB 2014B) array.
- **handles**—The `handles` structure contains the handles of all GUI components on the figure.

Note that each callback function has full access to the `handles` structure, and so each callback function can modify any GUI component in the figure. We took advantage of this structure in the callback function for the pushbutton in `MyFirstGUI`, where the callback function for the pushbutton modified the text displayed in the text field.

```
% Update the text field
set (handles.MyFirstText, 'String',str);
```

14.2.3 Adding Application Data to a Figure

It is possible to store any application-specific information needed by a GUI program in the `handles` structure instead of using global or persistent memory for that data. The resulting GUI design is more robust, since other MATLAB programs cannot accidentally modify the global GUI data, and since multiple copies of the same GUI cannot interfere with each other.

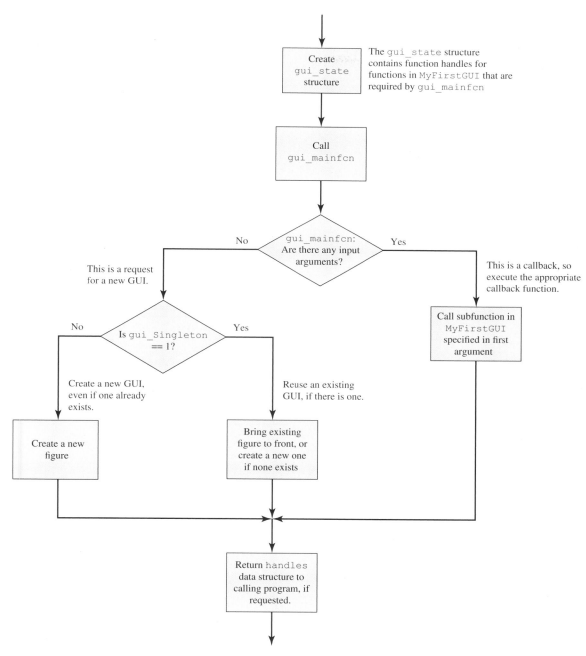

Figure 14.11 The operation of `MyFirstGUI`. If there are no calling arguments, it either creates a GUI or displays an existing GUI. If there are calling arguments, the first argument is assumed to be a callback function name, and `MyFirstGUI` calls the appropriate callback function.

To add local data to the `handles` structure, we must manually modify the M-file after it is created by `guide`. A programmer first adds the required local data to the `handles` structure, and then calls `guidata` to update the `handles` structure stored in the figure. For example, to add the number of mouse clicks `count` to the `handles` structure, we would modify the `MyFirstButton_Callback` function as follows:

```
function MyFirstButton_Callback(hObject, eventdata, handles)
% hObject    handle to MyFirstButton (see GCBO)
% eventdata reserved - to be defined in a future version of MATLAB
% handles    structure with handles and user data (see GUIDATA)

% Create the count field if it does not exist
if ~isfield(handles,'count')
   handles.count = 0;
end

% Update count
handles.count = handles.count + 1;

% Save the updated handles structure
guidata(hObject, handles);

% Create new string
str = sprintf('Total Clicks: %d',handles.count);

% Update the text field
set (handles.MyFirstText,'String',str);
```

🔲 Good Programming Practice

Store GUI application data in the `handles` structure, so that it will automatically be available to any callback function.

//

🔲 Good Programming Practice

If you modify any of the GUI application data in the `handles` structure, be sure to save the structure with a call to `guidata` before exiting the function where the modifications occurred.

//

14.2.4 A Few Useful Functions

Three special functions are occasionally used in the design of callback functions: gcbo, gcbf, and findobj. These functions are not actually needed with MATLAB GUIs, because the same information is available in the handles data structure. However, they were commonly used in earlier versions of MATLAB, and a programmer is sure to encounter them.

Function gcbo (*get callback object*) returns the handle of the object that generated the callback, while function gcbf (*get callback figure*) returns the handle of the figure containing that object. These functions can be used by a callback function to determine the object and figure producing the callback, so that it can modify objects on that figure.

Function findobj searches through all of the child objects within a parent object, looking for ones that have a specific value of a specified property. It returns a handle to any objects with the matching characteristics. The most common form of findobj is

```
Hndl = findobj(parent,'Property',Value);
```

where parent is the handle of a parent object such as a figure, 'Property' is the property to examine, and 'Value' is the value to look for.

For example, suppose that a programmer would like to change the color of all the lines in a plot on the callback figure. He or she could find the lines and change the line colors to red with the following statements

```
Hndl = findobj(gcbf,'Type','Line');
for ii = 1:length(Hndl)
    set( Hndl,'Color','r' );
end
```

14.3 Object Properties

Every GUI object includes an extensive list of properties that can be used to customize the object. These properties are slightly different for each type of object (figures, axes, uicontrols, and so forth). All of the properties for all types of objects are documented on the online Help Browser, but a few of the more important properties for figure and uicontrol objects are summarized in Tables 14.2 and 14.3.

Object properties can be modified using either the Property Inspector or the get and set functions. While the Property Inspector is a convenient way to adjust properties during GUI design, we must use get and set to adjust them dynamically from within a program, such as in a callback function.

Table 14.2: Important `figure` Properties

Property	Description
Color	Specifies the color of the figure. The value is either a pre-defined color such as `'r'`, `'g'`, or `'b'`, or else a three-element vector specifying the red, green, and blue components of the color on a 0-1 scale. For example, the color magenta would be specified by `[1 0 1]`.
CurrentCharacter	Contains the character corresponding to the last key pressed in this figure.
CurrentPoint	Location of the last button click in this figure, measured from the lower left-hand corner of the figure in units specified in the `Units` property.
Dockable	Specifies whether on not the figure can be docked to the desktop. Possible values are `'on'` or `'off'`.
MenuBar	Specifies whether on not the default set of menus appear on the figure. Possible values are `'figure'` to display the default menus or `'none'` to delete them.
Name	A string containing the name that appears in the title bar of a figure.
NumberTitle	Specifies whether or not the figure number appears in the title bar. Possible values are `'on'` or `'off'`.
Position	Specifies the position of a figure on the screen, in the units specified by the `'units'` property. This value accepts a four-element vector in which the first two elements are the x and y positions of the lower left-hand corner of the figure, and the next two elements are the width and height of the figure.
SelectionType	Specifies the type of selection for the last mouse click on this figure. A single click returns type `'normal'`, while a double click returns type `'open'`. There are additional options; see the MATLAB online documentation.
Tag	The "name" of the figure, which can be used to locate it.
Units	The units used to describe the position of the figure. Possible choices are `'inches'`, `'centimeters'`, `'normalized'`, `'points'`, `'pixels'`, or `'characters'`. The default units are `'pixels'`.
Visible	Specifies whether or not this figure is visible. Possible values are `'on'` or `'off'`.
WindowStyle	Specifies whether this figure is normal or modal (see discussion of Dialog Boxes). Possible values are `'normal'` or `'modal'`.

Table 14.3: Important `uicontrol` Properties

Property	Description
BackgroundColor	Specifies the background color of the object. The value is either a predefined color such as `'r'`, `'g'`, or `'b'`, or else a three-element vector specifying the red, green, and blue components of the color on a 0–1 scale. For example, the color magenta would be specified by `[1 0 1]`.
Callback	Specifies the name and parameters of the function to be called when the object is activated by a keyboard or text input.

(continued)

Table 14.3: Important `uicontrol` Properties (*Continued*)

Property	Description
Enable	Specifies whether or not this object is selectable. If it not enabled, it will not respond to mouse or keyboard input. Possible values are `'on'` or `'off'`.
FontAngle	A string containing the font angle for text displayed on the object. Possible values are `'normal'`, `'italic'`, and `'oblique'`.
FontName	A string containing the font name for text displayed on the object.
FontSize	A number specifying the font size for text displayed on the object.
FontUnits	The units in which the font size is defined. Possible choices are `'inches'`, `'centimeters'`, `'normalized'`, `'points'`, and `'pixels'`. The default font units are `'points'`.
FontWeight	A string containing the font weight for text displayed on the object. Possible values are `'light'`, `'normal'`, `'demi'`, and `'bold'`. The default font weight is `'normal'`.
ForegroundColor	Specifies the foreground color of the object.
HorizontalAlignment	Specifies the horizontal alignment of a text string within the object. Possible values are `'left'`, `'center'`, and `'right'`.
Max	The maximum size of the `value` property for this object.
Min	The minimum size of the `value` property for this object.
Parent	The handle of the figure containing this object.
Position	Specifies the position of the object on the screen, in the units specified by the `'units'` property. This value accepts a four-element vector in which the first two elements are the *x* and *y* positions of the lower left-hand corner of the object *relative to the figure containing it*, and the next two elements are the width and height of the object.
Tag	The "name" of the object, which can be used to locate it.
TooltipString	Specifies the help text to be displayed when a user places the mouse pointer over an object.
Units	The units used to describe the position of the figure. Possible choices are `'inches'`, `'centimeters'`, `'normalized'`, `'points'`, `'pixels'`, or `'characters'`. The default units are `'pixels'`.
Value	The current value of the `uicontrol`. For toggle buttons, check boxes, and radio buttons, the value is max when the button is on and min when the button is off. Other controls have different meanings for this term.
Visible	Specifies whether or not this object is visible. Possible values are `'on'` or `'off'`.

14.4 Graphical User Interface Components

This section summarizes the basic characteristics of common graphical user interface components. It describes how to create and use each component, as well as the types of events each component can generate. The components discussed in this section are:

- Static text fields
- Edit boxes
- Pushbuttons
- Toggle buttons
- Checkboxes
- Radio buttons
- Popup menus
- List boxes
- Sliders
- Tables

14.4.1 Static Text Fields

A **static text field** is a graphical object that displays one or more text strings, which are specified in the text field's `String` property. The `String` property accepts a string or a cell array of strings. If the input value is a string, it will be displayed on a single line. If the input value is a cell array of strings, the first element will be displayed on the first line of the text box, the second element will be displayed on the second line of the text box, and so forth. You can specify how the text is aligned in the display area by setting the horizontal alignment property. By default, text fields are horizontally centered. A text field is created by a `uicontrol` whose style property is `'text'`. A text field can be added to a GUI by using the text tool (▣) in the Layout Editor.

Text fields do not create callbacks, but the value displayed in the text field can be updated from another component's callback function by changing the text field's `String` property, as shown in program `MyFirstGUI` in Section 14.2.

14.4.2 Edit Boxes

An **edit box** is a graphical object that allows a user to enter one or more text strings. It is created by a `uicontrol` whose style property is `'edit'`. If the `min` property and `max` property are both set to 1, then *the edit box will accept a single line of text*, and it will generate a callback when the user presses the Enter key or the ESC key after typing the text.

Figure 14.12*a* shows a simple GUI containing an edit box named `'EditBox'` and a text field named `'TextBox'`. When a user presses Enter or ESC after typing a string into the edit box, the program automatically calls the function `EditBox_Callback`, which is shown in Figure 14.12*b*. This function locates the edit box, using the `handles` structure, and recovers the string typed by the user. Then, it locates the text field and displays the string in the text field. Figure 14.13 shows this GUI just after it had started and after the user had typed the word `'Hello'` in the edit box.

If the `max` property is set to a number greater than the `min` property, then *the edit box will accept as many lines of text as the user wishes to enter*. The textbox will include a vertical scrollbar to allow the user to move up and down through the data. Either the scrollbar or the up and down arrows can be used to move between

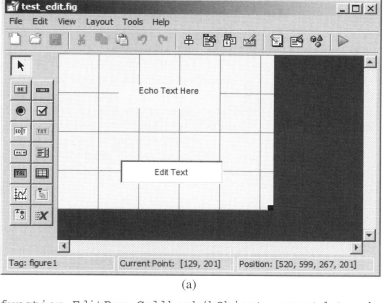

(a)

```
function EditBox_Callback(hObject, eventdata, handles)

% Find the value typed into the edit box
str = get (handles.EditBox,'String');

% Place the value into the text field
set (handles.TextBox,'String',str);
```

(b)

Figure 14.12 (a) Layout of a simple GUI with a single-line edit box and a text field. (b) The callback function for this GUI.

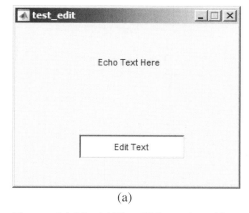

(a)

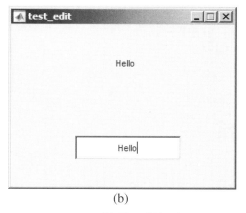

(b)

Figure 14.13 (a) The GUI produced by program `test_edit`. (b) The GUI after a user types 'Hello' into the edit box and presses Enter.

the input lines. If the user presses the Enter key in a multi-line edit box, the current line is finished and the cursor moves down to the next line for additional input. If the user presses the ESC key or clicks a point on the figure background with the mouse, a callback will be generated, and the data typed into the edit box will be available as a cell array of strings in the `uicontrol`'s `String` property.

Figure 14.14a shows a simple GUI containing a multi-line edit box named `'EditBox2'` and a text field named `'TextBox2'`. When a user presses ESC after typing a set of lines into the edit box, the program automatically calls the function `EditBox2_Callback`, which is shown in Figure 14.14b. This function locates the edit box using the `handles` structure, and recovers the strings typed by the user. Then, it locates the text field and displays the strings in the text field. Figure 14.15 shows this GUI just after it has started, and after the user has typed four lines in the edit box.

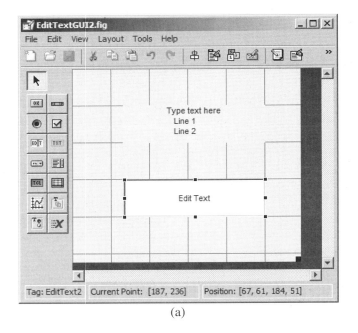

(a)

```
function EditBox2_Callback(hObject, eventdata, handles)

% Find the value typed into the edit box
str = get (handles.EditBox,'String');

% Place the value into the text field
set (handles.TextBox2,'String',str);
```
(b)

Figure 14.14 (a) Layout of a simple GUI with a multi-line edit box and a text field. (b) The callback function for this GUI.

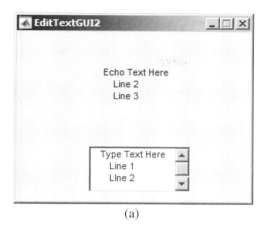

(a)

(b)

Figure 14.15 (a) The GUI produced by program `test_edit2`. (b) The GUI after a user types four lines into the edit box and presses ESC.

14.4.3 Pushbutton

A **pushbutton** is a component that a user can click on to trigger a specific action. The pushbutton generates a callback when the user clicks on it with the mouse. A pushbutton is created by creating a `uicontrol` whose style property is `'pushbutton'`. It can be added to a GUI by using the pushbutton tool (▨) in the Layout Editor.

Function `MyFirstGUI` in Figure 14.10 illustrates the use of pushbutton.

14.4.4 Toggle Buttons

A **toggle button** is a type of button that has two states: on (depressed) and off (not depressed). A toggle button switches between these two states whenever the mouse clicks on it, and it generates a callback each time. The `'Value'` property of the toggle button is set to max (usually 1) when the button is on, and min (usually 0) when the button is off.

A toggle button is created by a `uicontrol` whose style property is `'togglebutton'`. It can be added to a GUI by using the toggle button tool (▨) in the Layout Editor.

Figure 14.16*a* shows a simple GUI containing a toggle button named `'ToggleButton'` and a text field named `'TextBox'`. When a user clicks on the toggle button, it automatically calls the function `ToggleButton_Callback`, which is shown in Figure 14.16*b*. This function locates the toggle button, using the `handles` structure, and recovers its state from the `'Value'` property. Then, it locates the text field and displays the state in the text field. Figure 14.17 shows this GUI just after it has started, and after the user has clicked on the toggle button for the first time.

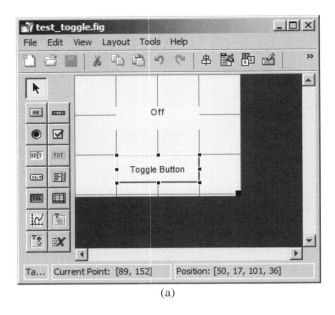

(a)

```
function ToggleButton_Callback(hObject, eventdata, handles)
% hObject     handle to ToggleButton (see GCBO)
% eventdata   reserved - to be defined in a future version of MATLAB
% handles     structure with handles and user data (see GUIDATA)

% Find the state of the toggle button
state = get(handles.ToggleButton,'Value');

% Place the corect value into the text field
if state == 0
   string ='Off';
else
   string ='On';
end
set (handles.TextBox,'String',string);
```

(b)

Figure 14.16 (a) Layout of a simple GUI with a toggle button and a text field. (b) The callback function for this GUI.

14.4.5 Checkboxes and Radio Buttons

Checkboxes and radio buttons are essentially identical to toggle buttons except that they have different shapes. Like toggle buttons, they have two states: on and off. They switch between these two states whenever the mouse clicks on them, generating a

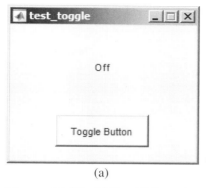 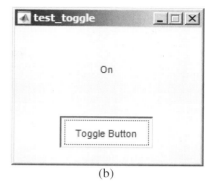

(a)   (b)

Figure 14.17 (a) The GUI produced by program `test_togglebutton` when the toggle button is off. (b) The GUI when the toggle button is on.

callback each time. The `'Value'` property of the checkbox or radio button is set to `max` (usually 1) when they are on, and `min` (usually 0) when they are off. Both checkboxes and radio buttons are illustrated in Figure 14.1.

A checkbox is created by a `uicontrol` whose style property is `'checkbox'`, and a radio button is created by a `uicontrol` whose style property is `'radiobutton'`. A checkbox can be added to a GUI by using the checkbox tool (☑) in the Layout Editor, and a radio button can be added to a GUI by using the radio button tool (⦿) in the Layout Editor.

Checkboxes are traditionally used to display on/off options, while groups of radio buttons are traditionally used to select among mutually exclusive options.

Figure 14.18*a* shows a simple GUI containing a checkbox named `'CheckBox'` and a text field named `'TextBox'`. When a user clicks on the checkbox, it automatically calls the function `CheckButton_Callback`, which is shown in Figure 14.18*b*. This function locates the checkbox, using the `handles` structure, and recovers its state from the `'Value'` property. Then, it locates the text field and displays the state in the text field. Figure 14.19 shows this GUI just after it has started and after the user has clicked on the toggle button for the first time.

Figure 14.20*a* shows an example of how to create a group of mutually exclusive options with radio buttons. The GUI in this figure creates three radio buttons, labeled Option 1, Option 2, and Option 3, plus a text field to display the currently selected results.

The corresponding callback functions are shown in Figure 14.20*b*. When the user clicks on a radio button, the corresponding callback function is executed. That function sets the text box to display the current option, turns on that radio button, and turns off all other radio buttons.

Figure 14.21 shows this GUI after Option 2 has been selected.

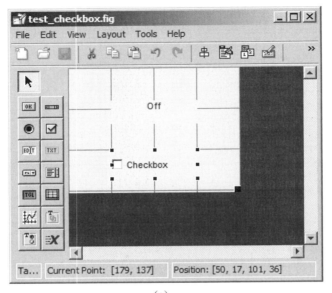

(a)

```
function CheckBox_Callback(hObject, eventdata, handles)

% Find the state of the checkbox
state = get(handles.CheckBox,'Value');

% Place the value into the text field
if state == 0
   set (handles.TextBox,'String','Off');
else
   set (handles.TextBox,'String','On');
end   ;
```

(b)

Figure 14.18 (a) Layout of a simple GUI with a CheckBox and a text field. (b) The callback function for this GUI.

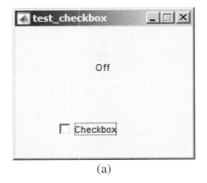

(a)

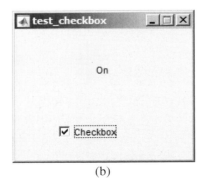

(b)

Figure 14.19 (a) The GUI produced by program `test_checkbox` when the toggle button is off. (b) The GUI when the toggle button is on.

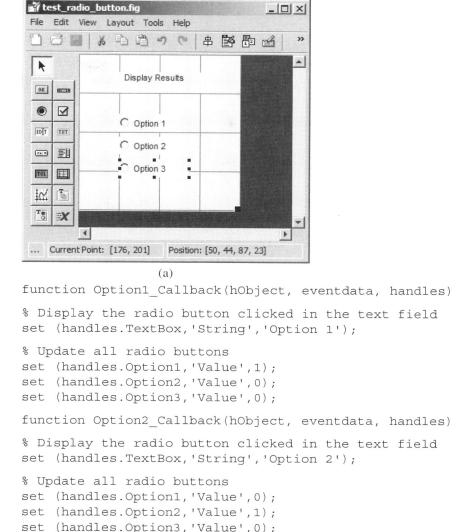

(a)

```
function Option1_Callback(hObject, eventdata, handles)

% Display the radio button clicked in the text field
set (handles.TextBox,'String','Option 1');

% Update all radio buttons
set (handles.Option1,'Value',1);
set (handles.Option2,'Value',0);
set (handles.Option3,'Value',0);

function Option2_Callback(hObject, eventdata, handles)

% Display the radio button clicked in the text field
set (handles.TextBox,'String','Option 2');

% Update all radio buttons
set (handles.Option1,'Value',0);
set (handles.Option2,'Value',1);
set (handles.Option3,'Value',0);

function Option3_Callback(hObject, eventdata, handles)

% Display the radio button clicked in the text field
set (handles.TextBox,'String','Option 3');

% Update all radio buttons
set (handles.Option1,'Value',0);
set (handles.Option2,'Value',0);
set (handles.Option3,'Value',1);
```

(b)

Figure 14.20 (a) Layout of a simple GUI with three radio buttons and a text field. (b) The callback functions for this GUI. When a user clicks on a radio button, it is set to "on" and all other radio buttons are set to "off."

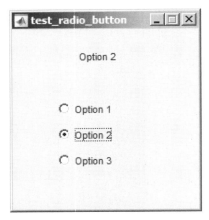

Figure 14.21 The GUI produced by program `test_radio_button` when Option 2 has been selected.

14.4.6 Popup Menus

Popup menus are graphical objects that allow a user to select one of a mutually exclusive list of options. The list of options that the user can select among is specified by a cell array of strings, and the `'Value'` property contains an integer indicating which of the strings is currently selected. A popup menu can be added to a GUI by using the popup menu tool (▥) in the Layout Editor.

Figure 14.22*a* shows an example of a popup menu. This GUI in this figure creates a popup menu with five options, labeled `Option 1`, `Option 2`, and so forth.

The corresponding callback function is shown in Figure 14.22*b*. The callback function recovers the selected option by checking the `'Value'` parameter of the popup menu, and creates and displays a string containing that value in the text field. Figure 14.23 shows this GUI after Option 4 has been selected.

14.4.7 List Boxes

List boxes are graphical objects that display many lines of text, and allow a user to select one or more of those lines. If there are more lines of text than can fit in the list box, a scroll bar will be created to allow the user to scroll up and down within the list box. The lines of text among which the user can select are specified by a cell array of strings, and the `'Value'` property indicates which of the strings are currently selected.

A list box is created by a `uicontrol` whose style property is `'listbox'`. A list box can be added to a GUI by using the listbox tool (▤) in the Layout Editor.

List boxes can be used to select a single item from a selection of possible choices. In normal GUI usage, a single mouse click on a list item selects that item

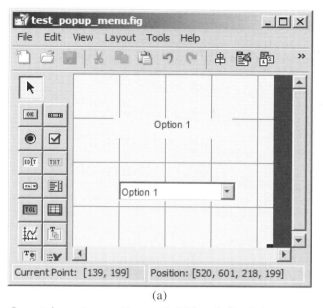

(a)

```
function PopupMenu_Callback(hObject, eventdata, handles)
% Find the value of the popup menu
value = get(handles.PopupMenu,'Value');

% Place the value into the text field
str = ['Option' num2str(value)];
set (handles.TextBox,'String',str);
```
(b)

Figure 14.22 (a) Layout of a simple GUI with a popup menu and a text field to display the current selection. (b) The callback functions for this GUI.

Figure 14.23 The GUI produced by program test_popup_menu.

but does not cause an action to occur. Instead, the action waits on some external trigger, such as a pushbutton. However, a mouse double-click causes an action to happen immediately. Single-click and double-click events can be distinguished using the `SelectionType` property of the figure in which the clicks occurred. A single mouse click will place the string `'normal'` in the `SelectionType` property, while a double mouse click will place the string `'open'` in the `Selection Type` property.

It is also possible for a list box to allow multiple selections from the list. If the difference between the `max` and `min` properties of the list box is greater than one, then multiple selections are allowed. Otherwise, only one item may be selected from the list.

Figure 14.24*a* shows an example of a single-selection list box. The GUI in this figure creates a list box with eight options, labeled `Option 1`, `Option 2`, and so forth. In addition, it creates a pushbutton to perform selection and a text field to display the selected choice. Both the list box and the pushbutton generate callbacks.

The corresponding callback functions are shown in Figure 14.24*b*. If a selection is made in the list box, then function `Listbox1_Callback` will be executed. This function will check the *figure producing the callback* (using function `gcbf`) to see if the selecting action were a single-click or a double-click. If it were a single-click, the function does nothing. If it were a double-click, then the function gets the selected value from the listbox and writes an appropriate string into the text field.

If the pushbutton is selected, then function `Button1_Callback` will be executed. This function gets the selected value from the listbox, and writes an appropriate string into the text field.

In an end-of-chapter exercise, you will be asked to modify this example to allow multiple selections in the list box.

14.4.8 Sliders

Sliders are graphical objects that allow a user to select values from a continuous range between a specified minimum value and a specified maximum value by moving a bar with a mouse. The `'Value'` property of the slider is set to a value between `min` and `max` depending on the position of the slider.

A slider is created by a `uicontrol` whose style property is `'slider'`. A slider can be added to a GUI by using the slider tool (▭) in the Layout Editor.

Figure 14.26*a* shows the layout for a simple GUI containing a slider and a text field. The `'Min'` property for this slider is set to zero, and the `'Max'` property is set to one. When a user drags the slider, it automatically calls the function `Slider1_Callback`, which is shown in Figure 14.26*b*. This function gets the value of the slider from the `'Value'` property, and displays the value in the text field. Figure 14.27 shows this GUI with the slider at some intermediate position in its range.

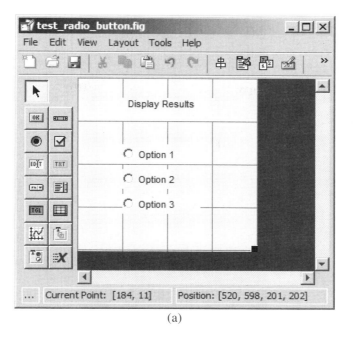

(a)

```
function Button1_Callback(hObject, eventdata, handles)

% Find the value of the popup menu
value = get(handles.Listbox1,'Value');

% Update text label
str = ['Option' num2str(value)];
set (handles.Label1,'String',str);

function Listbox1_Callback(hObject, eventdata, handles)

% If this was a double click, update the label.
selectiontype = get(gcbf,'SelectionType');
if selectiontype(1) =='o'

   % Find the value of the popup menu
   value = get(handles.Listbox1,'Value');

   % Update text label
   str = ['Option' num2str(value)];
   set (handles.Label1,'String',str);
end
```

(b)

Figure 14.24 (a) Layout of a simple GUI with a list box, a pushbutton, and a text field. (b) The callback functions for this GUI.

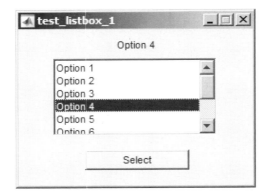

Figure 14.25 The GUI produced by program `test_listbox`.

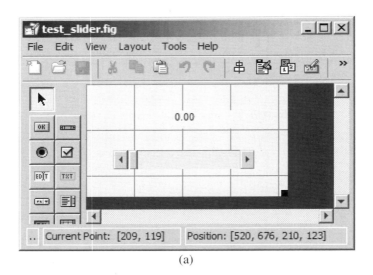

(a)

```
function Slider1_Callback(hObject, eventdata, handles)

% Find the value of the slider
value = get(handles.Slider1,'Value');

% Place the value in the text field
str = sprintf('%.2f',value);
set (handles.Label1,'String',str);
```
(b)

Figure 14.26 (a) Layout of a simple GUI with a slider and a text field. (b) The callback function for this GUI.

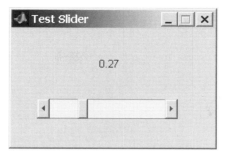

Figure 14.27 The GUI produced by program `test_slider`.

14.4.9 Tables

A table is a two-dimensional display of data with rows and columns. Sliders are graphical objects that allow a user to select values from a continuous range between a specified minimum value and a specified maximum value by moving a bar with a mouse. The `'Value'` property of the slider is set to a value between min and max depending on the position of the slider.

A table is created by a `uitable` object. A table can be added to a GUI by using the table tool (▦) in the Layout Editor. The `uitable` is different from the uicontrol used for the other GUI elements, and it has different properties that must be set to use it properly. The key properties needed for a `uitable` are:

1. `Data`—A two-dimensional cell array containing a collection of values to display in the table. The types of data displayed in the table can be mixed, with some columns being numeric, some logical, and others character. The size of the table is usually specified to be equal to the size of the data supplied to the table.
2. `ColumnName`—A one-dimensional cell array containing the labels for each column in the table.
3. `ColumnFormat`—A one-dimensional cell array containing the format of the data to display in each column (`'numeric'`, `'logical'`, and so forth).
4. `ColumnEditable`—A one-dimensional logical array of logical values indication whether each column can be edited or not.
5. `RowName`—A one-dimensional cell array containing the labels for each row in the table.
6. `CellEditCallback`—The specified function will be called when the data in a cell of the table is modified.
7. `CellSelectionCallback`—The specified function will be called when the cell selection is modified and will provide a list of all the currently selected cells.

Figure 14.28*a* shows the layout for a simple GUI containing a table and two text fields. The table is 3 × 4, because the initializing data was the cell array {1, 2, 3, 4; 5, 6, 7, 8; 9, 10, 11, 12}. The Property Inspector was used to specify that columns 1, 3, and 4 were editable and column 2 was not. Callback functions were defined for the `CellEditCallback` and the `CellSelectionCallback`.

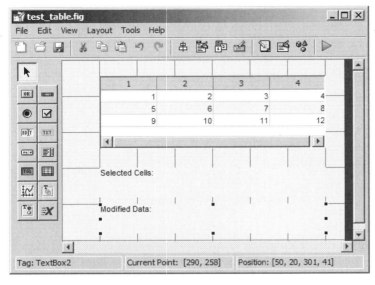

(a)

```
function uitable1_CellSelectionCallback(hObject, eventdata, handles)
% Get the cells selected
rows = eventdata.Indices(:,1);
columns = eventdata.Indices(:,2);

% List the (row,column) pairs selected
string = ['Selected Cells:'];
for ii = 1:length(rows)
  string = [string'(', int2str(rows(ii))',' int2str(columns(ii))')'];
end

% Set the list into the string
set (handles.TextBox1,'String', string);

% Clear the modified cell
set (handles.TextBox2,'String','Modified Cell:');

function uitable1_CellEditCallback(hObject, eventdata, handles)
% Get the cells selected
rows = eventdata.Indices(:,1);
columns = eventdata.Indices(:,2);

% Display the data change
string = ['Modified Cell: (', int2str(rows)',' int2str(columns)')'];
string = [string'; Old data =' num2str(eventdata.PreviousData)];
string = [string'; New data =' num2str(eventdata.NewData)];

% Set the list into the string
set (handles.TextBox2,'String', string);
```

(b)

Figure 14.28 (a) Layout of a sample table with two text fields. (b) The callback functions for this GUI.

The `eventdata` structure is returned to each callback, with information such as the cells currently selected and the old and new values when a cell is modified. The sample GUI displays the selected cell and any modified information. The callback functions to create the list of selected cells and to show the modified data in a cell are shown in Figure 14.28*b*. Figure 14.29 shows this GUI with selected and modified cells.

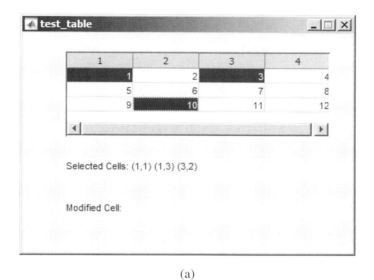

(a)

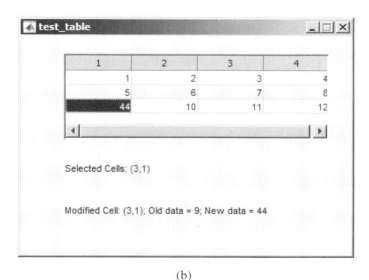

(b)

Figure 14.29 (a) The `test_table` GUI with multiple cells selected; (b) The `test_table` GUI with the data in cell (3,1) modified.

►Example 14.1—Temperature Conversion

Write a program that converts temperature from degrees Fahrenheit to degrees Celsius and vice versa over the range 0–100° C, using a GUI to accept data and display results. The program should include an edit box for the temperature in degrees Fahrenheit, an edit box for the temperature in degrees Celsius, and a slider to allow for the continuous adjustment of temperature. The user should be able to enter temperatures in either edit box or by moving the slider, and all GUI elements should adjust to the corresponding values.

Solution To create this program, we will need a text field and an edit box for the temperature in degrees Fahrenheit, another text field and an edit box for the temperature in degrees Celsius, and a slider. We will also need a function to convert degrees Fahrenheit to degrees Celsius, and a function to convert degrees Celsius to degrees Fahrenheit. Finally, we will need to write callback functions to support user inputs.

The range of values to convert will be 32–212° F or 0–100° C, so it will be convenient to set up the slider to cover the range 0–100, and to treat the value of the slider as a temperature in degrees C.

The first step in this process is to use guide to design the GUI. We can use guide to create the five required GUI elements and locate them in approximately the correct positions. Then, we can use the Property Inspector to perform the following steps:

1. Select appropriate names for each GUI element and store them in the appropriate Tag properties. The names will be 'Label1', 'Label2', 'Edit1', 'Edit2', and 'Slider1'.
2. Store 'Degrees F' and 'Degrees C' in the String properties of the two labels.
3. Set the slider's minimum and maximum limits to 0 and 100, respectively.
4. Store initial values in the String property of the two edit fields and in the Value property of the slider. We will initialize the temperature to 32° F or 0° C, which corresponds to a slider value of 0.
5. Set the Name property of the figure containing the GUI to 'Temperature Conversion'.

Once these changes have been made, the GUI should be saved to file temp_conversion.fig. This will produce both a figure file and a matching M-file. The M-file will contain stubs for the three callback functions needed by the edit fields and the slider. The resulting GUI is shown during the layout process in Figure 14.30.

The next step in the process is to create the functions to convert degrees Fahrenheit to degrees Celsius. Function to_c will convert temperature from degrees Fahrenheit to degrees Celsius. It must implement the equation

$$\deg C = \frac{5}{9}(\deg F - 32) \tag{14.1}$$

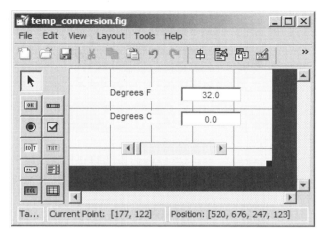

Figure 14.30 Layout of the temperature conversion GUI.

The code for this function is

```
function deg_c = to_c(deg_f)

% Convert degrees Fahrenheit to degrees C.
deg_c = (5/9) * (deg_f - 32);
end % function deg_c
```

Function to_f will convert temperature from degrees Celsius to degrees Fahrenheit. It must implement the equation

$$\deg F = \frac{9}{5}\deg C + 32 \tag{14.2}$$

The code for this function is

```
function deg_f = to_f(deg_c)

% Convert degrees Celsius to degrees Fahrenheit.
deg_f = (9/5) * deg_c + 32;
end % function deg_f
```

Finally, we must write the callback functions to tie it all together. The functions must respond to either edit box or to the slider, and must update all three components. (Note that we will update even the edit box that the user types into, so that the data can be displayed with a consistent format at all times and can correct errors if the user types an out-of-range input value.)

There is an extra complication here since the values entered into edit boxes are *strings* and we wish to treat them as *numbers*. If a user types the value 100 into an edit box, he or she has really created the string '100', not the number 100. The callback function must convert the strings into numbers so that the conversion can be calculated. This conversion is done with the str2num function that converts a string into a numerical value.

Also, the callback function will have to limit user entries to the valid temperature range, which is 0–100 °C and 32–212 °F.

The resulting callback functions are shown in Figure 14.31.

```
function Edit1_Callback(hObject, eventdata, handles)

% Update all temperature values
deg_f = str2num( get(hObject,'String'));
deg_f = max( [ 32 deg_f] );
deg_f = min( [212 deg_f] );
deg_c = to_c(deg_f);

% Now update the fields
set (handles.Edit1,'String',sprintf('%.1f',deg_f));
set (handles.Edit2,'String',sprintf('%.1f',deg_c));
set (handles.Slider1,'Value',deg_c);

function Edit2_Callback(hObject, eventdata, handles)

% Update all temperature values
deg_c = str2num( get(hObject,'String') );
deg_c = max( [ 0 deg_c] );
deg_c = min( [100 deg_c] );
deg_f = to_f(deg_c);

% Now update the fields
set (handles.Edit1,'String',sprintf('%.1f',deg_f));
set (handles.Edit2,'String',sprintf('%.1f',deg_c));
set (handles.Slider1,'Value',deg_c);

function Slider1_Callback(hObject, eventdata, handles)

% Update all temperature values
deg_c = get(hObject,'Value');
deg_f = to_f(deg_c);

% Now update the fields
set (handles.Edit1,'String',sprintf('%.1f',deg_f));
set (handles.Edit2,'String',sprintf('%.1f',deg_c));
set (handles.Slider1,'Value',deg_c);
```

Figure 14.31 Callback functions for the temperature conversion GUI.

The program is now complete. Execute it and enter several different values using both the edit boxes and the sliders. Be sure to use some out-of-range values. Does it appear to be functioning properly?

14.5 Additional Containers: Panels and Button Groups

MATLAB GUIs include two other types of containers: **panels** (created by the function `uipanel`) and **button groups** (created by the function `uibuttongroup`).

14.5.1 Panels

Panels are containers that can contain components or other containers, but they do *not* have a title bar and cannot have menus attached. A panel can contain GUI elements such as uicontrols, tables, axes, other panels, or button groups. Any elements placed in a panel will be positioned relative to the panel. If the panel is moved on the GUI, then all of the elements within it are moved as well. Panels are a great way to group related controls on a GUI.

A panel is created by a `uipanel` function. It can be added to a GUI by using the panel tool ( ) in the Layout Editor.

Each panel has a title and is usually surrounded by an etched or beveled line marking the edges of the panel. The title of a panel can be located at the left, center, or right side of either the top or bottom of the panel. Samples of panels with several combinations of title positions and edge styles are shown in Figure 14.32.

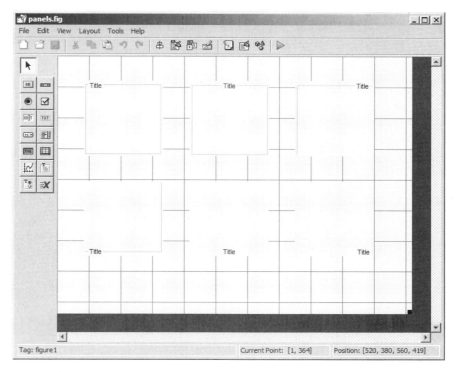

Figure 14.32 Examples of various panel styles.

Let's look at a simple example using panels. Suppose that we wanted to create a GUI to plot the function $y = ax^2 + bx + c$ between two specified values x_{min} and x_{max}. The GUI should allow the user to specify the values a, b, c, x_{min}, and x_{max}. In addition, it should allow the user to specify the style, color, and thickness of the line being plotted. These two sets of values (the ones specifying the line and the ones specifying what the line looks like) are logically distinct, so we can group them together in two panels on the GUI. One possible layout is shown in Figure 14.33. (You will be asked to finish this GUI and create an operational program in Exercise 14.7 at the end of the chapter.)

Table 14.4 contains a list of some important `uipanel` properties. These properties can be modified by the Property Inspector during the design phase, or they can be modified during execution with `get` and `set` functions.

14.5.2 Button Groups

Button groups are a special type of panel that can manage groups of radio buttons or toggle buttons to ensure that *no more than one button in the group is on at any time*. A button group is just like any other panel, except that the button group ensures that at most one radio button or toggle button is on at any given time. If one of them is turned on, then the button group turns off any buttons that were already on.

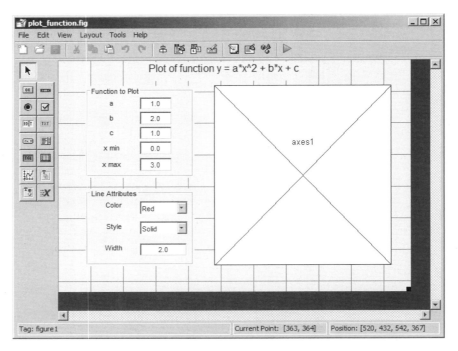

Figure 14.33 Layout of the Plot Function GUI, using panels to group related characteristics together.

Table 14.4: Important `uipanel` and `uibuttongroup` Properties

Property	Description
BackgroundColor	Specifies the color of the `uipanel` background. The value is either a predefined color such as `'r'`, `'g'`, or `'b'`, or else a three-element vector specifying the red, green, and blue components of the color on a 0–1 scale. For example, the color magenta would be specified by `[1 0 1]`.
BorderType	Type of border around the `uipanel`. Options are `'none'`, `'etchedin'`, `'etchedout'`, `'beveledin'`, `'beveledout'`, or `'line'`. The default border type is `'etchedin'`.
BorderWidth	Width of border around the `uipanel`.
FontAngle	A string containing the font angle for the title text. Possible values are `'normal'`, `'italic'`, and `'oblique'`.
FontName	A string containing the font name for the title text.
FontSize	A number specifying the font size for the title text.
FontUnits	The units in which the font size is defined. Possible choices are `'inches'`, `'centimeters'`, `'normalized'`, `'points'`, and `'pixels'`. The default font units are `'points'`.
FontWeight	A string containing the font weight for the title text. Possible values are `'light'`, `'normal'`, `'demi'`, and `'bold'`. The default font weight is `'normal'`.
ForegroundColor	Specifies the color of the title font and the border.
HighlightColor	Specifies the 3D border highlight color.
Position	Specifies the position of a panel relative to its parent `figure`, `uipanel`, or `uibuttongroup`, in the units specified by the `'units'` property. This value accepts a four-element vector in which the first two elements are the x and y positions of the lower left-hand corner of the panel, and the next two elements are the width and height of the panel.
ShadowColor	Specifies the color of the 3D border shadow.
Tag	The "name" of the `uipanel`, which can be used to access it.
Title	The title string.
TitlePosition	Location of the title string on the `uipanel`. Possible values are `'lefttop'`, `'centertop'`, `'righttop'`, `'leftbottom'`, `'centerbottom'`, and `'rightbottom'`. The default value is `'lefttop'`.
Units	The units used to describe the position of the `uipanel`. Possible choices are `'inches'`, `'centimeters'`, `'normalized'`, `'points'`, `'pixels'`, or `'characters'`. The default units are `'normalized'`.
Visible	Specifies whether or not this `uipanel` is visible. Possible values are `'on'` or `'off'`.

A button group is created by a `uibuttongroup` function. It can be added to a GUI by using the button group tool ([■]) in the Layout Editor.

If a radio button or a toggle button is controlled by a button group, then the user must attach the name of the function to execute when that button is selected

Figure 14.34 A button group controlling three radio buttons.

in a special button group property called `SelectionChangedFcn`. This callback is executed by the GUI whenever a radio button or toggle button changes state. Do *not* place the function in the usual `Callback` property, since the button group overwrites the callback property for every radio button or toggle button that it controls.

Figure 14.34 shows a simple GUI containing a button group and three radio buttons, labeled `'Option 1'`, `'Option 2'`, and `'Option 3'`. When a user clicks on one radio button in the group, the button is turned on and all other buttons in the group are turned off.

14.6 Dialog Boxes

A **dialog box** is a special type of figure that is used to display information or to get input from a user. Dialog boxes are used to display errors, provide warnings, ask questions, or get user input. They are also used to select files or printer properties.

Dialog boxes may be **modal** or **non-modal**. A modal dialog box does not allow any other window in the application to be accessed until it is dismissed, while a normal dialog box does not block access to other windows. Modal dialog boxes are typically used for warning and error messages that need urgent attention and cannot be ignored. By default, most dialog boxes are non-modal.

MATLAB includes many types of dialog boxes, the more important of which are summarized in Table 14.5. We will examine only a few types of dialog boxes here, but you can consult the MATLAB online documentation for the details of the others.

Table 14.5: Selected Dialog Boxes

Property	Description
dialog	Creates a generic dialog box.
errordlg	Displays an error message in a dialog box. The user must click the OK button to continue.
helpdlg	Displays a help message in a dialog box. The user must click the OK button to continue.
inputdlg	Displays a request for input data, and accepts the user's input values.
listdlg	Allows a user to make one or more selections from a list.
msgbox	Displays a message in a dialog box.
printdlg	Displays a printer selection dialog box.
questdlg	Asks a question. This dialog box can contain either two or three buttons, which by default are labeled Yes, No, and Cancel.
uigetdir	Displays a file open dialog box. This box allows a user to select a directory to open.
uigetfile	Displays a file open dialog box. This box allows a user to select a file to open *but does not actually open the file.*
uiputfile	Displays a file save dialog box. This box allows a user to select a file to save *but does not actually save the file.*
uisave	Save workspace variables to a file.
uisetcolor	Displays a color selection dialog box.
uisetfont	Displays a font selection dialog box.
waitbar	Displays or updates a wait bar dialog box.
warndlg	Displays a warning message in a dialog box. The user must click the OK button to continue.

14.6.1 Error and Warning Dialog Boxes

Error and warning dialog boxes have similar calling parameters and behavior. In fact, the only difference between them is the icon displayed in the dialog box. The most common calling sequence for these dialog boxes is

```
errordlg(error_string,box_title,create_mode);
warndlg(warning_string,box_title,create_mode);
```

The error_string or warning_string is the message to display to the user, and the box_title is the title of the dialog box. Finally, create_mode is a string that can be 'modal' or 'non-modal', depending on the type of dialog box you wish to create.

For example, the following statement creates a modal error message that cannot be ignored by the user. The dialog box produced by this statement is shown in Figure 14.35.

```
errordlg('Invalid input values!','Error Dialog Box','modal');
```

14.6.2 Input Dialog Boxes

Input dialog boxes prompt a user to enter one or more values that may be used by a program. They may be created with one of the following calling sequences.

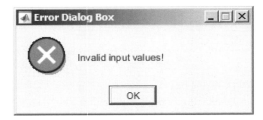

Figure 14.35 An error dialog box.

```
answer = inputdlg(prompt)
answer = inputdlg(prompt,title)
answer = inputdlg(prompt,title,line_no)
answer = inputdlg(prompt,title,line_no,default_answer)
```

Here, `prompt` is a cell array of strings, with each element of the array corresponding to one value that the user will be asked to enter. The parameter `title` specifies the title of the dialog box, while `line_no` specifies the number of lines to be allowed for each answer. Finally, `default_answer` is a cell array containing the default answers that will be used if the user fails to enter data for a particular item. Note that there must be as many default answers as there are prompts.

When the user clicks the `OK` button on the dialog box, his or her answers will be returned as a cell array of strings in variable `answer`.

As an example of an input dialog box, suppose that we wanted to allow a user to specify the position of a figure using an input dialog. The code to perform this function would be

```
prompt{1} = 'Starting x position:';
prompt{2} = 'Starting y position:';
prompt{3} = 'Width:';
prompt{4} = 'Height:';
title = 'Set Figure Position';
default_ans = {'50','50','180','100'};
answer = inputdlg(prompt,title,1,default_ans);
```

The resulting dialog box is shown in Figure 14.36.

14.6.3 The `uigetfile`, `uisetfile` and `uigetdir` Dialog Boxes

The `uigetfile` and `uisetfile` dialog boxes allow a user to interactively pick files to open or save. These functions use the standard file open or file save dialog boxes for the particular operating system that MATLAB is running on. They return strings containing the name and the path of the file but do not actually read or save it. The programmer is responsible for writing additional code for that purpose.

The form of these two dialog boxes is

```
[filename, pathname] = uigetfile(filter_spec,title);
[filename, pathname] = uisetfile(filter_spec,title);
```

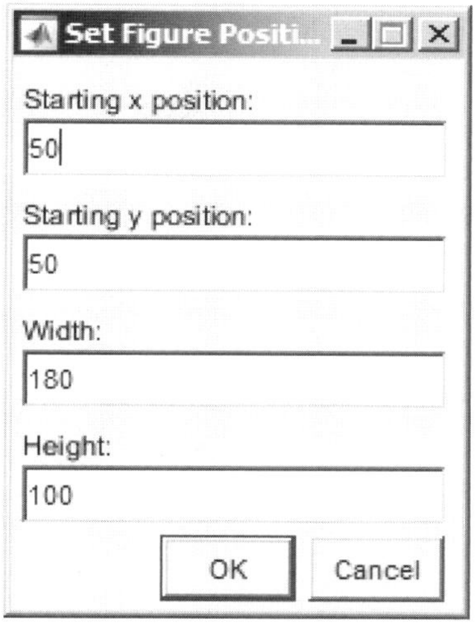

Figure 14.36 An input dialog box.

Parameter `filter_spec` is a string specifying the type of files to display in the dialog box, such as `'*.m'`,`'*.mat'`, and so forth. Parameter `title` is a string specifying the title of the dialog box. After the dialog box executes, `filename` contains the name of the selected file and `pathname` contains the path of the file. If the user cancels the dialog box, `filename` and `pathname` are set to zero.

The following script file illustrates the use of these dialog boxes. It prompts the user to enter the name of a mat-file, and then reads the contents of that file. The dialog box created by this code on a Windows 7 system is shown in Figure 14.37. (This is the standard open file dialog for Windows 7. It will appear slightly different on other Windows or Linux systems).

```
[filename, pathname] = uigetfile('*.mat','Load MAT File');
if filename ~= 0
   load([pathname filename]);
end
```

The `uigetdir` dialog box allows a user to interactively select a directory. This function uses the standard directory selection dialog box for the particular operating system that MATLAB is running on. It returns the name of the directory but does not actually do anything with it. The programmer is responsible for writing additional code to use the directory name.

The form of this dialog box is

```
directoryname = uigetdir(start_path, title);
```

Parameter `start_path` is the path of the initially selected directory. If it is not valid, then the dialog box opens with the base directory selected. Parameter `title` is a string specifying the title of the dialog box. After the dialog box executes, `directoryname` contains the name of the selected directory. If the user cancels the dialog box, `directoryname` is set to zero.

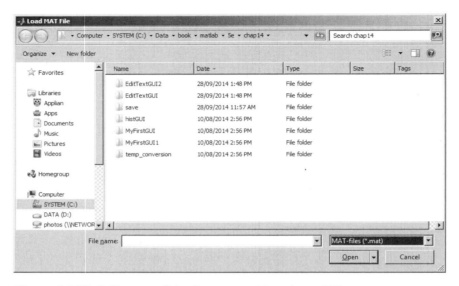

Figure 14.37 A file open dialog box created by `uigetfile`.

The following script file illustrates the use of this dialog box. It prompts the user to select a directory starting with the current MATLAB working directory. This dialog box created by this code on a Windows 7 system is shown in Figure 14.38. (This is the standard open file dialog for Windows 7. It will appear slightly different on other Windows or Linux systems).

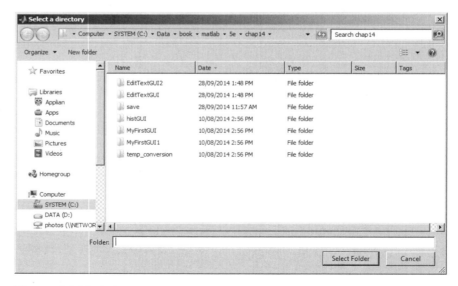

Figure 14.38 A directory selection dialog box created by `uigetdir`.

```
dir1 = uigetdir('C:\book\matlab\5e\chap14','Select a directory');
if dir1 ~= 0
    cd(dir1);
end
```

14.6.4 The `uisetcolor` and `uisetfont` Dialog Boxes

The `uisetcolor` and `uisetfont` dialog boxes allow a user to interactively select colors or fonts using the standard dialog boxes for the computer on which MATLAB is executing. The appearances of these boxes will vary for different operating systems. They provide a standard way to select colors or fonts within a MATLAB GUI.

Consult the MATLAB online documentation to learn more about these special-purpose dialog boxes. We will use them in some of the end-of-chapter exercises.

Good Programming Practice

Use dialog boxes to provide information or request input in GUI-based programs. If the information is urgent and should not be ignored, make the dialog boxes modal.

14.7 Menus

Menus can also be added to MATLAB GUIs. A menu allows a user to select actions without additional components appearing on the GUI display. They are useful for selecting less commonly used options without cluttering up the GUI with a lot of extra buttons.

There are two types of menus in MATLAB: **standard menus**, which are pulled down from the menu bar at the top of a figure, and **context menus**, which pop up over the figure when a user right-clicks the mouse over a graphical object. We will learn how to create and use both types of menus in this section.

Standard menus are created with `uimenu` objects. Each item in a menu is a separate `uimenu` object, including items in submenus. These `uimenu` objects are similar to `uicontrol` objects, and they have many of the same properties such as `Parent`, `Callback`, `Enable`, and so forth. A list of the more important `uimenu` properties is given in Table 14.6.

Each menu item is attached to a parent object, which is a figure for the top-level menus, or another menu item for submenus. All of the `uimenus` connected to the same parent appear on the same menu, and the cascade of items forms a tree of submenus. Figure 14.39*a* shows a typical MATLAB menu in operation, while Figure 14.39*b* shows the relationship among the objects making up the menu.

MATLAB menus are created using the Menu Editor, which can be selected by clicking the (▨) icon on the toolbar in the `guide` Layout Editor. Figure 14.39*c* shows the Menu Editor with the menu items that generate this menu structure. The additional properties in Table 14.6 that are not shown in the Menu Editor can be set with the Property Editor (`propedit`).

Table 14.6: Important `uimenu` Properties

Property	Description
Accelerator	A single character specifying the keyboard equivalent for the menu item. The keyboard combination CTRL + key allows a user to activate the menu item from the keyboard.
Callback	Specifies the name and parameters of the function to be called when the menu item is activated. It the menu item has a submenu, the callback executes *before the submenu is displayed.* If the menu item does not have submenus, then the callback executes when the mouse button is *released.*
Checked	When this property is `'on'`, a checkmark is placed to the left of the menu item. This property can be used to indicate the status of menu items that toggle between two states. Possible values are `'on'` or `'off'`.
Enable	Specifies whether or not this menu item is selectable. If it is not enabled, the menu item will not respond to mouse clicks or accelerator keys. Possible values are `'on'` or `'off'`.
ForegroundColor	Set color of text in the menu item.
Label	Specifies the text to be displayed on the menu. The ampersand character (&) can be used to specify a keyboard mnemonic for this menu item; it will not appear on the label. For example, the string `'&File'` will create a menu item displaying the text `'File'` and responding to the F key.
Parent	The handle of the parent object for this menu item. The parent object could be a figure or another menu item.
Position	Specifies the position of a menu item on the menu bar or within a menu. Position 1 is the left-most menu position for a top-level menu, and the highest position within a submenu.
Separator	When this property is `'on'`, a separating line is drawn above this menu item. Possible values are `'on'` or `'off'`.
Tag	The "name" of the menu item, which can be used to access it.
Visible	Specifies whether or not this menu item is visible. Possible values are `'on'` or `'off'`.

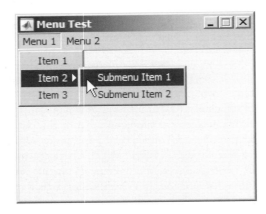

(a)

Figure 14.39 (a) A typical menu structure.

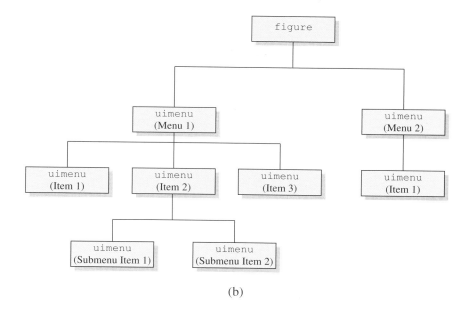

(b)

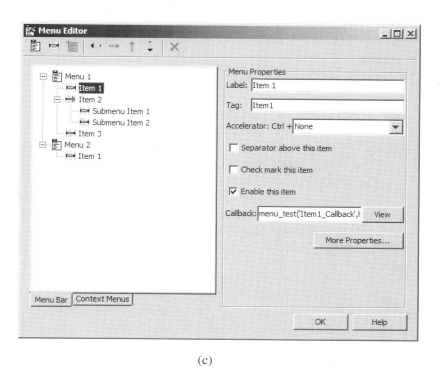

(c)

Figure 14.39 (*Continued*) (b) The relationships among the `uimenu` items creating the menu. (c) The Menu Editor structure that generated these menus.

Table 14.7: Important `uicontextmenu` Properties

Property	Description
Callback	Specifies the name and parameters of the function to be called when the context menu is activated. The callback executes before the context menu is displayed.
Parent	The handle of the parent object for this context menu.
Tag	The "name" of the context menu, which can be used to access it.
Visible	Specifies whether or not this context menu is visible. This property is set automatically and should normally not be modified.

Top-level context menus are created by `uicontextmenu` objects, and the lower level items within context menus are created by `uimenu` objects. Context menus are basically the same as standard menus, except that they can be associated with any GUI object (axes, lines, text, figures, and so forth).

14.7.1 Suppressing the Default Menu

Every MATLAB figure comes with a default set of standard menus. If you wish to delete these menus from a figure and create your own menus, you must first turn the default menus off. The display of default menus is controlled by the figure's `MenuBar` property. The possible values of this property are `'figure'` and `'none'`. If the property is set to `'figure'`, then the default menus are displayed. If the property is set to `'none'`, then the default menus are suppressed. You can use the Property Inspector to set the `MenuBar` property for your GUIs when you create them.

14.7.2 Creating Your Own Menus

Creating your own standard menus for a GUI is basically a three-step process.

1. First, create a new menu structure with the Menu Editor. Use the Menu Editor to define the structure, giving each menu item a `Label` to display and a unique `Tag` value. You can also specify whether or not there is a separator bar between menu items and whether or not each menu item has a check mark by it. A dummy callback function will be generated automatically for each menu item.

2. If necessary, edit the properties of each menu item using the Property Inspector. The Property Inspector can be started by clicking the More Options button on the Menu Editor. The most important menu item properties (`Label`, `Tag`, `Callback`, `Checked`, and `Separator`) can be set on the Menu Editor, so the Property Inspector is usually not needed. However, if you must set any of the other properties listed in Table 14.6, you will need to use the Property Inspector.

3. Third, implement a callback function to perform the actions required by your menu items. The prototype function is created automatically, but you must add the code to make each menu item behave properly.

The process of building menus will be illustrated in an example at the end of this section.

//

⊠ Programming Pitfalls

Only the `Label`, `Tag`, `Callback`, `Checked`, and `Separator` properties of a menu item can be set from the Menu Editor. If you need to set any of the other properties, you will have to use the Property Inspector on the figure, and select the appropriate menu item to edit.

//

14.7.3 Accelerator Keys and Keyboard Mnemonics

MATLAB menus support accelerator keys and keyboard mnemonics. **Accelerator keys** are "CTRL + key" combinations that cause a menu item to be executed *without opening the menu first*. For example, the accelerator key "o" might be assigned to the File/Open menu item. In that case, the keyboard combination CTRL + o will cause the File/Open callback function to be executed.

A few CRTL + key combinations are reserved for the use of the host operating system. These combinations differ between PC and Linux systems; consult the MATLAB online documentation to determine which combinations are legal for your type of computer.

Accelerator keys are defined by setting the `Accelerator` property in a `uimenu` object.

Keyboard mnemonics are single letters that can be pressed to cause a menu item to execute once the menu is open. The keyboard mnemonic letter for a given menu item is underlined[1]. For top-level menus, the keyboard mnemonic is executed by pressing ALT plus the mnemonic key at the same time. Once the top level menu is open, simply pressing the mnemonic key will cause a menu item to execute.

Figure 14.40 illustrates the use of keyboard mnemonics. The File menu is opened with the keys ALT + f, and once it is opened, the Exit menu item can be executed by simply typing "x"

Keyboard mnemonics are defined by placing the ampersand character (&) before the desired mnemonic letter in the `Label` property. The ampersand will not be displayed, but the following letter will be underlined, and it will act as a mnemonic key. For example, the `Label` property of the Exit menu item in Figure 14.40 is `'E&xit'`.

14.7.4 Creating Context Menus

Context menus are created in the same fashion as ordinary menus, except that the top-level menu item is a `uicontextmenu`. The parent of a `uicontextmenu`

---

[1]On Windows, the underlines are hidden until the ALT key is held down. This behavior can be modified. For example, the underlines can be made visible all the time in Windows 7 by selecting the "Underline keyboard shortcuts and access keys" option in the Ease of Access Center of the Control Panel.

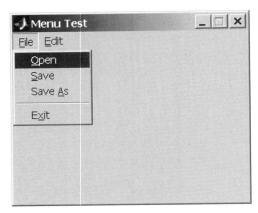

Figure 14.40 An example showing keyboard mnemonics. The menu shown was opened by typing the keys ALT+f, and the Exit option could be executed by simply typing "x".

must be a figure, but the context menu can be associated with and respond to right mouse clicks on any graphical object. Context menus are created using the "Context Menu" selection on the Menu Editor. Once the context menu is created, any number of menu items can be created under it.

To associate a context menu with a specific object, you must set the object's UIContextMenu property to the handle of the uicontextmenu. This is normally done using the Property Inspector, but it can be done with the set command as shown below. If Hcm is the handle to a context menu, the following statements will associate the context menu with a line created by a plot command.

```
H1 = plot(x,y);
set (H1,'UIContextMenu',Hcm);
```

We will create a context menu and associate it with a graphical object in the following example.

▶Example 14.2—Plotting Data Points

Write a program that opens a user-specified data file and plots the line specified by the points in the file. The program should include a File menu, with Open and Exit menu items. The program should also include a context menu attached to the line, with options to change the line style. Assume that the data in the file is in the form of (x,y) pairs, with one pair of data values per line.

Solution This program should include a standard menu with Open and Exit menu items, plus a set of axes on which to plot the data. It should also include a context menu specifying various line styles, which can be attached to the line after it is plotted. The options should include solid, dashed, dotted, and dash-dot line styles.

The first step in creating this program is to use guide to create the required GUI, which is only a set of axes in this case (see Figure 14.41*a*). Then, we must use

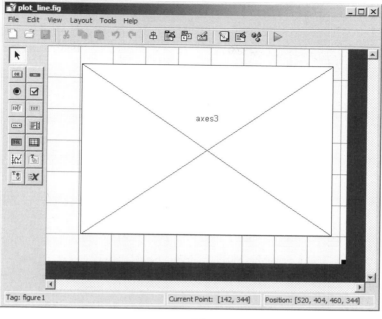

(a)

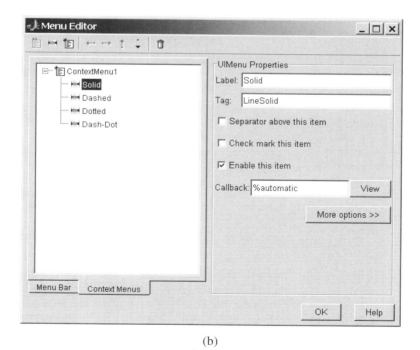

(b)

Figure 14.41 (a) The layout for `plot_line`. (b) The File menu in the Menu Editor.

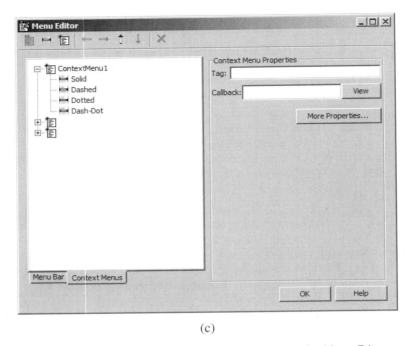

(c)

Figure 14.41 (*Continued*) (c) The context menu in the Menu Editor.

the Menu Editor to create the File menu. This menu will contain Open and Exit menu items, as shown in Figure 14.41*b*. Note that we must use the Menu Editor to set the `Label` and `Tag` and strings for each of these menu items. We will also define keyboard mnemonics "F" for File, "O" for Open and "x" for Exit, and place a separator between the Open and Exit menu items. Figure 14.41*b* shows the Exit menu item. Note that "x" is the keyboard mnemonic, and that the separator switch is turned on.

Next, we must use the Menu Editor to create the context menu. This menu starts with a `uicontextmenu` object, with four menu items attached to it (see Figure 14.41*c*). Again, we must set the `Label` and `Tag` strings for each of these menu items.

At this point, the GUI should be saved as `plot_line.fig`, and `plot_line.m` will be automatically created. Dummy callback functions will be automatically created for the menu items.

After the GUI is created, we must implement six callback functions for the Open, Exit, and linestyle menu items. The most difficult callback function is the response to the File/Open menu item. This callback must prompt the user for the name of the file (using a `uigetfile` dialog box), open the file, read the data, save it into `x` and `y` arrays, and close the file. Then, it must plot the line and save the line's handle as application data so that we can use it to modify the line style later. Finally, it must associate the context menu with the line. The `FileOpen_Callback` function is shown in Figure 14.42. Note that the function uses a dialog box to inform the user of file open errors.

```
function varargout = FileOpen_Callback(h, eventdata, ...
                                       handles, varargin)

% Get the file to open
[filename, pathname] = uigetfile ('*.dat','Load Data');
if filename ~= 0

    % Open the input file
    filename = [pathname filename];
    [fid,msg] = fopen(filename,'rt');

    % Check to see if the open failed.
    if fid < 0

        % There was an error--tell user.
        str = ['File ' filename ' could not be opened.'];
        title = 'File Open Failed';
        errordlg(str,title,'modal');

    else

        % File opened successfully. Read the (x,y) pairs from
        % the input file.  Get first (x,y) pair before the
        % loop starts.
        [in,count] = fscanf(fid,'%g',2);
        ii = 0;

        while ~feof(fid)
            ii = ii + 1;
            x(ii) = in(1);
            y(ii) = in(2);
            % Get next (x,y) pair
            [in,count] = fscanf(fid,'%g',2);
        end

        % Data read in.  Close file.
        fclose(fid);

        % Now plot the data.
        hline = plot(x,y,'LineWidth',3);
        xlabel('x');
        ylabel('y');
        grid on;

        % Associate the context menu with line
        set(hline,'Uicontextmenu',handles.ContextMenu1);
```

Get file name to open

Open file

Error message if open fails

Read data

Plot line

Set context menu

Figure 14.42 The File/Open callback function.

```
% Save the line's handle as application data
handles.hline = hline;
guidata(gcbf, handles);
```

Save handle as app data

```
    end
end
```

Figure 14.42 *(continued)*

The remaining callback functions are very simple. The `FileExit_Callback` function simply closes the figure, and the line style functions simply set the line style. When the user right-clicks a mouse button over the line, the context menu will appear. If the user selects an item from the menu, the resulting callback will use the line's saved handle to change its properties. These five functions are shown in Figure 14.43.

The output of the final program is shown in Figure 14.44. Experiment with it on your own computer to verify that it behaves properly.

```
function varargout = FileExit_Callback(h, eventdata, ...
                                      handles, varargin)
close(gcbf);

function varargout = LineSolid_Callback(h, eventdata, ...
                                       handles, varargin)
set(handles.hline,'LineStyle','-');

function varargout = LineDashed_Callback(h, eventdata, ...
                                        handles, varargin)
set(handles.hline,'LineStyle','--');

function varargout = LineDotted_Callback(h, eventdata, ...
                                        handles, varargin)
set(handles.hline,'LineStyle',':');

function varargout = LineDashDot_Callback(h, eventdata, ...
                                         handles, varargin)
set(handles.hline,'LineStyle','-.');
```

Figure 14.43 The remaining callback functions in `plot_line`.

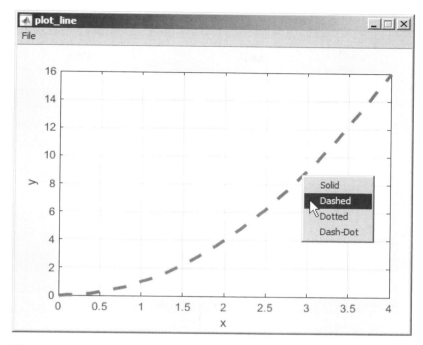

Figure 14.44 The GUI produced by program `plot_line`.

Quiz 14.1

This quiz provides a quick check to see if you have understood the concepts introduced in Sections 14.1 through 14.7. If you have trouble with the quiz, reread the section, ask your instructor, or discuss the material with a fellow student. The answers to this quiz are found in the back of the book.

1. List the types of graphical components discussed in this chapter. What is the purpose of each one?
2. List the types of containers discussed in this chapter. What are the differences among them?
3. What is a callback function? How are callback functions used in MATLAB GUIs?
4. Describe the steps required to create a GUI-based program.
5. Describe the purpose of the handles data structure.
6. How is application data saved in a MATLAB GUI? Why would you want to save application data in a GUI?

7. How can you make a graphical object invisible? How can you turn a graphical object off so that it will not respond to mouse clicks or keyboard input?

8. Which of the GUI components described in this chapter respond to mouse clicks? Which ones respond to keyboard inputs?

9. What are dialog boxes? How can you create a dialog box?

10. What is the difference between a modal and a non-modal dialog box?

11. What is the difference between a standard menu and a context menu? What components are used to create these menus?

12. What are accelerator keys? What are mnemonics?

14.8 Tips for Creating Efficient GUIs

This section lists a few miscellaneous tips for creating efficient graphical user interfaces.

14.8.1 Tool Tips

MATLAB GUIs support **tool tips**, which are small help windows that pop up beside a `uicontrol` GUI object whenever the mouse is held over the object for a while. Tool tips are used to provide a user with quick help about the purpose of each object on a GUI.

A tool tip is defined by setting an object's `TooltipString` property to the string that you wish to display. You will be asked to create tool tips in the end-of-chapter exercises.

Good Programming Practice

Define tool tips to provide users with helpful hints about the functions of your GUI components.

14.8.2 Toolbars

MATLAB GUIs can also support *toolbars*. A toolbar is a row of special pushbutton or toggle buttons along the top of a figure, just below the menu bar. Each button has a small figure or icon on it, representing its function. We have seen examples of toolbars in most of the MATLAB figures produced in this book. For example, Figure 14.45 shows a simple plot displaying the default toolbar.

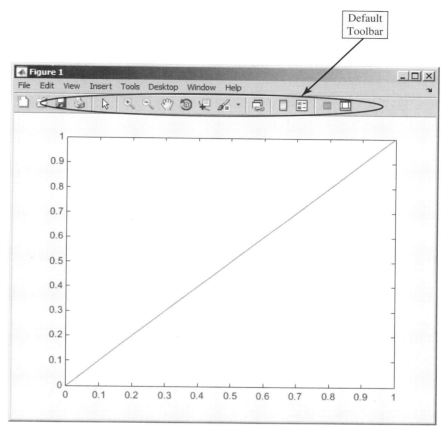

Figure 14.45 A MATLAB figure showing the default toolbar.

Every figure has a `ToolBar` property, which determines whether or not the default figure toolbar is displayed. The possible values of this property are `'none'`, `'auto'`, and `'figure'`. If the property is `'none'`, the default toolbar is not displayed. If the property is `'figure'`, the default toolbar is displayed. If the property is `'auto'`, the default toolbar is displayed unless the user defines a custom toolbar. If the property is `'auto'` and the user defines a custom toolbar, then it will be displayed instead of the default toolbar.

A programmer can create his or her own toolbar, using the `uitoolbar` function, and can add the toolbar equivalent of pushbutton and toggle buttons to the toolbar, using the `uipushtool` and `uitoggletool` functions. The user-defined toolbar can be displayed in addition to or instead of the default figure toolbar.

Toolbars are created and modified in guide by clicking on the Toolbar Editor (⬚).

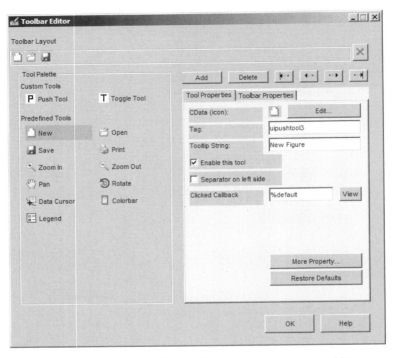

Figure 14.46 The Toolbar Editor allows a programmer to add items to the toolbar by dragging and dropping them from the Tool Palette to the Toolbar Layout.

14.8.3 Additional Enhancements

GUI-based programs can be much more sophisticated than we have described in this introductory chapter. In addition to the `Callback` property that we have been using in the chapter, `uicontrols` support four other types of callbacks: `CreateFcn`, `DeleteFcn`, `ButtonDownFcn`, and `KeyPressFcn`. MATLAB figures also support three important types of callbacks: `WindowButtonDownFcn`, `WindowButtonMotionFcn`, and `WindowButtonUpFcn`.

The **CreateFcn** property defines a callback that is automatically called whenever an object is created. It allows a programmer to customize his or her objects as they are created during program execution. Since this callback is executed before the object is completely defined, a programmer must specify the function to execute as a default property of the root before the object is created. For example, the following statement will cause the function `function_name` to be executed each time that a `uicontrol` is created. The function will be called after MATLAB creates the object's properties, so they will be available to the function when it executes.

```
set(groot,'DefaultUicontrolCreateFcn','function_name')
```

The **DeleteFcn** property defines a callback that is automatically called whenever an object is destroyed. It is executed before the object's properties are destroyed, so they will be available to the function when it executes. This callback provides the programmer with an opportunity to do custom clean-up work.

The **ButtonDownFcn** property defines a callback that is automatically called whenever a mouse button is pressed within a five-pixel border around a uicontrol. If the mouse button is pressed on the uicontrol, the Callback is executed. Otherwise, if it is near the border, the ButtonDownFcn is executed. If the uicontrol is not enabled, the ButtonDownFcn is executed even for clicks on the control.

The **KeyPressFcn** property defines a callback that is automatically called whenever a key is pressed while the specified object is selected or highlighted. This function can find out which key was pressed by checking the CurrentCharacter property of the enclosing figure, or else by checking the contents of the event data structure passed to the callback. It can use this information to change behavior, depending on which key was pressed.

The figure-level callback functions **WindowButtonDownFcn**, **WindowButtonMotionFcn**, and **WindowButtonUpFcn** allow a programmer to implement features such as animations and drag-and-drop, since the callbacks can detect the initial, intermediate, and final locations at which the mouse button is pressed. They are beyond the scope of this book, but are well worth learning about. Refer to the *Creating Graphical User Interfaces* section in the MATLAB user documentation for a description of these callbacks.

▶ Example 14.3—Creating a Histogram GUI

Write a program that opens a user-specified data file and calculates a histogram of the data in the file. The program should calculate the mean, median, and standard deviation of the data in the file. It should include a File menu, with Open and Exit menu items. It should also include a means to allow the user to change the number of bins in the histogram.

Select a color other than the default color for the figure and the text label backgrounds, use keyboard mnemonics for menu items, and add tool tips where appropriate.

Solution This program should include a standard menu with Open and Exit menu items, a set of axes on which to plot the histogram, and a set of six text fields for the mean, median, and standard deviation of the data. Three of these text fields will hold labels, and three will hold the read-only mean, median and standard deviation values. It must also include a label and an edit field to allow the user to selected the number of bins to display in the histogram.

We will select a light blue color [0.6 1.0 1.0] for the background of this GUI. To make the GUI have a light blue background, this color vector must be loaded into the 'Color' property of the figure and into the 'BackgroundColor' property of each text label with the Property Inspector during the GUI layout.

The first step in creating this program is to use guide to lay out the required GUI (see Figure 14.47a). Then, use the Property Inspector to set the properties of the seven text fields and the edit field. The fields must be given unique tags so that we can locate them from the callback functions. Next, use the Menu Editor to create the File menu (see Figure 14.47b). Finally, the resulting GUI should be saved as histGUI, creating histGUI.fig and histGUI.m.

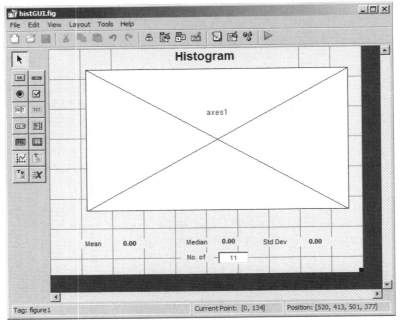

(a)

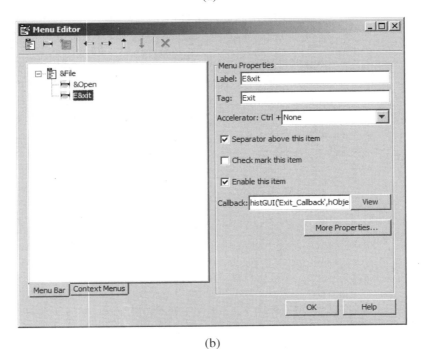

(b)

Figure 14.47 (a) The layout for `histGUI`. (b) The File menu in the Menu Editor.

After histGUI.m is saved, the function histGUI_OpeningFcn must be edited to initialize the background color of the figure, and to save the initial number of histogram bins in the handles structure. The modified code for the opening function is:

```
% --- Executes just before histGUI is made visible.
function histGUI_OpeningFcn(hObject, eventdata, handles, varargin)
% This function has no output args, see OutputFcn.
% hObject      handle to figure
% eventdata  reserved - to be defined in a future version of MATLAB
% handles      structure with handles and user data (see GUIDATA)
% varargin    command line arguments to histGUI (see VARARGIN)

% Choose default command line output for histGUI
handles.output = hObject;

% Set the initial number of bins
handles.nbins = 11;

% Update handles structure
guidata(hObject, handles);
```

Next, we must create callback functions for the File/Open menu item, the File/Exit menu item, and the "number of bins" edit box.

The File/Open callback must prompt the user for a file name and then read the data from the file. It must calculate and display the histogram and update the statistics text fields. Note that the data in the file must also be saved in the handles structure, so that it will be available for recalculation if the user changes the number of bins in the histogram. The callback function to perform these steps is shown below:

```
function Open_Callback(hObject, eventdata, handles)
% hObject      handle to Open (see GCBO)
% eventdata  reserved - to be defined in a future version of MATLAB
% handles      structure with handles and user data (see GUIDATA)

% Get file name
[filename,path] = uigetfile('*.dat','Load Data File');
if filename ~= 0

  % Read data
  x = textread([path filename],'%f');

  % Save in handles structure
  handles.x = x;
  guidata(gcbf, handles);

  % Create histogram
  hist(handles.x,handles.nbins);

  % Set axis labels
  xlabel('\bfValue');
  ylabel('\bfCount');
```

```
% Calculate statistics
ave = mean(x);
med = median(x);
sd  = std(x);
n   = length(x);

% Update fields
set (handles.MeanData,'String',sprintf('%7.2f',ave));
set (handles.MedianData,'String',sprintf('%7.2f',med));
set (handles.StdDevData,'String',sprintf('%7.2f',sd));
set (handles.TitleString,'String',['Histogram  (N = '  int2str(n)
')']);

end
```

The File/Exit callback is trivial. All it has to do is close the figure.

```
function Exit_Callback(hObject, eventdata, handles)
% hObject     handle to Exit (see GCBO)
% eventdata   reserved - to be defined in a future version of MATLAB
% handles     structure with handles and user data (see GUIDATA)

close(gcbf);
```

The NBins callback must read a numeric input value, round it off to the nearest integer, display that integer in the Edit Box, and recalculate and display the histogram. Note that the number of bins must also be saved in the handles structure, so that it will be available for recalculation if the user loads a new data file. The callback function to perform these steps is shown below:

```
function NBins_Callback(hObject, eventdata, handles)
% hObject     handle to NBins (see GCBO)
% eventdata   reserved - to be defined in a future version of MATLAB
% handles     structure with handles and user data (see GUIDATA)

% Get number of bins, round to integer, and update field
nbins = str2num(get(hObject,'String'));
nbins = round(nbins);
if nbins < 1
   nbins = 1;
end
set (handles.NBins,'String',int2str(nbins));

% Save in handles structure
handles.nbins = nbins;
guidata(gcbf, handles);

% Re-display data, if available
if handles.nbins > 0 & ~isempty(handles.x)

   % Create histogram
   hist(handles.x,handles.nbins);

end
```

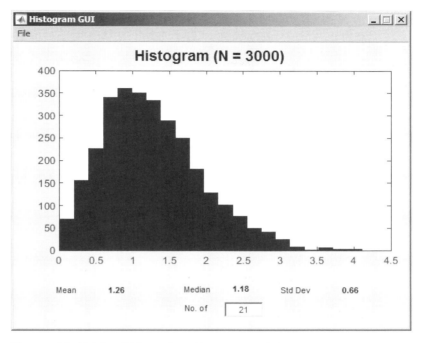

Figure 14.48 The GUI produced by program `histGUI`.

The final program is shown in Figure 14.48. Experiment with it on your own computer to verify that it behaves properly.

14.9 Summary

In Chapter 14, we learned how to create MATLAB graphical user interfaces. The three fundamental parts of a GUI are components (`uicontrols`, `uimenus`, `uicontextmenus`, `uitables`, toolbars, and axes), containers to contain them, and callbacks to implement actions in response to mouse clicks or keyboard inputs.

The standard GUI components created by `uicontrol` include text fields, edit boxes, pushbutton, toggle buttons, checkboxes, radio buttons, popup menus, list boxes, and sliders. The standard GUI component created by `uitable` is the MATLAB table. The standard GUI components created by `uimenu` and `uicontextmenu` are standard menus and context menus.

MATLAB containers consist of figures, panels, and button groups. Figures are created by the `figure` function. They are separate windows, complete with title bars, menus, and toolbars. Panels are created by the `uipanel` function. They are containers that reside within figures or other containers, and do not have title bars, menus, or toolbars. Panels can contain `uicontrol` components and other panels or button groups, and those items will be laid out with respect to the panel itself. If the panel is moved, all of its contents move with it. Button groups are created by the `uibuttongroup` function. They are special types of panels that control any radio buttons or toggle buttons contained within them to ensure that only one of them can be on at any time.

Any of these components and containers can be placed on a figure using `guide` (the GUI Development Environment tool). Once the GUI layout has been completed, the user must edit the object properties with the Property Inspector and then write a callback function to implement the actions associated with each GUI object.

Dialog boxes are special figures used to display information or to get input from a user. Dialog boxes are used to display errors, provide warnings, ask questions, or get user input. They are also used to select files or printer properties.

Dialog boxes may be modal or non-modal. A modal dialog box does not allow any other window in the application to be accessed until it is dismissed, while a normal (non-modal) dialog box does not block access to other windows. Modal dialog boxes are typically used for warning and error messages that need urgent attention and cannot be ignored.

Menus can also be added to MATLAB GUIs. A menu allows a user to select actions without additional components appearing on the GUI display. They are useful for selecting less commonly used options without cluttering up the GUI with a lot of extra buttons. Menus are created with the Menu Editor, and then the programmer must write a callback function to implement the actions associated with each menu item. For each menu item, the user must use the Menu Editor to set at least the `Label` and `Tag` properties.

Accelerator keys and keyboard mnemonics can be used to speed the operation of windows.

Compiling MATLAB functions to pcode can speed the execution of a program. It also protects your investment in your source code by allowing you to distribute the program to others in the form of pcode files. They may be freely executed, but it is not easy for someone to reengineer the files and take your ideas.

MATLAB uicontrol components have several additional properties for specifying less common types of callbacks, including `CreateFcn`, `DeleteFcn`, `ButtonDownFcn`, and `KeyPressFcn`. MATLAB figures also have several properties for specifying types of callbacks, including `WindowButtonDownFcn`, `WindowButtonMotionFcn`, and `WindowButtonUpFcn`. These various callbacks allow a user to customize the appearance and response of the MATLAB GUIs to various user inputs.

14.9.1 Summary of Good Programming Practice

The following guidelines should be adhered to when working with MATLAB GUIs.

1. Use `guide` to lay out a new GUI, and use the Property Inspector to set the initial properties of each component such as the text displayed on the component, the color of the component, and the name of the callback function, if required.

2. After creating a GUI with `guide`, manually edit the resulting function to add comments describing its purpose and components, and to implement the function of callbacks.
3. Store GUI application data in the `handles` structure, so that it will automatically be available to any callback function.
4. If you modify any of the GUI application data in the `handles` structure, be sure to save the structure with a call to `guidata` before exiting the function where the modifications occurred.
5. Use dialog boxes to provide information or request input in GUI-based programs. If the information is urgent and should not be ignored, make the dialog boxes modal.
6. Define tool tips to provide users with helpful hints about the functions of your GUI components.

14.9.2 MATLAB Summary

The following summary lists all of the MATLAB commands and functions described in this chapter, along with a brief description of each one. Also, refer to the summaries of Graphical object properties in Tables 14.2, 14.3, 14.4, 14.6, and 14.7.

`axes`	Function to create a set of axes.
`dialog`	Creates a generic dialog box.
`errordlg`	Displays an error message.
`helpdlg`	Displays a help message.
`findobj`	Finds a GUI object by matching one or more of its properties.
`gcbf`	Gets callback figure.
`gcbo`	Gets callback object.
`guidata`	Saves GUI application data in a figure.
`guihandles`	Gets the `handles` structure from the application data stored in a figure.
`guide`	GUI Development Environment tool.
`inputdlg`	Dialog to get input data from the user.
`printdlg`	Prints dialog box.
`questdlg`	Dialog box to ask a question.
`uibuttongroup`	Creates a button group container.
`uicontrol`	Function to create a GUI object.
`uicontextmenu`	Function to create a context menu.
`uigetdir`	Dialog box to select a directory.
`uigetfile`	Dialog box to select an input file.
`uimenu`	Function to create a standard menu or a menu item on either a standard menu or a context menu.
`uipanel`	Creates a panel.
`uipushtool`	Creates a pushbutton on a user-defined toolbar.
`uiputfile`	Dialog box to select an output file.

`uisetcolor`	Displays a color selection dialog box.
`uisetfont`	Displays a font selection dialog box.
`uitable`	Function to create a table.
`uitoggletool`	Creates a toggle button on a user-defined toolbar.
`uitoolbar`	Creates a user-defined toolbar.
`warndlg`	Displays a warning message.

14.10 Exercises

14.1 Explain the steps required to create a GUI in MATLAB.

14.2 What types of components can be used in MATLAB GUIs? What functions create them, and how do you select a particular component type?

14.3 What types of containers can be used in MATLAB GUIs? What function creates each of them?

14.4 How does a callback function work? How can a callback function locate the figures and objects that it needs to manipulate?

14.5 Create a GUI that uses a standard menu to select the background color displayed by the GUI. Include accelerator keys and keyboard mnemonics in the menu design. Design the GUI so that it defaults to a green background.

14.6 Create a GUI that uses a context menu to select the background color displayed by the GUI. Design the GUI so that it defaults to a yellow background.

14.7 Write a GUI program that plots the equation $y(x) = ax^2 + bx + c$. The program should include a set of axes for the plot and should include a panel containing GUI elements to input the values of a, b, c, and the minimum and maximum x to plot. A separate panel should contain controls to set the style, color, and thickness of the line being plotted. Include tool tips for each of your GUI elements.

14.8 Modify the GUI of Exercise 14.7 to include a menu. The menu should include two submenus to select the color and line style of the plotted line, with a check mark beside the currently selected menu choices. The menu should also include an Exit option. If the user selects this option, the program should create a modal question dialog box asking "Are You Sure?" with the appropriate responses. Include accelerator keys and keyboard mnemonics in the menu design. (Note that the menu items duplicate some GUI elements, so if a menu item is selected, the corresponding GUI elements must be updated as well, and vice versa.)

14.9 Modify the List Box example in Section 14.4.7 to allow for multiple selections in the list box. The text field should be expanded to multiple lines, so that it can display a list of all selections whenever the Select button is clicked.

14.10 **Random Number Distributions** Create a GUI to display the distributions of different types of random numbers. The program should create the distributions by generating an array of 1,000,000 random values from a distribution and using function `hist` to create a histogram. Be sure to label the title and axes of the histogram properly.

The program should support uniform, Gaussian, and Rayleigh distributions, with the distribution selection made by a popup menu. In addition, it should have an edit box to allow the user to select the number of bins in the histogram. Make sure that the values entered in the edit box are legal (the number of bins must be a positive integer).

14.11 Modify the temperature conversion GUI of Example 14.1 to add a "thermometer". The thermometer should be a set of rectangular axes with a red "fluid" level corresponding to the current temperature in degrees Celsius. The range of the thermometer should be 0–100 °C.

14.12 Modify the temperature conversion GUI of Exercise 14.11 to allow you to adjust the displayed temperature by clicking the mouse. (**Warning:** This exercise requires material not discussed in this chapter. Refer to the `CurrentPoint` property of `axes` objects in the online MATLAB documentation.)

14.13 Create a GUI that contains a title, and four pushbuttons grouped within a panel. The pushbutton should be labeled "Title Color," "Figure Color," "Panel Color," and "Title Font". If the Title Color button is selected, open a `uisetcolor` dialog box and change the title text to be in the selected color. If the Figure Color button is selected, open a `uisetcolor` dialog box and change the figure color and the title text background color to be the selected color. If the Panel Color button is selected, open a `uisetcolor` dialog box and change the panel background to be in the selected color. If the Title Font button is selected, open a `uisetfont` dialog box and change the title text to be in the selected font.

14.14 Create a GUI that contains a title and a button group. The button group will be titled "Style," and it should contain four radio buttons labeled "Plain," "Italic," "Bold," and "Bold Italic". Design the GUI so that the style in the currently selected radio button is applied to the title text.

14.15 **Least Squares Fit**   Create a GUI that can read an input data set from a file and perform a least-squares fit to the data. The data will be stored in a disk file in (x, y) format, with one x and one y value per line. Perform the least-squares fit with the MATLAB function `polyfit`, and plot both the original data and the least-squares fitted line. Include two menus: File and Edit. The File menu should include File/Open and File/Exit menu items, and the user should receive an "Are You Sure?" prompt before exiting. The Edit menu item should allow the user to customize the display, including line style, line color, and grid status.

14.16 Modify the GUI of the previous exercise to include an Edit/Preferences menu item that allows the user to suppress the "Are You Sure?" exit prompt.

14.17 Modify the GUI of the previous exercise to read and write an initialization file. The file should contain the line style, line color, grid choice (on/off), and exit prompt choice made by the user on previous runs. These choice should be automatically written out and saved when the program exits via the File/Exit menu item, and they should be read in and used whenever the program is started again.

Appendix A

UTF-8 Character Set

MATLAB strings use the UTF-8 character set, which contains many thousands of characters stored in a 16-bit field. The first 128 of the characters are the same as the ASCII character set, and they are shown in the table below. The results of MATLAB string comparison operations depend on the *relative lexicographic positions* of the characters being compared. For example, the character 'a' in the character set is a position 97 in the table, while the character "A" is at position 65. Therefore, the relational operator `'a' > 'A'` will return a 1 (true), since 97 > 65.

The table below shows the ASCII character set, with the first two decimal digits of the character number defined by the row, and the third digit defined by the column. Thus, the letter `'R'` is on row 8 and column 2, so it is character 82 in the ASCII character set.

	0	1	2	3	4	5	6	7	8	9
0	nul	soh	stx	etx	eot	enq	ack	bel	bs	ht
1	nl	vt	ff	cr	so	si	dle	dc1	dc2	dc3
2	dc4	nak	syn	etb	can	em	sub	esc	fs	gs
3	rs	us	sp	!	"	#	$	%	&	'
4	(	)	*	+	,	-	.	/	0	1
5	2	3	4	5	6	7	8	9	:	;
6	<	=	>	?	@	A	B	C	D	E
7	F	G	H	I	J	K	L	M	N	O
8	P	Q	R	S	T	U	V	W	X	Y
9	Z	[	\	]	^	_	`	a	b	c
10	d	e	f	g	h	I	j	k	l	m
11	n	o	p	q	r	s	t	u	v	w
12	x	y	z	{			}	~	del		

Answers to Quizzes

This appendix contains the answers to all of the quizzes in the book.

Quiz 1.1, page 22

1. The MATLAB Command Window is the window where a user enters commands. A user can enter interactive commands at the command prompt (») in the Command Window, and they will be executed on the spot. The Command Window is also used to start M-files executing. The Edit/Debug Window is an editor used to create, modify, and debug M-files. The Figure Window is used to display MATLAB graphical output.

2. You can get help in MATLAB by:

   - Typing `help <command_name>` in the Command Window. This command will display information about a command or function in the Command Window.

   - Typing `lookfor <keyword>` in the Command Window. This command will display in the Command Window a list of all commands or functions containing the keyword in their first comment line.

   - Starting the Help Browser by typing `helpwin` or `helpdesk` in the Command Window, by selecting "Help" from the Start menu, or by clicking on the question mark icon (?) on the desktop. The Help Browser contains an extensive hypertext-based description of all of the features in MATLAB, plus a complete copy of all manuals online in HTML and Adobe PDF formats. It is the most comprehensive source of help in MATLAB.

3. A workspace is the collection of all the variables and arrays that can be used by MATLAB when a particular command, M-file, or function is executing. All commands executed in the Command Window (and all script files executed from the Command Window) share a common workspace, so they can all share variables. The contents of the workspace can be examined with the `whos` command, or graphically with the Workspace Browser.

4. To clear the contents of a workspace, type `clear` or `clear variables` in the Command Window.

5. The commands to perform this calculation are:

```
» t = 5;
» x0 = 10;
» v0 = 15;
» a = -9.81;
» x = x0 + v0 * t + 1/2 * a * t^2
x =
   -37.6250
```

6. The commands to perform this calculation are:

```
» x = 3;
» y = 4;
» res = x^2 * y^3 / (x - y)^2
res =
    576
```

Questions 7 and 8 are intended to get you to explore the features of MATLAB. There is no single "right" answer for them.

Quiz 2.1, page 36

1. An array is a collection of data values organized into rows and columns, and known by a single name. Individual data values within an array are accessed by including the name of the array followed by subscripts in parentheses that identify the row and column of the particular value. The term "vector" is usually used to describe an array with only one dimension, while the term "matrix" is usually used to describe an array with two or more dimensions.

2. (a) This is a 3×4 array; (b) `c(2, 3)` $= -0.6$; (c) The array elements whose value is 0.6 are `c(1, 4)`, `c(2, 1)`, and `c(3, 2)`.

3. (a) 1×3; (b) 3×1; (c) 3×3; (d) 3×2; (e) 3×3; (f) 4×3; (g) 4×1.

4. `w(2,1)` $= 2$

5. `x(2,1)` $= -20i$

6. $y(2,1) = 0$

7. $v(3) = 3$

Quiz 2.2, page 44

1. (a) $c(2, :) = \begin{bmatrix} 0.6 & 1.1 & -0.6 & 3.1 \end{bmatrix}$

(b) $c(:, \text{end}) = \begin{bmatrix} 0.6 \\ 3.1 \\ 0.0 \end{bmatrix}$

(c) $c(1:2, 2:\text{end}) = \begin{bmatrix} -3.2 & 3.4 & 0.6 \\ 1.1 & -0.6 & 3.1 \end{bmatrix}$

(d) $c(6) = 0.6$

(e) $c(4, \text{end}) = \begin{bmatrix} -3.2 & 1.1 & 0.6 & 3.4 & -0.6 & 5.5 & 0.6 & 3.1 & 0.0 \end{bmatrix}$

(f) $c(1:2, 2:4) = \begin{bmatrix} -3.2 & 3.4 & 0.6 \\ 1.1 & -0.6 & 3.1 \end{bmatrix}$

(g) $c([1\ 3], 2) = \begin{bmatrix} -3.2 \\ 0.6 \end{bmatrix}$

(h) $c([2\ 2], [3\ 3]) = \begin{bmatrix} -0.6 & -0.6 \\ -0.6 & -0.6 \end{bmatrix}$

2. (a) $a = \begin{bmatrix} 7 & 8 & 9 \\ 4 & 5 & 6 \\ 1 & 2 & 3 \end{bmatrix}$ (b) $a = \begin{bmatrix} 4 & 5 & 6 \\ 4 & 5 & 6 \\ 4 & 5 & 6 \end{bmatrix}$ (c) $a = \begin{bmatrix} 4 & 5 & 6 \\ 4 & 5 & 6 \end{bmatrix}$

3. (a) $a = \begin{bmatrix} 1 & 0 & 0 \\ 1 & 2 & 3 \\ 0 & 0 & 1 \end{bmatrix}$ (b) $a = \begin{bmatrix} 1 & 0 & 4 \\ 0 & 1 & 5 \\ 0 & 0 & 6 \end{bmatrix}$ (c) $a = \begin{bmatrix} 1 & 0 & 0 \\ 0 & 1 & 0 \\ 9 & 7 & 8 \end{bmatrix}$

Quiz 2.3, page 51

1. The required command is `format long e`.

2. (a) These statements get the radius of a circle from the user, and calculate and display the area of the circle. (b) These statements display the value of π as an integer, so they display the string: `The value is 3!`.

3. The first statement outputs the value 12345.67 in exponential format; the second statement outputs the value in floating point format; the third statement outputs the value in general format; and the

fourth statement outputs the value in floating point format in a field 12 characters wide, with four places after the decimal point. The results of these statements are:

```
value = 1.234567e+004
value = 12345.670000
value = 12345.7
value = 12345.6700
```

Quiz 2.4, page 58

1. (a) This operation is illegal. Array multiplication must be between arrays of the same shape, or between an array and a scalar. (b) Legal matrix multiplication: result $= \begin{bmatrix} 4 & 4 \\ 3 & 3 \end{bmatrix}$ (c) Legal array multiplication: result $= \begin{bmatrix} 2 & 1 \\ -2 & 4 \end{bmatrix}$ (d) This operation is illegal. The matrix multiplication b * c yields a 1×2 array, and a is a 2×2 array, so the addition is illegal. (e) This operation is illegal. The array multiplication b .* c is between two arrays of different sizes, so the multiplication is illegal.

2. This result can be found from the operation x = A\B: $x = \begin{bmatrix} -0.5 \\ 1.0 \\ -0.5 \end{bmatrix}$

Quiz 3.1, page 114

1.
```
x = 0:pi/10:2*pi;
x1 = cos(2*x);
y1 = sin(x);
plot(x1,y1,'-ro','LineWidth',2.0,'MarkerSize',6,...
    'MarkerEdgeColor','b','MarkerFaceColor','b')
```

2. This question has no single specific answer; any combination of actions that changes the markers is acceptable.

3. `'\itf\rm(\itx\rm) = sin \theta cos 2\phi'`

4. `'\bfPlot of \Sigma \itx\rm\bf^{2} versus \itx'`

5. This string creates the characters: τ_m

6. This string creates the characters: $x_1^2 + x_2^2$ (units: $\mathbf{m^2}$)

7.
```
g = 0.5;
theta = 2*pi*(0.01:0.01:1);
r = 10*cos(3*theta);
polar (theta,r,'r-')
```

The resulting plot is shown below:

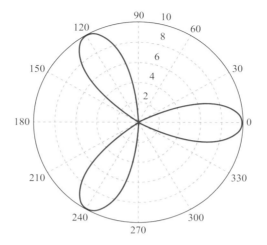

8.

```
figure(1);
x = linspace(0.01,100,501);
y = 1 ./ (2 *  x .^ 2);
plot(x,y);
figure(2);
x = logspace(0.01,100,101);
y = 1 ./ (2 *  x .^ 2)
loglog(x,y);
```

The resulting plots are shown below. The linear plot is dominated by the very large value at $x = 0.01$, and almost nothing is visible. The function looks like a straight line on the loglog plot.

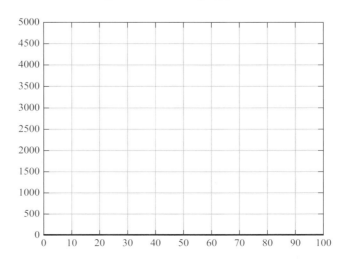

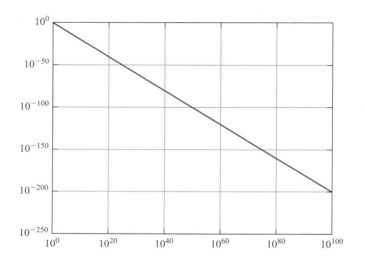

Quiz 4.1, page 140

	Expression	Result	Comment
1.	a > b	1 (logical true)	
2.	b > d	0 (logical false)	
3.	a > b && c > d	0 (logical false)	
4.	a == b	0 (logical false)	
5.	a & b > c	0 (logical false)	
6.	~~b	1 (logical true)	
7.	~(a > b)	$\begin{bmatrix} 0 & 0 \\ 0 & 1 \end{bmatrix}$ (logical array)	
8.	a > c && b > c	Illegal	The && and || operators only work between *scalar* operands.
9.	c <= d	Illegal	The <= operator must be between arrays of the same size, or between an array and a scalar.
10.	logical(d)	$\begin{bmatrix} 1 & 1 & 1 \\ 0 & 1 & 0 \end{bmatrix}$ (logical array)	

11. a * b > c $\begin{bmatrix} 1 & 0 \\ 0 & 1 \end{bmatrix}$
(logical array)

The expression a * b is evaluated first, producing the double array $\begin{bmatrix} 2 & -4 \\ 0 & 20 \end{bmatrix}$, and the logical operation is evaluated second, producing the final answer.

12. a * (b > c) $\begin{bmatrix} 2 & 0 \\ 0 & 2 \end{bmatrix}$
(double array)

The expression b > c produced the logical array $\begin{bmatrix} 1 & 0 \\ 0 & 1 \end{bmatrix}$, and multiplying that logical array by 2 converted the results back into a double array.

13. a*b^2 > a*c 0
(logical false)

14. d || b > a 1
(logical true)

15. (d | b) > a 0
(logical false)

16. isinf(a/b) 0
(logical false)

17. isinf(a/c) 1
(logical true)

18. a > b && 1
   ischar(d) (logical true)

19. isempty(c) 0
(logical false)

20. (~a) & b 0
(logical false)

21. (~a) + b -2
(double value)

~a is a logical 0. When added to b, the result is converted back to a double value.

Quiz 4.2, page 155

```
if x >= 0
    sqrt_x = sqrt(x);
else
    disp('ERROR: x < 0');
    sqrt_x = 0;
end
```

```
2. if abs(denominator) < 1.0E-300
       disp('Divide by 0 error.');
   else
       fun = numerator / denominator;
       disp(fun)
   end
3. if distance <= 100
       cost = 0.50 * distance;
   elseif distance <= 300
       cost = 50 + 0.30 * (distance - 100);
   else
       cost = 110 + 0.20 * (distance - 300);
   end
```

4. These statements are incorrect. For this structure to work, the second `if` statement would need to be an `elseif` statement.

5. These statements are legal. They will display the message "Prepare to stop."

6. These statements will execute, but they will not do what the programmer intended. If the `temperature` is 150, these statements will print out "Human body temperature exceeded." instead of "Boiling point of water exceeded.", because the `if` structure executes the first `true` condition and skips the rest. To get proper behavior, the order of these tests should be reversed.

Quiz 5.1, page 198

1. 4 times
2. 0 times
3. 1 time
4. 2 times
5. 2 times
6. `ires = 10`
7. `ires = 55`
8. `ires = 25;`
9. `ires = 49;`
10. With loops and branches:

```
for ii = -6*pi:pi/10:6*pi
    if sin(ii) > 0
        res(ii) = sin(ii);
    else
        res(ii) = 0;
    end
end
```

With vectorized code:
```
arr1 = sin(-6*pi:pi/10:6*pi);
res = zeros(size(arr1));
res(arr1>0) = arr1(arr1>0);
```

Quiz 6.1, page 251

1. Script files are collections of MATLAB statements that are stored in a file. Script files share the Command Window's workspace, so any variables that were defined before the script file starts are visible to the script file, and any variables created by the script file remain in the workspace after the script file finishes executing. A script file has no input arguments and returns no results, but script files can communicate with other script files through the data left behind in the workspace. In contrast, each MATLAB function runs in its own independent workspace. It receives input data through an input argument list and returns results to the caller through an output argument list.

2. The `help` command displays all of the comment lines in a function until either the first blank line or the first executable statement is reached.

3. The H1 comment line is the first comment line in the file. This line is searched by and displayed by the `lookfor` command. It should always contain a one-line summary of the purpose of a function.

4. In the pass-by-value scheme, a *copy* of each input argument is passed from a caller to a function, instead of the original argument itself. This practice contributes to good program design because the input arguments may be freely modified in the function without causing unintended side-effects in the caller.

5. A MATLAB function can have any number of arguments, and not all arguments need to be present each time the function is called. Function `nargin` is used to determine the number of input arguments actually present when a function is called, and function `nargout` is used to determine the number of output arguments actually present when a function is called.

6. This function call is incorrect. Function `test1` must be called with two input arguments. In this case, variable `y` will be undefined in function `test1`, and the function will abort.

7. This function call is correct. The function can be called with either one or two arguments.

Quiz 7.1, page 309

1. A local function is a second or subsequent function defined within a file. Local functions look just like ordinary functions, but they are accessible only to the other functions within the same file.

2. The scope of a function is defined as the locations within MATLAB from which the function can be accessed.

3. Private functions are functions that reside in subdirectories with the special name `private`. They are only visible to other functions in the `private` directory, or to functions in the parent directory. In other words, the scope of these functions is restricted to the private directory and to the parent directory that contains it.

4. Nested functions are functions that are defined entirely within the body of another function, called the host function. They are only visible to the host function in which they are embedded, and to other nested functions embedded at the same level within the same host function.

5. MATLAB locates functions in a specific order as follows:
   - MATLAB checks to see if there is a nested function within the current function with the specified name. If so, it is executed.
   - MATLAB checks to see if there is a local function within the current file with the specified name. If so, it is executed.
   - MATLAB checks for a private function with the specified name. If so, it is executed.
   - MATLAB checks for a function with the specified name in the current directory. If so, it is executed.
   - MATLAB checks for a function with the specified name on the MATLAB path. MATLAB will stop searching and execute the first function with the right name found on the path.

6. A function handle is a MATLAB data type that holds information to be used in referencing a function. When a function handle is created, MATLAB captures all the information about the function that it needs to execute it later on. Once the handle is created, it can be used to execute the function at any time.

7. The result returns the name of the function that the handle was created from:
```
>> myfun(@cosh)
ans =
cosh
```

Quiz 8.1, page 331

1. (a) true (1); (b) false (0); (c) 25

2. If `array` is a complex array, then the function `plot(array)` plots the real componets of each element in the array versus the imaginary components of each element in the array.

Quiz 9.1, page 365

1. These statements concatenate the two lines together, and variable `res` contains the string `'This is a test!This line, too.'`

2. These statements are illegal—there is no function `strcati`.

3. These statements are illegal—the two strings must have the same number of columns, and these strings are of different lengths.

4. These statements are legal—function `strvcat` can pad out input values of different lengths. The result is that the two strings appear on to different rows in the final result:
   ```
   » res = strvcat(str1,str2)
   res =
   This is another test!
   This line, too.
   ```

5. These statements return true (1), because the two strings match in the first 5 characters.

6. These statements return the locations of every 's' in the input string: 4 7 13.

7. These statements assign the character 'x' to every location in `str1` that contains a blank. The resulting string is `Thisxisxaxtest!xx`.

8. These statements return an array with 12 values, corresponding to the 12 characters in the input string. The output array contains 1 at the locations of each alphanumeric value and 0 at all other locations:
   ```
   » str1 = 'aBcD 1234 !?';
   » res = isstrprop(str1,'alphanum')
     Columns 1 through 5
        1      1      1      1      0
     Columns 6 through 10
        1      1      1      1      0
     Columns 11 through 12
        0      0
   ```

9. These statements shift all alphabetic characters in the first seven columns of `str1` to uppercase. The resulting string is `ABCD 1234 !?`.

10. `str1` contains `456` with three blanks before and after it, and `str2` contains `abc` with three blanks before and after it. String `str3` is the concatenation of the two strings, so it is 18 characters long: `   456      abc   `. String `str4` is the concatenation of the two strings with leading and trailing blanks removed, so it is 6 characters long: `456abc`. String `str5` is the concatenation of the two strings with only trailing blanks removed, so it is 12 characters long: `   456   abc`.

11. These statements will fail, because `strncmp` requires a length parameter.

Quiz 9.2, page 370

1. These statements are illegal, since it is not possible to add objects of these two classes.

2. These statements are illegal, since it is not possible to multiply objects of these two classes.

3. These statements are legal, since `single` and `double` objects can be multiplied using matrix multiplication. The result is a single array containing $\begin{bmatrix} 3 & 2 \\ -2 & 3 \end{bmatrix}$.

4. These statements are legal, since `single` and `double` objects can be multiplied using array multiplication. The result is a single array containing $\begin{bmatrix} 3 & 0 \\ 0 & 3 \end{bmatrix}$.

Quiz 10.1, page 403

1. A sparse array is a special type of array in which memory is only allocated for the non-zero elements in the array. Memory values are allocated for both the subscripts and the value of each element in a sparse array. By contrast, a memory location is allocated for every value in a full array, whether the value is 0 or not. Sparse arrays can be converted to full arrays using the `full` function, and full arrays can be converted to sparse arrays using the `sparse` function.

2. A cell array is an array of "pointers," each element of which can point to any type of MATLAB data. It differs from an ordinary array in that each element of a cell array can point to a different type of data, such as a numeric array, a string, another cell array, or a structure. Also, cell arrays use braces { } instead of parentheses ( ) for selecting and displaying the contents of cells.

3. *Content indexing* involves placing braces { } around the cell subscripts, together with cell contents in ordinary notation. This type of indexing defines the contents of the data structure contained in a cell. *Cell indexing* involves placing braces { } around the data to be stored in a cell, together with cell subscripts in ordinary subscript notation. This type of indexing creates a data structure containing the specified data, and then assigns that data structure to a cell.

4. A structure is a data type in which each individual element is given a name. The individual elements of a structure are known as fields, and each field in a structure may have a different type. The individual fields are addressed by combining the name of the structure with the name of the field, separated by a period. Structures differ from ordinary arrays and cell arrays in that ordinary arrays and cell

array elements are addressed by subscript, while structure elements are addressed by name.

5. Function `varargin` appears as the last item in an input argument list, and it returns a cell array containing all of the actual arguments specified when the function is called, each in an individual element of a cell array. This function allows a MATLAB function to support any number of input arguments.

6. (a) `a(1,1) = [3x3 double]`. The contents of cell array element `a(1,1)` is a 3×3 double array, and this data structure is displayed.

(b) $a\{1,1\} = \begin{bmatrix} 1 & 2 & 3 \\ 4 & 5 & 6 \\ 7 & 8 & 9 \end{bmatrix}$. This statement displays the *value* of the data structure stored in element `a(1,1)/`.

(c) These statements are illegal, since you can not multiply a data structure by a value.

(d) These statements are legal, since you *can* multiply the contents of the data structure by a value. The result is $\begin{bmatrix} 2 & 4 & 6 \\ 8 & 10 & 12 \\ 14 & 16 & 18 \end{bmatrix}$.

(e) $a\{2,2\} = \begin{bmatrix} -4 & -3 & -2 \\ -1 & 0 & 1 \\ 2 & 3 & 4 \end{bmatrix}$.

(f) This statement is legal. It initializes cell array element `a(2,3)` to be a 2×1 double array containing the values $\begin{bmatrix} -17 \\ 17 \end{bmatrix}$.

(g) `a{2,2}(2,2) = 0`.

7. (a) $b(1).a - b(2).a = \begin{bmatrix} -3 & 1 & -1 \\ -2 & 0 & -2 \\ -3 & 3 & 5 \end{bmatrix}$.

(b) `strncmp(b(1).b,b(2).b,6) = 1`, since the two structure elements contain character strings that are identical in their first 6 characters.

(c) `mean(b(1).c) = 2`

(d) This statement is illegal, since you cannot treat an individual element of a structure array as though it were an array itself.

(e) `b = 1x2 struct array with fields:`
     `a`
     `b`
     `c`

(f) `b(1).('b') = 'Element 1'`

(g) `b(1) =`
      `a: [3x3 double]`
      `b: 'Element 1'`
      `c: [1 2 3]`

Quiz 11.1, page 427

1. The `textread` function is designed to read ASCII files that are formatted into columns of data, where each column can be of a different type. This command is very useful for importing tables of data printed out by other applications, since it can handle data of mixed types within a single file.

2. MAT files are relatively efficient users of disk space, and they store the full precision of every variable—no precision is lost due to conversion to and from ASCII format. (If compression is used, MAT files take up even less space.) In addition, MAT files preserve all of the information about each variable in the workspace, including its class, name, and whether or not it is global. A disadvantage of MAT files is that they are unique to MATLAB and cannot be used to share data with other programs.

3. Function `fopen` is used to open files, and function `fclose` is used to close files. On PCs (but not on Linux or UNIX computers), there is a difference between the format of a text file and a binary file. In order to open files in text mode on a PC, a `'t'` must be appended to the permission string in the `fopen` function.

4. `fid = fopen('myinput.dat','at')`

5. ```
   fid = fopen('input.dat','r');
   if fid < 0;
       disp('File input.dat does not exist.');
   end
   ```

6. These statements are incorrect. They open a file as a text file, but then read the data in binary format. (Function `fscanf` should be used to read text data, as we see later in this chapter.)

7. These statements are correct. They create a 10-element array x, open a binary output file `file1`, write the array to the file, and close the file. Next, they open the file again for reading, and read the data into array `array` in a [2 Inf] format. The resulting contents of the array are $\begin{bmatrix} 1 & 3 & 5 & 7 & 9 \\ 2 & 4 & 6 & 8 & 10 \end{bmatrix}$.

Quiz 11.2, page 441

1. Formatted I/O operations produce formatted files. A formatted file contains recognizable characters, numbers, and so forth, stored as ASCII text. Formatted files have the advantages that we can readily see what sort of data they contain, and it is easy to exchange data between different types of programs using them. However, formatted I/O operations take longer to read and write, and formatted files take up more
   ```

space than unformatted files. Unformatted I/O operations copy the information from a computer's memory directly to the disk file with no conversions at all. These operations are much faster than formatted I/O operations because there is no conversion. In addition, the data occupies a much smaller amount of disk space. However, unformatted data cannot be examined and interpreted directly by humans.

2. Formatted I/O should be used whenever we need to exchange data between MATLAB and other programs, or when a person needs to be able to examine and/or modify the data in the file. Otherwise, unformatted I/O should be used.

3.
```
fprintf('     Table of Cosines and Sines\n\n');
fprintf(' theta       cos(theta)   sin(theta)\n');
fprintf(' =====       ==========   ==========\n');
for ii = 0:0.1:1
    theta = pi * ii;
    fprintf('%7.4f %11.5f %11.5f\n',theta,cos(theta),sin(theta));
end
```

4. These statements are incorrect. The `%s` descriptor must correspond to a character string in the output list.

5. These statements are technically correct, but the results are undesirable. It is possible to mix binary and formatted data in a single file the way that these statements do, but the file is then very hard to use for any purpose. Normally, binary data and formatted data should be written to separate files.

Quiz 12.1, page 523

1. A class is the software blueprint from which objects are made. It defines the properties, which are the data in the object, and the methods, which are the way in which the data is manipulated. When objects are instantiated (created), each object receives its own unique copy of the instance variables defined in the properties, but all share the same methods.

2. A user-defined class is created using the `classdef` structure. Properties and methods are declared in `properties` and `methods` blocks within the class definition. The basic structure of the class definition is

```
classdef (Attributes) ClassName < SuperClass

    properties (Attributes)
        PropertyName1
        PropertyName2
        . . .
    end
```

```
        methods (Attributes)
            function [obj = ] methodName(obj,arg1,arg2, ...)
                ...
        end
    end
```

3. The principal components in a class are:

   ■ **Properties.** Properties define the instance variables that will be created when an object is instantiated from a class. Instance variables are the data encapsulated inside an object. A new set of instance variables is created each time that an object is instantiated from the class.

   ■ **Methods.** Methods implement the behaviors of a class. Some methods may be explicitly defined in a class, while other methods may be inherited from superclasses of the class.

   ■ **Constructor.** Constructors are special methods that specify how to initialize an object when it instantiated.

   ■ **Destructor.** Destructors are special methods that clean up the resources (open files, etc.) used by an object just before it is destroyed.

4. Constructors are special methods that specify how to initialize an object when it instantiated. The arguments of the constructor include values to use in initializing the properties. Constructors are easy to identify because they have the same name as the class that they are initializing, and the only output argument is the object constructed. Note that constructors should always be built to accept the case with default inputs (no arguments) as well as the case with arguments because the constructor may be called without arguments when objects of subclasses are created.

5. Destructors are special methods that clean up the resources (open files, etc.) used by an object just before it is destroyed. Just before an object is destroyed, it makes a call to a special method named **delete** if it exists. The only input argument is the object to be destroyed, and there must be no output argument. Many classes do not need a delete method at all.

6. Events are notices that an object broadcasts when something happens, such as a property value changing or a user entering data on the keyboard or clicking a button with a mouse. Listeners are objects that execute a callback method when notified that an event of interest has occurred. Programs use events to communicate things that happen to objects and respond to these events by executing the listener's callback function. Events are triggered when a method calls the `notify` function on the event. A program can listen for and respond to an event by registering as a listener for that event using the `adlistener` function. (Listeners are objects that

execute a callback method when notified that an event of interest has occurred.)

7. Exceptions are interruptions to the normal flow of program execution due to errors in the code. When an error occurs that a method cannot recover from by itself, it collects information about the error (what the error was, what line it occurred on, and the calling stack describing how program execution got to that point). It bundles this information into a MException object and then throws the exception using the throw function. Programs handle exceptions by using try / catch structures. Code is executed in the try part of the structure, and errors that occur are trapped in the catch part of the structure, where they can be examined and efforts can be made to recover from the problem.

8. A subclass is a class that is derived from a parent class, called a superclass. The subclass inherits all the public or protected properties and methods of the parent class, and it can add additional properties and override the methods defined in the superclass. A subclass is created by specifying the superclass in the class definition.

```
classdef (Attributes) ClassName < SuperClass

end
```

Quiz 14.1, page 629

1. The types of graphical components discussed in this chapter are listed below, together with their purposes.

Table B.1: GUI Components Discussed in Chapter 14

Component	Created By	Description
		Graphical Controls
Pushbutton	uicontrol	A graphical component that implements a pushbutton. It triggers a callback when clicked with a mouse.
Toggle Button	uicontrol	A graphical component that implements a toggle button. A toggle button is either "on" or "off," and it changes state each time that it is clicked. Each mouse button click also triggers a callback.
Radio Button	uicontrol	A radio button is a type of toggle button that appears as a small circle with a dot in the middle when it is "on." Groups of radio buttons are used to implement mutually exclusive choices. Each mouse click on a radio button triggers a callback.
Check Box	uicontrol	A check box is a type of toggle button that appears as a small square with a checkmark in it when it is "on." Each mouse click on a check box triggers a callback.

(continued)

Table B.1: GUI Components Discussed in Chapter 14 (*Continued*)

Component	Created By	Description
Edit Box	uicontrol	An edit box displays a text string and allows the user to modify the information displayed. A callback is triggered when the user presses the Enter key.
List Box	uicontrol	A list box is a graphical control that displays a series of text strings. A user may select one of the text strings by single- or double-clicking on them. A callback is triggered when the user selects a string.
Popup Menus	uicontrol	A popup menu is a graphical control that displays a series of text strings in response to a mouse click. When the popup menu is not clicked on, only the currently selected string is visible.
Slider	uicontrol	A slider is a graphical control to adjust a value in a smooth, continuous fashion by dragging the control with a mouse. Each slider change triggers a callback.
Table	uitable	Creates a table of data.
Static Elements		
Frame	uicontrol	Creates a frame, which is a rectangular box within a figure. Frames are used to group sets of controls together. Frames never trigger callbacks. (This is a deprecated component, which should not be used in new GUIs.)
Text Field	uicontrol	Creates a label, which is a text string located at a point on the figure. Text fields never trigger callbacks.
Menus, Toolbars, Axes		
Menu Items	uimenu	Creates a menu item. Menu items trigger a callback when a mouse button is released over them.
Context Menus	uicontextmenu	Creates a context menu, which is a menu that appears over a graphical object when a user right-clicks the mouse on that object.
Toolbar	uitoolbar	Creates a toolbar, which is a bar across the top of the figure, containing quick-access buttons.
Toolbar Pushbutton	uipushtool	Creates a pushbutton to go in a toolbar.
Toolbar Toggle Button	uitoggletool	Creates a toggle button to go in a toolbar.
Axes	axes	Creates a new set of axes to display data on. Axes never trigger callbacks.

2. The types of containers discussed in this chapter are listed below, together with their differences.

Table B.2: GUI Components Discussed in Chapter 14

Component	Created By	Description
		Containers
Figure	`uicontrol`	Creates a figure, which is a container that can hold components and other containers. Figures are separate windows that have title bars and can have menus.
Panel	`uipanel`	Creates a panel, which is container that can hold components and other containers. Unlike figures, panels do not have title bars or menus. Panels can be placed inside figures or other panels.
Button Group	`uibuttongroup`	Creates a button group, which is a special kind of panel. Button groups automatically manage groups of radio buttons or toggle buttons to ensure that only one item of the group is on at any given time.

3. A callback function is a function that is executed whenever an action (mouse click, keyboard input, etc.) occurs on a specific GUI component. They are used to perform an action when a user clicks on or types in a GUI component. Callback functions are specified by the `'Callback'` property in a `uicontrol`, `uimenu`, `uicontextmenu`, `uipushtool`, or `uitoggletool`. When a new GUI is created, the callbacks are set automatically by `guide` to be `xxx_Callback`, where `xxx` is the value of the `Tag` property of the corresponding GUI component.

4. The basic steps required to create a MATLAB GUI are:

- Decide what elements are required for the GUI and what the function of each element will be. Make a rough layout of the components by hand on a piece of paper.

- Use a MATLAB tool called `guide` (GUI Development Environment) to lay out the components on a figure. The size of the figure, and the alignment and spacing of components on the figure, can be adjusted using the tools built into `guide`.

- Use a MATLAB tool called the Property Inspector (built into `guide`) to give each component a name (a "tag"), and to set the characteristics of each component, such as its color, the text it displays, and so forth.

- Save the figure to a file. When the figure is saved, two files will be created on disk with the same name but different extents. The `fig` file contains the actual GUI that you have created, and the M-file contains the code to load the figure, and also skeleton callbacks for each GUI element.

- Write code to implement the behavior associated with each callback function.

5. The `handles` data structure is a structure containing the handles of all components within a figure. Each structure element has the name of a component and the value of the component's handle. This structure is passed to every callback function, allowing each function to have access to every component in the figure.

6. Application data can be saved in a GUI by adding it to the handles structure, and saving that structure after it is modified using function `guidata`. Since the `handles` structure is automatically passed to every callback function, additional data added to the structure will be available to any callback function in the GUI. (Each function that modifies the handles structure must be sure to save the modified version with a call to `guidata` before the function exits.)

7. A graphical object can be made invisible by setting its `'Visible'` property to `'off'`. A graphical object can be disabled so that it will not respond to mouse clicks or keyboard input by setting its `'Enable'` property to `'off'`.

8. Pushbuttons, toggle buttons, radio buttons, check boxes, list boxes, popup menus, and sliders all respond to mouse clicks. Edit boxes respond to keyboard inputs.

9. A dialog box is a special type of figure that is used to display information or to get input from a user. Dialog boxes are used to display errors, provide warnings, ask questions, or get user input. Dialog boxes can be created by any of the functions listed in Table 14.5, including `errordlg`, `warndlg`, `inputdlg`, `uigetfile`, and so on.

10. A modal dialog box does not allow any other window in the application to be accessed until it is dismissed, while a normal dialog box does not block access to other windows.

11. A standard menu is tied to a menu bar running across the top of a figure, while a context menu can be attached to any GUI component. Standard menus are activated by normal mouse clicks on the menu bar, while context menus are activated by mouse right-clicks over the associated GUI component. Menus are built out of `uimenu` components. Context menus are built out of both `uicontextmenu` and `uimenu` components.

12. Accelerator keys are keys that may be typed on the keyboard to cause a menu item to be selected. Keyboard mnemonic keys are CTRL+key combinations that cause a menu item to be executed. The principal difference between accelerator keys and keyboard mnemonics is that accelerator keys only work to select a menu item if a menu has already been opened, while keyboard mnemonics can trigger an action even if a menu has not been opened.

Index

Note: **Boldface** numbers indicate illustrations or tables.